INTRODUCTION TO THE CONTROLLOGIX PROGRAMMABLE AUTOMATION CONTROLLER WITH LABS

Second Edition

Gary Dunning

Australia • Brazil • Japan • Korea • Mexico • Singapore • Spain • United Kingdom • United States

Introduction to the ControlLogix Programmable Automation Controller with Labs, Second Edition
Gary Dunning

Vice President, Careers and Computing: Dave Garza

Director of Learning Solutions: Sandy Clark

Associate Acquisitions Editor: Nicole Sgueglia

Director, Development – Careers and Computing: Marah Bellegarde

Managing Editor: Larry Main

Senior Product Manager: John Fisher

Editorial Assistant: Kaitlin Schlicht

Brand Manager: Kristin McNary

Market Development Manager: Erin Brennan

Senior Production Director: Wendy Troeger

Production Manager: Mark Bernard

Content Project Manager: Barbara LeFleur

Senior Art Director: David Arsenault

Technology Project Manager: Joe Pliss

Media Editor: Debbie Bordeaux

Cover Image(s): BACKGROUND IMAGE:
© iStockphoto/P. Wei
INSET IMAGES (left to right):
© iStockphoto/Bart Coenders
© iStockphoto/Joseph Gareri
© iStockphoto/Baran Ozdemir

© 2014, 2008 Delmar, Cengage Learning

ALL RIGHTS RESERVED. No part of this work covered by the copyright herein may be reproduced, transmitted, stored, or used in any form or by any means graphic, electronic, or mechanical, including but not limited to photocopying, recording, scanning, digitizing, taping, Web distribution, information networks, or information storage and retrieval systems, except as permitted under Section 107 or 108 of the 1976 United States Copyright Act, without the prior written permission of the publisher.

ControlLogix® is a registered trademark of Rockwell Automation, Inc. Rockwell Automation, Inc. has not sponsored or reviewed this book and is in no way affiliated with this publication. The statements, analysis, and opinions in this book are those of the author and not of Rockwell Automation, Inc.

> For product information and technology assistance, contact us at
> **Cengage Learning Customer & Sales Support, 1-800-354-9706**
> For permission to use material from this text or product,
> submit all requests online at **www.cengage.com/permissions**.
> Further permissions questions can be e-mailed to
> **permissionrequest@cengage.com**

Library of Congress Control Number: 2012948478

ISBN-10: 1-111-53929-4
ISBN-13: 978-1-111-53929-0

Delmar
5 Maxwell Drive
Clifton Park, NY 12065-2919
USA

Cengage Learning is a leading provider of customized learning solutions with office locations around the globe, including Singapore, the United Kingdom, Australia, Mexico, Brazil, and Japan. Locate your local office at: **international.cengage.com/region**

Cengage Learning products are represented in Canada by Nelson Education, Ltd.

To learn more about Delmar, visit **www.cengage.com/delmar**

Purchase any of our products at your local college store or at our preferred online store **www.cengagebrain.com**

Notice to the Reader
Publisher does not warrant or guarantee any of the products described herein or perform any independent analysis in connection with any of the product information contained herein. Publisher does not assume, and expressly disclaims, any obligation to obtain and include information other than that provided to it by the manufacturer. The reader is expressly warned to consider and adopt all safety precautions that might be indicated by the activities described herein and to avoid all potential hazards. By following the instructions contained herein, the reader willingly assumes all risks in connection with such instructions. The publisher makes no representations or warranties of any kind, including but not limited to, the warranties of fitness for particular purpose or merchantability, nor are any such representations implied with respect to the material set forth herein, and the publisher takes no responsibility with respect to such material. The publisher shall not be liable for any special, consequential, or exemplary damages resulting, in whole or part, from the readers' use of, or reliance upon, this material.

Printed in the United States of America
1 2 3 4 5   19 18 17 16 15

TABLE OF CONTENTS

Introduction v

Part 1: Introduction to the ControlLogix Programmable Automation Controller

Introduction 1

Chapter 1 Introduction to ControlLogix Hardware 3

Chapter 2 Introduction to RSLogix 5000 Software 47

Chapter 3 Number Systems 83

Chapter 4 RSLogix 5000 Project Organization 106

Chapter 5 Understanding ControlLogix I/O Addressing 126

Chapter 6 Modular ControlLogix I/O Configuration 148

Chapter 7 CompactLogix I/O Configuration 195

Chapter 8 Communication between Your Personal Computer and Your ControlLogix 214

Chapter 9 Creating and Monitoring RSLogix 5000 Tags 234

Chapter 10 Introduction to Logic 255

Chapter 11 The Basic Relay Instructions 283

Chapter 12 ControlLogix Timer Instructions 309

Chapter 13 Adding Ladder Rung Documentation 333

Chapter 14 ControlLogix Counter Instructions 350

Chapter 15 Comparison Instructions 385

Chapter 16 Data-Handling Instructions 410

Chapter 17 Introduction to the Get System Values (GSV) and Set Systems Values (SSV) Instructions 440

Chapter 18 Introduction to the RSLogix 5000 Function Block Programming Language ... 474

Part 2: Configuring RSLinx Drivers and Communication

Introduction 521

Chapter 19 Configuring a Serial Driver Using RSLinx 523

Chapter 20 Configuring the Keyspan by Tripp Lite, High-Speed USB to Serial Adapter Model USA-19HS 532

Chapter 21 Configuring an RSLinx Serial Driver Using a Rockwell Automation 9300-USBS USB to Serial Adapter 541

Chapter 22 Installing and Configuring a USB Driver for 1756-L7 Series Controllers 551

Chapter 23 Determine and Modify a Personal Computer's IP Address 560

Chapter 24 Configuring 1756 ControlLogix Modular Ethernet Hardware 579

Chapter 25 Configuring Ethernet IP Address for a CompactLogix 1769-L23E, 1769-L32E, or 1769-L35E 592

Chapter 26 Configuring a 1756-ENET Ethernet Driver Using RSLinx 600

Chapter 27 Configuring Ethernet/IP Drivers Using RSLinx Software 605

Chapter 28 Configuring a CompactLogix 1769-L23E, 1769-L32E, or 1769-L35E Ethernet/IP Driver Using RSLinx 611

Chapter 29 Configuring a USB Driver for a 1756 Ethernet Communications Module 618

Glossary 629

INTRODUCTION

Welcome to the world of the Allen-Bradley/Rockwell Automation ControlLogix family of programmable logic controllers (PLCs) and RSLogix 5000 software. ControlLogix is a much advanced PLC that has many features not found in traditional PLCs; as a result, it is referred to as a programmable automation controller, or PAC. Most people are familiar with the term PLC, we will be referring to the ControlLogix as a PAC in this text. There are three different packages of RSLogix software. RSLogix 5000 is for the ControlLogix family of PACs, which includes CompactLogix. RSLogix 5 software is for programming the Allen-Bradley PLC 5, and RSLogix 500 is for the SLC 500 family of PLCs, which also includes the MicroLogix. RSLogix 5 or RSLogix 500 cannot be used to program any ControlLogix family member.

This manual is an introductory text and lab manual to familiarize you with new terms and concepts associated with the ControlLogix Platform and RSLogix 5000 software. The first chapter introduces the ControlLogix hardware. Chapter 2 introduces the reader to the RSLogix 5000 software features and basic navigation. Other chapters guide the reader through RSLogix 500 project organization, understanding I/O addressing, performing an I/O configuration for digital as well as analog modules, understanding tag structure, creating tags, setting up a communications path, creating a new RSLogix 5000 project, understanding basic ladder programming and basic relay ladder instructions, editing, adding documentation, and downloading and monitoring the RSLogix 5000 project to either the modular ControlLogix or CompactLogix controller. The book also introduces the Function block programming language and provides the reader an opportunity to interpret and monitor basic function blocks.

Part 2 of the book covers setting up communications, including RSLinx drivers for serial, USB, and Ethernet/IP drivers.

To complete the lab exercises in this book, you need a personal computer running a recent version of Microsoft Windows, such as Windows XP, Vista, or Windows 7. Refer to RSLogix 5000 Release notes for information on Windows software that has been tested with the version of RSLogix 5000 being used and software that has not been tested but is expected to operate correctly. The labs were created using RSLogix 5000 version 18 software. Even though RSLogix 5000 version 21 was the current release of the RSLogix 5000 software at the time of printing, not everyone has the latest version of the software, so we use version 18 in this text. The projects can be upgraded to version 19, 20, or 21, if desired, and the labs should still work fine. Keep in mind that the controller major firmware level must match the project software level. For example, if the RSLogix 5000 project software major revision is version 18, the controller firmware major revision must also be 18.

WHAT'S NEW IN THE SECOND EDITION?

The second edition of *Introduction to the ControlLogix Programmable Automation Controller with Labs* has been updated and expanded to include many new topics and lab exercises.

- Chapter 1 was updated to include the new 1756-L7 series controllers as well as the 5370 L series of CompactLogix controllers released with software version 20.
- Chapter 6 has been expanded to include modular analog I/O configuration and labs.
- Chapter 8: Communication between your Personal Computer and Your PLC is new. Setting up communications was removed from the first edition timer chapter and moved forward so that students can download their projects and go online much earlier in the class.

The following new chapters were added:
- Chapter 14: ControlLogix Counter Instructions
- Chapter 15: Comparison Instructions
- Chapter 16: Data-Handling Instructions.
- Chapter 17: Introduction to the Get System Values (GSV) and Set System Values (SSV) Instructions
- Chapter 18: Introduction to the RSLogix 5000 Function Block Programming Language

Part 2: ControlLogix and RSLinx Communications

One area where most PLC users experience difficulty is in setting up communications between their personal computer and the PLC. The new chapters in Part 2 provide the student, as well as users in industry, a reference for configuring different communications options using Rockwell Software's RSLinx. Whereas RSLogix 5000 software is used to create, modify, and monitor the project to be downloaded into your controller, RSLinx, a second package of software, is used to configure the communications drivers between the personal computer and the PLC. After the required driver is configured, we look at setting up the communication path in the RSLogix 5000 project Path toolbar, identifying the link between the two pieces of hardware. The project can then be downloaded. The new chapters step the user through configuring different communications options including setting up USB drivers as well as determining or setting up the IP address for a personal computer, configuring or modifying the IP address of an Ethernet module using Rockwell Software's RSLinx, configuring an Ethernet/IP driver, and downloading the project.

- Introduction to Part 2 Chapters
- Chapter 19: Configuring a Serial Driver Using RSLinx
- Chapter 20: Configuring the Keyspan by Tripp Lite, High-Speed USB to Serial Adapter Model USA-19HS
- Chapter 21: Configuring an RSLinx Serial Driver Using a Rockwell Automation 9300-USBS USB to Serial Adapter
- Chapter 22: Installing and Configuring a USB Driver for 1756-L7 Series Controllers
- Chapter 23: Determining and Modifying a Personal Computer's IP address
- Chapter 24: Configuring 1756 ControlLogix Modular Ethernet Hardware
- Chapter 25: Configuring the Ethernet IP Address for a CompactLogix 1769-L23E, 1769-L32E, or 1769-L35E
- Chapter 26: Configuring a 1756-ENET Ethernet Driver Using RSLinx
- Chapter 27: Configuring Ethernet/IP Drivers Using RSLinx Software
- Chapter 28: Configuring CompactLogix 1769-L23E, 1769-L32E, or 1769-L35E Ethernet/IP Driver Using RSLinx
- Chapter 29: Configuring a USB Driver for a 1756-EN2T Ethernet Communications Module

Again, welcome to the world of ControlLogix, and good luck as you work through the exercises!

SUPPLEMENTS

Student Companion Site

A Student Companion Website is available containing six lab files for use in the lab exercises. Accessing a Student Companion Website site from CengageBrain:

1. Go to: hyperlink "http://www.cengagebrain.com/" \t "_blank" http://www.cengagebrain.com
2. Enter author, title or ISBN in the search window
3. **When you arrive at the Procuct page, click on the Access Now Tab**
4. Click on the resources listed under Book Resources in the left navigation pane to access the project files.

Instructor Companion Site

An Instructor Companion Web site containing supplementary material is available. This site contains answers to Review Questions, testbanks, an image gallery of text figures, and chapter presentations done in PowerPoint™. Contact Delmar Cengage Learning or your local sales representative to obtain an instructor account.

Accessing an Instructor Companion Web site from SSO Front Door

1. Go to http://login.cengage.com and log in using the Instructor e-mail address and password. Enter author, title, or ISBN in the **Add a title to your bookshelf** search box. Click **Search** and then click **Add to My Bookshelf** to add Instructor Resources.
2. At the Product page, click the **Instructor Companion site** link **New Users.**

If you're new to Cengage.com and do not have a password, contact your sales representative.

PART

1

INTRODUCTION TO THE CONTROLLOGIX PROGRAMMABLE AUTOMATION CONTROLLER

This is an introductory text and lab manual that introduces the reader to the Rockwell Automation ControlLogix Programmable Automation Controller and RSLogix 5000 software. The first chapters introduce ControlLogix and CompactLogix hardware, the RSLogix 5000 software, project organization, and tag formatting, as well as digital and analog I/O configuration. After learning what ControlLogix is and basic project organization, we move on to creating a new project. An introduction to basic ladder logic programming and knowledge of basic ladder instructions prepares you to create rungs. Once you have completed creating your first project, we download, run, and monitor the newly created project. We then move on to working with additional instructions such as timers and counters, comparison, data handling, and adding documentation. Chapter 17 introduces the Get System Values (GSV) and Set System Values (SSV) instructions. The last chapter in this section introduces the Function Block Diagram programming language, including interpretation labs.

CHAPTER 1

Introduction to ControlLogix Hardware

OBJECTIVES

After completing this lesson you should be able to

- Identify different members of the ControlLogix family.
- Understand available ControlLogix controllers and their features.
- Identify I/O modules by their part numbers.
- Understand the differences between the modular ControlLogix and the members of the CompactLogix family.
- Identify ControlLogix communications modules.

INTRODUCTION

This chapter introduces ControlLogix system hardware and the different members of the ControlLogix family. ControlLogix is the newest member of the Rockwell Automation and Allen Bradley PLC offerings. ControlLogix is not a regular *programmable logic controller* (PLC). It is much more, including many advanced features not available in the earlier PLC-5 or SLC 500 families of PLCs. These older PLCs are 16-bit computers. ControlLogix is a 32-bit computer. Because ControlLogix has many advanced features that are far beyond the capabilities of common PLCs, ControlLogix is classified as a *programmable automation controller* (PAC). ControlLogix can be programmed in four different programming languages: ladder logic, function block diagram, sequential function chart, and structured text.

The Rockwell Automation SLC 500 PLC can only be programmed in ladder logic, whereas the PLC-5 can be programmed in ladder logic, sequential function chart, and structured text. Function block diagram is a new programming language for Rockwell Automation. As an automation controller, ControlLogix can control most anything in a modern automated system, including

- Sequential control such as conveyor control
- Process control
- Batching using Phase Manager
- Motion Control using ladder, function block, or structured text-based programming (analog or digital motion control modules can be inserted directly into the modular ControlLogix chassis or 1768 Modular CompactLogix backplane)
- A communications gateway or a way to bridge from one industrial network to another

Figure 1-1 illustrates the members of the ControlLogix family, which include the Modular ControlLogix PAC and smaller versions with fewer capabilities, such as the FlexLogix and CompactLogix. The PowerFlex 700S drive is shown, as it can be part of the ControlLogix family and DriveLogix. SoftLogix is a software PAC when an industrial computer is used as the PAC controller instead of the traditional chassis-mounted controller. When an industrial computer is used as the controller, remote chassis of I/O is connected via networks such as Ethernet/IP, ControlNet, and DeviceNet.

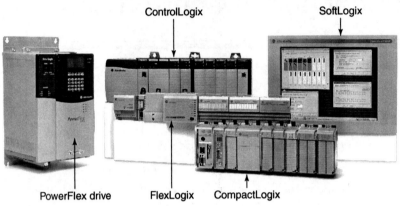

Figure 1-1 Members of the ControlLogix family.
Used with permission Rockwell Automation, Inc.

The ControlLogix modular PAC is a high-performance multicontroller system. Unlike traditional PLCs in which only a single controller is allowed in the chassis, ControlLogix provides the flexibility to add multiple controllers as system requirements and complexity dictate. The key is the backplane of the chassis, which is a ControlNet network. The backplane is called ControlBus. Data on a ControlNet network are transmitted at 5 million bits per second. Being a ControlNet network, the backplane allows the controller or controllers to be inserted in any slot of the chassis. This is contrary to traditional PLCs in which only one controller is allowed and must be inserted into the chassis in the far left slot of the chassis, which is known as *slot zero*.

ControlLogix chassis slot numbers are still numbered, starting with slot zero on the far left and incrementing in decimal numbers to the right. The modular ControlLogix chassis sizes are 4, 7, 10, 13, and 17 slots. Figure 1-2 shows two controllers installed in a 10-slot ControlLogix chassis, one controller in slot 3 and the other in slot 6. The controllers can be easily identified by their key switches. The third module from the left is a ControlNet bridge or communication

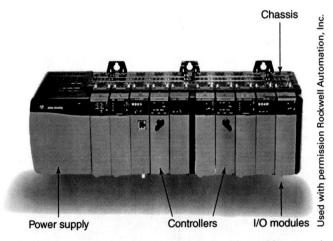

Figure 1-2 Example of a possible ControlLogix modular chassis and its associated power supply, controller, and I/O modules.

module. This is distinguishable by the RJ-45 *network access port* (NAP) or connection near the top of the gray portion of the module's faceplate. The remaining modules could be standard I/O, specialty, or other communications modules. The power supply is mounted on the far left. There are no chassis-mounted power supplies in a ControlLogix system, as might be found in a PLC-5 chassis.

CONTROLLOGIX CONTROLLER FEATURES

ControlLogix modular controllers are grouped into a series such as series 5, series 6, and series 7 controllers. For the most part, controllers within the same series are the same except for the amount of memory they contain. Modular controllers are available with only a serial port up through the series 6 controllers. Series 7 controllers contain a USB port. Communications to other networks, such as Ethernet/IP, ControlNet, DeviceNet, Foundation Field Bus, Data Highway Plus, Remote I/O, or DH-485, are made through purchasing and inserting the desired communications modules into the ControlLogix chassis. Figure 1-3 shows a 1756-L63 controller. Note the controller identification near the top of the module. The text Logix 5563 identifies the module in the figure as a 1756-L63 controller. Directly below the controller identification are the status LEDs (lights). Below the LEDs is the key switch, which is used to switch the controllers operating mode from Run to Remote or to Program mode. Behind the currently open door is the battery, serial port, and nonvolatile memory card slot. Each of these features is defined below. The first release of a piece of hardware such as a controller is a series A controller. As improvements are made, the module increments to a series B, whose features we discuss in this text. Refer to Figure 1-3 to identify 1756-L63 series B controller features.

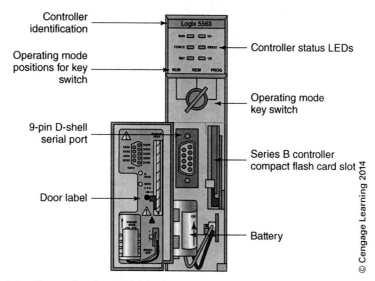

Figure 1-3 Modular ControlLogix 1756-L63 Series B controller.

Figure 1-4 describes the part numbers, series, memory, and minimum RSLogix 5000 software required to use the series 6 controller.

Series 7 Controllers

The L7 series of controllers were released in 2010. The 1756-L73 and 1756-L75 controllers came out in the spring of 2010 with the release of RSLogix 5000 software version 18. The 1756-L72 and 1765-L74 controllers became available in the fall of 2010 with the release of RSLogix software version 19. Some of the new features for the L7 series are listed here:

6 INTRODUCTION TO CONTROLLOGIX HARDWARE

Controller	Memory	Minimum RSLogix 5000 Software
1756-L61 Series A or B	2 MB	12
1756-L62 Series A or B	4 MB	12
1756-L63 Series A	8 MB	10
1756-L63 Series B		12
1756-L64	16 MB	16
1756-L65	32 MB	17

Figure 1-4 ControlLogix series 6 controller specifications.
© Cengage Learning 2014

- Dual core CPU
- USB communication port to replace the standard 9-pin D-shell serial port
- Compact Flash nonvolatile memory card replaced with the smaller secure digital (SD) nonvolatile memory card
- Scrolling data display on the front of the controller to provide the user more information as to the controller status
- Energy storage module (ESM) replacing the lithium battery

Figure 1-5 illustrates the series 7 controller features.

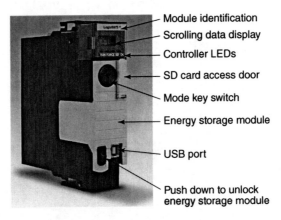

Figure 1-5 Series 7 controller features.
Used with permission Rockwell Automation, Inc.

The table in Figure 1-6 provides an overview for Series 7 controller memory specifications and minimum RSLogix 5000 software required to use the controller.

Controller	Memory	Minimum RSLogix 5000 Software
1756-L71	2 MB	20.11
1756-L72	4 MB	19.11
1756-L73	8 MB	18.11
1756-L74	16 MB	19.11
1756-L75	32 MB	18.11

Figure 1-6 Series 7 controller memory and minimum RSLogix software.
© Cengage Learning 2014

Controller Operating Modes

A controller has two modes of operation. The controller is in either Program mode or some variation of Run mode. The most common modes are Run and Program. The available operating modes when using the key switch include Run, Program, and Remote. The operating modes of the ControlLogix are explained in the following paragraphs. Figure 1-7 illustrates 1756-L63 controller features, including the operating mode key switch.

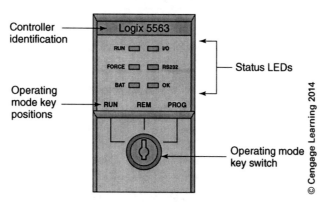

Figure 1-7 ControlLogix controller features.

Program Mode

When the controller is in Program mode, it is accepting instructions either as a new program or as changes to instructions (adding or deleting instructions, or *rungs,* of an existing program). Changing or adding instructions or rungs is typically accomplished in Program mode; editing an existing running program, however, is called "online editing."

Run Mode

After all instructions have been entered in a new program or all editing has been completed to an existing program, the controller is put in Run mode. While in Run mode, a controller is executing the instructions programmed on the rungs of ladder logic that make up the user program. In Run mode, the controller is in its operating cycle, which is called the *controller scan* or *sweep.*

Remote Run Mode

ControlLogix controllers have key switches built into the front of the controller module. This key switch can be used to put the controller in Run or Program mode. Using the key switch to change modes is called *operating in local mode.* If a PAC controller is put in Run mode when programming or monitoring a PAC program from an industrial or personal computer, the PAC is then in Remote Run mode (REM). The key switch is in REM.

Remote Program Mode

If the controller is put in Program mode from a remote programming device such as a personal computer, it is said to be in *Remote Program mode.* Any programming operations can be accomplished from a computer while it is connected to the controller that is in Remote Program mode.

Test Mode

After developing a user program or when done editing a user program, a programmer might wish to test program execution before allowing the PAC to operate the actual hardware. Most Test modes operate much like Run mode, with the exception of actually energizing real-world outputs. The controller still reads inputs, executes the ladder logic, and updates the output tags, but all without energizing output circuits. ControlLogix controllers have continuous-scan Test mode available for

8 INTRODUCTION TO CONTROLLOGIX HARDWARE

testing programs before they are put into production. The continuous-scan Test mode is used to run the program continuously to check out programs or for troubleshooting. The only difference between normal Run mode and continuous-scan Test mode is that although the output tags are updated in the latter, the programmer determines whether to energize the output points.

In Test mode, each ControlLogix output point state is determined by how the output state during Program mode parameter is set up in the RSLogix 5000 I/O configuration for the PLC project. Selections are "on," "off," or "hold" and are configured by the user as part of the software I/O configuration.

Operating modes can also be selected from the RSLogix 5000 software when online with the controller. When using software, Test mode can be selected in addition to the other three. Figure 1-8 shows the drop-down menu by which operating modes can be selected when using the RSLogix 5000 software.

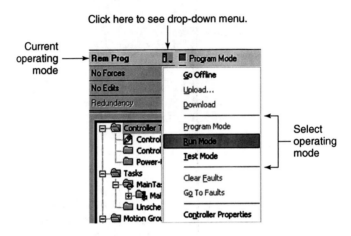

Figure 1-8 RSLogix 5000 Software remote operating mode selection drop-down menu.
Used with permission Rockwell Automation, Inc.

Controller LEDs

The specific controller you have determines which LEDs are available. Figure 1-7 shows a common ControlLogix controller and its controller status LEDs. Each LED is described here. Refer to the controller's installation guide for information on specific controllers.

RUN　　The controller is in Run mode, Program mode, or Test mode.

FORCE　Displays the status of I/O forces in the controller.

BAT　　This light alerts the user to the status of the controller battery. If the light is off, the battery is good. If the light is solid red, the battery is either not installed or needs replacement. Powering down the controller will most likely result in the loss of the project being stored in volatile memory.

I/O　　The I/O light supplies the following information regarding the controller:
- There are no devices in the I/O configuration.
- Controller memory is empty.
- The controller is communicating with all devices in its I/O configuration.
- One or more devices in the I/O configuration are not responding.

RS-232　Displays when there is serial communication to and from the RS-232 serial port.

OK　　The OK light supplies the following information regarding the controller:
- This is a new controller and needs firmware update.
- A major fault is on controller.
- Controller is OK.
- Controller is storing to or loading from nonvolatile memory.

Battery Backup for the Controller's Volatile Memory

The read–write memory (RAM) in the controller makes it possible to enter and change the user program and store data in memory. This memory is referred to as *volatile memory* because there must always be power supplied to the memory's integrated circuit chips. As soon as power is removed or lost through a power failure, volatile memory chips forget what was stored in them. A PAC would be of little value if every time it powered down—whether from power failure (e.g., blown fuses or tripped circuit breakers) or power interruptions (from a utility or weather-related power outages)—the operator had to reenter the user program and any data stored in the data tables. To circumvent memory failure problems, PAC manufacturers use lithium batteries to provide backup power for volatile memory. Battery life varies among controllers and is usually determined by the amount of memory that is backed up and the amount of time the battery is actually providing power to memory. The lithium batteries are not rechargeable. When the PAC is powered, the battery is not being used.

Most controllers have a battery condition status indicator light (an LED). Normally this light is off because the battery is OK. When the battery voltage falls below the acceptable threshold, the battery condition status indicator light comes on. When the indicating light comes on, the battery must be changed within a few days to a few weeks to avoid memory loss. Check the particular controller's specifications for the amount of time after the battery LED comes on to see how much time will transpire before memory loss. Typically, changing the battery is as simple as unplugging and replacing the old battery (consult the manufacturer's documentation and procedures for more information).

Series 6 Battery Backup for Volatile Memory

Series A 1756-L61, 1756-L62, and 1756-63 controllers and earlier use the controller battery 100 percent of the time to maintain power to controller volatile memory to maintain your project and associated data in memory when the controller is either shut down or as the result of a power loss. Because the ControlLogix controllers have a vast amount of memory, a lot of battery power is required to maintain the contents of volatile memory when the controller is powered down. As a result, the battery life is limited, and a series A controller battery if used 100 percent may drain to a critical level in as little as two or three weeks.

Series B L63 controllers and L64 and L65 controllers use their batteries differently. These controllers have a blue-colored battery that is used only to transfer the project to onboard nonvolatile memory when power is lost as the result of power loss.

Series 7 Controller Battery Backup for Volatile Memory

Series 7 controllers have a capacitor module that replaces the battery function and eliminates worrying about project loss from a dead battery on earlier controllers. The capacitor assembly is referred to as an energy storage module. The ESM is charged when the controller is powered and will provide the energy to transfer the contents of controller volatile memory to onboard nonvolatile memory at power down or power loss. Volatile memory transfer on power loss or power down works about the same as the series 6 series B controllers using the ESM rather than the battery.

Lithium Battery Handling and Disposal

Used lithium batteries are considered hazardous waste and require special precautions in handling, transporting, and disposal.

Follow the manufacturer's guidelines in handling or disposing of lithium batteries. Observe the following precautions:

1. These batteries are not rechargeable. Attempting to recharge a lithium battery could cause the battery to overheat and possibly explode.
2. Do not attempt to open, puncture, or crush a lithium battery. The possibility of an explosion or contact with corrosive, toxic, or flammable liquids could result.
3. Do not incinerate batteries. High temperatures could cause them to explode.
4. Severe burns can result if the positive and negative terminals are shorted together.

5. The U.S. Department of Transportation regulates the transportation of materials such as lithium batteries within the United States. As a general rule, lithium batteries cannot be transported on a passenger aircraft. They may be transported via motor vehicle, rail freight, cargo vessel, or cargo aircraft. Before shipping lithium batteries, check on the proper procedures, packaging, and hazard identification.
6. As a general rule, contact the city or county recycling authority for correct local disposal methods.

In many situations, a machine builder does not want to have to worry about whether the end customer will be responsible for replacing a low battery. In addition, if the end customer is negligent in replacing the battery and power to the PAC is lost or interrupted, causing the PAC program to be lost, then who will pay to reload the PAC program? Because most machine builders consider their user programs proprietary, they would not wish to mail the end user a CD-ROM to reload the program. To circumvent these types of problems, optional nonvolatile memory is available for many controllers.

Nonvolatile Memory

Nonvolatile memory is an option for some ControlLogix controllers. If available, nonvolatile memory will either be built into the ControlLogix controller or available as either a 64-megabyte (MB) or 128 MB compact flash card that is similar to a digital camera's memory card. There are many advantages to using a nonvolatile memory:

1. Nonvolatile memory requires no battery, so battery problems in some applications are eliminated.
2. If a machine builder needs to upgrade the program contained on a nonvolatile compact flash card, it is simple to produce a new card with the current version of the project and send it to the end user. The end user simply shuts down the PAC, removes the card from the controller, installs the new memory card, and powers up the unit. If programmed to do so, the controller will automatically upload the new project from the memory card into controller memory and go into Run mode. By receiving the old memory card from the end customer, the machine builder can feel confident that the program was successfully upgraded and can reuse the memory card for another upgrade.
3. Nonvolatile memory also offers automatic upload on startup, where memory has been corrupted, or when a project in the controller is missing because of a dead battery.

A major advantage when using nonvolatile memory in a PAC is the option for automatic project load on startup or RAM memory error. If the PLC is in a remote location such as an unmanned facility, an automatic project load can be a real benefit. This could eliminate a field service engineer from making a service call to reload memory and start a PAC. A machine builder's remotely installed PAC-controlled equipment can automatically load nonvolatile memory into the controller's user memory under the conditions listed here.

These automatic memory load functions can be set up in controller properties by setting the correct options from the RSLogix 5000's software controller properties nonvolatile memory tab under any of the following conditions:

1. When a controller memory error or memory corruption is detected
2. Whenever controller power is cycled and after the load the controller status is returned to Remote Program mode
3. Whenever controller power is cycled and after the load the controller automatically goes into REM
4. If the controller fails

In the last case, compact flash card can be removed from the failed controller and inserted into the new controller. On power up, the firmware and project can be loaded into the new controller. This feature requires RSLogix 5000 software version 12 or greater.

Some ControlLogix controllers have the option of also using the newer industrial compact flash cards for nonvolatile memory storage. ControlLogix controllers require a minimum firmware version of 11 to use the compact flash card. In the case of a 1756-L61, L62, or L63 controller, a 64-MB compact flash card can be installed in a series A controller as illustrated in Figure 1-9. The top portion of the figure shows the controller on its side and a wire locking clip that must be unlocked before the compact flash card can be installed. The center portion illustrates insertion of the compact flash card. Once inserted, the wire locking clip is moved downward until it snaps over the card, locking it into position. In Figure 1-3, note the compact flash card slot on the upper-right area behind the controller's door. The series B controller compact flash card can be inserted into the controller from the front of the controller without removing the controller from the chassis. Figure 1-10 shows a 64-MB compact flash card.

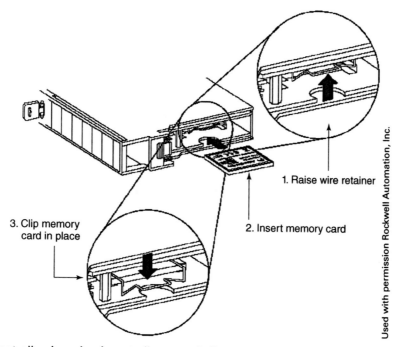

Figure 1-9 ControlLogix series A controller nonvolatile compact flash card insertion.

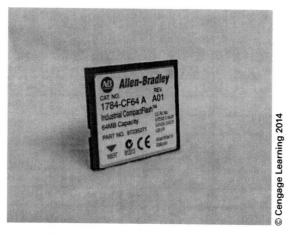

Figure 1-10 A 64-MB compact flash memory card.

Figure 1-11 lists current controllers with their memory size, battery part numbers, and nonvolatile project backup options.

INTRODUCTION TO CONTROLLOGIX HARDWARE

CONTROLLOGIX CONTROLLERS BATTERY AND NON-VOLATILE MEMORY					
Controller	Memory	1756-BA1	1756-BA2	Energy Storage Module	Nonvolatile Memory
Series 6 Controllers					
1756-L61	2 MB	Yes	No	No	Compact Flash
1756-L62	4 MB	Yes	No	No	Compact Flash
1756-L63	8 MB	Series A	Series B	No	Compact Flash
1756-L64	16 MB	No	Yes	No	Compact Flash
1756-L65	32 MB	No	Yes	No	Compact Flash
Series 7 Controllers					
1756-L71	2 MB	No	No	Yes	SD card
1756-L72	4 MB	No	No	Yes	SD card
1756-L73	8 MB	No	No	Yes	SD card
1756-L74	16 MB	No	No	Yes	SD card
1756-L75	32 MB	No	No	Yes	SD card

Figure 1-11 Current controllers with their memory size, battery part numbers, and nonvolatile project backup options.
© Cengage Learning 2014

SELECTING A CONTROLLOGIX MODULAR CONTROLLER

When selecting a ControlLogix controller here are some considerations:

1. How much memory is required for the current application as well as future expansion?
2. Is removable nonvolatile memory required?
3. Do you need a standard serial port or USB?
4. Battery backup or energy storage module (capacitor backup)
5. Major minimum and maximum RSLogix 5000 software revisions

Figure 1-12 lists basic modular controller specifications.

CONTROLLOGIX CONTROLLERS SPECIFICATIONS						
Controller	Memory	Standard Serial Port	USB Port	Energy Storage Module	Non Volatile Memory	Minimum / Maximum RSLogix 5000 Software
Series 6 Controllers						
1756-L61	2 MB	Yes	No	No	Compact Flash	12 / 20
1756-L62	4 MB	Yes	No	No	Compact Flash	12 / 20
1756-L63	8 MB	Yes	No	No	Compact Flash	10 / 20* 12 / 20**
1756-L64	16 MB	Yes	No	No	Compact Flash	16 / 20
1756-L65	32 MB	Yes	No	No	Compact Flash	17 / 20
Series 7 Controllers						
1756-L71	2 MB	No	Yes	Yes	SD card	18 / current
1756-L72	4 MB	No	Yes	Yes	SD card	19 / current
1756-L73	8 MB	No	Yes	Yes	SD card	18 / current
1756-L74	16 MB	No	Yes	Yes	SD card	19 / current
1756-L75	32 MB	No	Yes	Yes	SD card	20 / current

*Series A Controller
**Series B controller

Figure 1-12 Modular controller basic specifications.
© Cengage Learning 2014

COMPACTLOGIX

CompactLogix is also member of the ControlLogix family. CompactLogix was designed for low-end to medium-sized applications that require less memory, software capabilities, less I/O, and less communications capabilities than the modular ControlLogix. CompactLogix is used in many applications as an upgrade from a 16-bit SLC 500 controller to 32-bit ControlLogix hardware with its many additional features. Because CompactLogix is a member of the ControlLogix family, RSLogix 5000 software is used for programming. There are two groups of CompactLogix PLC hardware: the 1769 Modular CompactLogix and the 1768 Modular CompactLogix. We look at the 1769 Modular CompactLogix first.

COMPACTLOGIX 1769-L23

The 1769-L23 CompactLogix is a fixed or packaged controller where the controller, power supply, and I/O are contained in a single unit. A sample of some of the 1769-L23 series CompactLogix is illustrated in Figure 1-13. The L23 series provide a small-sized CompactLogix PAC for applications that need only a few I/O points and minimum memory.

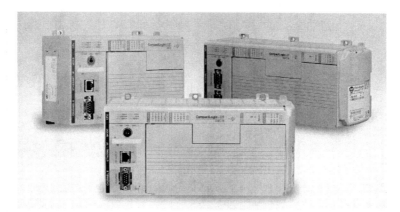

Figure 1-13 1769-L23 fixed I/O CompactLogix.
Used with permission Rockwell Automation, Inc.

Figure 1-14 lists part numbers and basic specification for the 1769-L23 CompactLogix.

Part Number 1769-E	Embedded I/O	1769 Expansion Modules	Memory	Embedded Communications Ports	Minimun / Maximum RSLogix 5000 Software Revision
L23QB1B	16 DC Inputs 16 DC Outputs	2 I/O modules or 2 communication	512K	Isolated Serial and Ethernet/IP	17 to current revision
L23E-QBFC1B	16 DC Inputs 16 DC Outputs 4 universal analog in 2 universal analog out 4 high-speed counters	2 I/O modules or 1 communication	512K	Isolated serial and Non-isolated serial	17 to current revision
L23E-QBFC1B	16 DC Inputs 16 DC Outputs 4 universal analog in 2 universal analog out 4 high-speed counters	2 I/O modules or 1 communication	512K	Isolated Serial and Ethernet/IP	17 to current revision

Figure 1-14 1769-L23 CompactLogix selected specifications.
© Cengage Learning 2014

COMPACTLOGIX 5370 COMPACTLOGIX L2 CONTROLLERS

CompactLogix 5370-L2 controllers with additional features are illustrated in Figure 1-15, were released in December 2011 with the release of RSLogix 5000 version 20 software. An overview of the L2 controller's features is provided here.

- Memory up to 3 MB
- USB port replaces serial communications port
- Internal energy storage (capacitor) eliminates lithium battery
- 1 GB SD card for nonvolatile memory
- Uses existing 1769 I/O modules
- Dual Ethernet ports to support Device level ring network topology
- Integrated motion control across Ethernet/IP

Figure 1-15 shows two of the CompactLogix 5370-L2 controllers

Figure 1-15 CompactLogix 5370-L2 fixed I/O programmable automation controller.
Used with permission Rockwell Automation, Inc.

Figure 1-16 lists selected specifications and part numbers for the newer CompactLogix 5370-L2 Controllers.

Part Number 1769-	Memory	I/O	1769 Expansion Modules	Memory Card	Minimun RSLogix 5000 Software Revision
L24ER-QB1B	.75 MB	16 DC Inputs 16 DC Outputs	4	1 GB included 2 GB optional	20
L24ER-QBFC1B	.75 MB	16 DC Inputs 16 DC Outputs 4 universal analog in 2 universal analog out 4 high-speed counters	4	1 GB included 2 GB optional	20
L27ERM-QBFC1B	1 MB	16 DC Inputs 16 DC Outputs 4 universal analog in 2 universal analog out 4 high-speed counters	4	1 GB included 2 GB optional	20

Figure 1-16 CompactLogix 5370-L2 controllers.
© Cengage Learning 2014

1769 MODULAR COMPACTLOGIX

1769 Modular CompactLogix does not have a chassis for inserting the controller and modules like the modular ControlLogix. CompactLogix is considered a rackless design: The modules simply snap together on a DIN rail or can be mounted directly to the panel backplate. Figure 1-17 shows a 1769 Modular CompactLogix with many of the pieces identified. Note that each module is separate and connected together to make a "bank." The power supply is always near the center of the bank. The 1769 controller must be the leftmost module in the first bank of the system. Three I/O modules then can be placed between the controller and the power supply. As many as eight I/O modules can be placed to the right of the power supply. The number of local I/O modules configurable depends on the controller selected. Figure 1-18 lists controller part numbers and the total number of local I/O modules allowed. The suffix at the end of the part number identifies the controller's communication options. As an example, a "C" at the end of the part number signifies integrated ControlNet communications, while a "CR" signifies ControlNet with redundancy. An "E" at the end of the part number signifies integrated Ethernet communications.

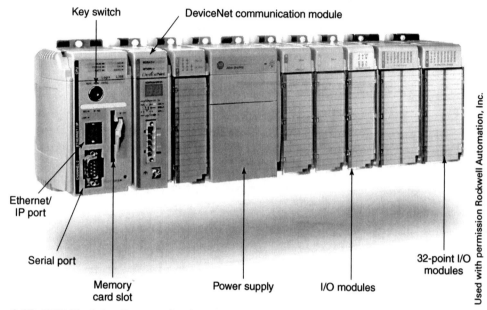

Figure 1-17 1769 Modular CompactLogix.

Controller Part Number	Controller Communications	Local Modules
1769-L31	Serial only	16
1769-L32C	ControlNet	16
1769-L32E	Ethernet/IP	16
1769-L35CR	ControlNet with redundancy	30
1769-L35E	Ethernet/IP	30

Figure 1-18 Total number of local modules configured into three banks.
© Cengage Learning 2014

Local I/O modules are modules that can be directly connected together, using expansion cables, in as many as three banks without using a separate communication module in the bank such as Ethernet or ControlNet, and going out on a network. When expanding the system into a second or third bank, each bank must have its own power supply. Expansion cables and endcaps are used to connect the banks either vertically or horizontally. The last bank requires an endcap to provide termination. Because the controller is the leftmost module, termination for the left side of the bank is built into the controller.

A simple application could include a single bank of I/O modules, a controller, and basic communications. Unlike ControlLogix controllers, CompactLogix controllers are available with integrated ControlNet (C), ControlNet with redundancy (CR), or Ethernet/IP (E) along with a serial port built into the controller. Figure 1-19 shows specifications available for selected 1769 Modular CompactLogix controllers. Each controller can have a 64-MB compact flash card installed.

Part Number	Memory (in Bytes)	Communication Ports	Total Local I/O Modules	Total Local I/O Banks	64-MB Compact Flash Nonvolatile Memory
1769-L31	512 K	2 serial only	16	3	Yes
1769-L32E	768 K	Ethernet/IP and serial	16	3	Yes
1769-L32C	768 K	ControlNet/serial	16	3	Yes
1769-L35E	1.5 M	Ethernet/IP and serial	30	3	Yes
1769-L35CR	1.5 M	ControlNet with redundancy and serial	30	3	Yes

Figure 1-19 CompactLogix controller specifications (part numbers and features).
© Cengage Learning 2014

CompactLogix 5370-L3 Programmable Automation Controllers

CompactLogix 5370-L3 controllers have additional features beyond those available from the 5370-L2 controllers. The 5370-L3 controllers were also released in December 2011. Minimum RSLogix 5000 version 20 software is required to use these controllers. An overview of the L3 controller's features is provided here:

- Memory up to 3 MB
- USB port replaces serial communications port
- Internal energy storage (capacitor) eliminates lithium battery
- 1 GB SD card for nonvolatiile memory
- Uses existing 1769 I/O modules
- Dual Ethernet ports to support Device level ring network topology
- Integrated motion control across Ethernet/IP

Figure 1-20 shows one of the CompactLogix 5370-L3 controllers with 1769 I/O modules. The I/O modules and power supply are the same as those used with the 1769 CompactLogix illustrated in Figure 1-17.

Figure 1-20 A CompactLogix 5370-L3 controller with 1769 I/O modules.
Used with permission Rockwell Automation, Inc.

Figure 1-21 lists selected specifications and part numbers for the newer CompactLogix 5370-L3 Controllers.

Part Number 1769-	Memory	Local 1769 I/O Modules	1769 Expansion I/O Points	Memory Card	Minimum RSLogix 5000 Software Revision	Integrated Motion
L30ER	1 MB	8	256	1 GB included 2 GB optional	20	No
L30ERM	1 MB	8	256	1 GB included 2 GB optional	20	Yes
L30ER-NSE	1 MB	8	256	1 GB included 2 GB optional	20	No
L33ER	2 MB	16	512	1 GB included 2 GB optional	20	No
L33ERM	2 MB	16	512	1 GB included 2 GB optional	20	Yes
L36ERM	3 MB	30	960	1 GB included 2 GB optional	20	Yes

Figure 1-21 CompactLogix 5370-L3 Controllers specifications.
© Cengage Learning 2014

1768 MODULAR COMPACTLOGIX

A newer member of the CompactLogix family is the 1768 Modular CompactLogix, which was designed for low-end to medium-sized applications that require motion control and network connectivity but have fewer software capabilities and fewer I/O capabilities than the modular ControlLogix. A unique feature of the 1768 Modular CompactLogix is that it uses the same I/O modules as the 1769 Modular CompactLogix. The 1768 can integrate as many as eight 1769 Modular CompactLogix I/O modules to the right of the controller. Figure 1-22 shows an example of a 1768 Modular CompactLogix. Note the 1769 Modular CompactLogix backplane to the right of the controller. The controller and 1768 backplane is to the left. The controller is near the center. Much like the 1769 Modular CompactLogix, this platform supports as many as three banks of 1769 I/O Modules in a local I/O configuration. As many as eight I/O modules can be in the same bank as the controller.

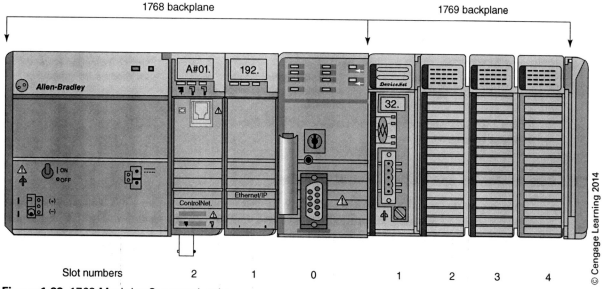

Figure 1-22 1768 Modular CompactLogix.

1768 Controller Properties

Currently, there are two 1768 controllers: the L43 and L45. The major differences between the modules include memory size, total I/O count, and the number of modules installable in the 1768 backplane. Figure 1-23 illustrates an overview of the controller capabilities.

Part Number	Memory	Maximum 1768 Modules	Communication Ports	Total Local I/O Modules	Total Local I/O Banks	64-MB Compact Flash Nonvolatile Memory
1768-L43	2 MB	2	Serial only	16	3	Yes
1769-L45	3 MB	4	Serial only	30	3	Yes

Figure 1-23 1768 Modular CompactLogix controllers (part numbers and features).
© Cengage Learning 2014

Controller features are much like the modular ControlLogix outlined earlier. The top portion of the controller contains status indicators, and the lower portion has the operating mode key switch, RS-232 serial port, and compact flash slot. Much like the modular ControlLogix controllers, the only communication port available on the L43 or L45 controller is an RS-232 serial port. Additional communications options are available as separate communication modules. These modules are introduced in this text. Figure 1-24 illustrates the features of the L43 and L45 CompactLogix controllers. Each feature also is explained. Refer to the 1768 Modular CompactLogix user's manual for additional information on controller features.

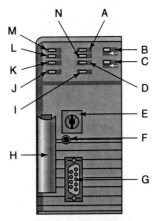

A. I/O
  What is the status of the I/O?
B. Power for 1768 backplane
C. Power for 1769 backplane
D. DCHO
  Is channel 0 configured same as default or differently?
E. Operating mode key switch
F. Button to reset communications to default
G. 9-Pin D-shell RS-232 serial port
H. Compact Flash nonvolatile memory slot and cover
I. OK
  Controller status. As an example, is controller OK or faulted?
J. CF
  Compact Flash card activity.
K. Memory Save
  User program being saved to flash memory?
L. FORCE
  Status of I/O forces
M. RUN
  The Controller operating mode status. Is controller in Run, Program, or Test mode?
N. CH0
  Channel 0 is the serial port. This LED displays status of serial communications.

Figure 1-24 L43 and L45 CompactLogix controller features.
© Cengage Learning 2014

1768 Communication and Motion Control Modules

Communication modules provide an interface to outside networks. The 1768 backplane can have as many as two communication modules and two motion control modules. The L43 controller supports a total of two modules in the 1768 backplane, whereas the L45 controller support as many as four. DeviceNet communications modules would be placed in the 1769 backplane. The available 1768 modules available for insertion in the 1768 backplane as of this writing are shown in Figure 1-25.

Module Part Number	Module Usage
1768-CNB	Interface to ControlNet network
1768-CNBR	Interface to ControlNet network with redundant network cabling
1768-ENBT	Interface to Ethernet/IP network
1768-EWEB	Web server module for remote monitoring and modification of data from an extensible markup language (XML) Web page
1768-M04SE	Digital motion control module for interface to serial real-time communication system (SERCOS) motion drives

Figure 1-25 Modules for insertion in 1768 backplane.
© Cengage Learning 2014

FLEXLOGIX

FlexLogix is the third member of the ControlLogix PAC family. It provides the option to integrate a 32-bit ControlLogix controller with Rockwell Automation's Flex I/O blocks. Flex I/O blocks and their terminal bases have been available for many years as an easy compact method to configure Remote I/O for platforms such as the PLC-5 or SLC 500 PLCs. Communication modules such as DeviceNet, Remote I/O, and ControlNet can serve as the communication interface back to the local PLC from the Flex I/O blocks. The selected communication module would be on the left side of the assembly, whereas as many as eight digital or analog I/O modules could be clipped together to the right of the communication module on a DIN rail to form Flex I/O. This configuration only provided a network interface back to the local PLC controller. There was no intelligence in the remote Flex I/O, only the selected communication interface. With the arrival of FlexLogix, the communication interface module could be replaced with a 32-bit ControlLogix controller. The same eight Flex I/O terminal bases and I/O modules could be used to create either a stand-alone small PAC or a small PAC with as many as two communication interface cards for communication back to a supervisory or other PAC. Figure 1-26 shows FlexLogix. Note the Flex I/O block and terminal base to the right of the controller. The controller is near the center of the figure. There are two slots for communication cards behind the front of the controller, referred to as "daughter cards." The communication cards in the figure inserted behind the controller faceplate are ControlNet network interface cards. Because each card has two bayonet-type connections, these cards provide ControlNet communications with redundant cabling. Figure 1-27 shows a DeviceNet daughter card for the FlexLogix. An Ethernet/IP daughter card is also available.

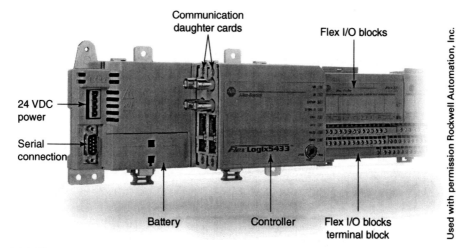

Figure 1-26 FlexLogix 5543.

20 INTRODUCTION TO CONTROLLOGIX HARDWARE

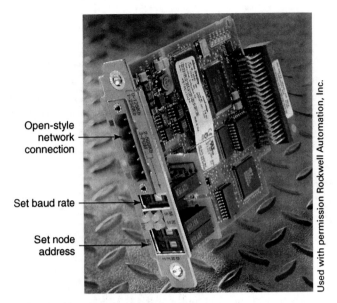

Figure 1-27 FlexLogix DeviceNet daughter card.

RSLogix 5000 software version 16 is the last version of RSLogix 5000 software the FlexLogix supports. In other words, as of this writing, FlexLogix is not upgradeable beyond RSLogix 5000 software version 16. RSLogix 5000 version 16 was released at the end of April 2007. FlexLogix cannot be upgraded to new features available when RSLogix 5000 version 17 or future software revisions are released. Now that we have introduced the controllers of the ControlLogix family, we start by looking at I/O modules.

CONTROLLOGIX DIGITAL I/O MODULES

ControlLogix digital input modules fall into two categories: (1) 1756 modules for the modular ControlLogix and (2) 1769 modules for the 1768 and 1769 Modular CompactLogix platform. Being the larger system, the modular or 1756 ControlLogix has more modules with additional features.

ControlLogix 1756 Digital I/O Modules

Each 1756 ControlLogix digital I/O module falls into one of four categories: standard, individually isolated, electronically fused, or diagnostic. We look closely at the diagnostic module features when we introduce configuring ControlLogix I/O in Chapter 6. Chapter 6 lab exercises provide the opportunity to configure diagnostic 1756 I/O modules. An overview of the module types is listed here.

- A standard input or output module is similar to a module found in a traditional PLC. The standard module does not have additional module-based or I/O point–based diagnostic features.
- Individually isolated I/O modules have individually isolated I/O points. Each point has its own separate common screw terminal.
- Electronically fused output modules have internal electronic fusing to protect the module output points from drawing too much current.
- Diagnostic input and output modules have additional I/O point–level diagnostic features such as open-wire detection for input modules and no load detection for an output module.

Figure 1-28 illustrates the format of modular ControlLogix I/O module part numbers. Modular ControlLogix part numbers begin with 1756.

INTRODUCTION TO CONTROLLOGIX HARDWARE 21

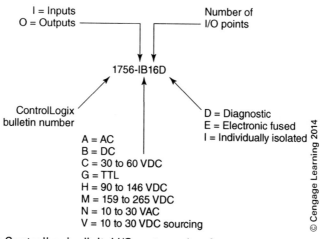

Figure 1-28 Modular ControlLogix digital I/O part number format.

1756 Module Features

Modular ControlLogix digital I/O modules contain two pieces: the module and the removable terminal block. The removable terminal block is where the input or output wiring is connected. The terminal block can be easily removed by using the module door as a handle. Figure 1-29 shows a diagnostic input module with the major features identified. Note the removable terminal block locking switch in the upper right of the module. The locking switch slides up or down to lock or unlock the removable terminal block. If the switch is in the up position, the terminal block can be easily removed by pulling it away from the module. Additional features are explained here.

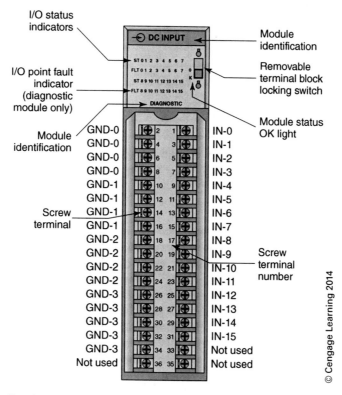

Figure 1-29 ControlLogix digital diagnostic input module.

- The very top of the module identifies the type of module.
- The state (ST) lights for each I/O point indicate the on or off status of each I/O point.
- Fault (FLT) lights or LEDs are only available on diagnostic modules. There is a fault LED for each input or output point. The numbered FLT LED alerts the user that the numbered input or output point that is illuminated has some type of point-level fault. We look at what causes the FLT lights to illuminate in the lesson for Chapter 6. If an output module were electronically fused such as a 1756-OA8E, it would have an electronic fuse LED for each point rather than a fault LED.
- Additional module identification is found below the module status LEDs. The illustrated module is a diagnostic module. A standard module has no text in this area.
- The OK LED is found to the left of the removable terminal block locking switch. This LED has five states and is used for troubleshooting module problems. Refer to the module user's manual for additional information on the OK LED.
- The bottom portion of the module is the removable terminal block for landing input or output wires. Note there is a screw number stamped into the terminal block plastic. Do not confuse this number with the actual input or output bit number. Because there are only two basic removable terminal block part numbers, these terminal blocks are used on many different I/O modules. A sticker included with each I/O module can be affixed to the inside of the terminal block's door to identify the actual screw terminal and the associated bit number. This information can also be found in the user's manual.

Figure 1-30 lists a few of the 1756 digital I/O modules showing the part number format.

Part Number	Voltage	Input/Output	Number of I/O Points	Module Type
1756-IA16	120 AC	Input	16	Standard
1756-IA16I	120 AC	Input	16	Individually isolated
1756-OA8	120 or 240 AC	Output	8	Standard
1756-OA8D	120 AC	Output	8	Diagnostic
1756-OA8E	120 AC	Output	8	Electronic fusing
1756-IB16	12 or 24 VDC	Input	16	Standard
1756-IB16D	12 or 24 VDC	Input	16	Diagnostic
1756-IB16I	12 or 24 VDC	Input	16	Individually isolated

Figure 1-30 Selected 1756 digital I/O part numbers (sample 1756 digital I/O modules).
© Cengage Learning 2014

I/O Module Wiring Options

ControlLogix I/O module input or output wiring can be connected directly to a module's removable terminal block. However, when ordering a new I/O module, the removable terminal block does not come with the module, so the user must order either the proper removable terminal block or the proper 1492 interface module (IFM) for the specific module. The 1492 IFMs can be ordered with a prewired cable and removable terminal block already connected. Installation is as easy as plugging the removable terminal block into the I/O module and the other end of the prewired cable into the IFM module. The IFM simply snaps to the DIN rail. Figure 1-31 is an example of a ControlLogix interfaced to a couple of IFM modules on a DIN rail. There are many variations of the IFM modules—for example, some IFM modules have LEDs and others have fuses.

The 1492-IFMs are extremely popular with machine builders because the IFM module has a large terminal strip for landing wires. It is much easier to wire an IFM with a large terminal block rather than attempting to wire a removable terminal block with either 16 or 32 I/O points.

INTRODUCTION TO CONTROLLOGIX HARDWARE 23

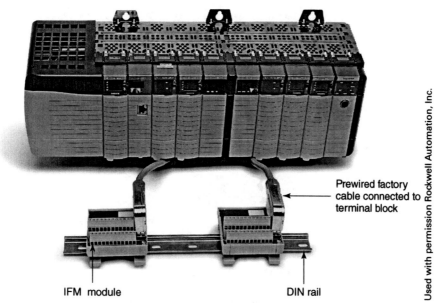

Figure 1-31 Examples of 1492-IFM modules and prewired cable and terminal block connected to ControlLogix I/O modules in a modular chassis.

CONTROLLOGIX ANALOG MODULES

Analog input and output signal connections to an analog module are referred to as *channels*. ControlLogix analog tags are either integers or real numbers. Tag data format is assigned in the analog modules I/O configuration. Typical analog signals could be 0–10 VDC, 0–5 VDC, –10 to +10 VDC, 0 to 20 milliamps (ma) or 4 to 20 ma. The module converts the varying voltage or current input signal to a value and stores the value in a buffer in the module until the signal is transferred to the controller input tag. Analog output values—as either integers or real numbers—are transferred from the controller's output tag to the module and converted to the desired varying voltage or current signal and output to the field device. The 1756 analog modules come as 4, 6, 8, or 16 channel modules. Figure 1-32 illustrates the 1756 analog module part number format.

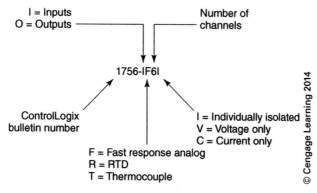

Figure 1-32 ControlLogix analog module part number format.

Sample 1756 analog part numbers are shown in Figure 1-33.

Figure 1-34 shows a typical 1756 modular ControlLogix analog module with many of the features identified.

Catalog Number	Description
1756-IF16	16-point nonisolated current/voltage inputs
1756-IF8	8-point nonisolated current/voltage inputs
1756-IF6I	6-point isolated current/voltage inputs
1756-IR6I	6-point isolated RTD inputs
1756-OF8	8-point nonisolated current/voltage outputs
1756-OF6VI	6-point isolated current/voltage outputs

Figure 1-33 Selected ControlLogix analog module part numbers.
© Cengage Learning 2014

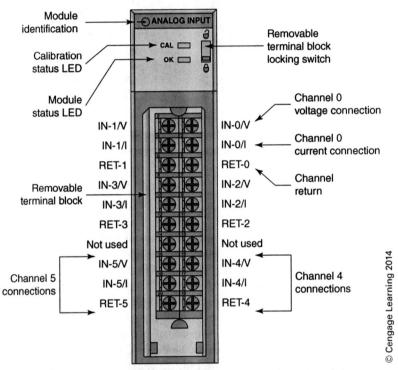

Figure 1-34 1756-IF6I ControlLogix six-channel isolated analog input module.

1756 Modular ControlLogix analog module features are listed here.

- The very top of the module identifies the type of module.
- The locking switch slides up or down to lock or unlock the removable terminal block. If the switch is in the up position, the terminal block can be easily removed by pulling it away from the module.
- The OK LED alerts the user as to the current state of the module. LEDs can be in one of four states: flashing green, steady green, flashing red, or solid red. This LED is used to troubleshoot module problems. Refer to the module user's manual for additional information on the OK LED.
- The CAL LED is flashing green when the module is being calibrated. Normal state is off.
- The bottom portion of the module is the removable terminal block for landing input or output wires. Note that a screw number is stamped into the terminal block plastic. Do not confuse this number with the actual analog channel number. Because there are only two basic removable terminal block part numbers, these terminal blocks are used on many different I/O modules. A sticker is included with each I/O module that

can be affixed to the inside of the terminal block's door to identify the actual screw terminal and the associated channel number. This information can also be found in the user manual.

COMPACTLOGIX 1769 I/O MODULES

Both the 1769 and 1768 Modular CompactLogix use the same 1769 digital and analog I/O modules. Because CompactLogix has fewer features than the modular ControlLogix, I/O module selections are more standard. There are no diagnostic or electronically fused modules for the CompactLogix platform. Here modules fall into two general categories: standard and individually isolated. For the most part, the module part-numbering format is the same except that CompactLogix modules start with 1769 as the beginning part number. Figure 1-35 provides a sample of selected CompactLogix I/O and analog modules.

Part Number	Voltage	Input/Output	Number of I/O Points	Module Type
1769-IA8I	120 VAC	Input	8	Individually isolated
1769-IA16	120 VAC	Input	16	Standard
1769-OA8	120 or 240 VAC	Output	8	Standard
1769-IQ16	24 VDC	Input	16	Sink or source
1769-OB8	24 VDC	Output	8	Sourcing
1769-OW8I	AC or DC	Relay outputs	8	Isolated
1769-IF4I	Volts or current	Inputs	4 channels	Analog isolated
1769-OF2	Volts or current	Outputs	2 channels	Analog

Figure 1-35 Selected 1756 digital I/O part numbers (selected 1769 digital and analog I/O modules).
© Cengage Learning 2014

HOW CONTROLLOGIX MODULES OPERATE

ControlLogix can be a simple stand-alone PAC or a highly complex PAC with thousands of rungs of sequential logic, process control, batching, and motion control. As many as 128,000 inputs and outputs can be controlled across multiple chassis and multiple networks. Information can be sent to multiple devices such as operator interfaces, other PLCs, variable frequency drives, motion control drives, and other ControlLogix controllers in the same or remote chassis. To increase the efficiency of this large controller, the ControlLogix scan has been modified from a traditional PLC. To start, we next review how a traditional PLC scans.

Traditional PLC Operating Cycle or Scan

When the traditional PLC is put into Run mode, the controller begins the following sequence. To assist understanding, here we only list the major pieces of the controller scan. We use a PLC-5 or SLC 500 in our example (see Figure 1-36).

1. When the controller enters Run mode, all of the inputs are read and the information is written in the input status table. The input status table now has fresh information as to the current state of the inputs for use as the ladder logic is solved.
2. The main ladder file is executed starting with the first rung, typically rung 0. Rungs are read left to right, starting with rung 0, to the highest numbered rung. As each rung is solved, the state of the output is written to the output status table. The output status table is updated as each rung is solved, so the information is current in case a later rung might wish to examine the output state of the earlier rung during the logic scan. Output information is not written to the output modules at this point.

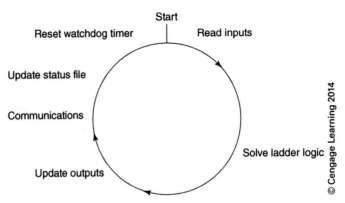

Figure 1-36 Typical PLC scan.

3. When all of the ladder rungs have been solved, one word from the output status table is sent—one word at a time—by way of the backplane to the respective output module, where field devices are updated.
4. The communication section contains three basic pieces:
   a. Communications update which includes
      - Updating the computer screen with current information if the user is online and monitoring the PLC
      - Sending any message instruction programmed on the ladder
   b. Updating the S2 status file
   c. Resetting the watchdog timer, which ensures that the scan is completed in a predetermined time and the scan is restarted (if the watchdog timer times out, the controller will fault)
5. The scan starting over

ControlLogix Operating Cycle

The ControlLogix operating cycle is entirely different from the step-by-step scan described for a traditional PLC. Inside a ControlLogix two separate 32-bit unsynchronized processes are going on simultaneously—that is, *asynchronously*. ControlLogix I/O updates occur asynchronously to the scan of the logic. This provides more efficiency and more control over when the input field device data are updated in the input tag and when the output data resulting from the solved logic are sent to the output modules and their respective field devices. The programmer can specify the time when the input or output data are to be transferred by entering the desired value in the *requested packet interval* (RPI) parameter when performing the I/O configuration for the module in the RSLogix 5000 software. The RPI specifies the rate at which the module being configured sends or multicasts its data. The time configured in the I/O configuration for a digital I/O module ranges from 200 microseconds to 750 milliseconds. For an input module, this is the time when input information is sent from the module across the backplane to the input tags. For an output module, this is the time when output tag information is transferred across the backplane to the specific output module and then on to the field devices. The default RPI for a digital module for RSLogix 5000 starting with version 16 is 20 milliseconds. Figure 1-37 illustrates the logic scan as being one separate operation, whereas the update of the I/O is a totally separate function.

The RPI is set up in the I/O configuration for each module. As long as the programmer is communicating with I/O modules in the local chassis, he or she can select any valid RPI value. However, when I/O modules are in a remote chassis, RPI values are determined by how the network is configured. As an example, if communicating with I/O modules in a remote chassis across a ControlNet network then the *network update time* (NUT) is one network parameter that must be set up. The NUT is simply one update or scan of the network. If the network is updated at

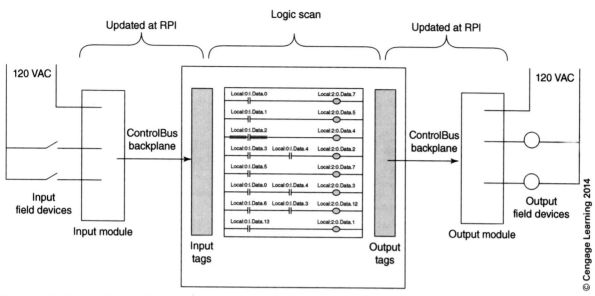

Figure 1-37 ControlLogix I/O update.

20 milliseconds, then one could not program an RPI for 10 milliseconds because data cannot be updated faster than the network can "see it." In Chapter 6, we set up an RPI for local I/O modules when we perform an I/O configuration.

The Communications Portion of the Scan

- Communications include updating the computer screen with current information if the user is online and monitoring the PAC, and sending any active message instructions programmed on the ladder.
- ControlLogix does not have a status file like a PLC-5 or SLC 500 PLC. Specific programming instructions are used to send and receive information to and from the controller.
- ControlLogix does not have a single watchdog timer as a traditional PLC. Each ControlLogix task has one watchdog timer. There may be as many as 32 watchdog timers in a ControlLogix project.
- The scan starts over.

COMMUNICATIONS MODULES

The common networks used with the ControlLogix include Ethernet/IP, ControlNet, DeviceNet, Data Highway Plus, Remote I/O, and serial. Before we explore the different communication modules, we first introduce some common network terminology.

network	A group of devices connected together using cable or fibers. Each network has its own unique cabling or fibers.
node or station	A physical device on a network.
node address	A unique address of a device on a network (much like a house on a street); also called a node or *station address*.
baud rate	The speed at which information is transferred in bits per second.

The ControlLogix 1756 modular controller and 1768 controller have built-in serial ports, so no additional hardware is required when using serial. Network communications other than serial will require the purchase of a separate communication module. Figure 1-38 lists selected ControlLogix network interface modules. Following, we introduce each network.

28 INTRODUCTION TO CONTROLLOGIX HARDWARE

Description	1756 Platform	1769 Platform	1768 Platform
Ethernet/IP (fiber module)	1756-EN2F		
Ethernet/IP	1756EN2T		
Ethernet/IP	1756-ENBT	1769-L32E/L35E*	1768-ENBT
Ethernet	1756-ENET		
Ethernet Web server	1756-EWEB		1768-EWEB
ControlNet bridge	1756-CNB	1769-L32C*	1768-CNB
ControlNet bridge with redundancy	1756-CNBR	1769-L35CR*	1768-CNBR
DeviceNet bridge	1756-DNB	1769-SDN	1769-SDN
Data Highway + Remote I/O	1756-DHRIO		

*Communication ports built into controller.

Figure 1-38 Communication module part number for the ControlLogix family of communication modules.
© Cengage Learning 2014

ETHERNET COMMUNICATIONS

Ethernet is the most popular network today because of its fast speed and readily available off-the-shelf components. The older version of Ethernet uses the 1756-ENET module with a speed of 10 million bits per second, half-duplex. Many of these older 1756-ENET modules are still used in industry. The 1756-ENBT is a commonly used module. The newer Ethernet industrial network, also know as Ethernet/IP, can communicate not only at 10 million bits per second (to interface to the older ENET module and slower networks) but also at 100 million bits per second, full-duplex. Ethernet/IP is an open industrial network standard that supports real-time I/O control as well as messaging among different pieces of hardware. Today Ethernet/IP is the most popular network for new applications for information transfer between controllers and human interface devices or variable frequency drives. Figure 1-39 shows the ENET as well as the ENBT communications modules. It is easy to identify which module is currently used in a system by looking at each module's features. The ENET module has no window near the top of the module; the newer ENBT module does, displaying the module's IP address as well as other

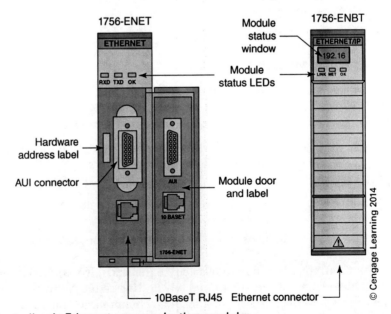

Figure 1-39 ControlLogix Ethernet communication modules.

module status information. The ENET not only has the 10 Base T-RJ45 Ethernet connector on the front of the module behind the module's door but also the older AUI connector found on older Ethernet systems, including most PLC-5 Ethernet controllers. There is no module door on the ENBT module, and the RJ45 connector is on the bottom of the module. Connecting devices on the Ethernet network is accomplished through standard industrial managed switches. The newer 1756-EN2T has two differences from the 1756-ENBT. First, the EN2T has twice the communications capacity. The newer module also has a built-in USB communications port for personal computer connectivity to access the RSLogix 5000 project residing in the controller.

The 1769 Modular CompactLogix Ethernet/IP interface is integrated on the controller, so no additional communications module is required (see Figure 1-40).

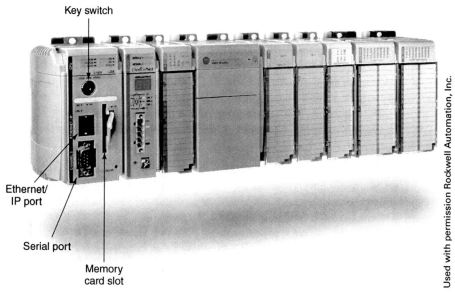

Figure 1-40 1769 Modular CompactLogix Ethernet controller with integrated Ethernet port.

THE CONTROLNET NETWORK

ControlNet is an open-control level network used for real-time data transfer of time-critical and non–time critical data among controllers or I/O on the same link. The network is basically a combination of Data Highway Plus and Remote I/O on the same network. ControlNet provides project upload, download, online monitoring, forcing of I/O, and online editing just like Data Highway Plus. Remote I/O with the ControlLogix platform typically use ControlNet as the communication link between the local chassis and remote chassis. An overview of network features is listed here:

- As many as 99 nodes addressed as decimal node addresses
- A baud rate fixed at 5 million bits per second
- Transfer of time-critical data in the scheduled bandwidth of the network
- Time-critical data with an RPI (if the network is working properly, ControlNet will meet or beat the RPI)
- Non–time critical data transfer via programming a message instruction on ladder rungs
- An open network's ability to have hardware from many different vendors purchased and connected as a node on the network
- Managed by ControlNet International (http://www.controlnet.org)

ControlNet Network Applications

Sharing data between a controller and I/O in a remote ControlLogix chassis is done over a ControlNet network. ControlNet is typically used in the following applications:

- ControlLogix processor, SLC 500, or PLC-5 processor-to-processor scheduled data exchange
- Local ControlLogix or PLC-5 chassis connection to a remote chassis for high-speed remote I/O connectivity
- Interlocking or synchronization of two nodes such as starting two variable-frequency drives
- Connecting two Data Highway Plus networks
- Connecting multiple DeviceNet networks

ControlNet Network Makeup

A ControlNet network is made of several nodes, each having a ControlNet communication interface. ControlNet nodes could include ControlLogix controller-to-controller communications, Remote I/O, and communication among PLC-5s, SLC 500s, variable-frequency drives, and operator interface, among others. A node is a physical device on the network; as many as 99 nodes are permitted on a ControlNet network. Each node must have a unique decimal node address within the range of 1 to 99. There is no node 0 in ControlNet, and duplicate node addresses are not allowed.

ControlNet Cable System

ControlNet is a trunk-line, drop-line type of network. The trunk cable is the bus or backbone of the network. The trunk cable may contain one or multiple sections called *segments*. Termination resistors are required at each end of a segment. As many as 250 meters of coaxial cable and 48 nodes may be on a segment before a repeater must be used. When using coaxial cable, fiber, and the appropriate repeaters, a ControlNet network could be as long as 18.6 miles (29.9 km). The trunk line is either quad-shielded RG-6 coaxial, fiber, or special-use cable, depending on the environment in which the cable will be installed. Cable can range from standard quad-shielded RG-6 coaxial to armored, fiber, and special-use cables such as flood burial, high-flex, or to a mixture of fiber and cable. A tap is required as the drop line off the trunk cable to each field device or node. Taps come in many configurations. The taps' drop-line length is fixed at 39.6 inches (100.6 cm). Figure 1-41

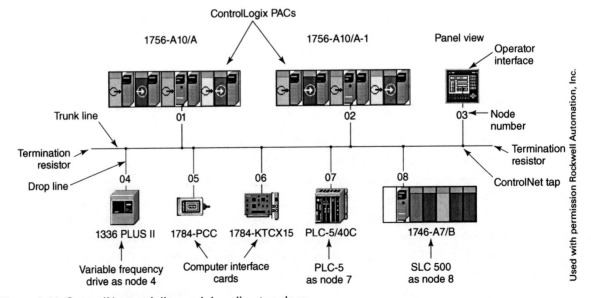

Figure 1-41 ControlNet trunk-line and drop-line topology.

shows all of the network elements just discussed, but only one segment and no repeaters. Figure 1-41 is a screen print from the graphical view of an actual network as displayed in RSNetworx for ControlNet. RSNetworx for ControlNet software is required to configure the network.

Figure 1-41 illustrates what a ControlNet network would look like as graphical view on RSNetworx for ControlNet software. Figure 1-42 shows a drawing of what an actual but different ControlNet network could look like. Note the identification of the components described above. This network has three segments connected together with a single repeater, and each segment is RG-6 coaxial. Note that the repeater has a left and a right side. When assembling a repeater, each side must be purchased separately. Each side of the repeater could be copper, fiber, or a combination.

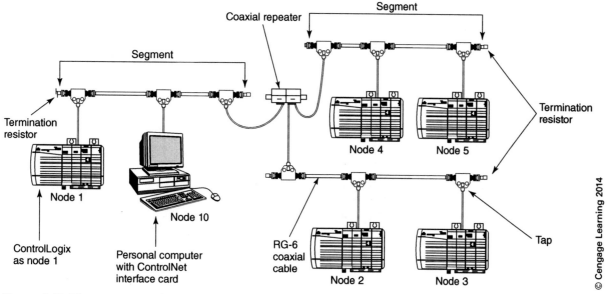

Figure 1-42 Three-segment ControlNet network.

CONTROLNET CABLE REDUNDANCY

When configuring a ControlNet network, a redundant ControlNet bridge module can be used to build a redundant cable network. With redundancy, if one cable fails or is damaged, the second cable can pick up communications seamlessly. When designing a redundant ControlNet network, redundant cables must be routed separately so that if one is damaged, the other cable remains intact and can continue communications. Of the networks introduced in this section, only ControlNet offers redundant cabling.

Notice in Figure 1-43 that there are two cables—that is, redundant media. Whereas a 1756-CNB, 1768-CNB, or 1769-L32C controller supports a single media network, the 1756-CNBR, 1768-CNBR, or 1769-L35CR controller supports redundant cabling, as illustrated in Figure 1-43.

ControlLogix ControlNet Interface Module

The 1756-CNB and 1756-CNBR are the ControlLogix PLC interface modules in the ControlNet network. CNB stands for ControlNet bridge module. The CNB has only one network connection, whereas the CNBR contains two connectors for network redundancy. Figure 1-44 illustrates the main features of a 1756-CNBR module. Notice the two BNC connections on the bottom of the module. One is channel A, the other channel B. The channel B connector is present only on a CNBR module and is used for redundancy. The NAP is an RJ-45 connector used as an interface between a notebook personal computer and the network, typically while working on the factory floor. A maintenance or electrical individual can walk up to any CNB

32 INTRODUCTION TO CONTROLLOGIX HARDWARE

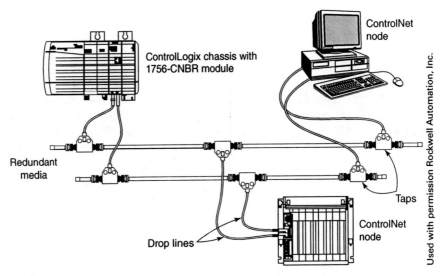

Figure 1-43 Three nodes on a redundant ControlNet network.

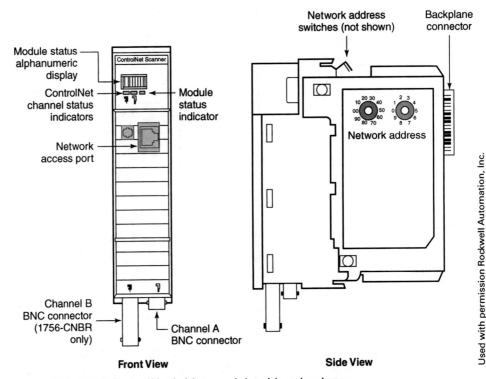

Figure 1-44 1756-CNBR ControlNet bridge module with redundancy.

or CNBR module with a personal computer and the appropriate interface card and cable and connect to the network at any NAP. The latest generation of ControlNet modules replace the RJ-45 connector with a USB port.

ControlNet channel status indicators and the module status indicator are used to view the health of the module and typically are used for troubleshooting. Also used for troubleshooting is the module status alphanumeric display. Here, module information such as firmware level can be viewed as the module is powered up. The module's node address and troubleshooting messages are displayed as the module is running. Located on the top of the module are the network address switches. These two rotary switches are used to set the ControlNet network node address for this module.

CompactLogix ControlNet Controller

The 1769-L32C controller can connect to a single-cable ControlNet network, whereas the 1769-L35CR can connect to a redundant network. Figure 1-45 shows a L35CR. Note the built-in serial port. The ControlNet network access port is used for personal computer interface to the controller and the ControlNet network. The bayonet connections are shown on the bottom of the controller module. Because one connector is in front of the other, you only see one bayonet connector in the figure.

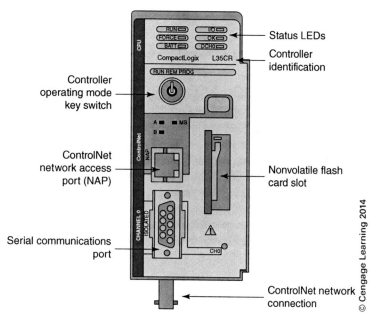

Figure 1-45 CompactLogix ControlNet controller.

DEVICENET COMMUNICATIONS

DeviceNet is also an open network intended to link low-level devices such as sensors, pushbutton stations, distributed I/O blocks, intelligent motor-starter overload devices, and variable-frequency drives to higher-level devices such as PLCs. One major advantage of DeviceNet is that a single four-conductor cable containing power and signal wires is used to connect the devices on the network and save wiring costs. Figure 1-46 illustrates one possible DeviceNet network. The figure is a screen print from Rockwell Software's RSNetworx for DeviceNet software. RSNetworx for DeviceNet software is required for network configuration.

The following is an overview of basic DeviceNet network specifications:

- Uses single cable with four wires—two power and two signal
- Has as many as 64 nodes, 0 to 63 as decimal node addresses
- Baud rates of 125 K, 250 K, or 500 K
- Maximum cable length at 125 K is 1,378 feet
- Needs DeviceNet interface module such as 1756-DNB or 1769-SDN in PAC chassis to access network
- Required RSNetworx for DeviceNet Software for network configuration

34 INTRODUCTION TO CONTROLLOGIX HARDWARE

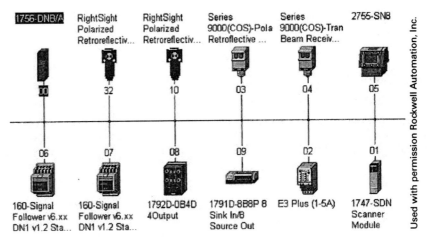

Figure 1-46 Graphical view of a DeviceNet network from RSNetworx for DeviceNet software.

The 1756-DNB communications module is inserted in a ControlLogix chassis to gain access to the network. Figure 1-47 illustrates a 1756-DNB. The very top of the module displays text that identifies the module type. This module also has an information window similar to the Ethernet and ControlNet communications modules. The DNB information window displays the module's node address, basic status information, and error codes for nodes on the network. Directly below the window are three status LEDs. Behind the door is a push button for changing the module's node address, a network cable wiring diagram, and an open-style terminal block connection to connect the network cable. The DNB module has no switches for configuration. Network configuration is accomplished using RSNetworx software. The latest generation of DeviceNet modules have a USB port providing an easy connection between a personal computer and DeviceNet network.

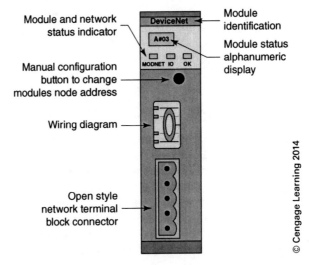

Figure 1-47 1756-DNB.

The upper-left corner of the Figure 1-46 shows a 1756-DNB module in RSNetworx for DeviceNet software. The network node address of the DNB is node 00. The node address of each device is listed directly above or below the device. The lower-right corner of the figure shows a 1747-SDN scanner or communication module as node 01. The number 1747 is the beginning part number of the SLC 500 family of PLCs. This could have been a PLC-5 or CompactLogix 1769-SDN DeviceNet scanner.

All of the devices in Figure 1-46 have one thing in common: They either send or receive a small amount of information across the network. This is a device-based network. Operator interface devices that typically exchange large amounts of information across the networks should be placed on an Ethernet/IP network rather than DeviceNet. DeviceNet was not designed with the bandwidth to accommodate devices that transfer large amounts of information. There are also two variable-frequency drives, one at node 6 and the other at node 7. Before DeviceNet, Remote I/O was an extremely popular network for variable-frequency drives. With its advent, DeviceNet quickly became the network of choice for many variable-frequency drive applications. With the high speeds that today's Ethernet/IP offers, it is currently most popular for newer drive and operator interface applications. Approximately 80 percent of modern drive applications include some type of network connectivity.

DATA HIGHWAY PLUS AND REMOTE I/O INTERFACE

Data Highway Plus (Data Highway + or DH+) is the native network for the Rockwell Automation PLC-5 family. Rockwell Automation's SLC 500 family SLC 5/04 processor can also connect directly to the Data Highway Plus network. Data Highway Plus cable can be connected either as daisy chain or in trunk-line or drop-line configurations. The network can have as many as 64 nodes numbered 0 to 77 in octal or base eight. DH+ is used for uploading, downloading, monitoring, forcing, and performing online editing to the PLC-5 family of processors or the 5/04.

Remote I/O is another network in which remote PLC-5 or SLC 500 chassis can be remotely mounted from the local chassis as a way to save wiring costs by communicating I/O data through the two-wire Belden 9463 cable, which is affectionately referred to as "blue hose." The network eliminates separate wiring runs to remote devices over long distances. RIO uses the same cable as the DH+ network. Both networks can have as many as 10,000 feet of cable at a default baud rate of 57,600 bits per second.

ControlLogix can easily be integrated into an existing DH+ or RIO system by using one or more DH+/RIO (1756-DHRIO) communications modules. In many manufacturing facilities, there may be a large installed base of PLC-5 and SLC 500 PLCs using DH+ and RIO, with the current supervisory PLC most likely to be a PLC-5. In many cases, there is the need to modernize the system to a faster controller with more memory, capabilities, and better communications. Rather than replacing all existing hardware with new ControlLogix hardware, the current PLC-5 supervisor can be replaced with a new modular ControlLogix. The new ControlLogix controller now supervises all existing hardware using the 1756-DHRIO module to interface current DH+ and RIO networks. This is a highly cost-effective way to upgrade the current system with a minimum hardware replacement and still gain access to a fully featured ControlLogix plus Ethernet/IP, ControlNet, and DeviceNet as well as DH+ and RIO network connectivity using the appropriate network modules discussed above. Figure 1-48 Illustrates an example of integrating an existing PLC-5 remote I/O chassis into a newer ControlLogix-supervised system using a 1756-DHRIO communications module.

To keep the figure simple, we included only one PLC-5 remote chassis; in many cases, there are multiple existing remote chassis. In many cases, existing Data Highway Plus networks are also be integrated into the newer ControlLogix-supervised system. Figure 1-49 is an example of a Data Highway Plus network containing PLC-5s, SLC 5/04s, and an operator interface. Note that each device on the network has a unique node address. Data Highway Plus node addresses are octal numbers.

1756-DHRIO Communications Module

The 1756-DHRIO communications module is inserted into the Modular ControlLogix chassis as a way to interface the ControlLogix chassis to Data Highway Plus or Remote I/O networks. Figure 1-50 shows a 1756-DHRIO module. The text at the very top of the module identifies the module as DH+/RIO. Below the module identification is a window that provides status information such as the module node number. Three status indicators are directly below the

36 INTRODUCTION TO CONTROLLOGIX HARDWARE

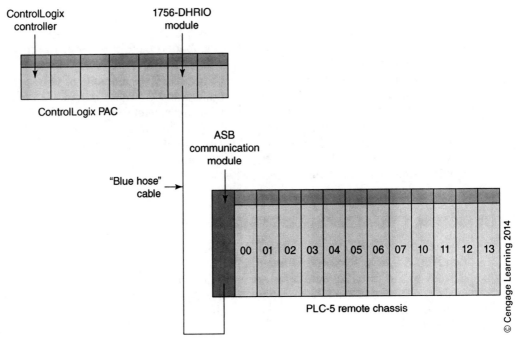

Figure 1-48 Existing PLC remote I/O chassis connected to modular ControlLogix and 1756-DHRIO communications card.

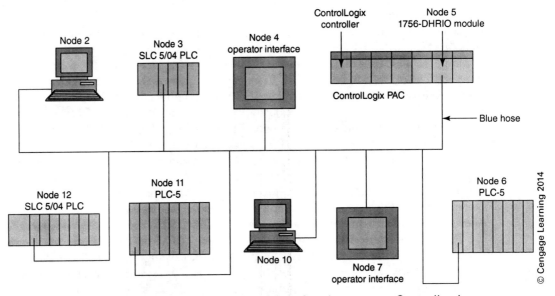

Figure 1-49 Existing Data Highway Plus network interfaced to a newer ControlLogix using the 1756-DHRIO module.

information window. The module in the figure has its door open and shows the two communications connections or channels: channels A and B. This module can be configured by the user to be two DH+ channels, two RIO channels, or one of each. On the top of the module are rotary switches that are used to configure the channels. If one or two DH+ channels were selected, then the node addresses are set using rotary switches on the bottom of the module. Labels on the side of the module explain how to set the switches.

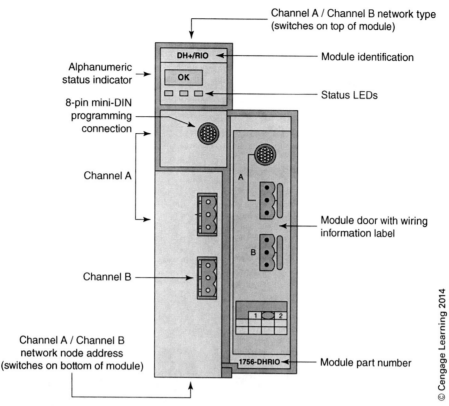

Figure 1-50 ControlLogix 1756-DHRIO communications module.

Refer to the figure and note that channel A has two connections, one round and the other somewhat rectangular. The round 8-pin mini-DIN connection is for personal computer interface to the DH+ network through the module. This is the same connection found on PLC-5 controllers and the SLC 5/04 controller. The rectangular connection is called a *three-pin Phoenix connection* and is where the blue hose is connected to the module. The round DIN connection and the three-pin Phoenix connection for channel A are internally connected. Anything connected to either connection is connected to the DH+ network as well as the ControlLogix by way of the module and backplane, assuming that the channel configuration switches are set to DH+. Channel A can also be used for remote I/O by setting the rotary switches for channel A to the appropriate setting. If using channel A for remote I/O, the round DIN connector is not used. Channel B also can be used to access either network. If channel B is used for Data Highway Plus, then there is no round DIN connection for direct computer access.

Now we consider taking the remote I/O network from Figure 1-48 and plugging the blue hose into channel B, and then taking the DH+ network from Figure 1-49 and plugging its blue hose connection into channel A. This shows how we can interface both networks using one DH+/RIO module. Next we introduce the concept of using the communication modules we have just discussed and the ControlLogix chassis as a way to bridge from one network to another.

CONTROLLOGIX AS A COMMUNICATIONS BRIDGE

ControlLogix can be used to move data from one network to a different network—for example, Data Highway Plus to Ethernet/IP. This is called *bridging*. Bridging from Data Highway Plus to Ethernet/IP is extremely popular. Assume we had two data Highway Plus networks at separate ends of a large facility and we wanted to share information between the two networks. The total network cable currently used plus the distance from one network to another exceeds the 10,000 cable feet allowed on a single Data Highway Plus network. Considering that a modular ControlLogix has already been installed on each end of the facility, we will use the DHRIO modules as a way to get the network connected to each ControlLogix. Because Ethernet/IP is already installed in the plant, we will use the Ethernet/IP network to connect the two networks together. DHRIO modules are used to bring the Data Highway Plus network into the ControlLogix chassis. The ControlLogix backplane is used to bridge from Data Highway Plus on each end to Ethernet, and Ethernet transports the information from one ENBT to the other. Figure 1-51 illustrates a conceptual example using Ethernet and two DHRIO modules to bridge between the networks.

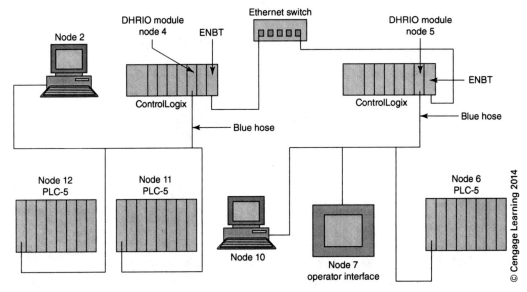

Figure 1-51 Ethernet bridging between two Data Highway Plus networks.

The figure is a typical example of integrating existing hardware to a ControlLogix system as well as using ControlLogix as a bridging vehicle to tie remote networks together.

Keep in mind that Figure 1-51 illustrates only one possible example of using ControlLogix as a bridge. Any of the communications modules mentioned in this section can be used to bridge to a different network.

SUMMARY

This chapter introduced ControlLogix system hardware and the different members of the ControlLogix family. ControlLogix is a 32-bit computer and the newest member of the Rockwell Automation/Allen Bradley PLC offerings. ControlLogix is not a regular PLC—it is much more and has many advanced features not available in the earlier PLC-5 or SLC 500 families of PLCs. Because ControlLogix has many advanced features that are far beyond the capabilities of common PLCs, ControlLogix is classified not as a PLC, but as a PAC. As an automation controller, ControlLogix can control most anything in a modern automated system. Examples are listed here:

- Sequential control such as conveyor control
- Process control
- Batching using a newer feature called Phase Manager
- Motion control using ladder, function block, or structured text-based programming (analog or digital motion control modules can be inserted directly into the modular ControlLogix chassis or the 1768 Modular CompactLogix backplane)
- A communications gateway or a way to bridge from one industrial network to another

ControlLogix family members include the Modular ControlLogix PAC, smaller versions with fewer capabilities such as the FlexLogix and the 1769 and 1768 Modular CompactLogix, along with PowerFlex 700S drive, and SoftLogix.

The modular ControlLogix is a high-performance multicontroller system. Unlike traditional PLCs in which only a single controller is allowed in the chassis, ControlLogix provides the flexibility to add multiple controllers as system requirements and complexity dictate. The key is that the backplane of the chassis is a ControlNet network called ControlBus. Data on a ControlNet network is transmitted at 5 million bits per second. Being a ControlNet network, the chassis backplane allows the controller or controllers to be inserted in any slot of the chassis. This is contrary to traditional PLCs in which only one controller is allowed, which must be inserted into the chassis in the far left slot of the chassis, known as *slot 0*. ControlLogix chassis slot numbers are still numbered, starting with slot 0 on the far left and incrementing in decimal numbers to the right. The modular ControlLogix chassis sizes are 4, 7, 10, 13, and 17 slots.

We also introduced the following:

- Digital and analog modules and their part numbers
- Volatile and nonvolatile memory
- How ControlLogix modules operate and the RPI
- 1769 Modular CompactLogix
- 1768 Modular CompactLogix
- FlexLogix Platform
- ControlLogix communication or bridge modules

Again, the intention of this chapter is only to introduce ControlLogix system hardware and different members of the ControlLogix family. A wealth of additional information is available that should be reviewed before installing, programming, using, or troubleshooting any of the mentioned networks or the ControlLogix itself. Visit Rockwell Automation's Web site—http://www.ab.com—which has a publications library and a literature library. Search through the available publications. A network specialist should always be consulted before installing, maintaining, troubleshooting, or modifying any network.

REVIEW QUESTIONS

Note: For ease of handing in assignments, students are to answer using their own paper.

1. ControlLogix is classified not as a PLC, but as a _____.
2. ControlLogix is a _____ bit computer.
3. For an output module, the _____ is the time that output tag information is transferred across the backplane to the specific output module and then on to the field devices.
4. _____ is a software PLC when an industrial computer is used as the PLC controller instead of the traditional chassis-mounted controller.
5. When an industrial computer is used as with SoftLogix, the remote chassis of I/O is connected via networks such as _____, _____, or _____.
6. List the four different programming languages ControlLogix can be programmed in.
7. The modular ControlLogix controller is available with one _____ communications port.
8. Communications to other networks such as Ethernet/IP, ControlNet, DeviceNet, Foundation Field Bus, Data Highway Plus, Remote I/O, or DH-485 are made through purchasing and inserting the desired _____ into the ControlLogix chassis.
9. Identify the components of the ControlLogix controller illustrated in Figure 1-52.
10. In Figure 1-52, is this a series A or series B ControlLogix controller? How do you know?

40 INTRODUCTION TO CONTROLLOGIX HARDWARE

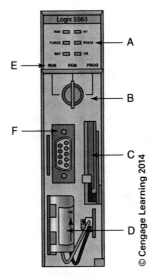

Figure 1-52 ControlLogix controller features.

11. The RPI specifies the rate at which the module being configured sends _____ or _____ its data.
12. ControlLogix chassis slot numbers are numbered starting with slot _____ on the far left and incrementing in _____ numbers to the right.
13. While in _____ mode, a controller is executing the instructions programmed on the rungs of ladder logic that make up the user project.
14. When the controller is in _____ mode, it is not running but can accept a new project or modifications to the existing project.
15. The modular ControlLogix chassis sizes are _____.
16. The controller battery is a nonrechargeable lithium battery used to keep the _____ memory alive when power is removed from the controller's chassis.
17. The ControlLogix modular chassis backplane is a ControlNet network called _____.
18. Data on a ControlNet network are transmitted at _____ bits per second.
19. The programmer can specify the time when the input or output data are transferred by entering the desired value in the _____ or _____ parameter when performing the I/O configuration.
20. The following questions refer to Figure 1-53. Identify and briefly explain each called out feature.
21. In Run mode, the controller or CPU is in its operating cycle, which is called the *controller* _____.

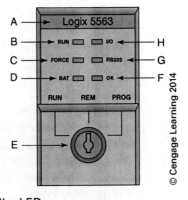

Figure 1-53 ControlLogix controller LEDs.

INTRODUCTION TO CONTROLLOGIX HARDWARE **41**

22. _____ or _____ PLCs were designed for low-end to medium-sized applications that require fewer software capabilities, less I/O, and fewer communications capabilities than the modular ControlLogix.
23. _____ memory is an option for some ControlLogix controllers.
24. Used lithium batteries are considered _____, and special precautions are needed in their handling, transportation, and disposal.
25. A unique feature of the 1768 Modular CompactLogix is that it uses the same I/O modules as the _____.
26. RSLogix 5000 software version _____ is the last version of the software that the FlexLogix will support.
27. Nonvolatile memory is either built into the ControlLogix controller or available as a _____ megabyte _____ card.
28. The 1768 Modular CompactLogix can integrate as many as _____ 1769 Modular CompactLogix I/O modules to the right of the controller as a local I/O configuration.
29. _____ provides the option to integrate a 32-bit ControlLogix controller with Rockwell Automation's Flex I/O blocks.
30. Identify the components of the input module shown in Figure 1-54.

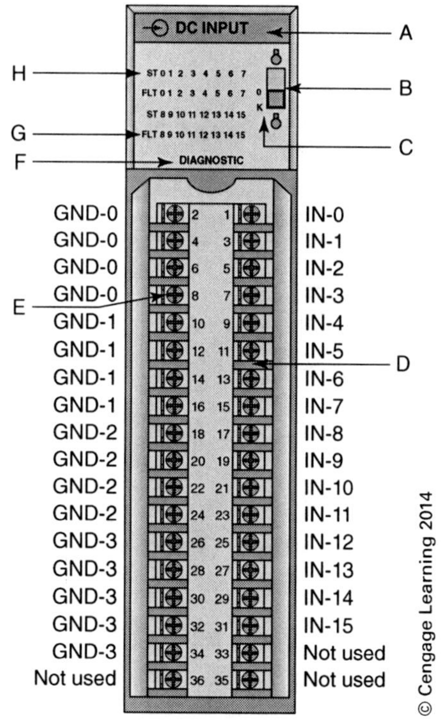

Figure 1-54 Module component identification.

31. ControlLogix I/O updates occur _____ to the scan of the logic scan.
32. Being a ControlNet network, the modular backplane allows the controller or controllers to be inserted in _____ slot of the chassis.
33. _____ software is used to configure I/O.
34. The RPI range for a digital I/O module ranges from _____ microseconds to _____ milliseconds.
35. For an input module, the _____ is the time when the input information is sent from the module across the backplane to the input tags.

42 INTRODUCTION TO CONTROLLOGIX HARDWARE

36. What are the most common networks used with the ControlLogix?
37. _____ is an open industrial network standard that uses standard off-the-shelf components, supports real-time I/O control, and uses messaging between different pieces of hardware.
38. _____ is an open-control network used for real-time data transfer of time-critical and non–time critical data between controllers and I/O on the same link.
39. _____ is an open network intended to link low-level devices such as sensors, push-button stations, distributed I/O blocks, intelligent motor-starter overload devices, and variable-frequency drives to higher-level devices such as PLCs.
40. Today the most popular network for new applications for information transfer between controllers, human interface devices, and variable-frequency drives is _____.
41. _____ is the only network that provides the option for redundant cabling.
42. One major advantage of _____ is to use a single cable to supply network power and communications to connect devices on a network saving wiring costs.
43. Refer to Figure 1-55 as you answer the following questions.
    a. What type of network is this?
    b. Define *segment*.
    c. How many segments are in the network represented by Figure 1-55?

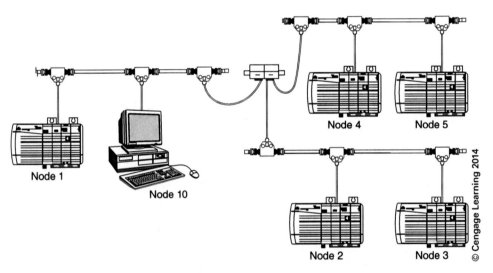

Figure 1-55 Network identification.

44. The following questions refer to Figure 1-56.
    a. Identify the module.
    b. What is the module's part number?
    c. Identify A.
    d. Identify B.
    e. Identify C.
    f. Identify D.
    g. Describe E.
    h. What does F do?
    i. Explain the purpose of item C.
    j. Is there a module similar to this that does not have both items D and E?
    k. What is that module's part number?
    l. Explain the difference between the two modules.

INTRODUCTION TO CONTROLLOGIX HARDWARE   43

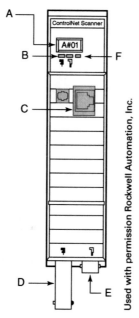

Figure 1-56 Module for identification for Question 44.

45. The following questions refer to the Figure 1-57.
   a. Identify this module.
   b. Identify A.
   c. Identify B.
   d. Identify C.
   e. What is D?
   f. Where are the actual switches for setting up the node address of this module?
   g. What is the range of valid network addresses for this module?

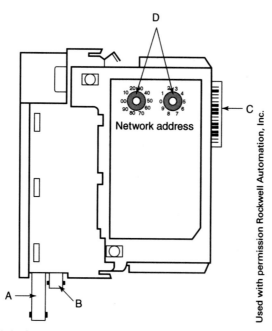

Figure 1-57 Network module for identification.

44 INTRODUCTION TO CONTROLLOGIX HARDWARE

46. Refer to Figure 1-58 for the following questions.
    a. Identify the module in Figure 1-58.
    b. Identify A.
    c. Identify B.
    d. Identify C.
    e. Identify D.
    f. Identify E.
    g. Identify F.
    h. Identify G.
    i. Identify H.

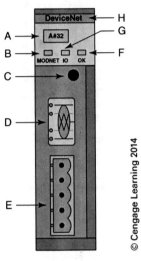

Figure 1-58 Module for identification.

47. The following questions refer to Figure 1-59.
    a. Identify the module in Figure 1-59.
    b. Identify A.
    c. Identify B.
    d. Identify C.
    e. Identify D.
    f. Identify E.
    g. Identify F.
    h. Identify G.
    i. Identify H.
    j. Callouts C and D in the figure are internally connected. Explain how these two ports work for this module.
    k. The DHRIO module has two channels: A and B. Explain what these can be configured to do.
48. Identify the module in Figure 1-60.
    a. Identify what this picture is.
    b. Identify A.
    c. Identify B.
    d. Identify C.
    e. Identify D.
    f. Identify E.
    g. Identify F.
    h. Identify G.

INTRODUCTION TO CONTROLLOGIX HARDWARE   45

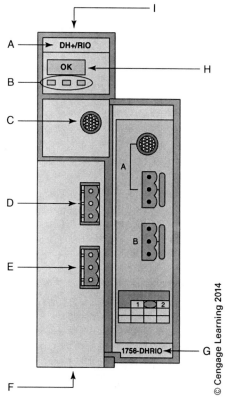

Figure 1-59 Module identification for Question 47.

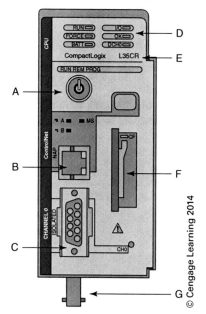

Figure 1-60 Module for identification.

46 INTRODUCTION TO CONTROLLOGIX HARDWARE

49. Identify the pieces in Figure 1-61.
   a. Describe the hardware assembly illustrated in the figure.
   b. Identify A.
   c. Identify B.
   d. Identify C.
   e. Identify D.
   f. Identify E.
   g. Identify F.
   h. Identify G.
   i. Identify H.
   j. Identify I.
   k. Identify J.

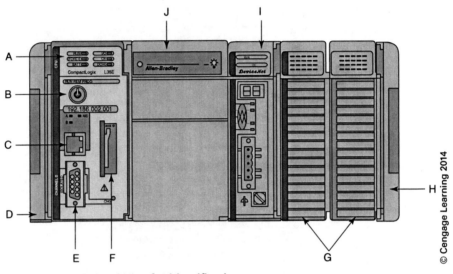

Figure 1-61 ControlLogix hardware for identification.

50. _____ is using ControlLogix to move data from one network to possibly a different network using communication modules and the ControlLogix backplane.

CHAPTER

2

Introduction to RSLogix 5000 Software

OBJECTIVES

After completing this lesson you should be able to

- Identify the major RSLogix 5000 software features.
- Identify RSLogix 5000 toolbars.
- Create and configure a new RSLogix 5000 project.
- Enable automatic project backup and project recovery.
- View and modify controller properties.
- Turn on and customize toolbars.
- Configure the way you want your ladder window to look.

INTRODUCTION

This chapter introduces only the basic features of the RSLogix 5000 software. As we work through future lessons, we will explore these features in more depth. Those who are familiar with Microsoft's Windows will find that many of the common Windows features and mechanics are the same. Those who have worked with RSLogix 5 or RSLogix 500 software will discover that the three RSLogix software packages are remarkably similar, especially the mechanics for programming ladder logic. Keep in mind that the ControlLogix has the same instruction set as the PLC-5. The PLC-5 is a 16-bit PLC, whereas ControlLogix is a 32-bit PAC; thus the RSLogix 5000 software has 32-bit instructions rather than the 16-bit instructions of the PLC-5 and SLC 500 PLCs.

The lab exercises provide an opportunity to work with, navigate through, and set up the RSLogix 5000 software. We look at the basic software features, including configuring the ladder window.

RSLOGIX 5000 MAIN PROJECT WINDOW

In Figure 2-1, the main RSLogix 5000 window displays an open project. Each major piece of the window is identified. We will look at each piece as we work through this lesson.

LOGIX 5000 LADDER LOGIC COMPONENTS

The ladder logic window is where the ladder rungs are displayed, monitored, and edited. The basic components of ladder logic are rungs, branches, and instructions. Figure 2-2 shows a sample ladder logic window. Notice the rung numbers in the upper-right corner. Rungs are the horizontal

48 INTRODUCTION TO RSLOGIX 5000 SOFTWARE

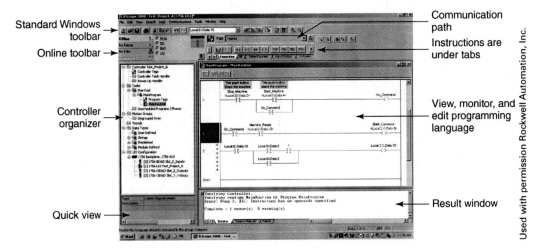

Figure 2-1 Major sections of the RSLogix 5000 software ladder logic program display and controller organizer.

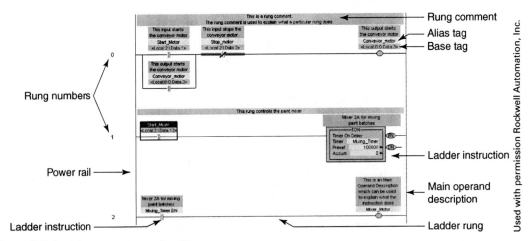

Figure 2-2 Ladder rung components.

lines; the power rails are the vertical lines. A ladder rung comprises instructions and associated documentation. Some normally open and normally closed instructions are shown on the left side of the rungs. These symbols represent incoming information, or *inputs*. *Output*-type instructions are on the right side of a ladder rung. The figure also shows both coil-type and box-type instructions; both types are output instructions. A *rung comment* is documentation that describes the function of the rung. A main operand description is documentation that identifies the function of the associated instruction.

Because ControlLogix is a tag-based PAC, each instruction has one or more tags associated with it. The beginning, or basic, tag is called a *base tag*. A base tag can have a more descriptive or more user-friendly name or tag associated with it, called an *alias tag*. Refer to the right side, of rung 0. We will work with all of these components in future lab exercises.

Controller Organizer

The controller organizer is a tree structure similar to Windows Explorer and used to organize the project. For now, we first define a project. An RSLogix project is the computer file that contains all of the information associated with the ladder logic to make it work. There is a lot more to what we commonly call a PLC program than just the ladder rungs. In addition, the controller organizer contains such items as communications, tags, data regarding how the application is to operate, and I/O configuration, among others. Figure 2-3 shows an example of a controller organizer for an RSLogix 5000 project. Keep in mind that the objects in the controller organizer change, depending on the specific application. At this point, do not be too concerned with all the objects listed in the figure; we discuss many of them in detail in future lessons.

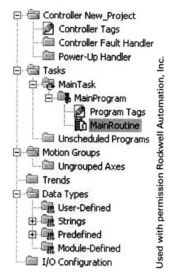

Figure 2-3 RSLogix 5000 controller organizer.

Menu Bar

As in Windows, the menu bar is located across the top of the window and contains all of the available menus (see Figure 2-4). Many of the menus are similar to the typical Windows menus that most of us are familiar with.

Figure 2-4 RSLogix 5000 menu bar.
Used with permission Rockwell Automation, Inc.

The following are samples of the items contained in the different menus:

File   The File drop-down menu includes New, Open, Close, Save, Save As, printing options, and options for closing the application.

Edit   The Edit drop-down includes Cut, Copy, Paste, Delete, and Insert, as well as controller properties.

View   The View drop-down includes toolbars, controller organizer options, errors, search results, watch window, and start page.

Search   Search includes features such as Find, Replace, Go To, Cross Reference, Find Next, and Find Previous.

Logic   The Logic menu allows the user to monitor tags, edit tags, perform verification, force I/O, and perform online edits.

50 INTRODUCTION TO RSLOGIX 5000 SOFTWARE

Communications The Communications menu includes Who Active, Go Online, Upload, Download, Go to Run Mode, Go to Test Mode, Go to Program Mode, Clear Faults, and Go to Faults.

Tools The Tools menu includes options for screen setup, Control Flash, Clear Keeper, Tag Data Monitor Tool, and a DeviceNet tag generator.

Window The Window menu includes New Window, Cascade, Tile, and Close.

Help Help options include Contents of Help, Instruction Help, Release Notes, Online Books, Quick Start, Learning Center, and About the RSLogix 5000 Software.

Standard Windows Toolbar

Figure 2-5 shows the standard Windows toolbar with icons identified.

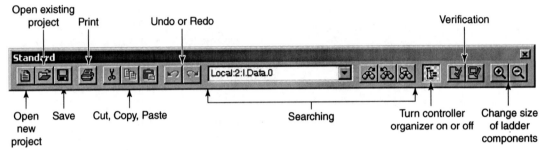

Figure 2-5 Standard Windows toolbar.
Used with permission Rockwell Automation, Inc.

Online Toolbar

The Online toolbar drop-down menu contains the selections shown in Figure 2-6. When online, the processor LEDs' current status and the processor key switch position are displayed. Clicking the icon in the bottom right corner takes us to the controller properties window.

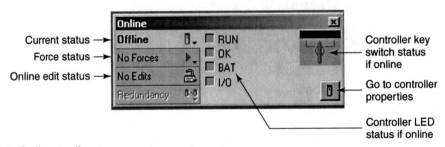

Figure 2-6 Online toolbar.
Used with permission Rockwell Automation, Inc.

Notice that the project's current status is Offline. Clicking Offline text displays the processor LEDs as shown in the figure. If the No Forces text were clicked, the processor LED object would be replaced with force status information and LEDs. Clicking the arrow icon to the right of the No Forces text provides another drop-down list with selections for forcing. Clicking the No Edits text displays information as to the status of the controller regarding online editing. Likewise, clicking the lock to the right of the No Edits text displays another drop-down list with online edit options.

Clicking the icon to the left of the Offline text displays the drop-down list as illustrated in Figure 2-7 and the desired action is simply selected from the drop-down list. This drop-down list is *dynamic*—that is, it changes, depending on the controller's current state. As an example, if the controller is faulted, the top two selections will change to Clear Faults and Go To Faults.

INTRODUCTION TO RSLOGIX 5000 SOFTWARE 51

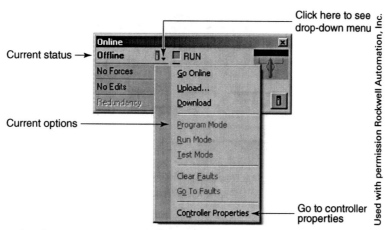

Figure 2-7 Example of a right-click drop-down menu.

Language Element Toolbar

One way to program ladder instructions is to drag and drop them from the Language Element toolbar to the desired position on a ladder rung or select where to place the instruction on the rung and simply click it on the toolbar. The Language Element toolbar organizes instructions under tabs, so timer and counter instructions can all be found under the Timer/Counter tab. The Favorites tab can be customized to contain regularly used instructions. Customizing the Favorites tab avoids the problem of switching back and forth between the different instruction tabs when programming. The Language Element toolbar is illustrated in Figure 2-8.

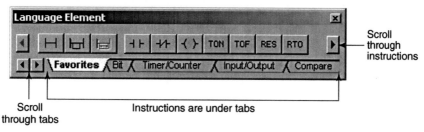

Figure 2-8 Language Element toolbar.
Used with permission Rockwell Automation, Inc.

Ladder Common Logic Toolbar

The Ladder Common Logic toolbar is an additional toolbar that may be customized to hold any desired specific instructions. Figure 2-9 shows how this toolbar might be customized.

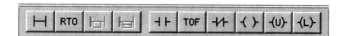

Figure 2-9 Customizable Ladder Common Logic toolbar.
Used with permission Rockwell Automation, Inc.

New Component Toolbar

As most of us know, Windows has many ways to do almost anything. The New Component toolbar is one way to create a new task, program, routine, tag, data type, and so on. Figure 2-10 shows the New Component toolbar and available selections. As an example, clicking the New Routine icon creates a new routine.

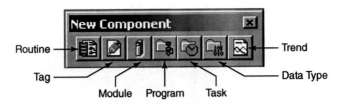

Figure 2-10 New Component toolbar.
Used with permission Rockwell Automation, Inc.

Path Bar Toolbar

The Path Bar toolbar is used to configure communication between a personal computer and the PAC. The toolbar has three parts: the current path, recent paths, and RSWho (see Figure 2-11).

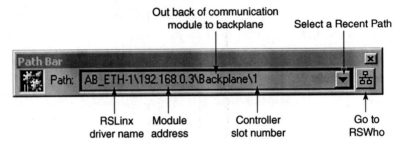

Figure 2-11 Path Bar toolbar.
Used with permission Rockwell Automation, Inc.

Current Path

The text displayed near the center of the toolbar is called the current path. If one attempted to upload, download, or go online at this time, this is the communications path that would be used. One might ask, "Is this path correct for what I want to do at this time?" Some users assume that whatever path is displayed is the correct one, but in some cases it is not. As a result, these users end up communicating with the wrong PAC. It is the responsibility of the individual uploading, downloading, or going online to make sure the path is correct. We discuss the pieces of the path and how to set it up in the lesson on setting up communications between the personal computer and PAC.

Select Recent Path

When setting up communications between a personal computer and the PAC processor, a communication path has to be determined and configured. The communication path is simply the route to send the project to the correct controller. First, the programmer must determine what the path will be. Second, the programmer must determine whether the path displayed in the path toolbar is correct. If it is correct, then the project can be downloaded. If the current path is not correct, then the programmer can click the Select a Recent Path arrow and see a drop-down list of previously used paths. If the correct path is found in the drop-down list, it can be selected and used for the current operation. If a usable path is not in the list, then a new one will have to be configured.

To verify the communication path to the desired PAC to communicate with, clicking the RSWho icon displays an RSLinx RSWho screen. We use RSLogix 5000 software to program, monitor, and modify our PAC project; a second package of software called RSLinx is used to set up communications drivers. RSLinx has an RSWho screen that can be used to view a graphical representation of users on the current network. The communication path can be set from that screen and communication established. This lesson introduces the toolbars. We will look at setting up communications, configuring the path, and downloading our RSLogix 5000 project in a future lesson.

TURNING TOOLBARS ON OR OFF

From the Menu bar, selecting View and then Toolbars displays a window similar to that shown in Figure 2-12. To turn a toolbar on, click its box to put a check in the box. Any checked box displays

INTRODUCTION TO RSLOGIX 5000 SOFTWARE

Figure 2-12 Setting Toolbars visibility.
Used with permission Rockwell Automation, Inc.

the selected toolbar; unchecked boxes leave those toolbars hidden. Toolbars turned on or off from this window determine which toolbars are displayed, regardless of which project is opened. Toolbar setup is not on a project-by-project basis but for the computer as a whole until someone changes the settings. Clicking Restore Factory Toolbar Layout restores the project's original toolbars.

CUSTOMIZING TOOLBARS

Only the Language Element toolbar and Ladder Common Logic toolbars can be customized. The Ladder Common Logic toolbar is a second easy-to-use toolbar used to configure desired specific programming instructions.

Upon clicking either of these toolbars, the Customize Toolbar button in the upper-right corner of the window becomes available. Refer to item 1 in Figure 2-13. The figure shows selecting the Language Element and Customize Toolbar options. The top screen is then displayed. Notice the arrow from item 2. All of the instructions are contained in folders in this window. The Timer/Counter folder is selected. Clicking the + in front of the Timer/Counter folder expands it to show the instructions inside. The arrow is pointing to the timer on delay (TON) instruction. By selecting the instruction and clicking the Add button as indicated by the arrow from item 3, the selected instruction will be displayed in the resulting Toolbar section on the right. The up and down arrow identified by item 4 provides the option to position the instruction anywhere desired in the list of icons that is displayed on the resulting toolbar. If there is an instruction to be removed from the Resulting Toolbar list, select the instruction and click the Remove button. Click OK to finish. The newly configured instruction will be displayed on the Language Element toolbar.

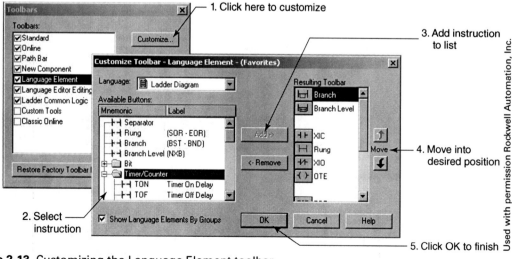

Figure 2-13 Customizing the Language Element toolbar.

HELP SCREENS

Some of the major Help features from RSLogix 5000 version 16 are described following. These features may be different or not available in software versions older than 16. Some of the described features and the screens displayed first became available with RSLogix 5000 version 16.

The help screens contain a lot of information to assist the user of the RSLogix 5000 software. After clicking Help, a menu bar drop-down similar to that in Figure 2-14 should be displayed. Numbered items in the following list help identify many of the items in the figure.

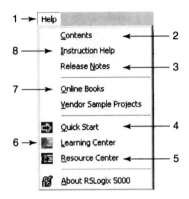

Figure 2-14 Help drop-down menu.
Used with permission Rockwell Automation, Inc.

1. Click Help from the menu bar to display the drop-down menu, as shown in Figure 2-14.
2. Selecting the Contents tab displays a table of contents in which one can select from a list resembling a book's table of contents (see Figure 2-15). Likewise, selecting the Index tab displays an alphabetical list of help topics similar to a book's index. The Find tab allows the user to enter multiple key words for help to find information.

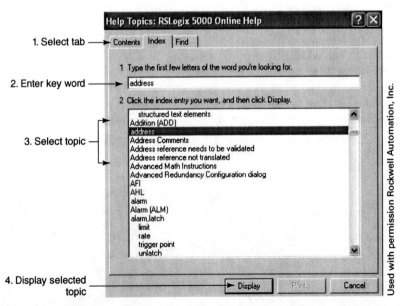

Figure 2-15 Help index entry.

3. The Release Notes selection provides information about the current software release, including installation information, past problems that have been fixed, and new features. When upgrading software from an older version, the release notes should always be read.

4. Quick Start (starting in version 16) provides a list of recent projects that have been opened and offers help in getting started and creating a first project.
5. The Resource Center (starting in version 16) provides information on downloads and online books. Figure 2-16 shows a portion of the Online books section of the Resource Center. The books can be referenced from this screen by simply clicking them. Note the list of manual categories listed to the left.

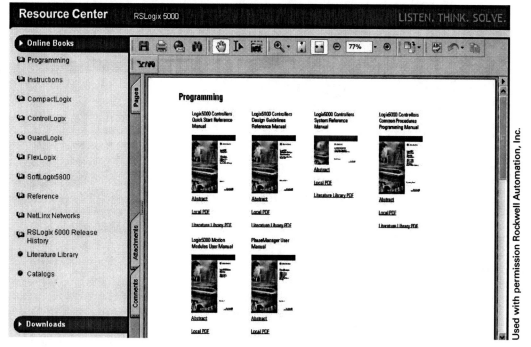

Figure 2-16 Online books available from the Resource Center.

6. The Learning Center, which first became available with RSLogix 5000 version 16, contains information under the questions "What's New?," "How Do I . . . ?," and "Did You Know . . . ?"
7. Clicking Online Books is another way to get to the screen shown in Figure 2-16.
8. If help with the Instruction Set is required, click Instruction Help. Figure 2-17 shows part of the Instruction Set Help.

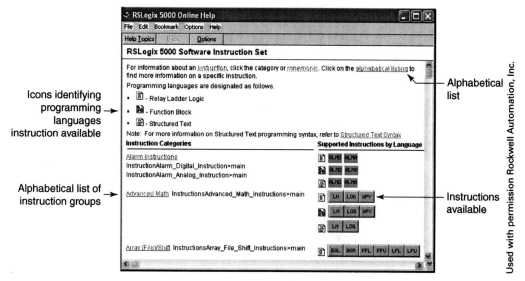

Figure 2-17 A portion of the Instruction Set Online Help screen.

DISPLAY TOOL TIPS

Using Windows's Tool Tips feature, users can hover over objects on their screens and obtain additional information regarding those objects. Figure 2-18 shows the Tool Tips for the Start Machine tag. The information displayed in this example includes the tag name, tag data type, the name of the base tag for which this tag is an alias, tag scope, and current value.

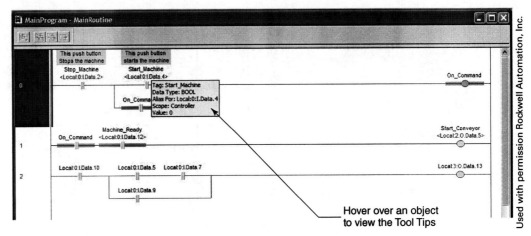

Figure 2-18 Tool Tips.

ROUTINE WINDOW COMPONENTS

The Routine window is where the logic contained in the selected routine is displayed. Remember that ControlLogix can be programmed in four different languages. A routine can be ladder logic, function block diagram, sequential function block, or structured text. The Routine window displays the routine in the language programmed. Figure 2-19 shows a ladder logic routine. Note the routine name in the upper-left corner of the window. Ladder rungs are displayed starting with rung 0. The rung numbers are listed to the left of the rung. The horizontal lines are the ladder rungs, whereas the vertical lines are the power rails. Note the tabs in the lower-left area of the window. Each time a different routine is opened, a tab is displayed for that routine. After clicking a different tab, a different routine is displayed.

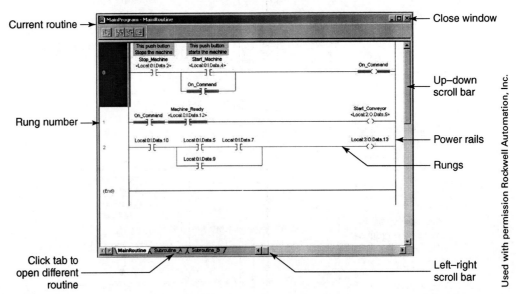

Figure 2-19 RSLogix 5000 Routine window.

Splitting Screen to View Multiple Ladders

A standard Windows feature is the ability to arrange open windows in different configurations. Figure 2-20 shows how a routine ladder window can be split into multiple views to monitor two routines at the same time. The figure shows the main routine on the top screen; the bottom screen is a subroutine.

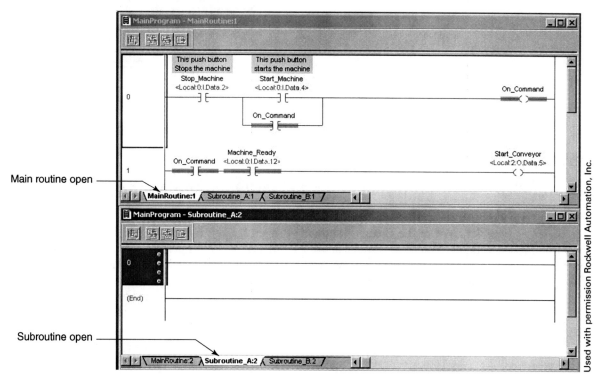

Figure 2-20 Split screen to monitor multiple routines.

CONFIGURING THE SOFTWARE DISPLAY

Workstation Options windows configure how the application and the different programming languages that are activated on a particular computer are displayed. Ladder programming is the most common programming language. The package of RSLogix 5000 software that was installed on a computer dictates which languages are available for programming. If a personal computer does not have the language installed and activated, the language selection will not be displayed in the categories listings. These configuration screens define how the routine screen is displayed for each of the different programming languages on a particular computer. Setting up these preferences is for this copy of RSLogix 5000 software loaded on this particular computer, no matter what project is opened. The Routine window can be customized regarding what is to be displayed, and the color in which objects are to be displayed. In this section, we introduce the major features for setting up software preferences. Selecting Tools and then Options from the RSLogix 5000 software menu bar displays Change general preferences of a RSLogix 5000 screen that is similar to Figure 2-21. The left side of the display lists the category listing.

The Application category is displayed in the left part of the figure. This category sets up such things as the default directory where projects are to be displayed and enables automatic project recovery and automatic project backup.

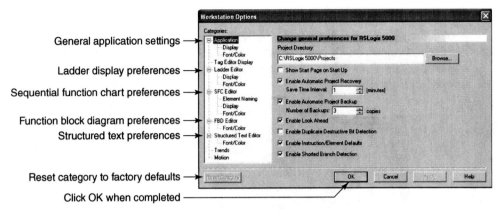

Figure 2-21 General preferences to configure the software display.
Used with permission Rockwell Automation, Inc.

AUTOMATIC PROJECT BACKUP

RSLogix 5000 software has two automatic project backup features. One is automatic recovery based on a user-configurable time. The second feature makes a backup of the project file each time a project is saved.

First, let's look at the Enable Automatic Project Backup feature. Figure 2-22 shows the Open Project window. Note the project file named "new.ACD." Assume this is our current project file. Directly below the new.ACD file are three backup files. The first is new_BAK000.acd. The name *new* is the project name, whereas BAK000 is the file backup number. The extension ".acd" is the file extension for a Windows RSLogix 5000 project file. The figure shows three backup files, 000 through 002. The number of backups is configurable between 1 and 999 in the General Applications window.

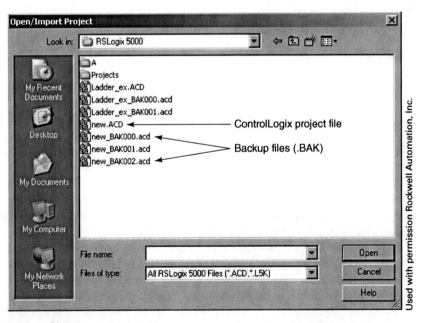

Figure 2-22 Backup files.

Here is how Automatic Project Backup feature works:

1. The current setting for Enable Auto Project Backup is set to three copies (refer to Figure 2-23).
2. If the project new.ACD is opened and edited, when the edited project is saved, the edited project will replace the original new.ACD file.

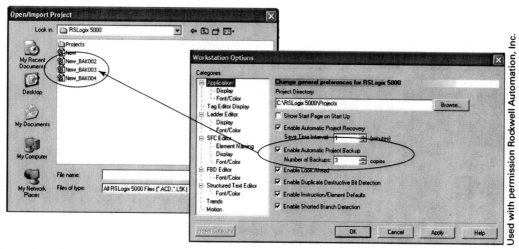

Figure 2-23 Workstation options.

3. The original new.ACD file becomes the next backup file, in this example, new_BAK003.
4. The oldest backup file, new_BAK000.acd, will be discarded.
5. When the project is saved again, the backup file is new_BAK004, as illustrated in Figure 2-23.
6. Backup files can be opened just like any other RSLogix 5000 file.

Configuring the Automatic Project Backup Feature

Selecting options from the Tools drop-down menu from the main menu bar displays a screen similar to that shown in Figure 2-23. Checking the Enable Automatic Project Backup selection enables this feature. Enter the number of backups desired. Valid entry is 1 to 999.

Configure Timed Automatic Project Recovery

Starting with RSLogix version 16, the timed backup feature that many users were familiar with in the RSLogix 500 or RSLogix 5 software became available for RSLogix 5000. This feature provides automatic backups on a timed basis. Figure 2-24 shows the Enable Automatic Project Recovery setup. The figure shows 1 minute as the time of the automatic backup. Valid entries are 1 to 1,440 minutes, or once every 24 hours.

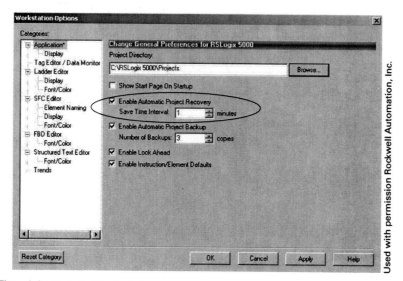

Figure 2-24 Timed Automatic Project Recovery.

60 INTRODUCTION TO RSLOGIX 5000 SOFTWARE

If a user's computer locks up, the window shown in Figure 2-25 will display during rebooting and ask how the user wishes to proceed. The message asks whether the user wants to attempt to recover the project file by using the timed backup file.

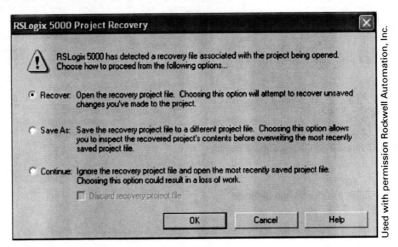

Figure 2-25 Automatic Project Recovery.

It might be best at this point to select Save As and save the recovery file as a new.acd file. How much data is lost is determined by the time interval selected and whether modifications have been made since the last automatic timed backup. To allow the user to view any differences between projects, RSLogix 5000 has a comparison utility that could be used to compare the last saved copy of a project to the saved copy of the recovery file. A decision could then be made as to how to proceed.

Next we look at modifying the colors used in the ladder window.

Configuring the Ladder Editor Display

Figure 2-26 shows the Display selection under Ladder Editor. Features such as rung numbers and text documentation such as rung comments, main operand descriptions, and tag alias information are displayed in this screen.

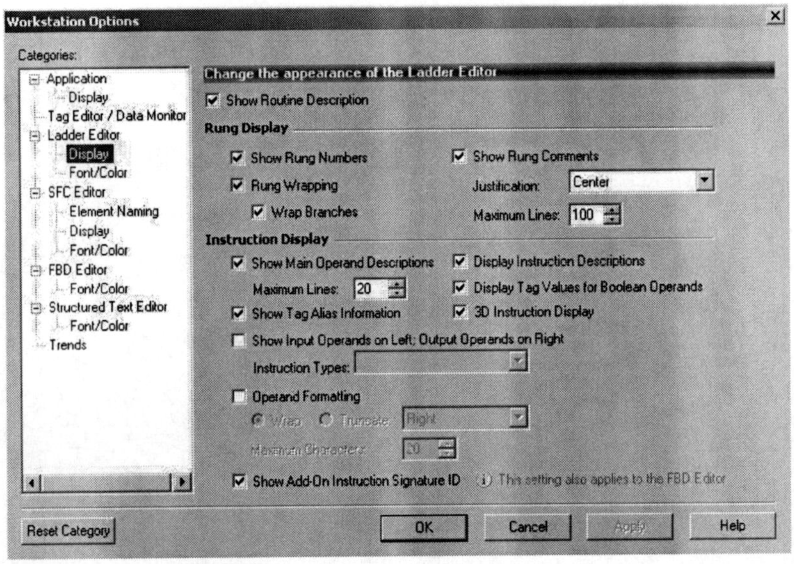

Figure 2-26 Ladder Editor appearance.

In the fonts and colors section, the user can select the colors and fonts in which the information will be displayed, as in the ladder routine window.

Configuring Ladder Editor Fonts and Colors

The Figure 2-27 has Ladder Editor's Fonts/Color category selected and displayed. Selections such the ladder window color, rung comment, descriptions, and alias tag information text and background colors are configured here. Refer to the figure for steps in selecting colors.

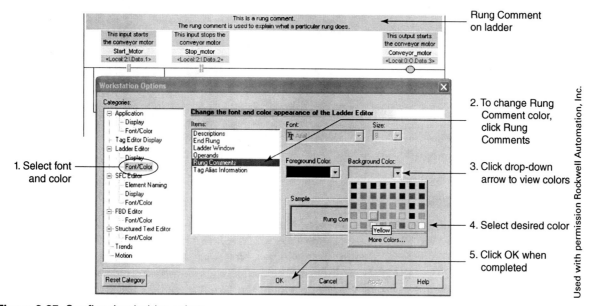

Figure 2-27 Configuring ladder colors.

1. From Ladder Editor under Workstation Options, select Font Color.
2. From the items list, select the color to be modified. The Rung Comment section is highlighted. A rung comment is identified near the top right of Figure 2-27.
3. Select Foreground Color or Background Color drop-down arrow. In the figure, the Background Color drop-down has been selected.
4. The Background Color palette is displayed. Hovering over a color selection displays Tool Tips that identify the color. If additional color selections are desired, select More Colors.
5. When the desired color has been selected, click OK.

This completes the Workstation options for the application and ladder display options. Next we look at the edit and monitor tag display options.

EDIT TAGS VIEW

ControlLogix does not have the predefined data files found in the RSLogix 5 or RSLogix 500 software. To edit or create a tag, go to the Edit Tags window. Monitoring tags is done in the Monitor Tags view. Neither window allows the user to do the opposite function. From the controller organizer double click Controller Tags to display the Edit Tags and Monitor Tags views. Figure 2-28 shows an Edit Tags window. Note that the Edit Tags tab as selected near the bottom left of the window. This window shows all tags in the project. By default, tags are arraigned alphabetically. They can be created or edited from this window. New tags are created also near the bottom left of the window on the blank row in the table. We examine the components of the window more closely in a later lesson.

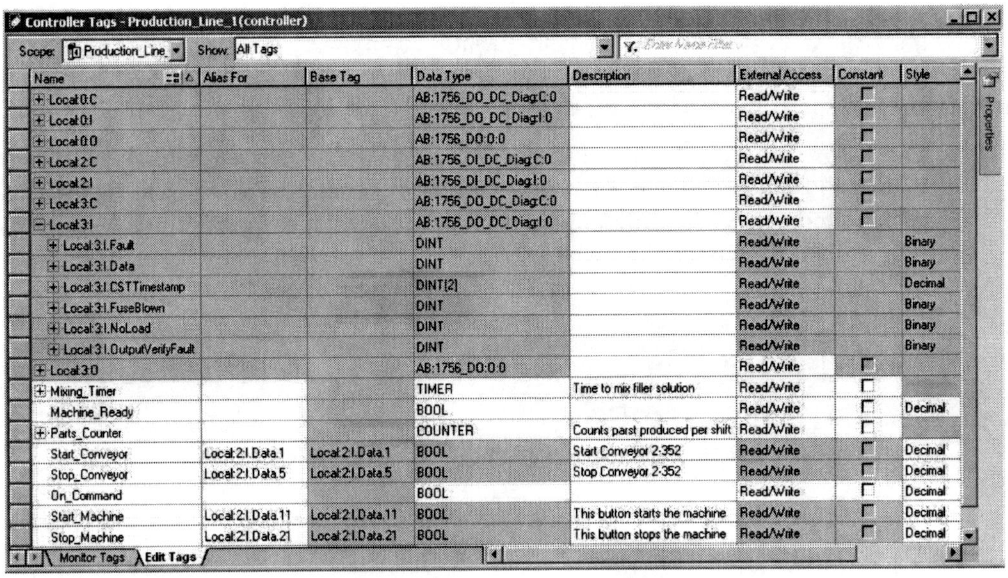

Controller Tags - Production_Line_1(controller)

Scope: Production_Line_ Show: All Tags

Name	Alias For	Base Tag	Data Type	Description	External Access	Constant	Style
+ Local:0:C			AB:1756_DO_DC_Diag:C:0		Read/Write	☐	
+ Local:0:I			AB:1756_DO_DC_Diag:I:0		Read/Write	☐	
+ Local:0:O			AB:1756_DO:O:0		Read/Write	☐	
+ Local:2:C			AB:1756_DI_DC_Diag:C:0		Read/Write	☐	
+ Local:2:I			AB:1756_DI_DC_Diag:I:0		Read/Write	☐	
+ Local:3:C			AB:1756_DO_DC_Diag:C:0		Read/Write	☐	
− Local:3:I			AB:1756_DO_DC_Diag:I:0		Read/Write	☐	
+ Local:3:I.Fault			DINT		Read/Write		Binary
+ Local:3:I.Data			DINT		Read/Write		Binary
+ Local:3:I.CSTTimestamp			DINT[2]		Read/Write		Decimal
+ Local:3:I.FuseBlown			DINT		Read/Write		Binary
+ Local:3:I.NoLoad			DINT		Read/Write		Binary
+ Local:3:I.OutputVerifyFault			DINT		Read/Write		Binary
+ Local:3:O			AB:1756_DO:O:0		Read/Write	☐	
+ Mixing_Timer			TIMER	Time to mix filler solution	Read/Write	☐	
Machine_Ready			BOOL		Read/Write	☐	Decimal
+ Parts_Counter			COUNTER	Counts parst produced per shift	Read/Write	☐	
Start_Conveyor	Local:2:I.Data.1	Local:2:I.Data.1	BOOL	Start Conveyor 2-352	Read/Write	☐	Decimal
Stop_Conveyor	Local:2:I.Data.5	Local:2:I.Data.5	BOOL	Stop Conveyor 2-352	Read/Write	☐	Decimal
On_Command			BOOL		Read/Write	☐	Decimal
Start_Machine	Local:2:I.Data.11	Local:2:I.Data.11	BOOL	This button starts the machine	Read/Write	☐	Decimal
Stop_Machine	Local:2:I.Data.21	Local:2:I.Data.21	BOOL	This button stops the machine	Read/Write	☐	Decimal

Monitor Tags | Edit Tags

Figure 2-28 Version 18's newer Edit Tags view.

MONITOR TAGS VIEW

Because ControlLogix does not have predefined data tables for monitoring PLC data as does the PLC-5 and SLC 500 PLCs, the Monitor Tags window is the place to monitor tags. Figure 2-29 is that of an RSLogix 5000 Monitor Tags window. Note that the Monitor Tags tab is selected at the bottom left. The tag names are listed on the far-left column, with the current value of that tag displayed in the value column. We examine all of the Monitor window features in a later lesson.

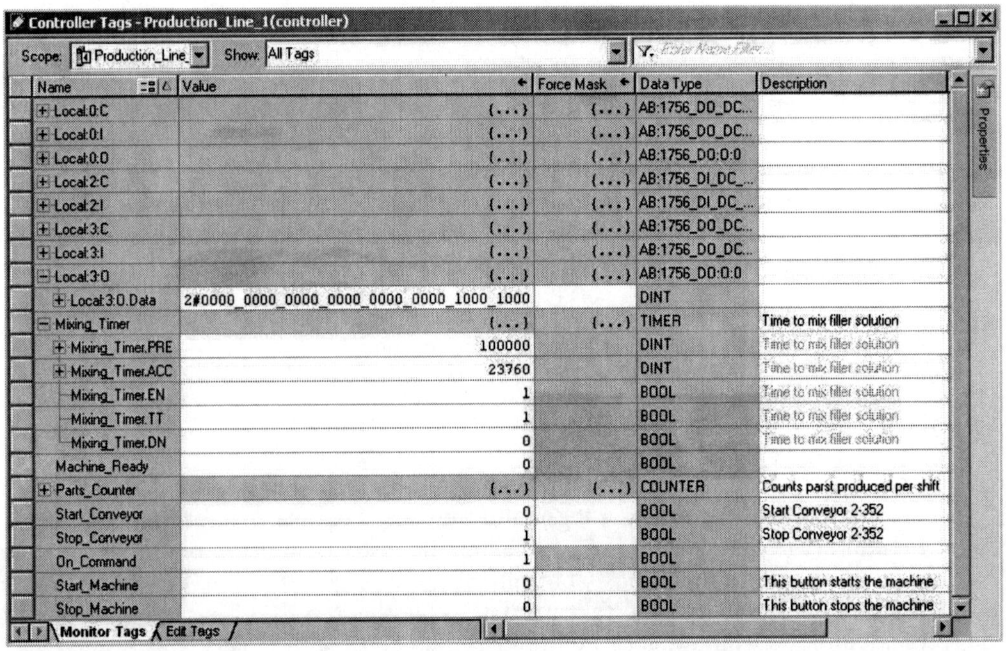

Controller Tags - Production_Line_1(controller)

Scope: Production_Line_ Show: All Tags

Name	Value	Force Mask	Data Type	Description
+ Local:0:C	{...}	{...}	AB:1756_DO_DC...	
+ Local:0:I	{...}	{...}	AB:1756_DO_DC...	
+ Local:0:O	{...}	{...}	AB:1756_DO:O:0	
+ Local:2:C	{...}	{...}	AB:1756_DI_DC_...	
+ Local:2:I	{...}	{...}	AB:1756_DI_DC_...	
+ Local:3:C	{...}	{...}	AB:1756_DO_DC...	
+ Local:3:I	{...}	{...}	AB:1756_DO_DC...	
− Local:3:O	{...}	{...}	AB:1756_DO:O:0	
+ Local:3:O.Data	2#0000_0000_0000_0000_0000_0000_1000_1000		DINT	
− Mixing_Timer	{...}	{...}	TIMER	Time to mix filler solution
+ Mixing_Timer.PRE	100000		DINT	Time to mix filler solution
+ Mixing_Timer.ACC	23760		DINT	Time to mix filler solution
Mixing_Timer.EN	1		BOOL	Time to mix filler solution
Mixing_Timer.TT	1		BOOL	Time to mix filler solution
Mixing_Timer.DN	0		BOOL	Time to mix filler solution
Machine_Ready	0		BOOL	
+ Parts_Counter	{...}	{...}	COUNTER	Counts parst produced per shift
Start_Conveyor	0		BOOL	Start Conveyor 2-352
Stop_Conveyor	1		BOOL	Stop Conveyor 2-352
On_Command	1		BOOL	
Start_Machine	0		BOOL	This button starts the machine
Stop_Machine	0		BOOL	This button stops the machine

Monitor Tags | Edit Tags

Figure 2-29 Version 18's newer Monitor Tags view.

CONTROLLER PROPERTIES SCREEN

The Controller Properties window displays basic information about the controller. Figure 2-30 shows some of the major features as follows:

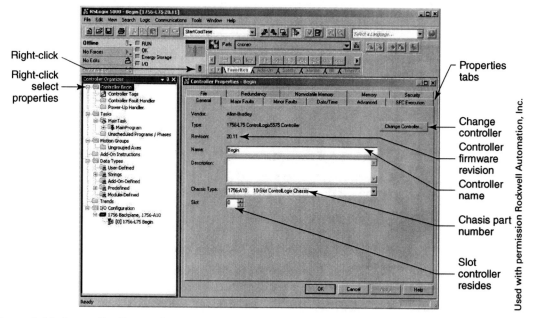

Figure 2-30 Controller Properties screen.

1. The controller is a 1756-L75 ControlLogix 5575 controller.
2. The software revision of this project is 20.11.
3. The name of the project is Begin.
4. The chassis used for this project is a 1756-A10. This is a 10-slot ControlLogix chassis. Remember, chassis are available with 4, 7, 10, 13, or 17 slots.
5. The slot is identified as 0. This specifies into which chassis slot controller this project will be allowed to download. Because ControlLogix is a multicontroller system, the programmer must specify which specific slot controller this project will be downloaded to.
6. Clicking the Major Faults tab shows information when a major fault occurs. A major fault shuts down the controller.
7. Clicking the Minor Faults tab shows information when a minor fault occurs. A minor fault is more like a warning and does not shut down the PAC.
8. The Date/Time tab is used to set up the date and time.
9. The estimated amount of memory used for this project along with estimated available memory can be viewed from the Memory tab.
10. Notice the Change Controller button. The project software version and controller part number can be changed from here. We talk about software versions and firmware levels in the next section.
11. The Serial Port tab is used to configure the serial port on the front of the controller.

The easiest way to display the controller properties is to click the controller icon in the bottom right of the online toolbar. Refer to Figure 2-30.

Figure 2-31 shows a CompactLogix Controller properties window.

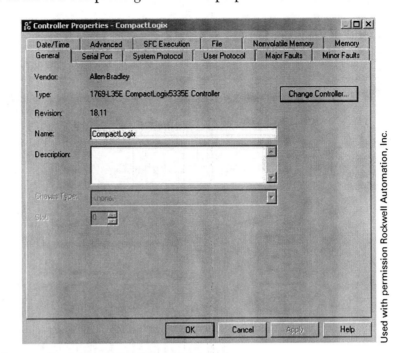

Figure 2-31 CompactLogix Controller Properties screen.

SOFTWARE REVISION LEVELS

The user's personal computer might have Windows 2000, Windows XP, or Windows Vista software loaded. If it had an older version of Windows loaded, then the user might wish to update the computer's software to Vista or Windows 7 in order to use new features contained in the newer software versions. Each major release of PAC programming software may provide new programming instructions, new features, fixes for old problems, and improvements. Rockwell Software products such as RSLogix 5000 software for programming the ControlLogix family of PACs have different versions levels or numbers. The first release of the software is revision 1.0. The major revision is 1, and the minor version is 0. A major revision change such as from version 1 to version 2 is made when new programming instructions or new features are introduced, somewhat like upgrading a personal computer from Windows XP to Vista or Windows 7. Minor modifications to PLC software, such as moving from version 2.0 to 2.1, include minor fixes to the software rather than the addition of new programming features. As of this writing, RSLogix 5000 software version 20.11 was current with version 21 tentatively scheduled for release in November 2012.

Loading RSLogix 5000 Software onto a Personal Computer

When loading RSLogix 5000 software onto a particular computer, determine which versions of RSLogix software could be used on the computer. As of this writing, a new set of software disks contains RSLogix 5000 software versions 12, 13, 15 through 20. Version 14 was a private release. Users can load any or all of the listed versions on their computers. To determine which versions to load, take an inventory of all ControlLogix family controllers and their project software versions in the facility and load the required versions.

Keep in mind that the personal computers on which a new project is being created will have different RSLogix 5000 software versions loaded. When the project is completed, it is transferred or downloaded into a ControlLogix controller, which must have the same major revision of RSLogix 5000 firmware. So if the software project is created in a major revision 20, then the controller into which the project is downloaded also must have major revision 20 firmware loaded. At this point, the minor revision number is not an issue.

Upgrading an Older Software Version

Assume an RSLogix 5000 project was created as version 18, but there were new features of newer version 20 that the user would like to incorporate into the original project. A project can be easily upgraded to a newer version of software that is loaded on a computer by simply going to Controller Properties by clicking the icon, as shown in Figure 2-32. This takes the user to the properties of this specific controller. In Figure 2-33, notice that number 1 identifies the current software version of this project as 18.11. To upgrade the project from version 18.11 to a newer version, click item number 2 to display the Change Controller window. Note that item number 3 also identifies the current project as version 18.11.

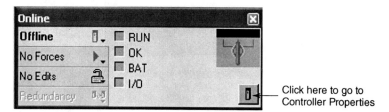

Figure 2-32 Click icon to go to Controller Properties.
Used with permission Rockwell Automation, Inc.

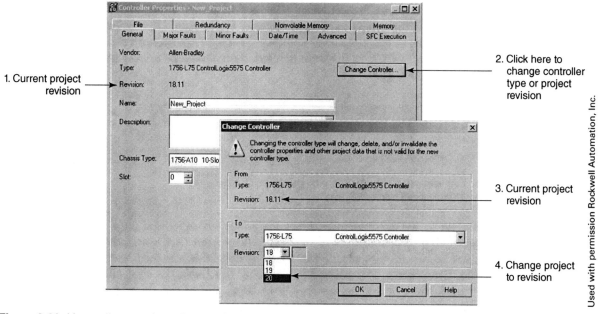

Figure 2-33 Upgrading version 18.11 project to version 20.

After clicking the drop-down arrow next to "Revision," a list of software versions available on this computer is displayed. The figure illustrates that versions 18, 19, and 20 are loaded. Remember that other older versions also could be loaded. When future versions are released, they will also be listed if installed on this computer. For this example, refer to item number 4; we click 20 and select OK. RSLogix upgrades the project from version 18.11 to version 20. Figure 2-34 shows the Controller properties screen with the project updated to 20.11.

66 INTRODUCTION TO RSLOGIX 5000 SOFTWARE

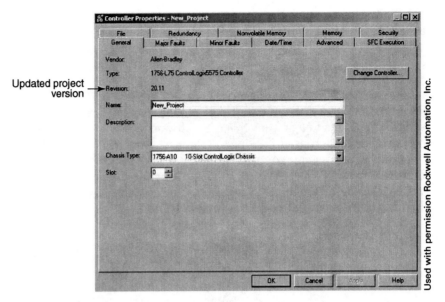

Updated project version

Figure 2-34 Project updated to version 20.11.

When updating the project to a newer version of software, the original project (version 18 in this example) is retained. Figure 2-35 shows part of the RSLogix 5000 files on a personal computer. Note the file named "New_Project"; this is the updated version 20 project. The actual file name is "New_Project.acd." The .acd extension signifies this is an RSLogix 5000 file. The original version 18 file named "New_Project.ACD_V18" is the original version file. To open this file, it must be renamed and the extension changed to .acd. Users should be careful as to what they do to the original version 18 file; the newly updated version 20 file cannot be changed back to version 18. If the original version 18 file were deleted, there would be no way to take the version 20 file and return to version 18.

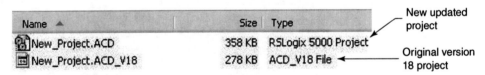

New updated project

Original version 18 project

Figure 2-35 Old version file retained with modified file extension.
Used with permission Rockwell Automation, Inc.

In our scenario, we just updated our version 18 project to version 20. Considering the controller the original project was running also had to be version 18, the updated project file as version 20 cannot be directly downloaded into the version 18 controller because the two are now incompatible. Even though it is beyond the scope of this book, the controller firmware version can be upgraded from its original version 18 to 20 as part of the downloading process of the version 20 project.

Creating a new RSLogix 5000 Project and Software Revision

When creating a new project, programmers select the software version of the project they are about to create. When determining the software version for the new project, first determine the firmware level of the ControlLogix controller into which the project will be downloaded. The major firmware revision level in the controller must match the major software level of the software project. As an example, a version 20 project can only be transferred to a ControlLogix controller that has a firmware version of 20. Figure 2-36 shows the New Controller—that is,

INTRODUCTION TO RSLOGIX 5000 SOFTWARE 67

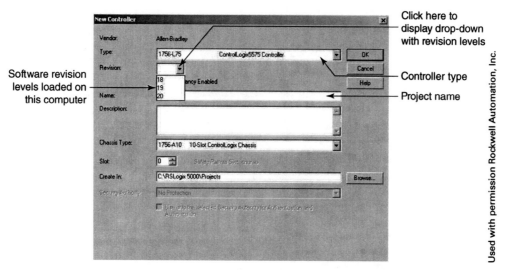

Figure 2-36 Drop-down menu showing available RSLogix software versions on a particular computer.

the create new project window—in the RSLogix 5000 software. When creating a new project, the controller type, project name, chassis size, and software version must be selected. In the figure, note the drop-down menu near the top left. The software revisions listed in this particular drop down are 18, 19, and 20. This particular computer has RSLogix versions 18, 19, and 20 loaded, so these are the options for creating a new RSLogix software project on this computer.

SUMMARY

This lesson introduced the basic RSLogix 5000 software features so that readers will be comfortable with the basic layout of the software and navigation. Future labs will use RSLogix 5000 software to create projects such as programming ladder rungs, editing and modifying ladder rungs, setting up the communication path for uploads and downloads, and going online.

Remember—there are three different packages of RSLogix software. RSLogix 5000 is for the ControlLogix family of programmable automation controllers. This is the software used for the lab exercises in this book. RSLogix 5 is for programming the Allen-Bradley PLC-5, whereas RSLogix 500 is for the SLC 500 family of PLCs, which also includes the MicroLogix.

REVIEW QUESTIONS

Note: For ease of handing in assignments, students are to answer using their own paper.

1. There are three packages of RSLogix software mentioned. Identify which PLC they are used with.
2. Define the term *project*.
3. Where would one view Major Fault information?
4. Refer to Figure 2-37 to identify the parts of the RSLogix 5000 main window.
5. RSLogix 5000 software has two project backup features. One is Automatic Project Recovery based on a user-configurable _____. The second Automatic Program Backup makes a backup of the project file each time the project is _____.
6. The _____ is a tree structure similar to Windows Explorer and is used to organize the project.
7. Refer to Figure 2-38 to identify the ladder components.
8. Like Microsoft Windows, the _____ is located across the top of the window and contains all of the available menus.

68 INTRODUCTION TO RSLOGIX 5000 SOFTWARE

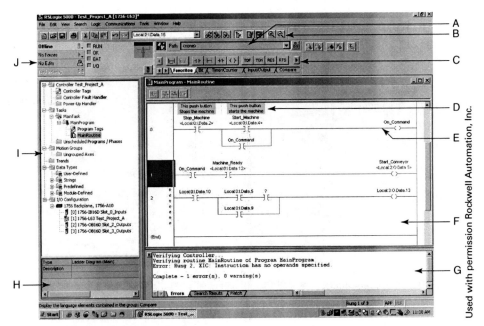

Figure 2-37 RSLogix 5000 Main Window.

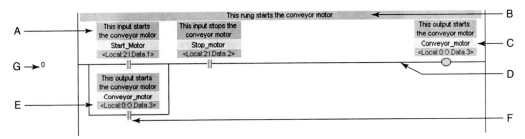

Figure 2-38 Figure for Question 7.
Used with permission Rockwell Automation, Inc.

9. Where would one check to determine the version of software used to create the current project displayed on a computer?
10. One way to program ladder instructions is to drag them from the _____ toolbar to the desired position on a ladder rung and drop them.
11. The _____ toolbar and _____ toolbar can be customized to hold specific programming instructions.
12. The software version and controller part number can be changed from the _____.
13. The software version of a project can only increase in number to a newer revision. Changing a project's software revision back to an _____ software version is not allowed.
14. The _____ toolbar is one way to create a new task, program, routine, tag, and data type.
15. How does a user change the software revision of the current project from RSLogix 5000 version 18 to version 20?
16. The _____ toolbar is used to configure communication between a personal computer and the PAC.
17. There are three parts to the Path Bar toolbar: _____, _____, and _____.

INTRODUCTION TO RSLOGIX 5000 SOFTWARE 69

18. It is _____ of the individual who is uploading, downloading, or going online to make sure the path is correct.
19. The _____ feature is a Windows feature in which users can hover a cursor over objects on their screens to obtain additional information regarding that object.
20. What is the name of the default directory where RSLogix 5000 projects are stored?
21. Selecting the Help _____ tab displays a window that looks like a table of contents, where the user can select from a list that resembles a book's table of contents.
22. The Help _____ tab displays an alphabetical list of help topics similar to a book's index.
23. The Help _____ tab allows entering multiple key words for help to find information.
24. The _____ window is where the logic contained in the selected routine is displayed.
25. ControlLogix can be programmed in four different languages. A routine can be ladder logic, function block diagram, sequential function block, or structured text. The _____ window displays the routine in the language programmed.

LAB EXERCISE 1: RSLogix 5000 Software Navigation

Note: For ease of handing in assignments, students are to answer using their own paper.

This lab exercise provides an opportunity to work with the RSLogix software to navigate through the basic windows, identify the basic features, and find and work with many of them.

1. _____ Open the RSLogix 5000 software.
2. _____ Open the Software_Introduction project.
3. _____ In the Controller Organizer, locate the tasks folder.
4. _____ Click + to expand it to show the Main Task folder.
5. _____ Click + to expand and see the Main Program folder.
6. _____ Click + to expand and see the Main Routine.
7. _____ Double-click Main Routine text.
8. _____ The Main Routine ladder rungs should be displayed.

LOCATING SOFTWARE COMPONENTS

1. _____ Locate each of the software components, and check them off here.
   - _____ Standard Windows toolbar
   - _____ Online toolbar
   - _____ Path Bar toolbar
   - _____ Language Element toolbar
   - _____ Ladder window
   - _____ Controller Organizer
   - _____ Rung Comments
   - _____ Ladder rungs
   - _____ Ladder rung numbers

VIEWING CONTROLLER PROPERTIES

1. _____ How do you get to the Controller Properties screen?
2. _____ Open the Controller Properties.

70 INTRODUCTION TO RSLOGIX 5000 SOFTWARE

3. _____ Write down the type of controller this project is using.
4. _____ List the revision level of this project.
5. _____ What is the name of this project?
6. _____ The slot is identified as _____.
7. _____ What does the slot number signify?
8. _____ The chassis used for this project is a 1756- _____.
9. _____ Chassis are available as _____ slots.
10. _____ Click the Major Faults tab. By clicking this tab, the user can view information when a major fault occurs.
11. _____ Click the Minor Faults tab and notice the Recent Faults area and the fault bits. This information can be used to help diagnose current minor faults.
12. _____ Click the Date/Time tab to set up the date and time. When the computer is online with the controller, the date and time can be set by clicking Set Date Time and Zone from the Workstation button. This synchronizes a personal computer's date and time information with the controller's.
13. _____ Click the Memory tab. View the estimated amount of memory used for this project along with estimated available memory.

LAB EXERCISE 2: Customizing the Language Element Toolbar

This exercise steps you through customizing the Language Element toolbar. Only the Language Element and Ladder Common Logic toolbars can be customized to suit the programmer. Refer to Figure 2-39 for the resulting Favorites tab.

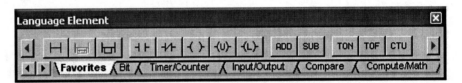

Figure 2-39 Resulting Favorites tab on Language Element toolbar.
Used with permission Rockwell Automation, Inc.

1. _____ On the RSLogix 5000 menu bar, click View.
2. _____ Click Toolbars.
3. _____ Select the Language Element toolbar.
4. _____ Click Customize.
5. _____ All of the instructions are contained in folders. Select the Timer/Counter folder.
6. _____ Clicking + in front of the Timer/Counter folder expands it to show the instructions inside.
7. _____ Select the TON (timer on delay) instruction.
8. _____ Click the Add button; the selected instruction is displayed in the resulting Toolbar section on the right.
9. _____ Select the TOF (timer off delay) instruction.
10. _____ Click the Add button; the selected instruction is displayed in the resulting Toolbar section on the right.
11. _____ Select the CTU (count up counter) instruction.
12. _____ Click the Add button; the selected instruction is displayed in the resulting Toolbar section on the right.
13. _____ Scroll down to the Compute/Math folder.
14. _____ Place the ADD (addition instruction) on the resulting toolbar.

15. _____ Also include the SUB (subtraction instruction).
16. _____ Use the up and down arrows to the right of the resulting toolbar to move the two math instructions before the timer and counter instructions. Refer again to Figure 2-39.
17. _____ At the top of the Available Buttons list is a separator or spacer. Add one spacer to separate the math instructions from the relay ladder logic instructions.
18. _____ Position a second separator between the math instructions and timer and counter group.
19. _____ Click OK to finish.
20. _____ Refer to Figure 2-39 to see the resulting toolbar. If not all of your favorite instructions are displayed, use the right or left arrows at each end of the icon list to move to the right or left to bring them into view.

LAB EXERCISE 3: Configuring the RSLogix 5000 Display

In this lab exercise, we practice modifying the configuration for our routine window. Because this is a beginning ControlLogix class, we only look at some of the basic features in this lab.

1. _____ Open the Software Introduction project if it is not already open from the last lab.
2. _____ From the RSLogix 5000 menu bar, select Tools.
3. _____ Click Options. You should see a window similar to that in Figure 2-21, with the Application category displayed.
4. _____ Note the Project Directory section. This is where RSLogix 5000 projects are stored. By default, the directory should read "C:\RSLogix5000\Projects." If your path is different, do not change it without obtaining your instructor's permission.
5. _____ RSLogix 5000 version 16 has a Startup page that is displayed when the software is launched. Uncheck the box to disable this feature.
6. _____ Verify that the check box is checked for the Enable Automatic Project Backup feature. Change the number of backups to "3."
7. _____ Verify that the check box is checked for the Enable Automatic Project Recovery feature. Change the time value to the time period you desire.
8. _____ Leave the remaining options as the default.
9. _____ From the categories list, select Ladder Editor.
10. _____ Select Display. You should see a window similar to that shown in Figure 2-26.
11. _____ Uncheck Show Rung Numbers.
12. _____ Uncheck Show Rung Comments.
13. _____ Uncheck 3D Instruction Display.
14. _____ Click Apply.
15. _____ Notice how your routine window has changed. The three items you unchecked should not be displayed.
16. _____ Go back and recheck the three options you just unchecked.
17. _____ Click Apply; they should be displayed again.
18. _____ Click Font/Color, which is also under the Ladder Editor category.
19. _____ From the Items list, select Rung Comments.
20. _____ Click the down arrow associated with Background Color. You should see a color selection drop-down menu similar to that shown in Figure 2-27.
21. _____ Select whichever color you want for your Rung Comment background. Additional colors are available by clicking the More Colors button.
22. _____ Change the Foreground Color to something you like.
23. _____ Experiment in changing the ladder window color.
24. _____ Change the End Rung color.

25. _____ Change the Descriptions colors to whatever you like.
26. _____ Click Apply to see the changes.
27. _____ To return the colors to the factory defaults, click the Reset Category button in the bottom left of the Workstation Options window.
28. _____ Look at the Categories list. Are languages other than ladder installed on your computer?

    How do you know? *Hint:* Look at Figure 2-27. If the SFC Editor is listed as in the figure, then the sequential function chart language is installed on your computer. If the FBD Editor is listed as in Figure 2-27, then the function block diagram language is installed. If Structured Text Editor is listed as in the figure, then the structured text language is installed on your computer.
29. _____ When done, click OK to exit the Workstation Options window.
30. _____ Save the project.

LAB EXERCISES 4 AND 5: Creating the Begin Project Using RSLogix 5000 Software

The following lab exercises create a new RSLogix 5000 project that we will name "Begin." We will use this project also in future exercises. After creating the project, we configure our display and customize the Favorites tab on the Language Element toolbar.

Lab Exercise 4 creates a new project for a modular ControlLogix, whereas Lab Exercise 5 is for a CompactLogix. Select the lab depending on the hardware you have.

LAB EXERCISE 4: Creating the Begin Project in RSLogix 5000 Software for a Modular ControlLogix Programmable Automation Controller

Note: For ease of handing in assignments, students are to answer using their own paper.

INTRODUCTION

In this lab, you create a basic project in RSLogix 5000 software for a ControlLogix. This configures the specific ControlLogix hardware you will be using in your lab exercises. To complete this lab, you should have a ControlLogix modular PAC similar to that shown in Figure 2-40. Your specific hardware is probably different from that in the figure.

Figure 2-40 Modular ControlLogix programmable automation controller.
Used with permission Rockwell Automation, Inc.

CREATING A CONTROLLOGIX MODULAR PAC BEGIN PROJECT

This lab exercise steps you through creating the Begin project for your specific modular ControlLogix. Before we start, we have to determine the specific hardware you are using.

Determining Your Hardware

1. _____ What size chassis do you have?
2. _____ What is the chassis part number?
3. _____ Determine how many controllers are in the chassis and their part numbers. Refer to Chapter 1 for controller information. List your controller(s) part numbers in the Figure 2-41. Make sure the controller(s) are listed in the correct chassis slot.

MY MODULAR CHASSIS HARDWARE CONFIGURATION		
Slot Number	Module Part Number	Description of Module
0		
1		
2		
3		
4		
5		
6		
7		
8		
9		

Figure 2-41 Begin project ControlLogix chassis configuration.
© Cengage Learning 2014

4. _____ How much memory does your controller(s) have?
5. _____ What is the firmware revision of your processor(s)?
6. _____ What is the part number of your power supply?

Determining I/O Modules

7. _____ What does it mean if an I/O module part number has a "D" suffix, as in a 1756-OB16D?
8. _____ What does the "O" in the same part number signify?
9. _____ What does the "B" in the part number signify?
10. _____ What does the "16" represent?
11. _____ If a module part number had an "I" suffix such as a 1756-IB16I, what would this tell you about the module?
12. _____ If a module part number had an "E" suffix such as a 1756-OA8E, what would this tell you about the module?
13. _____ List the I/O modules in your chassis in the table in Figure 2-41.

Determining Communication Modules

14. _____ Define the type of communication module from the part numbers listed here:
    a. 1756-ENBT
    b. 1756-ENET
    c. 1756-CNB
    d. 1756-CNBR
    e. 1756-DHRIO
    f. 1756-DNB
15. _____ Are there any communication modules in your chassis? List them in the table in the correct slot.

Figure 2-42 illustrates a possible completed table representing the modules in your chassis.

74 INTRODUCTION TO RSLOGIX 5000 SOFTWARE

MODULAR CHASSIS HARDWARE CONFIGURATION		
Slot Number	Module Part Number	Description of Module
0	1756-OB16D	16-point diagnostic output module
1	1756-L63 processor	L63 processor
2	1756-IB16D	16-point diagnostic input module
3	1756-CNB	ControlNet bridge module
4	1756-DNB	DeviceNet bridge module
5	1756-ENBT	Ethernet communications module
6	Empty slot	
7	Empty slot	
8	Empty slot	
9	Empty slot	

Figure 2-42 Completed table example.
© Cengage Learning 2014

CREATING THE BEGIN PROJECT

1. _____ Open your RSLogix 5000 software.
2. _____ Click New, as illustrated in Figure 2-43, to create a new RSLogix 5000 project.

Figure 2-43 Creating a new project.
Used with permission Rockwell Automation, Inc.

3. _____ You should see a window similar to that in Figure 2-44.

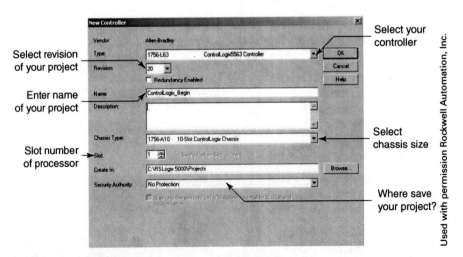

Figure 2-44 Creating a new project in RSLogix 5000 software.

4. _____ Select your controller by clicking the drop-down arrow and selecting the controller part number from the list.
5. _____ Likewise select the project software revision type. Exercises in this book assume you are using RSLogix 5000 version 18.
6. _____ Enter the name as ControlLogix Begin.

INTRODUCTION TO RSLOGIX 5000 SOFTWARE 75

7. _____ You can enter a description if you wish. The description is optional.
8. _____ Select the chassis type.
9. _____ Select the slot number in the chassis where the controller resides into which this project will be downloaded.
10. _____ The "Create In" selection is the directory on your personal computer where this project will be saved. We will accept the default directory, as shown in the figure.
11. _____ Click OK when completed.
12. _____ When the new project is created, your controller organizer should look similar to that shown in Figure 2-45.
13. _____ Save this project as ControlLogix Begin on your computer hard drive.

The main components of your project are identified in Figure 2-45 and described here.

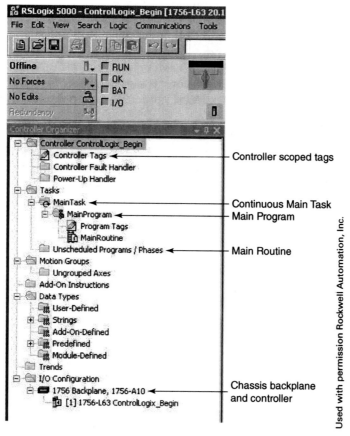

Figure 2-45 RSLogix 5000 Controller Organizer.

Controller Scoped Tags

ControlLogix is a tag-based PLC. A tag is simply a name. There is no predefined ladder or data files like those found in a traditional PLC such as the Allen-Bradley PLC-5 or SLC 500. Tags are simply arranged alphabetically. The arrow is pointing to the controller scoped tags collection. This is the primary area where tags are created, edited, and stored. We will complete a lab exercise a little later in which you will create new tags, modify existing tags, and monitor data in the controller scoped tags collection.

Continuous Main Task

Four types of tasks are possible within an RSLogix 5000 project. A task is simply an organizational mechanism. There is no executable code contained directly within it. Executable code is contained in a routine, which is a subset of a task and a program. You will be creating tasks, programs, and routines as we move into future lab exercises.

Main Program

A ControlLogix project can comprise many programs. A program is only an organizational mechanism. There is no executable code directly stored in one. Programs are organized or listed in the order which they are to be executed.

Main Routine

Routines are where the project's executable code is stored. A single routine must be all the same programming language. Because ControlLogix supports four programming languages—ladder diagram, function block diagram, sequential function chart, and structured text—these are the available routine types. As an example, a ladder diagram routine must be 100 percent ladder logic. Languages cannot be mixed in the same routine. Each program has a main routine and possibly subroutines. In Figure 2-45, notice the ladder-type icon to the left of the Main Routine text. This signifies the routine as ladder logic. There is a specific icon for each programming language. Also note a piece of paper with a "1" on it sits on top of the ladder icon. This identifies this as the Main Routine, or where the controller starts executing the logic within each program. Additional routines are designated as subroutines. You will create programs and routines in future lessons.

I/O Configuration, Chassis Backplane, and Controller

This is the I/O configuration section of the project. Notice the little chassis icon and the text "1756 Backplane, 1756-A10." This identifies the chassis part number and the number of slots in the chassis. A future lab exercise will step you through performing an I/O configuration that is using the RSLogix 5000 software to populate the chassis with the specific modules used for this application.

When creating a new project, one item specified in the new controller window is the slot of the controller into which this project is to be downloaded. Notice there is a controller icon with a left pointing arrow and "[1] 1756-L63 ControlLogix_Begin." This identifies the controller into which this project will be downloaded. The controller is in slot 1, and it is a 1756-L63 with a project name of "ControlLogix_Begin." We will work with all of these project components in greater depth in a future lesson.

14. _____ Click Tools and then Options to get to Workstation options.
15. _____ Set up your Automatic Project recovery with a time you are comfortable with.
16. _____ Set up your Automatic Project Backup with the number of backups you wish. Remember, these are the BAK files. For this project, the files will be saved starting as "ControlLogix_Begin.BAK000."
17. _____ Go to the Ladder Editor's Font/Color section and set the desired colors for the routine window.
18. _____ Remember that how you set these workstation options will pertain to any project opened on this computer until someone changes them.
19. _____ This completes this lab. You have created a new RSLogix 5000 project, which we will build on in later lessons.
20. _____ Save your project.

LAB EXERCISE 5: Creating the Begin Project in RSLogix 5000 Software for a CompactLogix Programmable Automation Controller

Note: For ease of handing in assignments, students are to answer using their own paper.

INTRODUCTION

This lab creates a basic project in RSLogix 5000 software for a CompactLogix. In this lab, you determine and configure the specific hardware you are using in your lab exercises. Complete this lab if you are using a CompactLogix PAC for your lab exercises. Figure 2-46 illustrates a CompactLogix PAC.

Figure 2-46 Example of a 1769 CompactLogix PAC.
Used with permission Rockwell Automation, Inc.

Determining your CompactLogix Hardware

This lab exercise guides you through creating the Begin project for your specific modular CompactLogix PAC. CompactLogix does not have a chassis for the modules to slide into. Modules simply clip together on a DIN rail to make an assembly called a *bank*.

Examine your CompactLogix and list your hardware here.

1. _____ Are you using a 1768 or 1769 CompactLogix?
2. _____ List your controller part number: _____.
3. _____ How much memory does your controller have? _____
4. _____ What communications ports does your controller have? _____
5. _____ Enter the controller information in the table in Figure 2-49.
6. _____ What is the firmware revision of your controller? _____
7. _____ List the part number of your power supply: _____
8. _____ If you are using the 1768 CompactLogix, do you have an Ethernet, a ControlNet, or a motion control module in the 1768 bus?

Determining I/O Modules

9. _____ What does it mean if an I/O module part number has an "I" suffix, as in 1769-I8I?
10. _____ What does the first "I" in this part number signify?
11. _____ What does the "8" in the part number signify?
12. _____ What does the "16" represent?
13. _____ Identify the specifics of an I/O module if a module's part number is 1769-IQ16.
14. _____ List the I/O modules in your PLC in the table in Figure 2-49.
15. _____ Are there any communication modules on your chassis?

The table in Figure 2-47 is an example of what a completed table for a 1769 CompactLogix hardware configuration could look like. Your specific CompactLogix hardware is listed in the table shown in Figure 2-49.

The table in Figure 2-48 shows what a completed table for a 1768 CompactLogix hardware configuration could look like.

Figure 2-49 is the table for you to fill in for the specific CompactLogix you will use in future lab exercises.

1769 COMPACTLOGIX HARDWARE CONFIGURATION		
Slot Number	Module Part Number	Description of Module
0	1769-L35E	CompactLogix Ethernet processor
1	1769-IQ16	16 inputs, 24 volts DC
2	1769-OB16	16 outputs, 24 volts DC
3	1769-IF4	4-channel analog inputs
4	1769-OF2	2-channel analog outputs
5		
6		
7		
8		
9		
10		

Figure 2-47 Completed table example.
© Cengage Learning 2014

1768 COMPACTLOGIX HARDWARE CONFIGURATION		
Slot Number	Module Part Number	Description of Module
1768 Bus		
1	1768-ENBT	Ethernet communications module
0	1768-L43	CompactLogix controller
1769 Bus		
1	1769-IQ16	16 inputs, 24 volts DC
2	1769-OB16	16 outputs, 24 volts DC
3	1769-IF4	4-channel analog inputs
4	1769-OF2	2-channel analog outputs

Figure 2-48 1768 CompactLogix hardware configuration example.
© Cengage Learning 2014

17 _____ COMPACTLOGIX HARDWARE CONFIGURATION		
Slot Number	Module Part Number	Description of Module

Figure 2-49 My Begin project CompactLogix configuration.
© Cengage Learning 2014

INTRODUCTION TO RSLOGIX 5000 SOFTWARE 79

LAB EXERCISE 6: Creating a 1768 or 1769 CompactLogix Begin Project

1. _____ Open your RSLogix 5000 software.
2. _____ Select the new icon as shown in Figure 2-50.

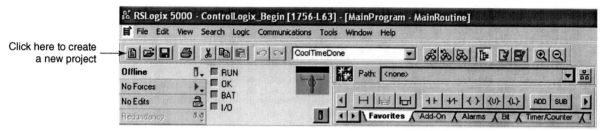

Figure 2-50 Create new RSLogix 5000 project.
Used with permission Rockwell Automation, Inc.

3. _____ The window in Figure 2-51 should be displayed.

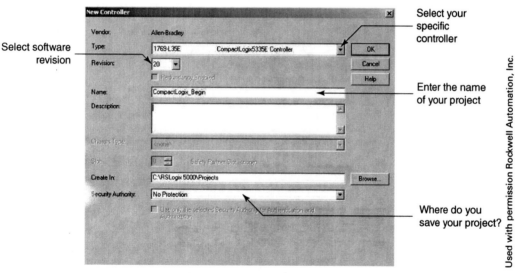

Figure 2-51 New Controller.

4. _____ Select the proper 1768 or 1769 controller.
5. _____ Select the software version you are using.
6. _____ Name the project "CompactLogix Begin."
7. _____ Adding a description is optional.
8. _____ Leave "Create In" as the default RSLogix 50000 directory.
9. _____ Click OK when you have completed filling in the new project information.
10. _____ Figure 2-52 illustrates how the controller organizer should look for the new 1769 CompactLogix Begin project just created.
11. _____ Save this project as "ControlLogix_Begin" on your computer's hard drive.

The main components of your project are identified in Figure 2-52 and described following.

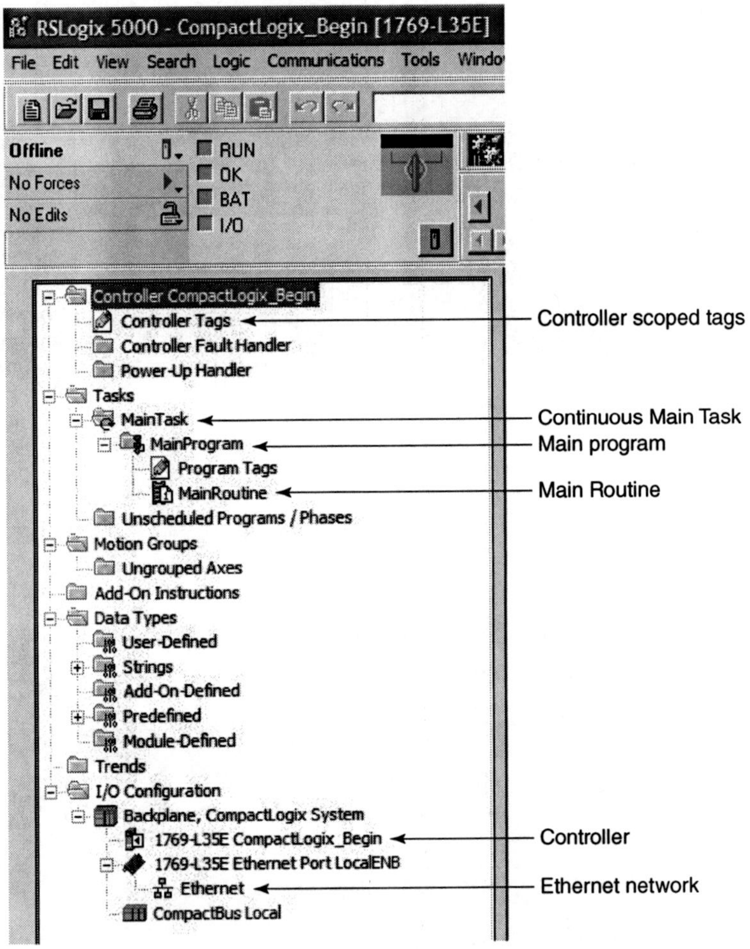

Figure 2-52 Controller organizer of your CompactLogix 1769 new project.

Controller Scoped Tags

ControlLogix is a tag-based PLC. A tag is simply a name. There is no predefined ladder or data files such as one would find in a traditional PLC such as the Allen-Bradley PLC-5 or SLC 500. Tags are simply arranged alphabetically. The top arrow in Figure 2-52 is pointing to the controller scoped tags collection. This is the primary area where tags are created, edited, and stored. Later, we will complete a lab exercise in which you will create new tags, modify existing tags, and monitor data in the controller scoped tags collection.

Continuous Main Task

Four types of tasks are possible within an RSLogix 5000 project. A task is simply an organizational mechanism. There is no executable code contained directly within one. Executable code is contained in a routine, which is a subset of a task and a program. You will be creating tasks, programs, and routines as we move into future lab exercises.

Main Program

A CompactLogix project can have as many as 100 programs per task. A program is only an organizational mechanism. There is no executable code directly stored in one. Programs are organized or listed in the order which they are to be executed.

Main Routine

Routines are where the project's executable code is stored. A single routine must be entirely in the same programming language. Because ControlLogix supports four programming languages—ladder diagram, function block diagram, sequential function chart, and structured text—these are the available routine types. As an example, a ladder diagram routine must be 100 percent ladder logic. Languages cannot be mixed in the same routine. Each program has a main routine and possibly several or many subroutines. In Figure 2-52, notice the ladder-type icon to the left of the Main Routine text. This signifies the routine as ladder logic. Each programming language has a specific icon. Also note a piece of paper with a "1" on it sits on top of the ladder icon, identifying this as the main routine, or where the controller starts in executing the logic within each program. Additional routines are designated as subroutines. You will create programs and routines in future lessons.

I/O Configuration, Chassis Backplane, and Controller

This is the I/O configuration section of the project. Notice the little chassis icon and the text "Backplane, CompactLogix System." A future lab exercise will guide you through performing and I/O configuration that uses RSLogix software to populate the bank with specific modules used for this application.

Note that there is a controller icon with a left-pointing arrow and "1769-L35E CompactLogix_Begin." This identifies the controller into which this project will be downloaded. The controller is a 1769-L35E Ethernet controller with a project name of "CompactlLogix_Begin." We will work with all of these project components in greater depth in a future lesson.

If you are using a 1768 CompactLogix, Figure 2-53 illustrates the new "1768CompactLogix_Begin" project that you have just created. The other parts of the figure are the same for both the 1768 and 1769 CompactLogix.

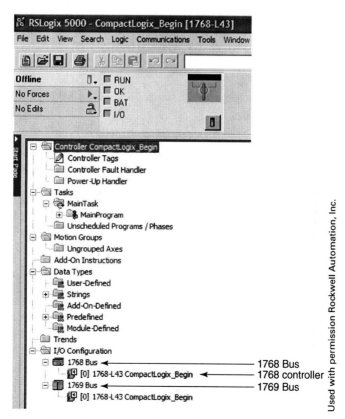

Figure 2-53 CompactLogix 1768 hardware configuration.

82 INTRODUCTION TO RSLOGIX 5000 SOFTWARE

12. _____ Click Tools and then Options to get to Workstation Options.
13. _____ Set up your Automatic Project recovery with a time you are comfortable with.
14. _____ Set up your Automatic Project Backup with the number of backups you wish. Remember that these are the BAK files. For this project, the files will be saved starting as "ControlLogix_Begin.BAK000."

Selecting Ladder Window Colors

The user can also modify the colors of the ladder window components. Rung Comments, ladder window background color, and other documentation contained in the window can be modified as desired. Here we modify some of the ladder window colors. Refer to Figure 2-54 through the procedure.

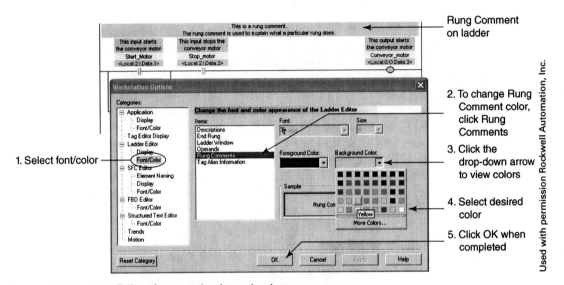

Figure 2-54 Ladder Editor fonts and color selections.

15. _____ Select Tools from the RSLogix 5000 menu bar.
16. _____ Select Options.
17. _____ Under the Ladder Editor category, select Font/Color.
18. _____ Select Rung Comments.
19. _____ Click the down arrow to change the background color.
20. _____ Select the desired color.
21. _____ If desired, change the foreground color for Rung Comments.
22. _____ Experiment with changing other color options in this section.
23. _____ Save your project.

CHAPTER

3

Number Systems

OBJECTIVES

After completing this chapter, you should be able to

- Understand decimal, binary, octal, and hexadecimal numbers.
- Convert from one number system to another.
- Understand new terms associated with ControlLogix data formatting.

INTRODUCTION

To effectively choose, install, maintain, and troubleshoot today's programmable logic controllers (PLCs) you must possess basic PLC programming skills. To program the modern PLC, you must understand the different methods by which internal data is represented. Typical data representation inside a PLC can include binary, binary-coded decimal, decimal, octal, hexadecimal, or some combination of these. In this chapter we are going to use the common term PLC to refer to either a programmable logic controller or programmable automation controller, as the chapter information pertains equally to both.

Programmable controllers do not understand our everyday human languages such as spoken words or numbering conventions. Thus, programmable controllers cannot add, subtract, multiply, divide, or otherwise manipulate the decimal numbers to which we are accustomed. The PLC, like any microprocessor-based device, works in the world of zeros and ones. PLCs process ones and zeros using the binary number system and binary arithmetic.

Alphanumeric characters that are entered into a computer from a keyboard are converted into unique binary patterns representing each keyboard character. The computer manipulates this binary data as directed by the user program and converts it back to decimal data for output to a monitor or printer. The PLC works the same way when a human inputs decimal data through an operator interface device or keyboard to change the preset values. The PLC reconverts this internal binary data to decimal data before outputting the modified values to the operator interface device for human viewing.

The decimal, binary, octal, and hexadecimal systems are common number systems used with PLCs. Binary coding of decimal numbers (BCD) and binary coding of seven or eight bits to represent alphanumeric characters, as in the American Standard Code for Information Interchange (ASCII), are common coding techniques. This chapter investigates most of these different data representations.

Upon beginning to read this chapter, you are probably asking why you should study and understand different number systems. Some important reasons follow:

1. Internal PLC data tables represent data in binary format.
2. Some PLCs use decimal numbering to represent addresses.
3. Other PLCs use octal numbering to represent addresses.

4. Operator interface devices, in older systems such as thumbwheels and seven-segment output displays, have their own binary coding, which allows for an easy human-machine electrical interface.
5. Analog input data is converted to numeric data as it is entered into the PLC.
6. Analog output data must be converted from numeric data as the signal is prepared for output from the PLC to the analog output hardware device.
7. Many PLCs display error codes in hexadecimal.
8. Some PLCs have multiple instructions that require hexadecimal masks for filtering data.
9. Hexadecimal error codes may need to be converted to decimal as part of a fault routine.
10. When configuring, some PLC analog I/O cards require a 16-bit word to be constructed from a table of operating options. This 16-bit word is converted to decimal before being programmed on the ladder rung that sends the configuration information to the analog module.

NUMBER SYSTEM CHARACTERISTICS

There are many ways to use numbers to count. In our everyday lives, we depend on the decimal number system to count everything from quantity, weight, and speed to money. Our method of counting is based on 10, which is why it is referred to as base 10. The base, or radix, of a number simply identifies how many unique symbols are used in that particular number system. Decimal has 10 unique symbols used for counting, starting with 0 and ending with 9. Notice that the largest-value symbol is one value less than the base. The number systems typically encountered when working with PLCs include binary (base 2), decimal (base 10), octal (base 8), and hexadecimal (base 16). Figure 3-1 lists each number system, its base, how many digits (or valid symbols) are available, and the range of the available symbols.

NUMBER SYSTEM CHARACTERISTICS			
System	Base	Digits	Range
Binary	2	2	0 and 1
Octal	8	8	0 through 7
Decimal	10	10	0 through 9
Hexadecimal	16	16	0 through 9 and A through F

Figure 3-1 Number systems typically used with PLCs.
© Cengage Learning 2014

Ask yourself, where are these different number systems used and how do they affect me?

As a PLC programmer, or electrical or maintenance individual working with PLCs, you will someday find yourself attempting to understand information stored inside the PLC in different number systems, program or configure PLC hardware, understand or program instructions that use masks, or set up or troubleshoot a network. Listed here are a few other reasons specific number system understanding is important:

1. Computers and PLCs only understand binary information. In many cases humans have to understand and convert decimal information to and from binary to understand what is happening inside the PLC.
2. If using an octal-based PLC such as a PLC-5, I/O addresses are octal-formatted.
3. When using a PLC-5 or SLC 5/04 processor and Data Highway Plus as your communication network, station addresses are octal.
4. Even though decimal numbers are familiar to humans, PLCs for the most part only understand binary. In many cases information must be converted back and forth.
5. Hexadecimal information is found in processor or I/O module error codes, instruction masking, and DeviceNet ControlNet and Ethernet/IP electronic data sheet file identification, to mention a few.

6. You might use a ControlLogix PAC to communicate with an existing Data Highway Plus or Remote I/O network, both of which are addressed in octal.

In this chapter we explore the world of numbers as used by different PLCs. We introduce number systems typically used by PLCs, examine methods of conversion from one to the other, and look at examples of how selected number systems are used in PLC applications. Let us start by looking closely at the decimal number system, with which we all are familiar.

THE DECIMAL NUMBER SYSTEM

In our everyday life, we are accustomed to counting in the decimal system; that is, 0, 1, 2, 3, 4, 5, 6, 7, 8, 9, 10, 11, and so forth. We use 10 symbols, 0 through 9. After the number 9, we use various combinations of these symbols to express numerical information. Our decimal system is configured in what is called base 10. The characteristic that distinguishes different number systems is the base, or radix, which simply identifies the number of symbols (numbers) used to represent a given quantity. The base, or radix, also tells us what weight each digit position represents in relation to the other digits.

Let us look at an example. In base 10, the number 123 in reality represents three values:

Here we have a 1 in the 100s place. This equals:	100
We have a 2 in the 10s place. This equals:	20
We have a 3 in the 1s place. This equals:	3
Adding these values, we get the following:	123

We call this number "one hundred twenty-three."

To signify that the number 123 is in base 10, it can also be written as 123_{10}. The subscript, 10, indicates that the numerical value is represented in base 10. In our daily life it is accepted that the numbers we encounter are decimal, so the base, or radix, is usually not included.

DECIMAL PLACE VALUES

Each physical number position has a weighted value. Each successive decimal number position, from right to left, is 10 times greater than the previous position, as we illustrate here. We calculate the powers of 10 to arrive at each decimal place value:

Powers of 10:	→	10^4	10^3	10^2	10^1	10^0
Place value:	→	10,000	1,000	100	10	1

The first placeholder, or least significant digit, is 10^0, which equals 1. This is our 1s place. When counting, we count from 0 through 9. When we reach 9, we run out of 1s. The next number after 9 is indicated by placing a 1 in the second number position, the 10s place, while our 9 advances one count, to 0. The second place value equals 10^1 or $10 \times 1 = 10$. This is the 10s place. This leaves us with a number that looks like 10. We call this "10" because we have one 10 and no 1s.

As we count, we have one 10 and one 1, which we call "11"; one 10 and two 1s which we call "12"; and so on, until we reach one 10 and nine 1s. We have run out of numbers in the 1s place, and so, once again, we carry our count into the 10s place. The next number, then, is two 10s and zero 1s. We call this number "20." Eventually, when we reach nine 10s and nine 1s (99), we have to go into the third (100s) place to continue counting.

The third place value, the 100s place, is written 10^2, or $10 \times 10 = 100$. The number in this place tells us how many 100s we have. The third place is 10^3 or $10 \times 10 \times 10 = 1,000$. This is the 1,000s place. The fourth place is 10^4, or $10 \times 10 \times 10 \times 10 = 10,000$. This is the 10,000s place. Thus, the number 12,345 should be easily identified, as shown in Figure 3-2.

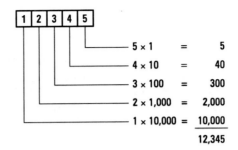

Figure 3-2 Derivation of the decimal number 12,345.
© Cengage Learning 2014

BINARY NUMBERS

In the next few pages we are going to introduce the world of computer numbers: the binary number system.

Binary Numbers

Binary numbering uses only two digits, 0 and 1. Although humans understand and use the decimal number system, microprocessor-controlled devices such as digital computers and programmable controllers understand binary signals instead. Computer-based devices are designed to understand and operate with binary information, as it is easier to design a machine to distinguish between 2 conditions rather than 10, which would be needed if the decimal system were used. Also, most basic industrial control devices operate on two conditions, such as open or closed, on or off. As a result, to design a machine to effectively control two-state devices using a two-state signal such as binary was an easy decision.

Computers do accept decimal numbers as inputs and also produce decimal outputs to satisfy the human operator; however, values are converted by the input circuitry to a binary format acceptable to the central processing unit (CPU). The output circuitry then converts binary signals from the CPU to decimal values recognized by humans.

Digital computers and other microprocessor-controlled equipment, such as PLCs, only understand and operate on binary values (1s and 0s). These binary digits are called bits. The term *bit* is derived from BInary digiT. A single bit is a single 1 or 0.

BINARY DATA REPRESENTATION

In our everyday lives, we communicate to others using groups of letters called words. In a similar fashion, computer data are organized into words and stored in PLC memory.

Computers work with groups of binary digits (bits) organized into words to either store or manipulate information. For many years, the most modern PLCs were 16-bit machines. The SLC 500, MicroLogix, and PLC-5 are examples of a 16-bit PLC. A 16-bit PLC works with information in a group of 16 bits called a word. The only differences between words in the English language and computer words are the length and the available characters. English words can be of varying lengths and can contain any of 26 alphabet characters, but computer words are always the same length. Rather than letters making up a computer word, binary bits are used. Unlike in English, when working with computers, there are only two characters available, 1 or 0, each of which is called a bit. A single bit is the smallest unit of data that can be represented.

One measure of a PLC's capabilities is the length of the data words on which it can operate. Many current PLCs use 16 bits to represent a word. Figure 3-3 illustrates two 16-bit words. Notice that only 1 bit, either a 1 or 0, is permitted in each of the 16 available positions.

NUMBER SYSTEMS 87

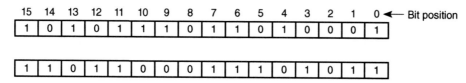

Figure 3-3 Information represented by arranging the 1s and 0s in different combinations within a 16-bit word.
© Cengage Learning 2014

- A 16-bit word is composed of the following parts:
- 16 single bits
- *nibble:* the lower (lower nibble) or upper (upper nibble) 4-bit group of a byte
- *byte:* the group of the lower (lower byte) or upper (upper byte) 8 bits of a 16-bit word
- *word:* a group of 16 bits stored and used together

Each part is illustrated in Figure 3-4.

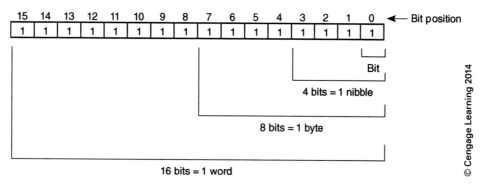

Figure 3-4 Parts of a 16-bit word.

Each 16-bit word is made up of two bytes, an upper and a lower byte. Each **byte** contains two nibbles. Each byte is broken down into an upper and a lower nibble, and each **nibble** is broken down into four individual bits. Figure 3-5 illustrates the breakdown of a 16-bit word.

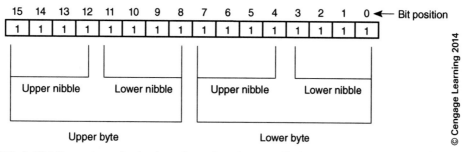

Figure 3-5 A 16-bit word can be broken down into bytes, nibbles, and bits.

A newer PLC like ControlLogix is a 32-bit computer. Even though a ControlLogix programmer has the option to store information in 16-bit words like other PLCs, a 32-bit controller such as ControlLogix has a basic memory allocation of 32 bits, or a double word. Figure 3-6 illustrates a double word and its associated pieces.

88 NUMBER SYSTEMS

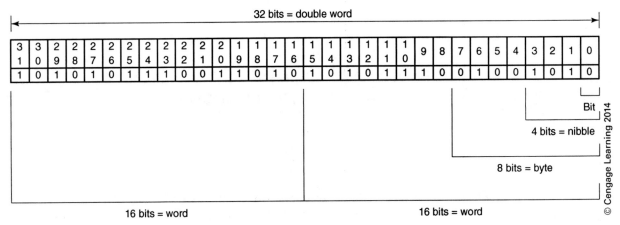

Figure 3-6 A 32-bit double word makeup.

Binary Bit Position and Weighting

The physical position of the bits in a 16-bit word or 32-bit double word is important. Each physical position has a weighted value. Because computers only understand binary, or base 2, each successive position from right to left is two times greater than the previous position. Figure 3-7 illustrates the bit positions for a 16-bit word.

| 2^{31} | 2^{30} | 2^{29} | 2^{28} | 2^{27} | 2^{26} | 2^{25} | 2^{24} | 2^{23} | 2^{22} | 2^{21} | 2^{20} | 2^{19} | 2^{18} | 2^{17} | 2^{16} | 2^{15} | 2^{14} | 2^{13} | 2^{12} | 2^{11} | 2^{10} | 2^{9} | 2^{8} | 2^{7} | 2^{6} | 2^{5} | 2^{4} | 2^{3} | 2^{2} | 2^{1} | 2^{0} |

Figure 3-7 A 32-bit binary double word bit weighting.
© Cengage Learning 2014

The first position, the least significant bit (or LSB), is the 1s place, or 2^0. The next place is the 2s place, or 2^1 ($2 \times 1 = 2$). The third place is the 4s place, or 2^2 ($2 \times 2 = 4$). The fourth place is the 8s place, or 2^3 ($2 \times 2 \times 2 = 8$), and so on. Figure 3-8 illustrates the decimal-weighting value of each bit position.

BINARY PLACE VALUE WEIGHTINGS							
Value	**Decimal**	**Value**	**Decimal**	**Value**	**Decimal**	**Value**	**Decimal**
2^0	1	2^8	256	2^{16}	65,536	2^{24}	16,777,216
2^1	2	2^9	512	2^{17}	131,072	2^{25}	33,554,432
2^2	4	2^{10}	1,024	2^{18}	262,144	2^{26}	67,108,864
2^3	8	2^{11}	2,048	2^{19}	524,288	2^{27}	134,217,728
2^4	16	2^{12}	4,096	2^{20}	1,048,576	2^{28}	268,435,456
2^5	32	2^{13}	8,192	2^{21}	2,097,152	2^{29}	536,870,912
2^6	64	2^{14}	16,384	2^{22}	4,194,304	2^{30}	1,073,741,824
2^7	128	2^{15}	32,768	2^{23}	8,388,608	2^{31}	2,147,483,647

Figure 3-8 Binary place values converted to decimals for 32-bit double words.
© Cengage Learning 2014

Remember how decimal numbers such as 12,345 from Figure 3-2 went together? Using the same basic method, it should not be too difficult to convert binary to decimal. As an example, let's look at the binary value 0000000010101010. If we determine each bit's value similar to the way we did for the decimal number in Figure 3-2, it should be easy to determine the decimal value of this binary value.

The value of the binary word 0000 0000 1010 1010 is equal to 170 in decimal. Refer to Figure 3-9.

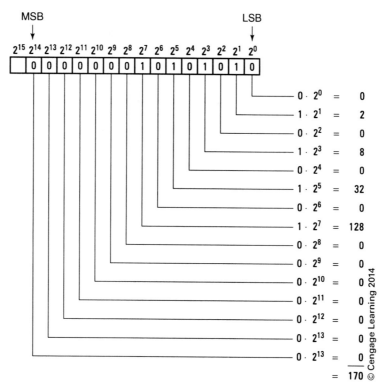

Figure 3-9 Decimal place value equivalent.

Figure 3-10 shows a sample of decimal values and their binary equivalents. Notice that the number 170 from Figure 3-9 is included in the table.

Powers of two	2^9	2^8	2^7	2^6	2^5	2^4	2^3	2^2	2^1	2^0
Place Value	512	256	128	64	32	16	8	4	2	1
Decimal 1 =										1
Decimal 10 =							1	0	1	0
Decimal 37 =					1	0	0	1	0	1
Decimal 52 =					1	1	0	1	0	0
Decimal 74 =				1	0	0	1	0	1	0
Decimal 124 =				1	1	1	1	1	0	0
Decimal 170 =			1	0	1	0	1	0	1	0
Decimal 263 =		1	0	0	0	0	0	1	1	1
Decimal 312 =		1	0	0	1	1	1	0	0	0
Decimal 832 =	1	1	0	1	0	0	1	0	1	0

Figure 3-10 Examples of decimal numbers and their binary equivalents.
© Cengage Learning 2014

Referring back to Figure 3-9, notice the identification of the *least significant bit* (LSB). This is the lowest value bit placeholder, or the 1s place. The *most significant bit* (MSB) notation is the highest value bit placeholder, or the 16,384 place. The LSB is bit 0, whereas the MSB is bit 14. Bit 15 is left blank for now. We will discuss its function next.

16-BIT SIGNED INTEGERS

When working with decimal numbers, we place either a plus or minus in front of a decimal value to identify a positive or negative value. Unfortunately, the PLC does not have the ability to use a plus or minus sign, as it only understands 1s and 0s. We take the left-most bit position, and use this bit to represent the sign of the number. Let's use a 0 in bit position 15 to represent a positive number and a 1 to represent a negative number. Data represented in this fashion are called a 16-bit signed integer. Figure 3-11 is similar to Figure 3-9, except now we have a way to identify whether the value is a positive or negative value.

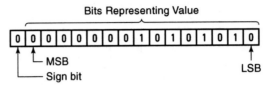

Figure 3-11 A 16-bit signed integer where bit 15 represents the sign of the binary value.
© Cengage Learning 2014

The lower 15 bits (0–15) represent the value. Figure 3-11 illustrates a 16-bit signed integer.

The maximum value that can be stored in a 16-bit signed integer is 32,767. Refer to Figure 3-12. The range of data table information stored in a 16-bit signed integer is −32,768 up to +32,767. It is important that you be familiar with this concept because this is the basic format in which information is stored in a ControlLogix if the data type of integer is selected.

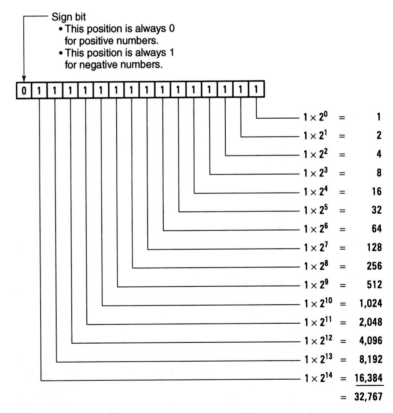

Figure 3-12 A 16-bit signed integer format and bit values.
© Cengage Learning 2014

32-BIT SIGNED INTEGERS

The ControlLogix, being a 32-bit PAC, uses data in the same general format, except that groups of 32 bits are used to store or manipulate information. Because 32 bits is 2 times 16, 32-bit information is known as a double word, or double integer (DINT). Figure 3-13 shows a 32-bit signed double integer for a ControlLogix PAC. The data range for a 32-bit signed integer is −2,147,483,648 to +2,147,483,647.

NUMBER SYSTEMS 91

Figure 3-13 A 32-bit signed integer with a data range of $-2{,}147{,}483{,}648$ to $+2{,}147{,}483{,}647$.
© Cengage Learning 2014

Information is still represented as a signed integer, now a 32-bit signed integer. Figure 3-14 illustrates the format of a 32-bit signed integer and its bit values.

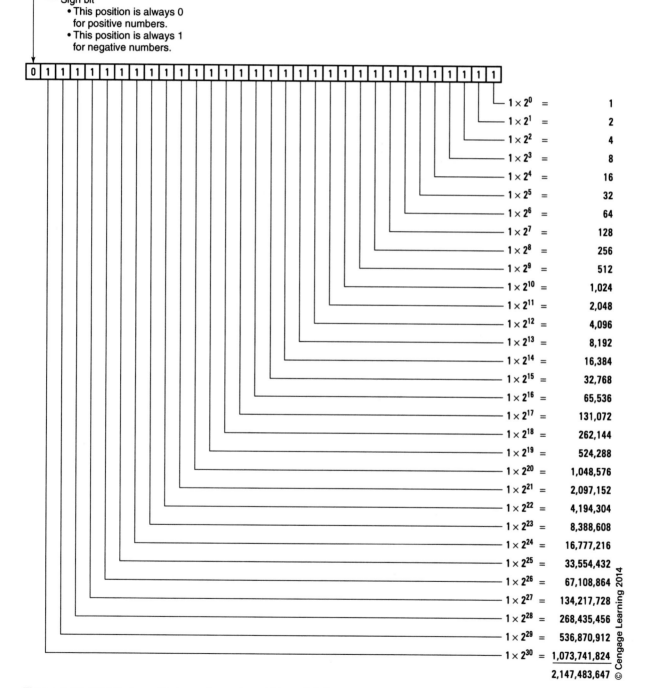

Figure 3-14 32-bit signed integer format and bit weightings.

TAGS AND DATA TYPES

Traditional PLCs allocate a portion of the controller's physical memory for the storage of program-related information. The memory is segmented into data files, making it is easier for the programmer to work with them. As an example, the PLC-5 and SLC 500 PLCs have the following data files:

File 0	Input status table
File 1	Output status table
File 2	Status file
File 3	Binary file
File 4	Timer file
File 5	Counter file
File 6	Control file
File 7	Integer file
File 8	Floating-point file

Each data table memory location has an associated physical address. These are used to address the ladder logic instructions as they are being programmed. To simplify things, a text reference called a *symbol* can be associated with the physical address. Figure 3-15 shows an Allen-Bradley SLC 500 PLC and an RSLogix 500 software ladder rung. The instruction shown in the figure is a counter that counts up from 0 to the preset value of 10,000 in this example. Note that the counter's physical address is C5:0. The "C" identifies the file type as counter, while the file number is 5. The zero specifies the counter. Refer to the list of files above; the default counter file for a SLC 500 or PLC-5 is C5. The symbol associated with that physical address is "Bad_Parts_Counter."

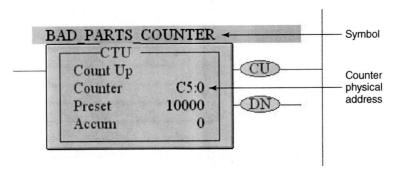

Figure 3-15 Symbolic addressing using RSLogix 500 software.
Used with permission Rockwell Automation, Inc.

ControlLogix is a tag- or name-based PAC. ControlLogix and its associated RSLogix 5000 software have no predefined data files as found in traditional PLCs. When creating a tag in RSLogix 5000 software, first it is assigned a unique name called a *tag*, and then the type of data the particular tag will be is selected. Figure 3-16 shows the same counter instruction as illustrated above for the SLC 500 PLC, except here the instruction is in the RSLogix 5000 software. Note that there is no separate file identification, only a tag name associated with the counter instruction. Tool Tips can be viewed by hovering over the tag, as illustrated in the bottom left of the instruction. The Tool Tips display the tag name, data type, and scope of the tag. Note that the data type—specifically, what this tag is—is counter.

Figure 3-17 lists the common terms used to identify the data format within a traditional PLC; the ControlLogix data type is listed in the next column. When creating tags, assign the tag name and select the data type. ControlLogix has many data types; only the basic data types are listed in Figure 3-17.

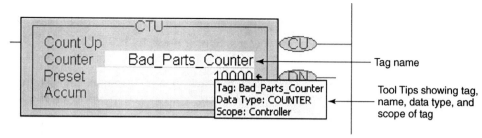

Figure 3-16 ControlLogix count up counter.
Used with permission Rockwell Automation, Inc.

CONTROLLOGIX BASIC TAG TERMINOLOGY			
Common Term	ControlLogix Data Type		Number of Bits	Data Range
Bit	BOOL	Boolean	1	1 or 0
Nibble		Nibble	4	
Byte	SINT	Short integer	8	−128 to +127
Word	INT	Integer	16	−32,768 to +32,767
Double word	DINT	Double integer	32	−2,147,483,648 to +2,147,483,647
Floating point	REAL		32	Very small or large numbers with decimal point
	LINT	Long integer	64	Currently only used for system timer

Figure 3-17 ControlLogix data-type terminology.
© Cengage Learning 2014

Figure 3-18 illustrates a portion of the RSLogix 5000 controller scoped tags collection in Monitor Tags view. Notice the tag named Total: Its data type is *integer* (INT) with a value of 22316. Selecting integer as the data type results in the data being a 16-bit signed integer. The tag named Production Total is a *double integer* (DINT), and its current value is 75499876. Temperature is a real number data type with the value of 234.8.

Name	Value	Data Type
Start	0	BOOL
Stop	1	BOOL
+ Production_Total	75499876	DINT
Temperature	234.8	REAL
+ Total	22316	INT
+ Value_1	43	SINT

Figure 3-18 Portion of an RSLogix 5000 controller scoped tags collection.
Used with permission Rockwell Automation, Inc.

DATA-TYPE SELECTION FOR MATH INSTRUCTIONS

As mentioned earlier, as we create a ControlLogix tag, we assign the tag a unique name and then specify its data type. When programming instructions to add, subtract, multiply, divide, or compute, the expected size and format of the source tag or tags and answer tag must be taken into consideration. Ask the following questions:

- Are any of the tags greater than 32,767?
- Are any tags less than or equal to 32,767?

- Are any tags less than −32,768?
- Do any tags require fractional accuracy?
- Is the answer (destination) tag less than or equal to 32,767?
- Does the destination tag require fractional accuracy (a decimal point)?

Figure 3-19 shows a multiplication instruction as an example. The programmer can mix or match the instruction source and destination data types, depending on the size and format of the tag.

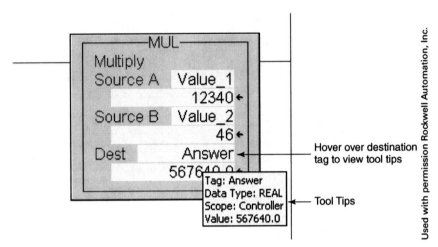

Figure 3-19 RSLogix 5000 multiply instruction with tool tips identifying the Answer tag.

Figure 3-19 also shows the Tool Tips displayed by hovering over a tag. Note that the Tool Tips identify the destination tag as Answer, data type as real, and tag scope as controller scoped, in addition to the current tag value. Figure 3-20 is a portion of the controller scoped tag collection for this RSLogix 5000 project, displaying the tags used in the multiply instruction. Note that the three tags used in the multiply example above are of three different data types:

- Source A tag named Value 1 is a DINT.
- Source B tag, named Value 2, is an INT.
- The destination tag, Answer, is a Real number.

Name	Value	Style	Data Type
Answer	567640.0	Float	REAL
+ Value_1	12340	Decimal	DINT
+ Value_2	46	Decimal	INT

Figure 3-20 Section of controller scoped tags monitor view.
Used with permission Rockwell Automation, Inc.

Figure 3-21 shows some of the possible data-type combinations when programming math instructions. Consider the questions above as you examine the destination data types in the figure.

Note: The result from math or data manipulation instructions that are stored in a smaller data type than the source tags may either be truncated or rounded. Refer to the Instruction Set Help or the user's manual for specific instructions.

MATH CALCULATIONS DATA TYPES		
Source A	Source B	Destination
INT	INT	INT
INT	INT	DINT
DINT	INT	DINT or real
INT	Real	INT, DINT, or real
Real	INT	INT, DINT, or real
DINT	REAL	DINT or real
DINT	Any	INT (see note)
Real	Any	INT (see note)

Figure 3-21 Examples of math instruction data types.
© Cengage Learning 2014

THE OCTAL NUMBER SYSTEM

In many cases, existing PLC systems are updated by replacing the current 16-bit supervisory PLC with a newer 32-bit PAC such as ControlLogix. The newer ControlLogix has more computer power, additional programming instructions, and advanced communication capabilities. Rather than replacing the entire PLC system and existing network or networks, only the supervisory PLC is replaced. Typical Allen-Bradley existing systems have PLC-5s, SLC 500s, or even some older PLC-2s, and possibly PLC-3s. These PLCs are connected to older networks such as Data Highway Plus or Remote I/O. Older PLC networks such as DH+ use octal node or station addresses. PLC-5 I/O addressing is octal. Any SLC 500 chassis on Remote I/O is addressed using the PLC-5 octal addressing format.

ControlLogix has a Data Highway Plus and Remote I/O (1756-DHRIO) communication module that can be placed into the ControlLogix chassis to directly communicate with DH+ or remote chassis over Remote I/O. As an example, when updating an older PLC-5 system to interface with ControlLogix, the current communication cable, known as the "blue hose," is simply unplugged from the current PLC-5 communication port(s) and plugged into the ControlLogix DH+/RIO ports. As discussed in Chapter 1, each DH+/RIO module has two channels that can be configured in any combination of DH+ and RIO. It is common to have multiple DH/+RIO modules in a single ControlLogix chassis to interface with numerous DH+ or RIO networks. As in introduction to the octal numbers encountered while working on these systems, we introduce the world of octal numbers. Although you probably will not be doing octal to other radix conversions, you do need to know how the number system is formatted.

Computers cannot work directly with decimal numbers and binary numerical representations, due to the long, random 1/0 sequences, which become difficult for humans to manage. To help humans work with computers, shorthand methods representing these binary values have been developed, one of which is the octal number system (base 8). Being base 8, valid octal numbers are 0 through 7. Remember, the span of valid numbers for any specific base is always one less than the base.

The octal number system is related to the binary system in that base 2 numbers and base 8 numbers are both based on powers of two. As a result, binary bits can also be used to represent octal numbers. Three binary bits can be used to represent the values 0 through 7, as illustrated in Figure 3-22.

Decimal	Binary	Decimal	Binary
0	000	4	100
1	001	5	101
2	010	6	110
3	011	7	111

Figure 3-22 Decimal numbers 0 through 7 represented with three binary bits.
© Cengage Learning 2014

In octal, the place values are $8^0 = 1$, $8^1 = 8$, $8^2 = 64$, $8^3 = 512$, $8^4 = 4,096$, and so on. Figure 3-23 illustrates the respective weights of each octal digit position.

OCTAL PLACE VALUES				
Powers of eight	8^3	8^2	8^1	8^0
Place value	512	64	8	1

Figure 3-23 Octal number system digit place values.
© Cengage Learning 2014

Decimal and octal numbers are equal for the values 0 through 7. When we reach 7 in octal, we run out of digits in the 1s place, much like we encountered when we reached 9 in the decimal number system. In both cases, we need to advance our count into the next place value. In octal, we advance from 7 to 10. A 10 in octal equals one 8 and zero 1s.

VALID OCTAL NUMBERS

When addressing nodes or stations on a DH+ network, PLC-5 I/O addresses in local or remote chassis, or SLC 500 I/O addresses in remote I/O chassis, the octal number format listed in Figure 3-24 is used to identify stations of nodes on the network or chassis slots for I/O addressing. Figure 3-24 lists the octal numbers from 0 to 277. Notice that no 8s or 9s are listed.

0–7	100–107	200–207
10–17	110–117	210–217
20–27	120–127	220–227
30–37	130–137	230–237
40–47	140–147	240–247
50–57	150–157	250–257
60–67	160–167	260–267
70–77	170–177	270–277

Figure 3-24 Octal numbers from 0 to 277.
© Cengage Learning 2014

Figure 3-25 illustrates a block diagram of a DH+ network. Each block represents a computer, an operator interface device, or a PLC chassis. Each block is assigned an octal address, which is referred to as either a *node* or *station* address. In the PLC world, the terms *node* and *station* are used interchangeably to represent a device on a network.

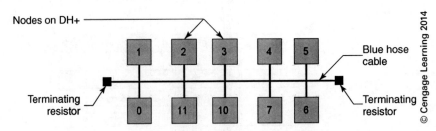

Figure 3-25 Block diagram of a DH+ network with node or station addresses.

Figure 3-26 shows a ControlLogix connected to a remote PLC-5 chassis via a remote I/O network. The ControlLogix chassis has a 1756-DH/RIO communication module to provide the interface from the ControlLogix backplane to the Remote I/O network. The PLC-5 chassis has a remote I/O communication module (1771-ASB) installed as its interface to the remote I/O network. Note that the remote PLC-5 chassis slots in this example are identified with octal numbers 0 to 7 and 10 to 13. This is a 12-slot PLC-5 chassis. These octal numbers are part of the I/O addressing for each slot in the PLC-5 chassis. In many applications, there may have been many existing PLC-5 chassis either on Data Highway Plus or remote I/O communicating with a PLC-5 controller. After the upgrade, the ControlLogix, by way of the 1756-DH/RIO module, is the new supervisory controller. Even though the figure shows a single PLC-5 chassis, there could be multiple chassis connected on Remote I/O. One or more of the existing remote chassis could have been SLC 500 instead.

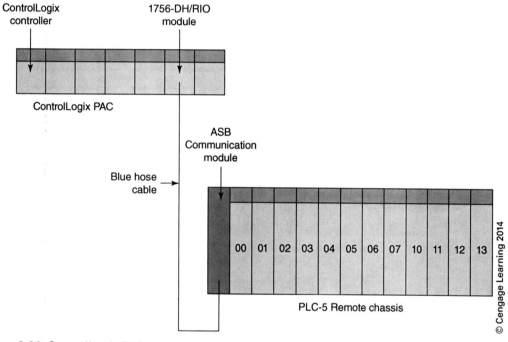

Figure 3-26 ControlLogix PLC connected to remote PLC-5 chassis via remote I/O.

INTRODUCTION TO HEXADECIMAL

When programming or maintaining modern PLCs, you are likely to come across hexadecimal numbers in many different situations. Some common instances of using hexadecimal numbers are

- Interpreting controller error codes
- Interpreting faulted I/O modules error codes
- Programming or interpreting instructions that use hexadecimal masking
- Selecting, determining, or downloading *electronic data sheet* (EDS) files from manufacturers' Web sites when installing or upgrading hardware on a DeviceNet, ControlNet, or Ethernet/IP network.

Hexadecimal, or *hex* as it is sometimes called, is base- or radix-16. The hexadecimal and binary number systems are also related because they are both based on a power of 2 (i.e., $2 \times 2 \times 2 \times 2 \times 16$). Sixteen different states or bit patterns can be represented using four bits. As a result, binary equivalent information is represented in groups of four bits. Figure 3-27 illustrates a DINT broken up into four-bit groups.

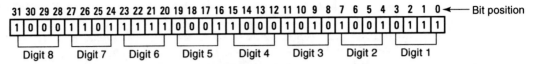

Figure 3-27 32-bit DINT hexadecimal bit patterns.
© Cengage Learning 2014

Hex numbering format is 0 through 9 and then 10 through 15; these are represented by the first letters of the alphabet, A through F. Figure 3-28 lists decimal numbers 0 through 15, their binary equivalents, and their hexadecimal values.

Decimal	Binary	Hexadecimal	Decimal	Binary	Hexadecimal
0	0000	0	8	1000	8
1	0001	1	9	1001	9
2	0010	2	10	1010	A
3	0011	3	11	1011	B
4	0100	4	12	1100	C
5	0101	5	13	1101	D
6	0110	6	14	1110	E
7	0111	7	15	1111	F

Figure 3-28 Decimal, hexadecimal, and binary comparisons.
© Cengage Learning 2014

Figure 3-29 shows examples of selected binary bit patterns, their decimal equivalents, and the corresponding hexadecimal values.

BINARY TO HEX COMPARISON					
Binary Weighting				Decimal	Hex
8	4	2	1		
0	1	1	1	7	7
1	0	0	0	8	8
1	0	0	1	9	9
1	0	1	0	10	A
1	0	1	1	11	B
1	1	0	0	12	C
1	1	0	1	13	D
1	1	1	0	14	E
1	1	1	1	15	F

Figure 3-29 Binary weighting, showing selected bit patterns and hex equivalents.
© Cengage Learning 2014

Let us look at two examples of how hexadecimal information can be represented as a word, an INT data type in ControlLogix, or as 32 bits—that is, as a DINT. Because a word is 16 bits, four groups of hexadecimal information can be represented in one word, as illustrated in Figure 3-30. Refer back to Figure 3-28 as a reference in converting the binary bit patterns into hex. When creating tags and selecting data types with RSLogix 5000 software, an INT data type could be selected to represent four hexadecimal characters.

NUMBER SYSTEMS 99

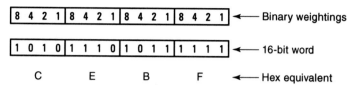

Figure 3-30 A 16-bit word representing hexadecimal data.
© Cengage Learning 2014

When creating tags and selecting data types with RSLogix 5000 software, a DINT data type could be selected to represent eight hexadecimal characters. Because a double word (DINT) is 32 bits wide, eight groups of hexadecimal information can be represented in one DINT, as shown in Figure 3-31. Refer back to Figure 3-28 as a reference when converting the binary bit patterns into hex. Referring back to Figure 3-27, can you determine the hexadecimal value represented?

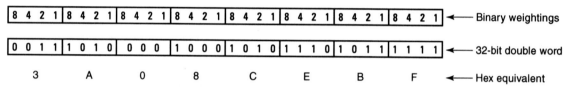

Figure 3-31 A 32-bit double integer (DINT) representing hexadecimal data.
© Cengage Learning 2014

HEXADECIMAL CONVERSION

Like other number systems, hexadecimal is based on the progression of powers of its base, which is 16. Hexadecimal place value weights are illustrated in Figure 3-32.

HEXADECIMAL PLACE VALUES				
Powers of 16	16^3	16^2	16^1	16^0
Place value	4,096	256	16	1

Figure 3-32 Place values of hexadecimal as a result of progression of powers of base 16.
© Cengage Learning 2014

Hexadecimal place values are base 16. The place value weights are determined as a result of the progression of powers of base 16: 16^0, 16^1, 16^2, 16^3, 16^4, and so on. The highest single-digit value allowed in any position is 15; then, the numbering rolls over to 0 and starts again.

Example 1:

Convert 123_{16} into its decimal equivalent, shown in Figure 3-33.

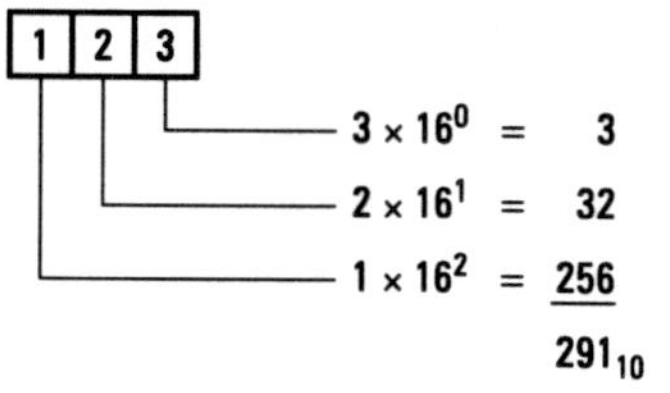

$$3 \times 16^0 = 3$$
$$2 \times 16^1 = 32$$
$$1 \times 16^2 = \underline{256}$$
$$291_{10}$$

Figure 3-33 Hexadecimal 123 converted to decimal.
© Cengage Learning 2014

Binary-to-Hexadecimal Conversion

With four bits, we can make 16 different bit combinations. One method of converting binary to hexadecimal is to start with the LSB and break the binary number into groups of four bits.

$1001000110110101_2 = $ _____?_____ base 16

Start with the least significant bit. Separate the bit pattern into groups of four bits. Add 0s if the most significant bit group does not fill out the required four bits. Second, assign the hexadecimal equivalent to each 4-bit code. Third, fill in the hexadecimal equivalent of each 4-bit nibble, as illustrated in Figure 3-34.

Binary Value	1001000110110101			
Grouped Binary	1001	0001	1011	0101
Result	9	1	B	5

Figure 3-34 Breaking the 16 bits up into 4-bit groups.
© Cengage Learning 2014

Example 2:

$1101101101111010_2 = $ _____?_____ base 16. Figure 3-35 illustrates the conversion process.

Binary Value	1101101101111010			
Grouped Binary	1101	1011	0111	1010
Result	D	B	7	A

Figure 3-35 Converting 4-bit groups into their respective hexadecimal equivalents.
© Cengage Learning 2014

Example 3:

$0101010110111111_2 = $ _____?_____ base 16. Figure 3-36 illustrates the conversion process.

Binary Value	0101010110111111			
Grouped Binary	0101	0101	1011	1111
Result	5	5	B	F

Figure 3-36 Converting binary to hexadecimal.
© Cengage Learning 2014

Example 4:

Newer 32-bit PLCs, such as the ControlLogix, use groups of bits that are 32 bits wide. Converting 10101111000110100011001011101000 to hex is the same basic procedure except that there are more bits. This is illustrated in Figure 3-37.

Binary Value	1010 1111 0001 1011 0011 0010 1110 1000							
Grouped Binary	1010	1111	0001	1011	0011	0010	1110	1000
Result	A	F	1	B	3	2	E	8

Figure 3-37 32-bit binary conversion to hexadecimal.
© Cengage Learning 2014

Hexadecimal-to-Binary Conversion

Conversion of hexadecimal to binary values is easily accomplished by assigning a 4-bit value to each hexadecimal digit. Convert the following:

$1234_{16} = $ _____?_____ base 2

First, separate the hexadecimal digits. Second, convert each separated hexadecimal digit into its 4-bit binary equivalent. Last, combine all 4-bit binary equivalents to form a 16-bit word. The conversion is now complete. Figure 3-38 illustrates the hexadecimal value of 1,234 by breaking it up into 4-bit groups and combining the groups to get the 16-bit binary answer.

Hex Value	1	2	3	4
Grouped Binary	0001	0010	0011	0100
Binary Result	\multicolumn{4}{c}{0001001000110100_2}

Figure 3-38 Converting hexadecimal to binary.
© Cengage Learning 2014

Example 5:

Convert $92AE_{16}$ to binary. Figure 3-39 illustrates the procedure.

Hex Value	9	2	A	E
Grouped Binary	1001	0010	1010	1110
Binary Result	\multicolumn{4}{c}{1001001010101110_2}

Figure 3-39 Converting hexadecimal 92AE to binary.
© Cengage Learning 2014

Example 6:

Newer 32-bit PACs, such as the ControlLogix, use groups of bits that are 32 bits wide. Converting AF1B32E8 to binary is the same basic procedure as with 16-bit words, except that there are more bits. This is illustrated in Figure 3-40.

Hex Value	A	F	1	B	3	2	E	8
Grouped Binary	1010	1111	0001	1011	0011	0010	1110	1000
Result	\multicolumn{8}{c}{1010 1111 0001 1011 0011 0010 1110 1000}

Figure 3-40 Convert AF1B32E8 to binary.
© Cengage Learning 2014

SUMMARY

This chapter introduced the various number systems used in today's computers and programmable controllers. To work with PLCs successfully, you need to have a solid grasp of binary fundamentals.

Similarly, a knowledge of the hexadecimal system will come in handy when working with PLC error codes as well as PLC instructions that use hex masks to control data flow. For example, the Allen-Bradley SLC 500, PLC-5, and ControlLogix use hex masks, with the masked move instruction, masked equal, immediate input with mask, and sequencer output instruction serving as examples. As an example, the masked move instruction is used to move data from a source location to a specified destination through a hexadecimal mask. The mask's function is to allow userselected destination data to be masked out. The sequencer instruction is used to transfer a 32-bit sequence step data through a hexadecimal mask to an output tag for the control of sequential machine operations.

In some instances, you may need to work with a PLC such as the Rockwell Automation PLC-5 that uses octal addressing; in such cases, you need to understand the octal number system to effectively work with the system.

In many cases, existing PLC systems are updated by replacing a 16-bit supervisory PLC with a newer 32-bit PLC, such as ControlLogix. The newer ControlLogix has more computer power, additional programming instructions, and advanced communication capabilities. Rather than replacing the entire PLC system and existing network or networks, only the supervisory PLC is replaced. These PLCs are connected to older networks such as Data Highway Plus or Remote I/O. Older PLC networks, such as DH+, use octal node or station addresses. PLC-5 I/O addressing is octal. Any SLC 500 chassis on remote I/O is addressed using the PLC-5 octal addressing format.

REVIEW QUESTIONS

Note: For ease of handing in assignments, students are to answer using their own paper.

1. As humans, we understand and use the _____, or base _____, number system.
2. Microprocessor-controlled devices and digital computers use the _____, or base _____, number system.
3. Where does the term *bit* come from?
4. What is a bit?
5. Match the following number systems to their corresponding bases:
   a. Decimal   _____ Base 2
   b. Binary   _____ Base 16
   c. Octal   _____ Base 10
   d. Hexadecimal   _____ Base 8
6. 100_{10} is equal to what in decimal?
7. 100_2 is equal to what in decimal?
8. Define LSB.
9. What position in a binary number holds the LSB position?
10. Define MSB.
11. What position in a binary number holds the MSB position?
12. From memory, convert the following binary numbers to decimal:
    a. 1010 = _____   f. 1111 = _____
    b. 1101 = _____   g. 1001 = _____
    c. 0110 = _____   h. 1011 = _____
    d. 1110 = _____   i. 0011 = _____
    e. 0111 = _____   j. 11010 = _____
13. List the binary numbers in order from 1 to 20.
14. Convert ABC hexadecimal into its decimal equivalent.
15. Convert 1011011101011101 base 2 into _____ base 16.
16. Complete Figure 3-41 by inserting the correct numbers for the corresponding bases as you work across each row.

Decimal	Binary	Hexadecimal
10	1010	A
		1C
74		
		2A3F
	1011101	

Figure 3-41 Question 16's conversion exercise.
© Cengage Learning 2014

17. A ControlLogix tag that is of the integer data type has a data range of _____ to positive _____.
18. Fill in the missing information in Figure 3-42.
19. A ControlLogix could be used to communicate with either an existing Data Highway Plus or a Remote I/O network through a _____ communications module. The node addresses on DH+ and the I/O addresses on the remote I/O network are _____ addressed.

NUMBER SYSTEMS 103

Number System	Number of Digits	Base	Range
Decimal			0 through 9
	2		
Octal			
		16	

Figure 3-42 Number systems characteristics.
© Cengage Learning 2014

20. Fill in the missing information in Figure 3-43.

Common Term	ControlLogix Data Type	Definition	Data Range
	BOOL		1 or 0
		Short integer	
			−32,768 to +32,767
Double word			
	Real		
		Long integer	

Figure 3-43 Data-type terminology.
© Cengage Learning 2014

21. A DINT is comprised of _____ words and _____ bytes.
22. When creating a tag for a ControlLogix, a _____ name is given to the tag and then the _____.
23. Hexadecimal, or hex as it is sometimes called, is base-or _____ 16.
24. Hexadecimal binary equivalent information is represented in groups of _____ bits.
25. When working with signed integers, a 0 represents a _____ value, whereas a 1 represents a _____ value.
26. Sixteen different states or bit patterns can be represented using _____ bits.
27. A LINT or _____ integer is comprised of 64 bits.
28. Hex information is found in processor of I/O module _____, instruction _____ and DeviceNet, ControlNet and Ethernet/IP _____ file identification.
29. A ControlLogix tag that is of the DINT data type has a data range of _____ to positive _____.
30. Hex numbering format is 0 through 9; then _____ through _____ are simply represented by the first letters of the alphabet, A through F.
31. A _____ or double integer is comprised of 32 bits.

LAB EXERCISE 1: Using the Windows Calculator to Convert between Binary and Decimal

Note: For ease of handing in assignments, students are to answer using their own paper.

An easy way to convert from one number system to another is to use the Windows calculator. Here we use the calculator to do binary to decimal and decimal to binary conversions. We start with the binary value 1100 1010 1010 1111 and convert it to a decimal value.

1. _____ In Windows, click Start.
2. _____ Click Programs.
3. _____ Click Accessories.
4. _____ Click Calculator.
5. _____ When the calculator appears, click View.
6. _____ Select Scientific to display the scientific calculator. The calculator should display as Figure 3-44.
7. _____ Make sure the Binary (Bin) radio button is selected.

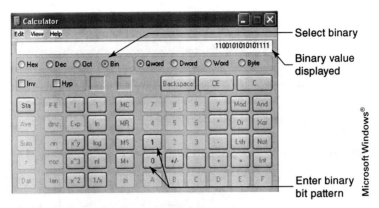

Figure 3-44 Entering binary value to be converted.

8. _____ Enter the binary bit pattern using the 1 or 0 keys.
9. _____ The binary value should be displayed.
10. _____ Notice there are radio buttons for hexadecimal (Hex), decimal (Dec), and octal (Oct).
11. _____ Converting a binary value to any other radix is as easy as clicking the desired button. Because we wish to convert this binary value to decimal, click the Dec button, as shown in Figure 3-45.

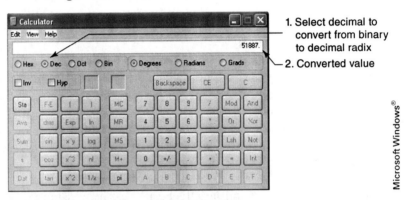

Figure 3-45 Select the Dec button to make the conversion.

12. _____ The converted decimal value should be displayed.
13. _____ Using the calculator, convert the following binary values to decimal:
   a. 1100 1111 0001 1001 _____
   b. 1010 0000 1100 0011 _____
   c. 0111 0000 1100 1111 0001 0001 0000 1111 _____
   d. 0000 1101 0011 0000 1010 1010 0001 0000 _____
14. _____ Convert the following decimal values to binary:
   a. 12,345 _____
   b. 32,767 _____
   c. 1,123,234 _____
   d. 1,756,234,110 _____

LAB EXERCISE 2: Using the Windows Calculator to Convert between Binary and Decimal and Hexadecimal

Note: For ease of handing in assignments, students are to answer using their own paper.

An easy way to convert from one number system to another is to use the Windows calculator. Here we use the calculator to convert among binary, decimal, and hexadecimal. We start with the hexadecimal value CAAF and convert it to a decimal value.

1. _____ In Windows, click Start.
2. _____ Click Programs.
3. _____ Click Accessories.
4. _____ Click Calculator.
5. _____ When the calculator appears, click View.
6. _____ Select Scientific to display the scientific calculator. See Figure 3-46.

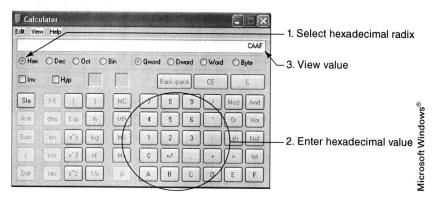

Figure 3-46 Entering binary value to be converted.

7. _____ Make sure the Hex radio button is selected.
8. _____ Enter the hex value using the numeric and alpha keys.
9. _____ The value should be displayed.
10. _____ Notice there are radio buttons for binary (Bin), decimal (Dec), and octal (Oct).
11. _____ Converting the hex value to any of the other radix is as easy as clicking the desired button. Because we wish to convert this hex value to decimal, click the Dec button, as shown in Figure 3-47.

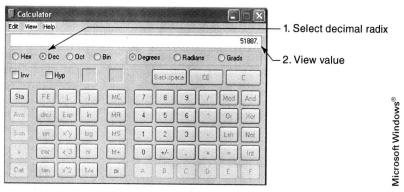

Figure 3-47 Select the Dec button to make the conversion

12. _____ The converted decimal value should be displayed.
13. _____ Using the calculator, convert the following binary values to hex:
    a. 1100 1111 _____
    b. 1010 1100 0011 1101 _____
    c. 0111 0000 0001 0001 1011 0011 0000 1111 _____
    d. 0000 1101 0011 0000 1010 1010 0001 0000 _____
14. _____ Convert the following hex values to binary.
    a. 1B2C _____
    b. 00FF _____
    c. ABCD 24C3 _____
    d. A6B2 102C _____

CHAPTER

4

RSLogix 5000 Project Organization

OBJECTIVES

After completing this lesson, you should be able to

- Understand task, program, and routine.
- Create a new RSLogix project.
- Create tasks, programs, and routines.
- Create a fault routine.
- Modify the program schedule.
- Unschedule a program.
- Assign a main routine and a fault routine.
- Create subroutines.

INTRODUCTION

Users who are familiar with the Allen-Bradley SLC 500 or PLC-5 are familiar with ladder files. The main ladder file for a SLC 500 or PLC-5 is ladder file 2. The SLC 500 has ladder files 2 through 255, and the PLC-5 has ladder files 2 through 999. Ladder logic is programmed within these ladder files. ControlLogix does not have ladder files. This chapter introduces what ControlLogix has instead of ladder files and how ControlLogix project files are organized.

CONTROLLOGIX PROJECT ORGANIZATION

The main components of a ControlLogix project consist of tasks, programs, and routines. The number of tasks a particular project can have is determined by the specific controller assigned to the project when the project was created.

Tasks

Tasks come in four types: continuous, periodic, event, and safety. The event task was first available in software version 13. The safety task was first introduced to the general public in RSLogix 5000 version 15 and is only available when a project is configured using ControlLogix as a safety PAC known as GuardLogix. Task configuration illustrated in this section is not the same for the safety task and is not covered in this text. Figure 4-1 shows a portion of an RSLogix 5000 project controller organizer. For illustrative proposes, this particular project has one of each kind of task. Note the icons that identify each type of task. Next, we introduce each one.

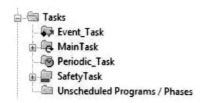

Figure 4-1 Portion of a project controller organizer showing different types of tasks.
Used with permission Rockwell Automation, Inc.

Continuous Task

A project can have only one continuous task. It can be replaced with a periodic or event task if desired. A continuous task runs continuously. When the task has finished executing, it completes the controller's housekeeping chores and starts over again. For those who are familiar with the Allen-Bradley SLC 500/MicroLogix or PLC-5, a ControlLogix continuous task is very similar to ladder file 2 in either of these other platforms. Figure 4-2 shows the continuous task icon. Here the continuous task is named "MainTask."

Figure 4-2 RSLogix 5000 continuous task icon.
Used with permission Rockwell Automation, Inc.

Periodic Task

A periodic task is a timed interrupt. If a user were working on something and a supervisor interrupted and asked the user to do something more important, then the basic principle of how an interrupt works would become apparent. When finished with the more important task, the user would return and pick up where he or she had left off with the original task. In the case of a periodic task, it is a timed interrupt.

As an example, assume the programmer wanted to update a *proportional integral derivative* (PID) loop every 1 second. Because the continuous task is being executed continuously, it will now be interrupted by the timed or periodic task. When the PID loop has been updated, the periodic task is completed, control would be returned to the continuous task and the continuous task will pick up where it left off. The continuous task continues executing the logic in its routines until interrupted again. Periodic tasks are assigned a priority of between 1 and 15. Priority 1 is highest, and priority 15 is lowest. One important concept is that the continuous task has no priority. It will always be interrupted by the other tasks. Keep in mind that priorities are relative. In other words, if multiple periodic tasks are assigned priorities of 3, 4, and 5, these priorities are the same as 1, 2, and 3. Those who are familiar with the SLC 500's or the PLC-5's *selectable timed interrupt* (STI) will note that it is very similar to the RSLogix 5000 periodic task. Figure 4-3 shows the periodic task icon. Although it is named "Periodic_Task" here, the task can be named as desired.

Figure 4-3 RSLogix 5000 Periodic Task icon.
Used with permission Rockwell Automation, Inc.

Event Task

An event task is also an interrupt, but rather than being on a timed basis, this task is triggered by something that happened or failed to happen, called an *event*. Each event task is also assigned a priority of between 1 and 15. Priority 1 is highest, and priority 15 is lowest. The continuous task is always interrupted by an event or periodic task. Priorities again are relative; if we assign event tasks priorities of 3, 4, and 5, then these priorities are the same as 1, 2, and 3. For those who are familiar with the SLC 500's *discrete input interrupt* (DII) or the PLC-5's *processor input interrupt* (PII), these are very similar to the RSLogix 5000 event task. As an example, assume we are bottling water. The bottles pass by a sensor that tests to see whether each one has a cap installed. A bottle without a cap passes the sensor; this is an example of an event. In this case, it would be classified as an input data state change. This input state change could trigger the event task. Depending on the priority assigned, this event task could interrupt other periodic or event tasks with lower priority that might be currently executing. The event task, regardless of its priority, always interrupts an executing continuous task. Figure 4-4 shows the event task icon. Here the task is named "Event_Task."

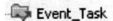

Figure 4-4 RSLogix 5000 event task icon.
Used with permission Rockwell Automation, Inc.

Application Scenario

Let us assume the following.

1. Our project has a continuous task.
2. There is a periodic task for PID update that is to execute every 20 milliseconds; the periodic task priority is 3.
3. A priority 1 event task is to be triggered by an input data state change from the sensor testing the bottles for caps.

Figure 4-5 illustrates what the controller organizer would look like for this scenario.

Figure 4-5 Tasks for this example.
Used with permission Rockwell Automation, Inc.

When the continuous task is executing, the time period for updating the PID loop expires. The periodic task, which has a higher priority than the continuous task, interrupts the continuous task so as to update the PID loop. Approximately halfway through the execution of the PID update, a bottle with no cap is detected by the sensor. The input data state change triggers the event task. Because the event task has a priority of 1, it interrupts the periodic task. The periodic task stops executing and gives control to the event task. The event task executes and then gives control back to the periodic task. The periodic task completes updating the PID loop. Assuming the periodic task is not interrupted again, control is given back to the continuous task. The continuous task continues from where it left off and continues executing its logic until interrupted again. Figure 4-6 is a graphical representation of our task scenario.

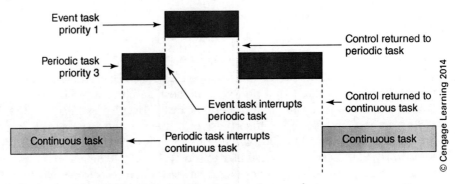

Figure 4-6 Sequence of operations for bottling interrupt scenario.

If any two tasks are assigned the same priority number, then they simply alternate, a situation known as *time slicing*. Task priority is set up in the task properties. Remember, the continuous task has no priority.

Safety Task

Starting with the RSLogix 5000 software version 15 release, the safety task is available for use in an RSLogix 5000 project. The safety task is automatically created when the programmer creates the project and selects one of the GuardLogix safety controllers. A safety task cannot be added using the methods outlined here for a periodic or event-driven task.

Figure 4-7 shows the safety task icon. Here the task is named "SafetyTask."

Figure 4-7 RSLogix 5000 safety task icon.
Used with permission Rockwell Automation, Inc.

The number of tasks a controller can support depends on the specific controller, controller firmware level, and RSLogix software version. Figure 4-8 lists the number of tasks for specific controllers beginning with firmware level 16.

ControlLogix Family Member	Number of Tasks Supported
ControlLogix Modular Controllers	
ControlLogix modular	32
GuardLogix	32
SoftLogix	100
DriveLogix	8
FlexLogix	8
CompactLogix Controllers	
1769-L31	4
1769-L32E	6
1769-L32C	6
1769-L35E	8
1769-L35CR	8

Figure 4-8 Controller tasks supported.
© Cengage Learning 2014

LAB EXERCISE 1: Tasks Programming

This lab exercise provides you with the opportunity to create new periodic and event tasks similar to those shown in Figure 4-5.

1. _____ Start your RSLogix 5000 software.
2. _____ Open the Begin project you created in Chapter 2's lab exercise. In this exercise, you will create a periodic and event task in the Begin project.
3. _____ Expand the Tasks folder. You should see the continuous task. Your project should look similar to that shown in Figure 4-9.
4. _____ To create a new task, right-click the Tasks folder. Select New Task (see Figure 4-10).
5. _____ From the Tasks dialog box, enter the task name. Refer to Figure 4-11.
6. _____ Enter a description if you wish.
7. _____ From the drop-down box, select Periodic.
8. _____ Enter "20 milliseconds" and the period or interrupt time.

110 RSLOGIX 5000 PROJECT ORGANIZATION

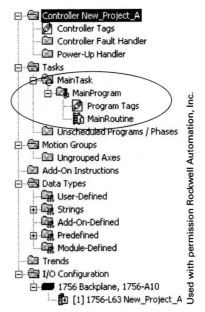

Figure 4-9 Controller organizer showing continuous task.

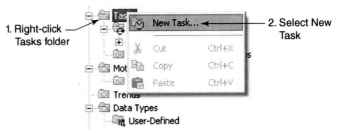

Figure 4-10 Creating a new task.
Used with permission Rockwell Automation, Inc.

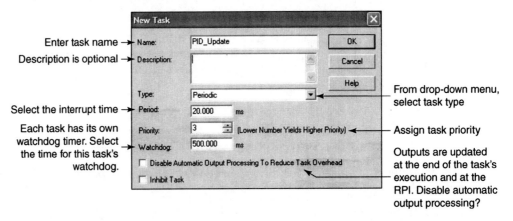

Figure 4-11 Configuring new task specifications.
Used with permission Rockwell Automation, Inc.

9. _____ For our example, enter priority of 3.
10. _____ Leave the Watchdog at its default value.
11. _____ Leave the Disable Automatic Output Processing and Inhibit Task boxes unchecked.
12. _____ Click OK when done.
13. _____ Your new task should be listed just below the continuous task in the Tasks folder.
14. _____ Next we create the event task for the missing bottle cap tracking. Right-click the Tasks folder.

15. _____ Select New Task (see Figure 4-12).

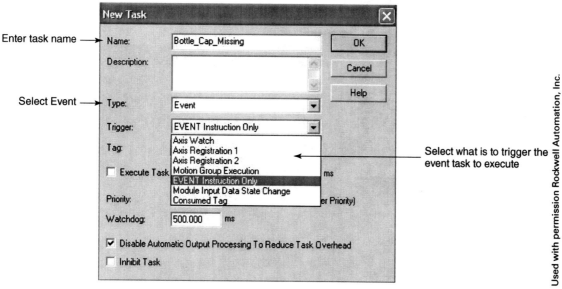

Figure 4-12 Event task trigger selections.

16. _____ Enter the task name.
17. _____ Select Event as the task type.
18. _____ From the trigger drop-down menu, select what will trigger task execution. Trigger selections fall into three groups:
    1. Input triggers
       Event instruction
       Module input data state change
    2. Network communications
       Consumed tag arrival
    3. Motion-control selections
       Axis registration 1
       Axis registration 2
       Axis watch
       Motion group execution
19. Input triggers fall into two groups: (1) a module input data state change and (2) a new instruction for ControlLogix called the *event instruction*. The event instruction became available in the RSLogix 5000 software with the introduction of the event task in version 13 of the software. The event instruction is programmed on a ladder rung as an output instruction. Any input logic can be programmed to trigger the event instruction to call the event task. Figure 4-13 illustrates the event instruction. Notice the task name programmed inside the event instruction. If the trigger event task input is true, the event instruction will be true and the event task will be called.

Figure 4-13 The event instruction calling the bottle cap missing task.
Used with permission Rockwell Automation, Inc.

20. For our example, we have a sensor inputting bottle cap information whether the cap is present into an input module. To keep our application simple, assume that we simulate this as input on a ladder rung.
21. _____ For our example, select trigger as "EVENT Instruction Only," as shown in Figure 4-14.

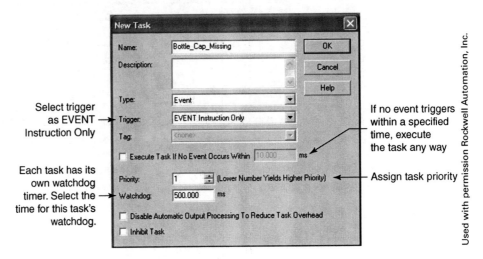

Figure 4-14 Event task configurations.

22. _____ The user has the option to have the task execute anyway if no trigger is detected within a specified time. Refer to Figure 4-14. For our example, do not check the box in front of Execute task if no event occurs within.
23. _____ The priority for our scenario was priority 1 for the bottle cap sensor. Select 1 for priority.
24. _____ Leave the Watchdog timer at default for our example.
25. _____ Select OK when done with configuring the task.
26. _____ Your task folder in the controller organizer should look like that shown in Figure 4-15.

Figure 4-15 Completed task configuration exercise.
Used with permission Rockwell Automation, Inc.

27. _____ This completes this section of the lab exercise. Be sure to save your project as "Tasks." You will add to it in future labs.

RSLOGIX 5000 PROGRAMS

An RSLogix 5000 program is the second level of scheduling within the RSLogix 5000 project. There is no executable code within a task or program. The function of programs is to provide the programmer an opportunity to determine and specify the order in which the programs execute. With RSLogix 5000 software version 13, as many as 32 programs could be scheduled within a task. Starting with software version 15, as many as 100 programs could be scheduled within each task. Execution is determined by the list position in which each program is placed, as illustrated in Figure 4-16. Under the MainTask are four programs. Because of the listed order,

RSLOGIX 5000 PROJECT ORGANIZATION   113

Figure 4-16 RSLogix 5000 controller organizer showing Main Task and four associated programs.
Used with permission Rockwell Automation, Inc.

the MainProgram is scheduled to execute first, then Program_A second, Program_B third, and Program_C last. In many cases, the specific order in which the programs are executed may not be important for a specific application. In this case, the programs are listed or scheduled by the order in which they were created. The order of execution can be modified by the programmer, as we see here.

LAB EXERCISE 2: Creating and Scheduling Programs

In this portion of the lab, we create and schedule the four programs in the main task, as shown in Figure 4-16, to provide you with an understanding of how to create and schedule programs. We will add these programs to your last lab RSLogix project, which you saved as "Tasks."

1. _____ Open the Tasks project you completed in the last lab exercise.
2. _____ Right-click on the MainTask folder. You should see a dialog box similar to that shown in Figure 4-17.

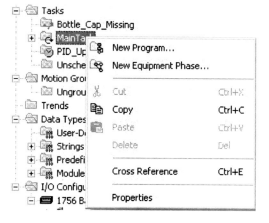

Figure 4-17 MainTask folder and opening dialog box.
Used with permission Rockwell Automation, Inc.

3. _____ To create a new program, select New Program from Figure 4-17.
4. _____ Enter the name "Program_A" as illustrated in Figure 4-18.

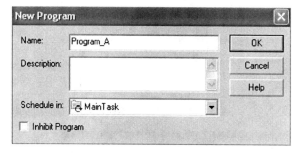

Figure 4-18 Creating a new program.
Used with permission Rockwell Automation, Inc.

5. _____ A description is optional.
6. _____ Leave Schedule in as MainTask because this is where we want to create this new program.
7. _____ When completed, click OK.
8. _____ Create Program_B in the same manner.
9. _____ Create Program_C in the same manner.
10. _____ Figure 4-19 shows the three new programs—A, B, and C—that you created within the continuous task. Notice that the first listed, or scheduled, program was created for you and named "MainProgram." The name can be changed by going to the program's properties dialog box and changing the name. The main program was created automatically as part of a new project's default configuration—that is, a new project, by default, creates a main continuous task and main program for you. Any other tasks, such as the periodic and event tasks, must be created by the programmer. All additional programs have to be created by the programmer.

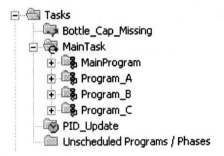

Figure 4-19 Newly created programs within the continuous task.
Used with permission Rockwell Automation, Inc.

The main purpose of programs is to determine which program executes first, second, and so on. The order in which the programs are created dictates where they fall in the list or schedule. In Figure 4-19, the MainProgram will execute first, and then Program_A second, Program_B third, and Program_C fourth. Next, we modify the program schedule.

11. _____ To view or modify the program schedule, right-click MainTask to show the drop-down menu. Refer to Figure 4-20.

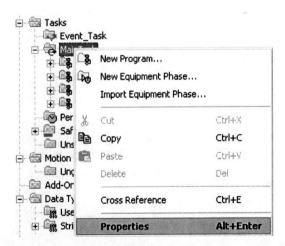

Figure 4-20 Right-click MainTask and select Properties.
Used with permission Rockwell Automation, Inc.

12. _____ Select Properties. The screen in Figure 4-21 should be displayed.

RSLOGIX 5000 PROJECT ORGANIZATION 115

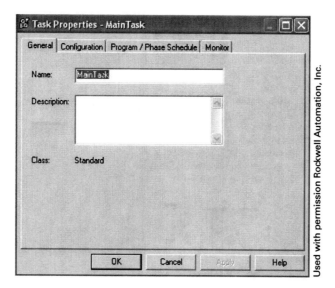

Figure 4-21 MainTask's Properties screen.

13. _____ Click the Program/Phase Schedule tab to view the program schedule as shown in Figure 4-22. The program schedule can be modified by clicking on the desired program, and by using the up or down arrows to the right, move the program to the desired position.

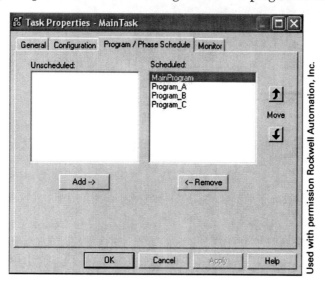

Figure 4-22 Program schedule for MainTask.

UNSCHEDULED PROGRAMS

Programs can be scheduled or unscheduled. A scheduled program will be allowed to execute, whereas any program that is unscheduled will not be allowed to execute.

14. _____ Use the up or down arrow buttons to select Program_C.
15. _____ Click Remove. Program_C should be moved to the unscheduled window as illustrated in Figure 4-23. Notice that the Tasks folder now shows Program_C under the Unscheduled folder. Remember that unscheduled programs are not executed by the controller.
16. _____ Select Program_C.
17. _____ Click Add to schedule Program_C.
18. _____ Click Apply.

116 RSLOGIX 5000 PROJECT ORGANIZATION

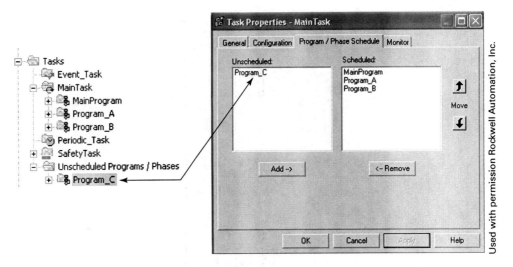

Figure 4-23 Unscheduled Program C will not execute.

19. _____ Use the move up or down arrows to move Program_C so that it is directly under the MainProgram, as illustrated in Figure 4-24. The resulting MainTask program schedule is illustrated on the right side of the figure.

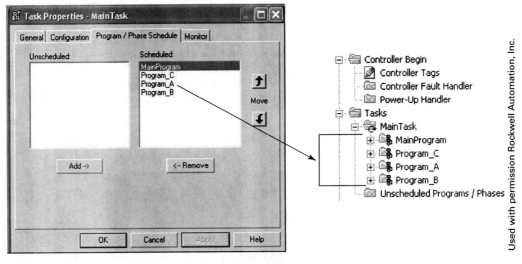

Figure 4-24 Program execution order is MainProgram, and then Program_C, Program_A, and Program_B.

20. _____ Click OK.
21. _____ View your project's Continuous Task folder to verify that the programs are in the correct order.
22. _____ Save your project.
23. _____ This completes this lab exercise.

RSLOGIX 5000 ROUTINES

RSlogix 5000 routines are where the programming code inside a program resides. There are three types of routines:

- A single main routine, which is required
- One or more subroutines, which are optional
- One fault routine per program, which is also optional

Routines can be any one of the four programming languages available for ControlLogix. A routine can be either: ladder logic, function block diagram, structured text or sequential function chart. Any routine must be entirely the same language. Each program will have one main routine that is typically followed by several subroutines. Figure 4-26 shows the Tasks folder from an RSLogix 5000 project. Under the Main Program, note that Main Routine icon is identified with a piece of paper with a "1" written on top of the ladder icon. The ladder icon signifies this routine is ladder logic.

Main Routine

The main routine is the starting point for program code execution. Every program must have a main routine. When the program is executed, the main routine if a ladder logic routine, will automatically execute the rungs starting with rung zero, left to right one rung at a time until the highest numbered rung has been executed. If familiar with the SLC 500 or PLC-5, the main routine is similar to ladder File 2.

Subroutine

Subroutines are blocks of logic than can be incorporated into a program to achieve efficiency or for organizational purposes. A program can be broken up into logical groups of rungs called subroutines. One of the principles of subroutines is to execute only the rungs needed currently in your process. Referring to Figure 4-25 note the Paint Part Blue, Paint Part Green, and Paint Part Red subroutines. If painting parts blue for the next four hours there is no need to execute the rungs for painting other colors. This is one method a programmer can use to gain efficiency and thus improve scan time in a PAC project.

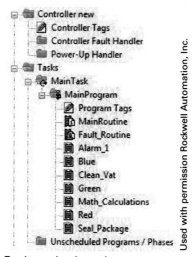

Figure 4-25 RSLogix 5000 Main, Fault, and subroutines.

The first subroutine in Figure 4-26 is a function block subroutine. Note that the icon identifies it as a function block diagram subroutine. Like any routine, a subroutine must be completely in the same programming language. This subroutine must be entirely function block diagram. The next two subroutines are sequential function chart and structured text. Subroutines A and B are ladder logic. Subroutines are accessed using the *jump to subroutine* (JSR) instruction.

Fault Routine

Faults fall into two general categories, major and minor. When a controller encounters a major fault, the controller stops (is faulted) and real-world outputs will be changed to their fault mode state as defined in the module's I/O configuration. On the other hand, a minor fault only sets a bit, and the controller will continue to run. An example of a minor fault would be the battery light alerting that the battery needs to be changed.

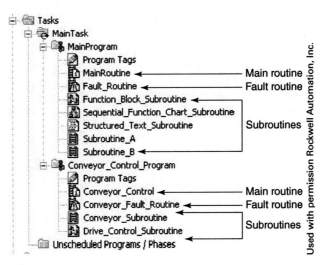

Figure 4-26 Main program showing subroutines.

Major faults can be further separated into major recoverable and major non recoverable. If a major non recoverable fault occurs, the controller will fault. If a major recoverable fault occurs, the programmer can create a fault routine and attempt to recover from the fault. Refer to Figure 4-25 and identify the fault routine for the main program and the fault routine for the conveyor control program. The fault routine is identified by an icon with a yellow cautionary triangle with an exclamation mark.

An instruction execution problem is an example of a major recoverable fault. This could be the result of bad programming or invalid data entry by an operator. As an example, an operator entering data on a touch screen operator interface device could mistakenly enter a negative value in a timer. This is defined as an instruction execution problem, or type 4 fault. If the fault is not cleared, the controller will shut down. If desired, a major recoverable fault such as this can be trapped in and recovered from in fault routine. By definition, a type 4 fault should be handled by a local fault routine. With a type 4 major recoverable fault declared, the controller will automatically enter the local fault routine if it exists. If after executing the logic in the local fault routine, the controller sees the problem has been corrected, the controller will continue to run as if the problem never existed. If it is desired not to fix the problem, or logic has not been written to trap the specific problem, the controller will exit the fault routine with the fault still active. If the controller exits the local fault with the fault still present, the Controller Fault Handler will be executed next to see if the problem can be recovered from there. Upon exiting the Controller Fault Handler with the problem resolved, the controller will continue to run. If the problem persists, controller faults. Refer back to Figure 4-9 and note the Controller Fault handler folder near the top of the controller organizer. Keep in mind that someone has to create and program a fault routine. Even though the Controller Fault handler has been created for you, no code exists to recover from any fault and will have to be created. We will work with understanding and programming a local fault routine in Chapter 17.

LAB EXERCISE 3: Creating Routines

This lab exercise steps you through creating routines and assigning a main routine and a fault routine.

1. _____ Open the Begin project. Right-click MainProgram.
2. _____ Select New Routine, as shown in Figure 4-27.
3. _____ Enter the routine name, as seen in Figure 4-28.
4. _____ Enter a description for the routine if you wish.

RSLOGIX 5000 PROJECT ORGANIZATION   119

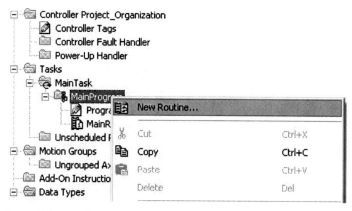

Figure 4-27 Creating a new routine.
Used with permission Rockwell Automation, Inc.

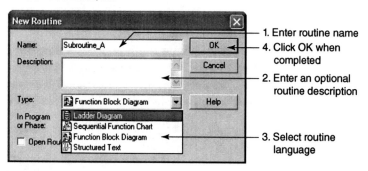

Figure 4-28 Creating a new routine.
Used with permission Rockwell Automation, Inc.

5. _____ Click the down arrow for the type window and select Ladder Diagram as the language for this routine.
6. _____ Behind the language-selection drop-down box is an In Program or Phase entry. This provides the programmer the opportunity to select in which program this routine will be created. For this lab, MainProgram is OK.
7. _____ On your own, create a new subroutine called "Subroutine_B."
8. _____ Next, we will create a new program and its associated routines. Right-click MainTask as shown in Figure 4-29.

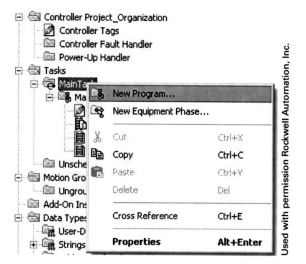

Figure 4-29 Creating a new program.

9. _____ Select New Program.
10. _____ Enter the program name, as in Figure 4-30. Notice the "Schedule in" section. This allows you to select in which task this program will be scheduled. For this lab, leave it as MainTask.

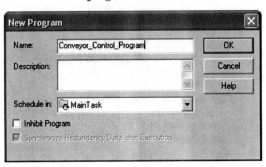

Figure 4-30 Creating a new program.
Used with permission Rockwell Automation, Inc.

11. _____ Click OK when completed.
12. _____ Create a new ladder diagram routine called "Conveyor_Control."
13. _____ Create another ladder diagram routine called "Conveyor_Subroutine."
14. _____ Create another ladder diagram routine called "Conveyor_Fault_Routine." When completed, your MainTask should look like that in Figure 4-31.

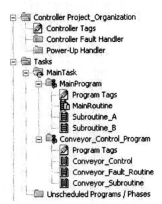

Figure 4-31 Completed Conveyor_Control_Program.
Used with permission Rockwell Automation, Inc.

Notice that each of the three routines we have just created has a ladder icon that signifies it is in ladder diagram language. Remember, a routine must be entirely in the same language—in this case, ladder logic. Refer back to the MainProgram and its associated MainRoutine. You can see that this main routine is identified by the ladder icon's piece of paper with a "1" on top of the ladder icon. In our newly created Conveyor_Control_Program, our main routine called "Conveyor_Control" does not have the piece of paper and 1 signifying it as the main routine. Our current configuration for the Conveyor_Control_Program has three subroutines, no main routine, and no fault routine. All routines are created as subroutines until they are assigned otherwise by the programmer. We will now assign our main routine and fault routine.

15. _____ Right-click Conveyor_Control_Program and select Properties from the drop-down, as shown in Figure 4-32.
16. _____ Select the Configuration tab, as in Figure 4-33.
17. _____ From the drop-down menu, select the Conveyor_Control routine that is to be the main routine.
18. _____ From the drop-down menu, select the Conveyor_Fault_Routine that is to be the fault routine.

RSLOGIX 5000 PROJECT ORGANIZATION 121

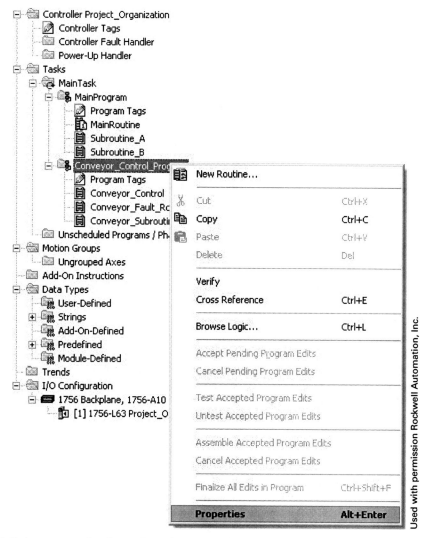

Figure 4-32 Select properties from drop-down menu.

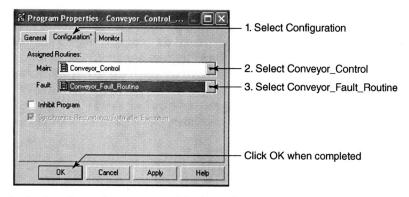

Figure 4-33 Assigning the main routine and fault routine.
Used with permission Rockwell Automation, Inc.

19. _____ When done, click OK.
20. _____ Your Conveyor_Control_Program should look like that shown in Figure 4-34. Save your project with a name of your choice.

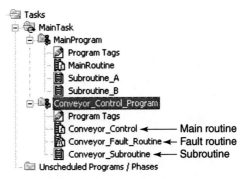

Figure 4-34 Conveyor_Control_Program with routines assigned.
Used with permission Rockwell Automation, Inc.

SUMMARY

Tasks come in four types: continuous, periodic, event, and safety. The event task was first available in software version 13. The safety task was first introduced to the general public in RSLogix 5000 version 15. The safety task is only available when the project is configured using a safety controller.

A project can have only one continuous task. The task can be replaced with a periodic or event task if desired. A periodic task is an interrupt based on time. As an example, assume the programmer wanted to update a PID loop every second. Because the continuous task is being executed continuously, it will be interrupted by the timed interrupt or periodic task. The event task, rather than being an interrupt on a timed basis, is a task triggered by something that happened or failed to happen. The trigger, or event, is definable by the programmer. As an example, assume you are bottling water. The bottles pass by a sensor that tests to see whether each one has a cap installed. At a certain moment, a bottle without a cap passes the sensor. This input state change could call the event task. Depending on the priority assigned, this task could interrupt other periodic or event tasks with lower priority that might be currently executing. Periodic and event tasks can be given a priority between 1 and 15. Regardless of its priority, the event or periodic task will always interrupt an executing continuous task.

An RSLogix 5000 program is the second level of scheduling within the RSLogix 5000 project. There is no executable code within a program. The function of programs is to provide the programmer the opportunity to specify the order in which the programs execute. Up through RSLogix 5000 software version 13, as many as 32 programs could be scheduled within a task. Starting with software version 15, as many as 100 programs can be scheduled within each task. Execution is determined by the position at which each program is placed within the list.

Inside RSLogix 5000 routines are where the programming code resides. Routines can be one of four types: ladder logic, function block diagram, structured text, or sequential function chart. Any one routine must be completely in the same language. Each program has one main routine typically followed by several or many subroutines. Each program also can have one fault routine, if desired. The main routine is identified by a piece of white paper with a "1" on top of the routine icon; the fault routine is signified by a yellow cautionary sign with an exclamation mark.

REVIEW QUESTIONS

1. The main components of a ControlLogix project consist of _____, _____, and _____.
2. Up through RSLogix 5000 software version 13, as many as ____ programs could be scheduled within a task.
3. Starting with RSLogix 5000 software version 15, as many as ____ programs could be scheduled within each task.
4. Tasks come in four types: _____, _____, _____, and _____.

5. An RSLogix 5000 _____ is the second level of scheduling within the RSLogix 5000 project.
6. There is no executable code within a _____ or _____.
7. The number of tasks a particular project can have is determined by the specific controller you assigned to the project when the project was created.
8. The _____ task was first available in software version 13.
9. The _____ task was first introduced to the general public in RSLogix 5000 version 15.
10. Periodic and event tasks are assigned a priority of between 1 and ____.
11. Task priority of ____ is highest, and priority ____ is lowest.
12. The _____ task has no priority.
13. Rather than being an interrupt on a timed basis, the _____ task is triggered by something that happens or fails to happen.
14. The function of ControlLogix _____ is to provide the programming individual an opportunity to determine and specify the order in which they execute.
15. Depending on the priority assigned, an event task could interrupt another periodic or event task with _____ priority that might be currently executing.
16. Regardless of its priority, an _____ or _____ task will always interrupt an executing continuous task.
17. _____ can be scheduled or unscheduled. A scheduled _____ will be allowed to execute, whereas any program that is _____ will not be allowed to execute.
18. The main purpose of having _____ is to determine which executes first, second, and so on.
19. A subroutine must be entirely in the same _____.
20. Subroutines are accessed using the _____ instruction.
21. The order in which the programs are created dictates where they fall in the list or _____.
22. The RSLogix 5000 _____ are where the programming code resides.
23. Routines can be one of four types: _____, _____, _____, or _____.
24. Any one _____ must be entirely in the same language.
25. Each _____ will have one main routine typically followed by one or more subroutines.
26. Each _____ can also have a fault routine if desired.
27. The main routine is identified with a ____ on top of the ladder icon.
28. The _____ is signified by a yellow cautionary sign with an exclamation mark.

EXERCISE A: Tasks, Programs, and Routines

Note: For ease of handing in assignments, students are to answer using their own paper.
Refer to Figure 4-35 as you answer the following questions.

1. What type of task is the main task?
2. Explain how this type of task works.
3. How many tasks can we have in a modular ControlLogix version 16 project?
4. The routines under the main program are what programming language?
    a. How do you know this?
    b. Can we mix programming languages within the same routine?
    c. How many routines can we have in a ControlLogix program?
5. What is the name of the fault routine?
    a. How is the fault routine identified?

124 RSLOGIX 5000 PROJECT ORGANIZATION

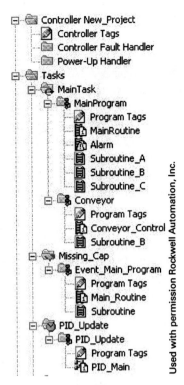

Figure 4-35 RSLogix screen printout showing project structure.

6. What type of task is the missing cap task?
   a. Explain how this task works.
7. What type is the PID task?
   a. Explain how this task works.
8. The PID main routine is in what programming language?
9. When creating a new routine that is to be a main routine, how do you assign this routine as a main routine?
10. In the current project, what is the order of execution of the programs under the continuous task?
11. How can the order of program execution be modified?
12. What does it mean to unschedule a program?
13. Why would you unschedule a program?
14. How is a program unscheduled?

LAB EXERCISE A: Project Creation and Organization

For this lab exercise, you will create a new RSLogix 5000 project in order to acquaint you with project structure. Programs will be created and scheduled. Main and fault routines will be created and assigned. Subroutines will be created within the newly created programs. A debug program will be created and unscheduled. We will not program any ladder rungs in our routines at this time. For this lab, you need only a computer with RSLogix 5000 software installed. Figure 4-35 illustrates the initial project to be created for this lab exercise. Once the project has been created, we will modify it.

1. _____ Open your RSLogix 5000 software.
2. _____ Create a new project and name it "Task Lab."

3. _____ Select the appropriate controller.
4. _____ Select the appropriate chassis size.
5. _____ If using a modular ControlLogix, fill in the slot number of the controller.
6. _____ Create the project tasks, programs, and routines, as illustrated in Figure 4-35. Specifications are listed here.
7. _____ PID update periodic task parameters:
   Task priority = 3
   Period = 250 milliseconds
   Leave the remaining parameters at default.
8. _____ Missing cap event task parameters:
   Task priority = 1
   Trigger = EVENT Instruction Only
   Leave remaining parameters at default.

LAB EXERCISE B: Modifying a Project

When done with creating the project as illustrated in Figure 4-35, modify the project as directed here.

1. _____ Right-click Subroutine A and go to the routine properties. Rename Subroutine A in the main program as "Math Calculations."
2. _____ Rename Subroutine B in the Conveyor program as "Convert Drive Temperature."
3. _____ Right-click Event_Main_Program. Rename it "Energize Diverter."
4. _____ Add a new program named "Mixer" to the main task.
5. _____ Schedule the mixer program to execute second.
6. _____ Create a ladder logic main routine called "Clean Mixer" within the mixer program.
7. _____ Create a new program called "Debug" within the main task.
8. _____ Create the main ladder logic routine in the debug program called "Main Debug Routine."
9. _____ Unschedule the debug program.
10. _____ Save your project.
11. _____ Ask your instructor to check your project or if it needs to be printed out and handed in.

CHAPTER

5

Understanding ControlLogix I/O Addressing

OBJECTIVES

After completing this lesson, you should be able to

- Explain format of ControlLogix I/O tags.
- Determine ControlLogix and CompactLogix tags.
- Identify local discrete and analog I/O tags.
- Identify remote I/O tags on ControlNet or Ethernet/IP networks.
- Interpret I/O configuration for local and remote ControlLogix hardware.

INTRODUCTION

When installing or troubleshooting a ControlLogix, one must be able to identify the tags assigned to the input and output field device wiring attached to an input or output module. The I/O tags are assigned to instructions as rungs are created. If we are interpreting ladder logic that has already been written, then we must be able to correlate I/O tags assigned to programming instructions and their respective I/O module screw terminals and associated field devices. Installers must be able to properly wire inputs and outputs to their respective module terminal points as specified in system wiring diagrams.

The ControlLogix modular PLC has physical chassis slots that are assigned slot numbers. Even though CompactLogix modules are not housed in a physical chassis, the I/O modules still have slot numbers assigned to them as part of their addressing or I/O tag identification. This lesson introduces both digital and analog tag formatting.

I/O TAG FORMAT

An I/O tag comprises several parts that help identify the specific I/O or module information we may be looking for. A typical digital I/O tag is made up of the following parts:

location: slot number: connection type. member name. bit number

Notice that each part of the tag is separated by either a colon or a period. The first part of the tag is the location of the module.

LOCATION

If a ControlLogix consists of a single chassis, as shown in Figure 5-1, then it is referred to as the *local chassis*. Once the local chassis slots have been filled, additional modules are placed into a remote chassis in relation to the local chassis. ControlLogix applications' remote chassis are connected to the local chassis by either a ControlNet or Ethernet/IP network connection. Each remote chassis has a communication module that provides network communications to the local chassis.

Figure 5-1 ControlLogix modular chassis.
Used with permission Rockwell Automation, Inc.

Figure 5-2 illustrates a local chassis as node 1 on a ControlNet network. In this example, there are two remote chassis identified as nodes 2 and 3. There is a 1756-CNB module in the local chassis as well as in the node 2 and node 3 chassis. Communication modules are referred to as *adapters*. When configuring the I/O for the local chassis, any I/O in remote chassis we wish to communicate with also must be configured. Figure 5-3 illustrates a sample I/O configuration that represents the network illustrated in Figure 5-2. Local and two remote chassis are identified. Notice the node addresses of the two remote chassis. The node 2 1756-CNB module's name is "Remote_CNB," and the node 3 CNB is named "Tank1."

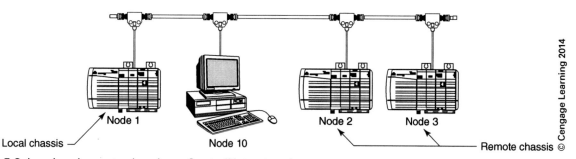

Figure 5-2 Local and remote chassis on ControlNet network.

The first part of an I/O tag is the location of the module. In this example, if the I/O module is in the local chassis, then the first word of the tag will be "Local." If the module resides in the node 2 chassis, then the first word of the tag will be "Remote_CNB." Modules in node 3's chassis are identified as "Tank1."

Figure 5-4 shows another I/O configuration with ControlNet as well as Ethernet networks configured in the I/O configuration. Slot 9 of the local chassis contains the local Ethernet (1756-ENBT) communications module. The chassis slot number is identified by [9] to the right of the module part number. The remote ENBT module's name is "Labeling_and_Packing." The first word of I/O tags in the remote Ethernet chassis are "Labeling_and_Packing."

Allen-Bradley refers to its communication modules as *adapters*, so the company's literature states that the first word of an I/O tag is either local or the adapter name. Figure 5-5 shows our beginning I/O tags as we start with the location of the module, followed by a colon.

128 UNDERSTANDING CONTROLLOGIX I/O ADDRESSING

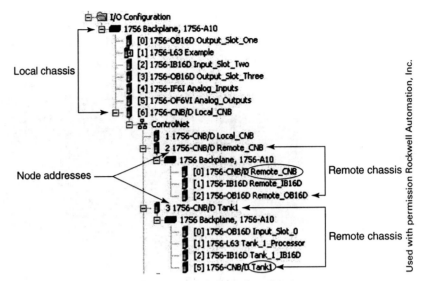

Figure 5-3 ControlLogix I/O configuration showing ControlNet network.

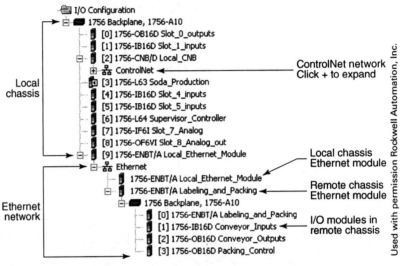

Figure 5-4 RSLogix 5000 software ControlLogix I/O configuration containing ControlNet and Ethernet communications.

Local:

Tank1:

Remote_CNB:

Labeling_and_Packing:

Figure 5-5 Sample tags showing the module location.
© Cengage Learning 2014

CHASSIS SLOT NUMBER

The second part of an I/O tag is the slot number where the module resides in the chassis. ControlLogix modular chassis slot numbers start numbering with 0 in the left-most slot, with increments in decimal numbers moving to the right. Chassis sizes for the modular ControlLogix are 4, 7, 10, 13, and 17 slots. Valid values for the chassis slot number portion of the tag are 0 through 16. Figure 5-6 illustrates slot numbers for a 10-slot chassis.

CompactLogix slot numbers start with slot 1 to the right of the controller, with increments in decimal numbers to the right-most slot. As many as three modules may be placed between the

UNDERSTANDING CONTROLLOGIX I/O ADDRESSING   129

Figure 5-6 ControlLogix modular chassis slot numbering.
Used with permission Rockwell Automation, Inc.

controller and the power supply. Depending on a specific controller, as many as 30 modules can be configured into groups referred to as banks. Figure 5-7 shows an example of a 1769 CompactLogix with slot numbers identified. The power supply does not consume a slot number.

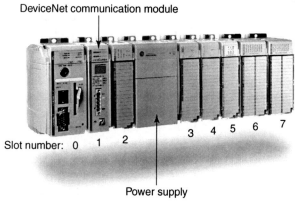

Figure 5-7 CompactLogix slot numbers.
Used with permission Rockwell Automation, Inc.

Figure 5-8 shows the RSLogix 5000 I/O configuration for the 1769 CompactLogix in Figure 5-7. The controller is in slot 0. Module slot numbers are in square brackets followed by the module part number and name.

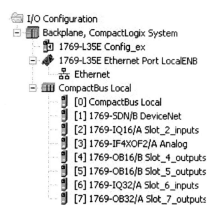

Figure 5-8 RSLogix 5000 CompactLogix I/O for Figure 5-7 hardware.
Used with permission Rockwell Automation, Inc.

Figure 5-9 shows the location and slot number portions of our sample tags I/O address and illustrates the location and slot number added to the tag string. In this case, the module is in slot 2 of the chassis.

130 UNDERSTANDING CONTROLLOGIX I/O ADDRESSING

Local :2:

Tank1:2:

Remote_CNB:2:

Labeling_and_Packing:2:

Figure 5-9 Location and slot number portions of sample tags we are building.
© Cengage Learning 2014

CONNECTION TYPE

The third member of an I/O tag is the connection type. There are three options for the connection portion of the I/O tag: I for input, O for output, and C for module configuration tags.

Connection Type C

The backplane in a ControlLogix chassis is a ControlNet network called "ControlBus." Each module in a ControlLogix chassis is intelligent, an entity in itself, or a separate node on the ControlBus network. After the I/O configuration is complete and the project is downloaded, the controller placed into run mode. The I/O configuration information is now transferred to each module the controller owns. In order to operate, every module must have an owner. The owner controller contains the I/O configuration for the module.

Each I/O module is in waiting mode until it receives its I/O configuration. Until that time, the module does not know how it is to operate. Several configuration tags are created as the result of the programmer performing the I/O configuration for each module. The I/O configuration is manually set up using RSLogix 5000 software. Configuration tags are of connection type C and stored in the controller scoped tags collection. Typically, configuration tags are not monitored or edited by a programmer at this level. If an I/O configuration has to be modified, then the programmer would right-click the specific module in the I/O configuration list and go to the module's properties. Figure 5-10 is an example of a module's I/O configuration tags set up during the configuration process and stored in the configuration tags.

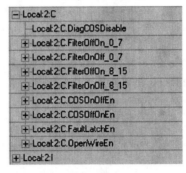

Figure 5-10 Portion of RSLogix 5000 I/O configuration showing an example of configuration tags in the controller scoped tags collection.
Used with permission Rockwell Automation, Inc.

Connection Type O

Output modules have an output tag that represents output information to be sent to the field devices. Figure 5-11 shows part of the controller scoped tags collection and illustrates the output tag for an output module in slot 0 of the local chassis. As I/O modules are configured, their I/O tags

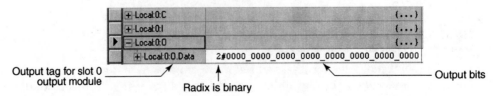

Figure 5-11 Monitor tags view showing output tag for a module residing in slot 0 of the local chassis.
Used with permission Rockwell Automation, Inc.

are automatically stored in the controller scoped tags collection. Controller scoped tags are global tags—that is, they can be used anywhere in the RSLogix 5000 project. Notice the 32 bits used to represent the output points. Output bit 0 is the right-most bit. Thirty-two bits are displayed regardless of how many output points the module actually contains. Only bits that represent actual output points are used. The "2#" to the left of the output bits identifies the radix of this tag as base-2, or binary. The figure shows a portion of the controller scoped tags collection monitor screen.

Figure 5-12 adds connection type O to our tag string. Notice that the capital O is followed by a dot or period.

Local:2:O.

Tank1:2:O.

Remote_CNB:2:O.

Labeling_and_Packing:2:O.

Figure 5-12 Adding the output connection type to a ControlLogix tag.
© Cengage Learning 2014

Input Module Connection Type I

Input modules have input tags that represent the status of their respective field input devices. An input module also has input status tags that provide information to the controller about the module's current status. The specific input tags assigned to an input module depend on the type of input module being configured. Standard modules have module status information or tags that represent the modules' status, whereas diagnostic modules have additional input status tags that represent module status to the point level. A diagnostic module such as a 1756-IB16D has more features and thus more tags than a standard module such as a 1756-IB16. Figure 5-13 illustrates the input tags from a 1756-IB16D module. Notice the 32 bits representing input bits and the radix identifier. Remember, all I/O modules are assigned 32 bits in their I/O tags, even though the bits with actual I/O points are used. The Local:2:I.Data tag illustrated in the figure represents input data. We will define this tag and other input tags from Figure 5-13 when we introduce the I/O configuration lesson.

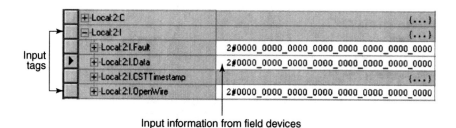

Figure 5-13 Portion of the controller scoped tags collection monitor view showing connection type I for the local input module in slot 2.
Used with permission Rockwell Automation, Inc.

Figure 5-14 illustrates our tags as we add the connection type I to the string number. Note the colon between the location and slot number. There is also a colon between the slot number and connection type. A period is placed after the connection type.

Tank1:2:I.

Remote_CNB:2:I.

Labeling_and_Packing:2:I.

Figure 5-14 Local I/O module tag showing location, slot number, and I connection type.
© Cengage Learning 2014

Output Module Connection Type I

Output modules also have input tags to provide the controller with module status information. Standard modules provide the controller with module status information, whereas diagnostic modules provide additional status information to the I/O point level. As an example, an electronically fused output module such as a 1756-OB16E has a fuse-blown input tag to provide the controller with the status of each electronically fused output point. The 1756-OB16D module also has electronic fusing in addition to the point-level diagnostic tags. Figure 5-15 is part of the controller scoped tags collection monitor view showing the tags of a 1756-OB16D module. Notice the Local:2:I.FuseBlown tag. We will define this tag and other input tags from Figure 5-15 when we introduce the I/O configuration lesson. The figure also shows the output modules configuration tags (Local:2:C), output tags (Local:2:O), and input tags (Local:2:I).

+ Local:2:C	(...)
+ Local:0:I	(...)
− Local:2:I	(...)
+ Local:2:I.Fault	2#0000_0000_0000_0000_0000_0000_0000_0000
+ Local:2:I.Data	2#0000_0000_0000_0000_0000_0000_0000_0000
+ Local:2:I.CSTTimestamp	(...)
+ Local:2:I.FuseBlown	2#0000_0000_0000_0000_0000_0000_0000_0000
+ Local:2:I.NoLoad	2#0000_0000_0000_0000_0000_0000_0000_0000
+ Local:2:I.OutputVerifyFault	2#0000_0000_0000_0000_0000_0000_0000_0000

Figure 5-15 Input tags from a 1756-OB16D output module.
Used with permission Rockwell Automation, Inc.

Figure 5-16 shows the tag string specifying the fuse-blown information. Note the period after the FuseBlown portion of the tag.

> Local :2:I.FuseBlown.
>
> Tank1:2:I.FuseBlown.
>
> Remote_CNB:2:I.FuseBlown.
>
> Labeling_and_Packing:2:I.FuseBlown.

Figure 5-16 FuseBlown added during tag string building.
© Cengage Learning 2014

THE MEMBER NAME

The next piece of the tag is the member name. The available members are specific to each type of module. Refer back to Figure 5-15. The groups of configuration and output tags have "+" signs in front of them. Tags within the group can be displayed by clicking the + sign. Note that the Local:2:I tag has been expanded to show all tags or members within that group. The input tags for this module include a fault, data, *coordinated system timer* (CST) time stamp, fuse blown, no load, and output verify fault. These are the members. We will define these member tags and their function in the I/O configuration lesson. I/O information is referred to as *data*. Figure 5-17 is a local input discrete tag member that specifies input information or data in our string. Note the period after the specified member name. We discuss the significance and use of these member tags in a later lesson. Figure 5-18 illustrates an electronically fused output module fuse-blown tag member added to the tag string.

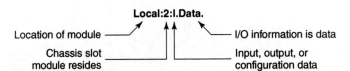

Figure 5-17 Tag showing the data member being added to the tag.
© Cengage Learning 2014

UNDERSTANDING CONTROLLOGIX I/O ADDRESSING **133**

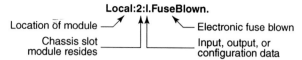

Figure 5-18 Electronically fused module fuse-blown member contained in the tag string.
© Cengage Learning 2014

BIT NUMBER

The last piece of the I/O tag string is the bit number. I/O modules are available as 8, 16, or 32 points. Valid entries are from 0 to 31. Figure 5-19 is a sample local input module discrete tag. This tag identifies the module as in the local chassis, chassis slot 2, input information. Data signifies I/O information, or data input bit number 3.

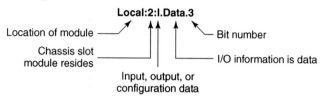

Figure 5-19 Completed I/O tag string for the local chassis slot 2.
© Cengage Learning 2014

Figure 5-20 illustrates the completed sample I/O tags we have been building in this lesson.

Local :2:I.Data.3

Tank1:2:I.Data.3

Remote_CNB:2:I.Data.3

Labeling_and_Packing:2:I.Data.3

Figure 5-20 Completed I/O tags.
© Cengage Learning 2014

Figure 5-21 is a sample local output discrete tag. The tag represents a module in the local chassis, slot 2, output information, data or I/O information, output bit 11.

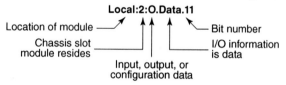

Figure 5-21 Completed output tag, local chassis slot 2.
© Cengage Learning 2014

Figure 5-22 illustrates the tags created for a slot 2 1756-OB16D module.

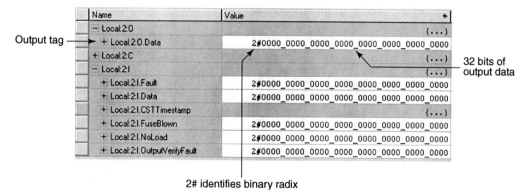

Figure 5-22 Configuration, input, and output tags displayed for 1756-OB16D.
Used with permission Rockwell Automation, Inc.

ANALOG I/O TAGS

Although I/O points are represented as bits, analog information is represented as a value. Analog module ports are referred to as *channels*. When an analog module is configured by the programmer, it elects that ControlLogix analog I/O data be represented either as an integer or as a real number. Figure 5-23 is a sample local analog tag. The tag identifies this analog module as in the local chassis, slot 4, input information, and channel 0 data. There is no period and bit number specified because the information is a number, not a single bit.

Figure 5-23 Local analog input tag, slot 4 module, channel 0 data.
© Cengage Learning 2014

Figure 5-24 illustrates a portion of the controller scoped tags monitor view that shows the analog input tag from Figure 5-23 as well as the other channels on that module. The module represented in the figure is a six-channel analog input module. The analog signal being input into channel zero is 345.7. This is a real number and is also referred to as *floating point*.

Tag	Value	Style	Type
Local:4:I.Ch0Data	345.7	Float	REAL
Local:4:I.Ch1Data	78.9	Float	REAL
Local:4:I.Ch2Data	6754.4	Float	REAL
Local:4:I.Ch3Data	0.0	Float	REAL
Local:4:I.Ch4Data	0.0	Float	REAL
Local:4:I.Ch5Data	0.0	Float	REAL

Figure 5-24 Controller scoped tags monitor view for a six-channel analog input module in chassis slot 4.
Used with permission Rockwell Automation, Inc.

VIEWING I/O TAGS IN CONTROLLER SCOPED TAGS COLLECTION

I/O configuration for RSLogix 5000 applications is manually configured by the individual doing the programming. When the I/O configuration is completed, I/O tags will be found in the controller scoped tags collection. RSLogix 5000 software version 13 and earlier identify controller scoped tags viewed in either the monitor or edit tags tab by placing the word *controller* in parentheses directly to the right of the project name, as illustrated in Figure 5-25. Click the + sign to expand a group and view the associated tags.

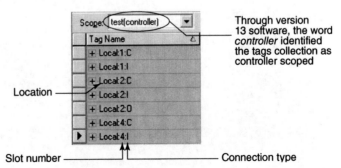

Figure 5-25 Controller scoped tags collection monitor view showing I/O modules in slots 1, 2, and 4.
Used with permission Rockwell Automation, Inc.

Figure 5-26 is the I/O configuration portion of the controller organizer for the tags displayed in Figure 5-25. The number in the square bracket is the chassis slot number where the module resides.

```
□ I/O Configuration
    [1] 1756-IB16D Inputs_Slot_1
    [2] 1756-OB16D Outputs_Slot_2
    [4] 1756-IF6I Analog_Inputs
```

Figure 5-26 ControlLogix I/O configuration showing modules configured in slots 1, 2, and 4.
Used with permission Rockwell Automation, Inc.

Starting with RSLogix 5000 Software version 15, identification of the controller scoped tags collection changes from the word *controller* in parentheses to a controller icon displayed to the left of the project name, as illustrated in Figure 5-27.

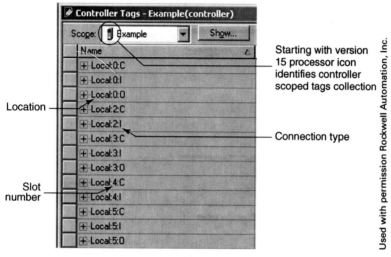

Figure 5-27 Identified controller scoped tags as controller icon displayed to the left.

Figure 5-28 is the I/O configuration for version 15 software and newer versions, for the modules displayed in Figure 5-27. Starting with version 15, the local controller is displayed in the I/O configuration. Note the icon with the left-pointing arrow for slot 1 and the 1756-L63 controller. The left-pointing arrow and controller icon identify the controller into which this project will be downloaded. In multicontroller systems, this is especially important because the arrow and icon identify in which controller the project resides. Earlier versions of software, like that shown in Figure 5-26, do not display the local controller in the I/O configuration.

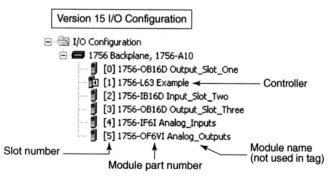

Figure 5-28 RSLogix 5000 version 15 and later ControlLogix I/O configuration.
Used with permission Rockwell Automation, Inc.

As the chassis modules are configured, they are assigned a name. Note the names assigned to each module after the module's part number. Programmers can name modules whatever they wish as long as each name is unique. The module name cannot contain the word local. The module's name is not used as part of the module's tags.

SUMMARY

When installing or troubleshooting a ControlLogix, one has to be able to identify the tags assigned to the input and output field device wiring. The I/O tags are assigned to instructions as rungs are created. In many cases, the ladder logic has already been written, so troubleshooters must be able to correlate I/O tags assigned to programming instructions and their respective I/O module screw terminals and associated field devices. Programmers must be able to correlate programming with prewired or soon-to-be-wired digital as well as analog field devices. Installers must be able to properly wire inputs and outputs to their respective module terminal points as specified in a system's wiring diagrams.

The modular ControlLogix has physical chassis slots that are assigned slot numbers. CompactLogix is a rackless design; however, I/O modules still have slot numbers assigned to them as part of their addressing or I/O tag identification. ControlLogix and CompactLogix tag formats are the same. An I/O tag comprises several parts to help identify the specific I/O or module information. A typical digital I/O tag is made up of the following parts:

> location: slot number: connection type. member name. bit number.

Analog input and output ports are referred to as *channels*. An analog tag is an integer or a real number rather than an I/O bit. Analog tags comprise "location: slot number: connection type. member name."

This lesson provided information on identifying digital as well as analog tag formatting and interpreting RSLogix 5000 I/O configurations for local as well as ControlNet and Ethernet/IP remote chassis.

REVIEW QUESTIONS

Note: For ease of handing in assignments, students are to answer using their own paper.

1. The _____ PAC has physical chassis slots that are assigned slot numbers.
2. The I/O configuration is _____ set up using RSLogix 5000 software.
3. Even though the CompactLogix is a _____ design, the I/O modules still have slot numbers assigned to them as part of their addressing or I/O tag identification.
4. An I/O tag comprises several parts to help identify the specific I/O or module information the user is looking for. A typical digital I/O tag is made up of the following parts:

    _____
    _____
    _____
    _____
    _____

5. Chassis sizes for the modular ControlLogix are _____ and 17 slots.
6. _____ bits are displayed in the I/O tag regardless of how many output points the module actually contains.
7. When performing an _____, the chassis modules are assigned a name. Module names are not used as part of the modules' tags.
8. Standard I/O modules have module status information or tags that represent the modules' status, whereas _____ modules have additional input status tags representing module status to the point level.
9. Although I/O points are represented as bits, analog information is represented as a real _____ or integer.
10. A _____ module such as a 1756-OB16D has more features and thus more tags than a standard module such as a 1756-OB16.
11. Analog module input ports are referred to as _____.
12. The first part of the tag is the _____ of the module.
13. If the ControlLogix consists of a single chassis, it is referred to as the _____ chassis.

14. Until a module receives its _____, the module does not know how it is to operate.
15. Once the local chassis slots have been filled, additional modules are placed into a _____ chassis in relation to the local chassis.
16. Newer applications' remote chassis are connected to the local chassis by either a _____ or _____ network connection.
17. Several configuration tags are created as the result of the programmer performing the _____ for each module.
18. Each remote chassis has a _____ that provides network communications to the local chassis.
19. As many as _____ modules may be placed between the 1769 CompactLogix controller and the power supply.
20. Remote communication modules are referred to as _____.
21. Each module in a ControlLogix chassis is intelligent, an entity in itself, or a separate _____ on the ControlBus network.
22. When the _____ controller goes into Run mode, I/O configuration information is sent to the respective modules across the backplane.
23. Each I/O module is in _____ mode until receiving its I/O configuration.
24. The "2#" to the left of the output bits identifies the _____ of this tag as base-2, or binary.
25. The modular ControlLogix platform is a _____ controller system in which a controller can be placed into _____ slot.
26. CompactLogix slot numbers start with slot _____ to the right of the controller and increment in decimal numbers to the right-most slots.
27. Because CompactLogix is a _____ design, modules are clipped together on a DIN rail.
28. Because Allen-Bradley refers to its communication modules as adapters, the company's literature states that the first word of a Remote I/O tag is the _____.
29. Depending on a specific CompactLogix controller, as many as _____ modules can be configured in three banks.
30. When the I/O configuration is completed, I/O tags are found in the _____ scoped tags collection.
31. When an analog module is configured by the programmer, the selected ControlLogix analog I/O data is represented either as an integer or as a _____.

LAB A: Determining ControlLogix I/O Tags

Note: For ease of handing in assignments, students are to answer using their own paper.

The next questions provide you with experience in identifying ControlLogix I/O tags by looking at the physical hardware configuration.

1. The following questions refer to the ControlLogix illustrated in Figure 5-29. Enter the I/O tag for the listed devices below.
   a. start conveyor _____
   b. conveyor running _____
   c. start dryer _____
   d. detergent flow valve _____
   e. start process _____
2. The following questions refer to the ControlLogix illustrated in Figure 5-30. Fill in the I/O tags for the devices listed.
   a. reject solenoid _____
   b. increment speed _____
   c. start mixer _____
   d. jog _____
   e. start dryer _____

138 UNDERSTANDING CONTROLLOGIX I/O ADDRESSING

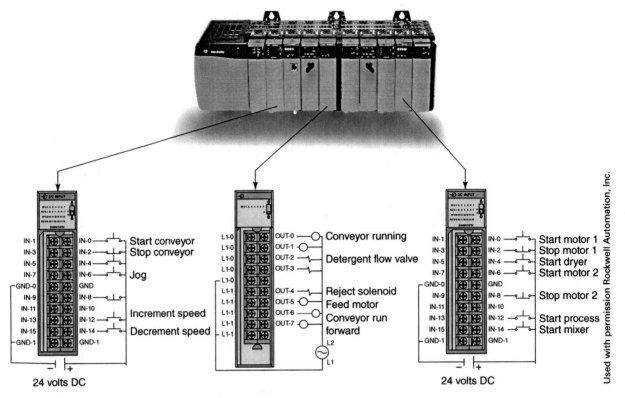

Figure 5-29 ControlLogix hardware for review question 1.

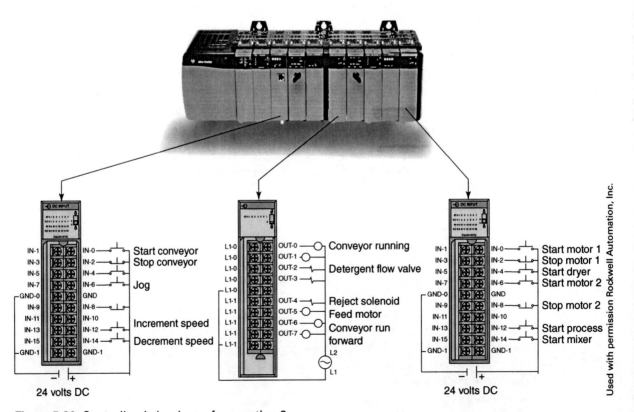

Figure 5-30 ControlLogix hardware for question 2.

3. The following questions provide you with experience in identifying I/O tags by looking at the physical hardware configuration. Refer to Figure 5-31 for your answers.

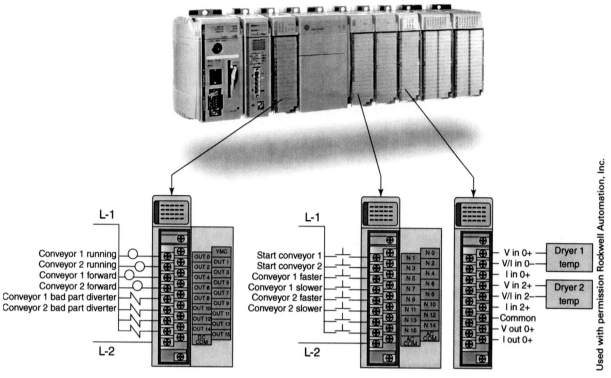

Figure 5-31 ControlLogix hardware for question 3.

  a. What member of the ControlLogix family is pictured in Figure 5-31?
  b. Write the tag for conveyor 1 running. _____
  c. What is to the immediate right of the module used for the answer to question 3b?
  _____
  d. The tag for start conveyor 2 is _____.
  e. Dryer temp 1 tag name is _____.
  f. Conveyor 2 bad part diverter tag name is _____.
  g. Dryer temp 2 tag name is _____.

4. The following questions refer to the ControlLogix and ControlNet network illustrated in Figure 5-32. Fill in the correct tag for the I/O devices listed. Assume node 1 is the local chassis. For this question determine the I/O tags as they would be referenced from a controller in the same chassis the module resides.
  a. conveyor running _____
  b. jog _____
  c. start conveyor _____
  d. start process _____

5. The following questions refer to the ControlLogix and ControlNet network illustrated in Figure 5-33. Fill in the requested tags below. Assume node 1 is the local chassis. For this question determine the I/O tags in the local chassis assuming they are owned by one of the controllers in the local chassis. Assume the node 2 module in question is remote I/O from the local chassis.
  a. start motor 1 _____
  b. decrement speed _____
  c. conveyor run forward _____
  d. detergent flow valve _____
  e. jog _____

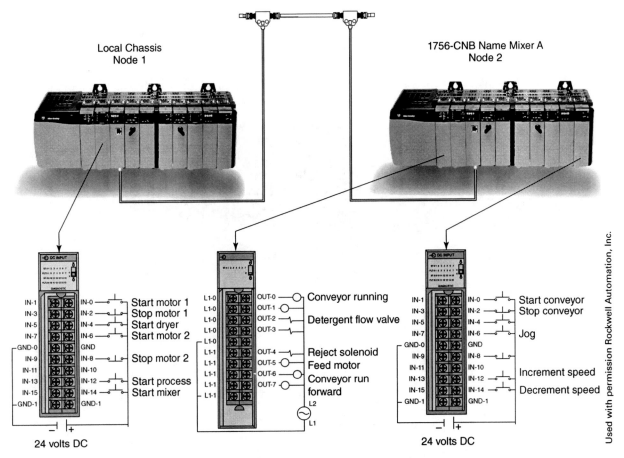

Figure 5-32 ControlLogix hardware configuration for question 4.

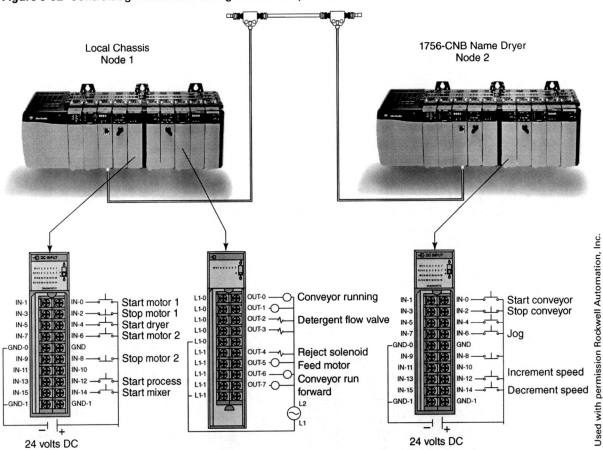

Figure 5-33 ControlLogix hardware configurations for question 5.

6. The following questions refer to the ControlLogix illustrated in Figure 5-34. Fill in the requested I/O tags. For this question determine the I/O tags in the local chassis assuming the modules in question are owned by one of the controllers in the local chassis. Assume the node 4 modules in question are remote I/O from the local chassis.

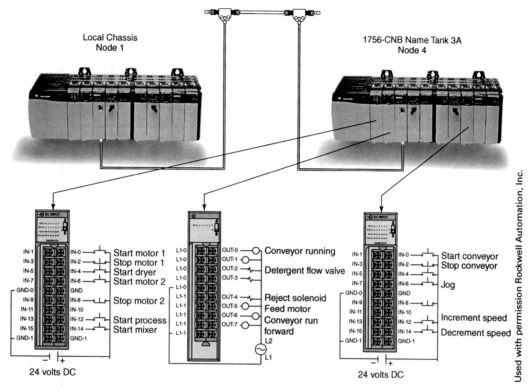

Figure 5-34 ControlLogix hardware configuration for question 6.

   a. feed motor _____
   b. reject solenoid _____
   c. increment speed _____
   d. start mixer _____

7. The following questions refer to the ControlNet network in Figure 5-35 and the RSLogix 5000 I/O configuration from Figure 5-36.
   a. What chassis slot is the local ControlNet communications module in? _____
   b. List the part number of the ControlNet communications module. _____
   c. What is the name of the remote ControlNet communications module? _____
   d. What is the node address of the remote ControlNet communications module? _____
   e. Write down the conveyor running tag. _____
   f. What is the start mixer tag? _____
   g. What is the stop conveyor tag? _____

8. The following questions provide you practice in interpreting an I/O configuration and how it associates to the physical hardware. Refer to the network in Figure 5-37 and RSLogix 5000 I/O configuration from Figure 5-38.
   a. List the names of the adapter modules:
      node 1: _____
      node 2: _____
      node 3: _____
      node 4: _____
      node 5: _____

142 UNDERSTANDING CONTROLLOGIX I/O ADDRESSING

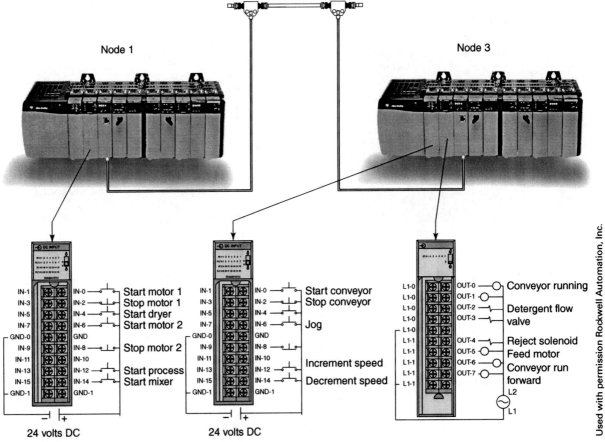

Figure 5-35 ControlLogix hardware configuration for question 7.

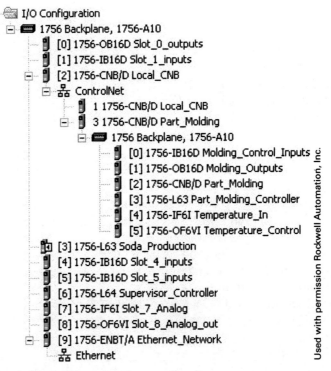

Figure 5-36 ControlLogix I/O configuration for question 7.

UNDERSTANDING CONTROLLOGIX I/O ADDRESSING 143

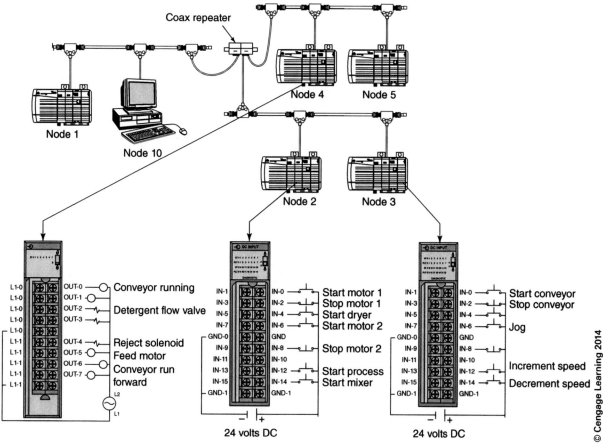

Figure 5-37 ControlLogix hardware for question 8.

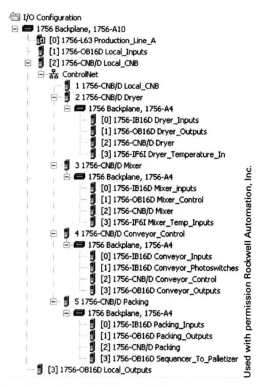

Figure 5-38 I/O configuration for network in Figure 5-37.

b. What network are we working with? _____
c. What is the feed motor tag? _____
d. What is the start process tag? _____
e. The name of the chassis slot 0 input module at node 5 is "Packing Inputs." Is that part of the input tag names for that module?
f. What is the stop conveyor tag? _____
g. What type of module resides in the node 3 chassis in slot 3?
h. Write a sample tag for the module you identified in question 8g.

9. The following questions refer to the network in Figure 5-39. Use the I/O configuration from Figure 5-38.

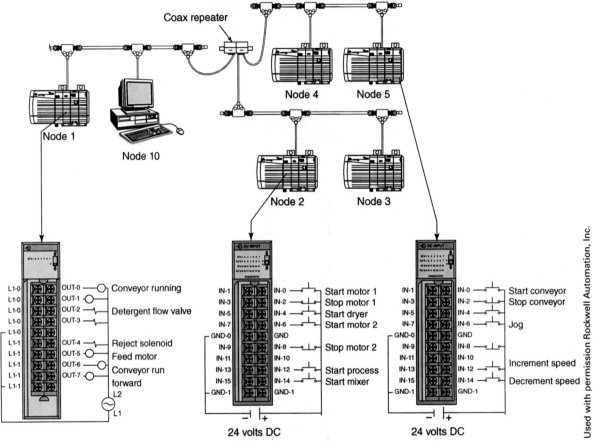

Figure 5-39 Hardware configuration for network in question 9.

a. What is the part number for the node 1 network communication module identified in Figure 5-39? _____
b. List the part number of the module identified from node 2. _____
c. The detergent flow valve tag is _____.
d. The start dryer tag is _____.
e. The jog tag is _____.

10. The following questions provide practice in interpreting an I/O configuration. Refer to the RSLogix 5000 software I/O configuration in Figure 5-40. Read each question and write down the requested information for your answer.
a. What is the name of the ControlNet bridge module in node 4? _____
b. What is the size of the local chassis? _____

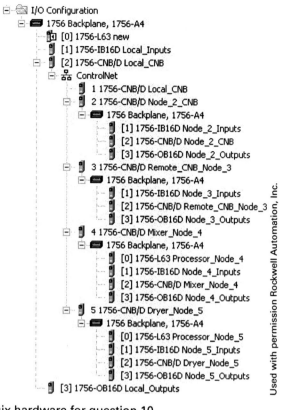

Figure 5-40 ControlLogix hardware for question 10.

    c. List the local chassis controller's part number. _____
    d. You are troubleshooting this system. The module named "Node 3 Outputs" has the suspected problem devices wire-connected to I/O point 10. What is the tag? _____

11. The following questions pertain to the RSLogix 5000 I/O configuration in Figure 5-41.
    a. What network or networks are represented in this I/O configuration? _____
    b. The module with the name "Temperature Control" is at which node? _____
In what chassis slot does the module reside? _____
List the name of the communication adapter. _____
Write down an I/O tag for this module. _____
    c. What is the name of the remote 1756-ENBT module? _____
What chassis slot is this module in? _____
Write down the input tag for I/O point 7 for the 1756-IB16D module in this chassis.
_____

Write down the output tag for I/O point 12 for the 1756-OB16D module in this chassis. _____
    d. Write down a possible I/O tag for the module named "Temperature_In."
_____

    e. While troubleshooting, you find that the device connected to I/O point 2 of the module named "Molding Control Inputs" is possibly causing problems. You must determine its tag so you can search for this point in your RSLogix 5000 software. After finding the tag, you will monitor this point in the monitor tags collection of the software. What is this device's tag? _____

146 UNDERSTANDING CONTROLLOGIX I/O ADDRESSING

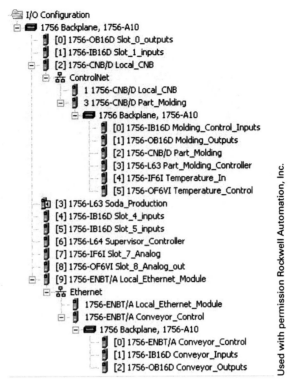

Figure 5-41 I/O configuration for question 11.

12. The following questions pertain to the RSLogix 5000 I/O configuration in Figure 5-42.
    a. What member of the ControlLogix family is this hardware? _____
    b. What controller is being used in this hardware configuration? _____
       What communication is available between your personal computer and this controller?
       _____
    c. If you were troubleshooting a problem and module D's I/O point 3 were connected to the field device, what would the tag be? _____
    d. If you were troubleshooting a problem and module A's I/O point 9 were connected to the field device, what would the tag be? _____
    e. If you were troubleshooting a problem and module E's I/O channel 0 were connected to the field device, what would the tag be? _____

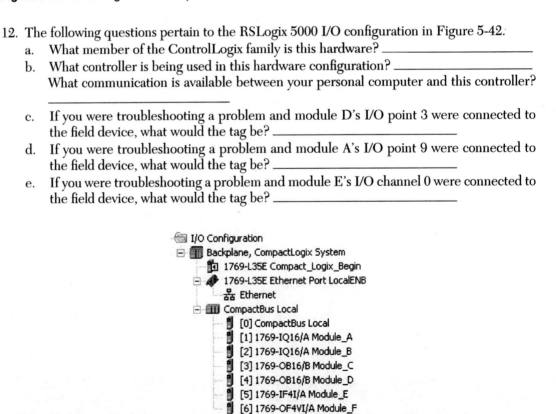

Figure 5-42 ControlLogix hardware for question 12.
Used with permission Rockwell Automation, Inc.

UNDERSTANDING CONTROLLOGIX I/O ADDRESSING   147

13. Refer to Figure 5-43 for the following questions.

Name	Value
+ Local:0:C	{...}
− Local:0:I	{...}
+ Local:0:I.Fault	2#0000_0000_0000_0000_0000_0000_0000_0000
+ Local:0:I.Data	2#0000_0000_0000_0000_0000_0000_0000_0000
+ Local:0:I.CSTTimestamp	{...}
+ Local:0:I.FuseBlown	2#0000_0000_0000_0000_0000_0000_0000_0000
+ Local:0:I.NoLoad	2#0000_0000_0000_0000_0000_0000_0000_0000
+ Local:0:I.OutputVerifyFault	2#0000_0000_0000_0000_0000_0000_0000_0000
− Local:0:O	{...}
+ Local:0:O.Data	2#0000_0000_0000_0000_0000_0000_0000_0000

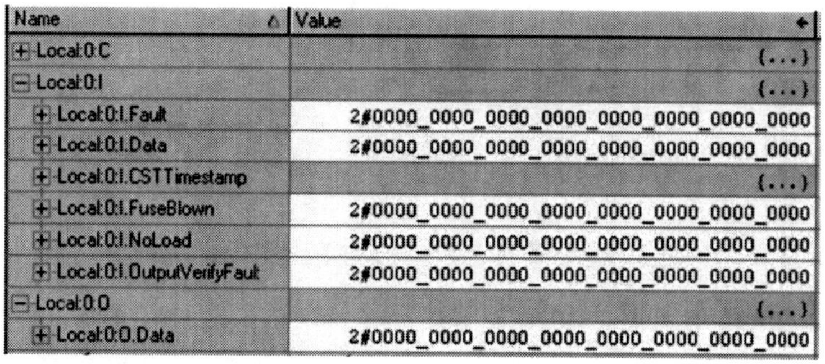

Figure 5-43 ControlLogix hardware for question 13.
Used with permission Rockwell Automation, Inc.

  a. What is being illustrated in Figure 5-43? _____
  b. Is an input or output module being represented here? _____
  c. What does "2#" signify? _____
  d. Because there are 32 bits listed in the value column for this module, must this be a 32-point I/O module? _____ Explain your answer. _____

14. The following questions pertain to the RSLogix 5000 tag collection in Figure 5-44.
  a. Looking at Figure 5-44, is this a controller scoped or programmed scoped tags collection? _____
      Explain how you determined your answer. _____

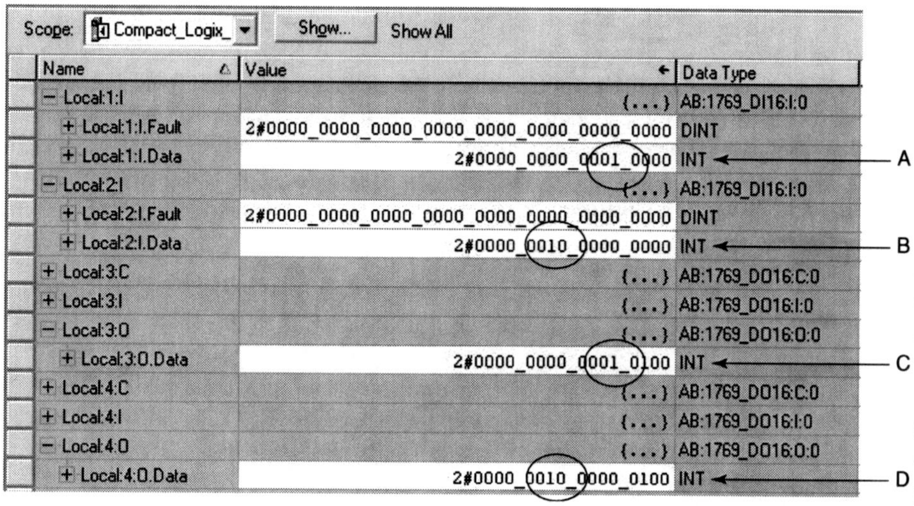

Figure 5-44 ControlLogix hardware for question 14.

  b. Are the following modules illustrated input or output modules?
      module slot 1 _____
      module slot 2 _____
      module slot 3 _____
      module slot 4 _____
  c. What is the tag for 1 bit identified as A?
  d. What is the tag for 1 bit identified as B?
  e. What is the tag for 1 bit identified as C?
  f. What is the tag for 1 bit identified as D?

CHAPTER

6

Modular ControlLogix I/O Configuration

OBJECTIVES

After completing this lesson, you should be able to

- Understand terminology associated with modules and their configuration.
- Perform a digital diagnostic input and output-module configuration for a modular ControlLogix.
- Perform an analog input and output-module configuration for a modular ControlLogix.
- Download your completed project and monitor tags.

INTRODUCTION

When we create a new project, the I/O modules we use must be manually configured by performing an I/O configuration. Until the modules are configured, no I/O tags are available in the monitor or edit tags views of the controller scoped tags collection. If we add a new I/O module to an existing RSLogix 5000 project, then we must configure the module before its I/O tags will be available for programming. If we are using diagnostic modules and wire additional inputs or outputs, the I/O configuration may need to be updated to turn on the newly installed I/O devices' diagnostic features.

LAB EXERCISE 1: ControlLogix Modular Digital I/O Configuration

EQUIPMENT NEEDED TO COMPLETE THIS LAB

This lab exercise steps through the completion of an I/O configuration using a ControlLogix modular programmable automation controller that uses RSLogix 5000 software. For this lab, we assume you are using the following ControlLogix hardware and software:

- RSLogix 5000 software version 18 or newer
- 1756-OB16D output module slot 0
- 1756-L63 controller in chassis slot 1
- 1756-IB6D input module in slot 2

See your instructor for specific instructions if you do not have this particular hardware.

THE LAB

1. _____ Open your RSLogix 5000 software.
2. _____ Create a new project.
3. _____ Select your specific controller.
4. _____ Assign your controller to chassis slot 1.

MODULAR CONTROLLOGIX I/O CONFIGURATION

5. _____ Select your software version.
6. _____ Give the project the name "I_O_Configuration."
7. _____ Look near the bottom of the controller organizer for the I/O configuration. Notice that the chassis size is identified and the controller is shown in chassis slot 1.
8. _____ Right-click 1756 Backplane, as in Figure 6-1, to start your configuration.

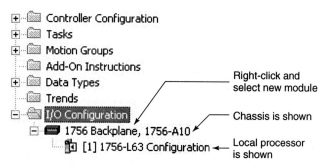

Figure 6-1 Beginning I/O configuration RSLogix 5000 version 15 or newer shows controller in slot 1. Used with permission Rockwell Automation, Inc.

The screen shown in Figure 6-2 appears. There are three ways to continue from here. First, you can select the module type from the By Category tab, as illustrated in the figure. Second, you can select the By Vendor tab to select by manufacturer. The third option is to select the Favorites tab. Figure 6-2 shows the Favorites tab selected and the current modules listed as favorites. Beginning with software version 20 the I/O configuration screens will look a bit different than earlier versions and the figures here. Since not everyone uses the latest version of software we are assuming you are using version 18. Even though different, one should have no trouble navigating through the newer screens.

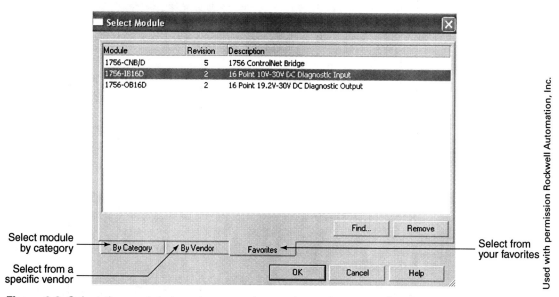

Figure 6-2 Select the module by category or by vendor or from your favorites.

1756-OB16D CONFIGURATION

For this portion of the lab, we configure the 1756-OB16D module residing in chassis slot 0.

9. _____ Select the By Category tab.
10. _____ Click + and Digital, as shown in Figure 6-3.

150 MODULAR CONTROLLOGIX I/O CONFIGURATION

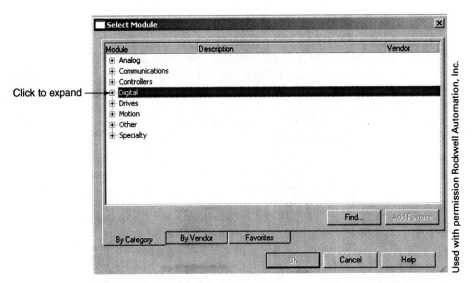

Figure 6-3 Click + to expand the type of modules group.

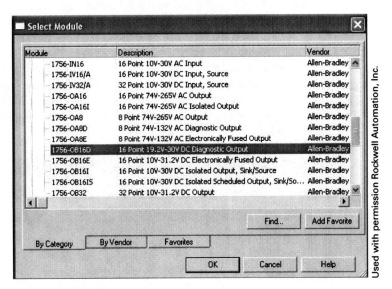

Figure 6-4 Select the 1756-OB16D module.

11. _____ Select the 1756-OB16D module, as in Figure 6-4, and click OK.
12. _____ Select the module's major revision level, as shown in Figure 6-5.
13. _____ Click OK.
14. _____ Give the module a name. For this lab, you can call the module "Outputs_Slot_0." Refer to Figure 6-6.
15. _____ A description of the module can be added if desired, although it is optional.
16. _____ Enter the slot number where the module resides in the chassis. For our example, the module is in chassis slot 0.
17. _____ Select the module's communication format. For our lab, select Full Diagnostics – Output Data. The section that follows provides a general description of the different communication formats. The specific communication formats depend on the selected I/O module. Some modules have only a couple of selections, whereas others may have many.

MODULAR CONTROLLOGIX I/O CONFIGURATION   **151**

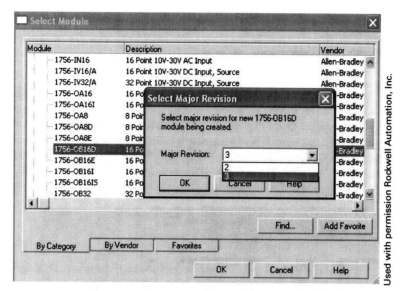

Figure 6-5 Select Major Revision.

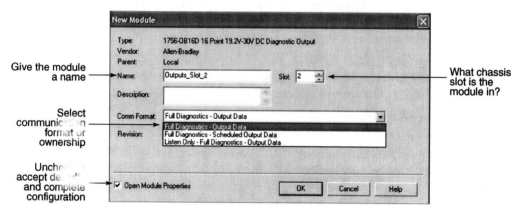

Figure 6-6 Select the module's ownership or communication format.
Used with permission Rockwell Automation, Inc.

OVERVIEW: I/O MODULE OWNERSHIP

The modular ControlLogix allows a programmer to assign ownership to I/O modules as part of the I/O configuration. Ownership is determined in this step of the I/O configuration. The owner controller holds the I/O configuration for each module it owns. When a project is downloaded into the controller and the controller is put into Run mode, the owner controller also establishes communication, or "connection," to the module and then sends each module it owns the module's configuration. Until a module receives its configuration, it does not know how to behave and is in waiting mode.

An input module may have multiple owners, whereas an output module can only have one owner. Because input modules may have multiple owners, the owner or owners of any input module could be in either the same chassis as the module or in a remote chassis. An output module can have only one owner. This owner could be in the same chassis as the output module or in a remote chassis. Another option for communicating with a module is a listen only connection.

Listen Only Connection

A listen only connection provides the programmer the option of listening to the input information or tags returning to the controller from the module. Remember: The owner controller provides the connection and configuration data for the module. A controller with a listen only connection can only monitor or listen to the module only as long as the owner has communication with the module. If the owner loses the connection to the module, the listen only connection is also broken.

Output-Module Input Tags

In addition to output tags, an output module has input tags "feeding back" status information to the controller. As an example, an electronically fused output module has information on the status of the electronic fuses on the module. This information is contained in the electronic fuse tag. The module sends the electronic fuse tag to the controller. From the controller's perspective, this is data being input from the output module. The electronic fuse tag is an input tag. A controller with a module configured as a listen only connection does not contain the I/O configuration for the module. If this controller is not the owner of the output module, it cannot control the output points.

Ownership Options

The list below provides a general list of the options for setting up ownership in the I/O configuration.

1. This controller has the module in its I/O configuration. This controller is the owner.
2. This controller has the module in its I/O configuration. This controller is not the owner and has a listen only connection.
3. This controller has the module in its I/O configuration, and the module is inhibited in the I/O configuration.
4. This controller does not have the module in the I/O configuration.

Using the 1756-OB16D as an example, the communication format—or ownership options—are listed below with explanation. Refer to Figure 6-6. Keep in mind that the specific communication format options are module specific.

Full Diagnostics – Output Data	This controller holds the I/O configuration for the module. This controller is the owner.
Listen Only Full Diagnostics – Output Data	This controller has an I/O configuration for the module; the module communications format is Listen Only Full Diagnostics–Output Data. This controller is not the owner of this module.
Full Diagnostics – Scheduled Output Data	This controller holds the I/O configuration for the module. This controller is the owner. This communications choice uses the controller's system clock to schedule a time delay before an output point turns on.

ELECTRONIC KEYING

The user specifies the slot number for the module as part of the I/O configuration. Electronic keying provides a safeguard: If the wrong type of module were to be placed in the slot, the controller would reject the connection or not communicate with the incorrect module. If the controller rejects the connection to the module, then the module will not work and the I/O LED on the front of the controller will alert the user that by flashing green there is a problem. When selecting electronic keying, a user needs to determine how closely matched a module replacement must be to the original part number, manufacturer, and major and minor revision. As shown in Figure 6-7, there are three choices: Exact Match, Compatible Module, and Disable Keying. Each option is explained below.

Exact Matches

For an exact match, the replacement module must be the same part number, manufacturer, and major and minor revision. An example of an exact match could be a module used by a pharmaceutical company, which is regulated by the Food and Drug Administration (FDA).

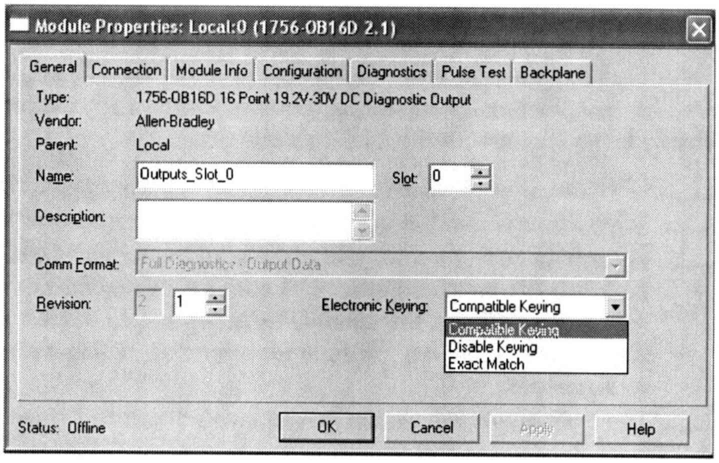

MODULAR CONTROLLOGIX I/O CONFIGURATION

Figure 6-7 Select the desired type of electronic keying for this module.

If the module fails and is replaced with an exact match, then typically only a notation needs to be made to satisfy the FDA. If the replacement is not an exact match, then the system will probably need to be revalidated, which could mean the system would be down for revalidation for days, weeks, or months.

Compatible Modules

A compatible module must be of the same part number, manufacturer, and major revision. The minor revision must be equal to or greater than the value entered into the right-hand revision box, as in Figure 6-7. A compatible module is typically the most common selection, but that is up to the programmer to determine. Refer to the specific module's user manual pages for an exact definition for your module.

Disable Keying

When keying is disabled, the module type of the replacement module must match that to be replaced, but a module with any firmware level can be placed in the slot. If keying is disabled, then the controller will attempt to communicate with the module. However, if the module is not the same type, then the connection will be rejected. As an example, if a standard input module were configured for the slot but a different part number but standard module were inserted in the slot, the connection would be established because the module type is the same. If, however, one attempted to put an output module in a slot configured for an input module, the connection would be rejected by the controller because the module type would not match. If a standard module were configured for the slot and a diagnostic module were inserted in the slot, then the connection would be rejected because the modules are of different types.

Selecting disable keying is not used under normal conditions. Typically, disable keying is used by project developers in the early stages of a project when the specific modules have not been installed.

The end user should not use disable keying as a shortcut in lieu of deleting the original module from the I/O configuration and redoing the I/O configuration when a newer replacement module would have a newer major firmware revision and thus would not be accepted by the PLC as a compatible module. In many cases, a replacement module with a higher major revision level would have newer features than the older module. As a result, the amount and format of the data sent by the newer module to the controller and the amount of and format of the data received from the controller will probably not be compatible, and the module will probably not work properly. When keying is disabled, the controller uses the major and minor revision information contained in the I/O configuration as its reference about what to send to and receive from

154 MODULAR CONTROLLOGIX I/O CONFIGURATION

the module regardless of what the module actually sends and expects to receive. In other words, if the user were to replace a module that was configured as a version 2.1 with a version 8.0 and keying were disabled, then the controller would treat the module as if it were a 2.1 because it was configured for that. In a case such as this, the data format and amount will not match, and there will probably be problems with the module's operation.

18. _____ Click the drop-down arrow in the Electronic Keying box (see Figure 6-7).
19. _____ Select Compatible Keying because it is used in most cases.
20. _____ Refer back to Figure 6-7 and make sure the Open Module Properties check box in the lower left-hand corner is checked. Unchecking the box and clicking OK completes the configuration of this module by accepting all default settings. Checking the box displays the remaining configuration screens, which we explore as we continue the configuration.
21. _____ The Connection tab should be displayed. Refer to Figure 6-8.

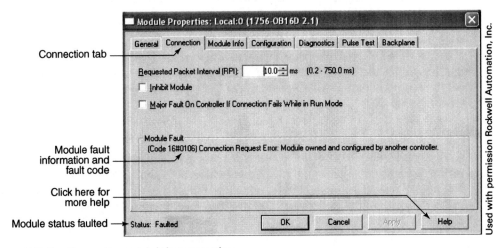

Figure 6-8 Configure the module's properties.

22. _____ Click the check box if you want a major fault on the controller if communication (the connection) fails to this module in Run mode.
23. _____ Click Inhibit Module if you wish to turn the module off as far as this controller is concerned. Inhibiting a module is typically used during startups or debugging.
24. _____ The *requested packet interval* (RPI) is user-selectable between 0.2 and 750 milliseconds—the time period when output information is sent from the output tags to this output module.
25. _____ Even though your module is not faulted, let us see how the screen might look if the module were faulted. Next, we look at how to obtain additional information, the hexadecimal error code, and the fault explanation in the module fault box. Refer back to Figure 6-8 to view the fault code and fault text. For our example, the error code is listed as 16#0106. The 16# signifies the radix of the error code is hexadecimal, or base 16. The error explanation states, "Connection Request Error: Module owned or configured by another controller." More information about the fault can be found by clicking the Help button.
26. _____ Assume your module was faulted as illustrated in Figure 6-8. You could click the Help button to display a screen like that shown in Figure 6-9. Notice that "module faults" is underlined.
27. _____ Click the underlined "module faults" text to display a screen with error-range information similar to that shown in Figure 6-10. Because our error code was 16#0106, it is in the second range listing.

MODULAR CONTROLLOGIX I/O CONFIGURATION   155

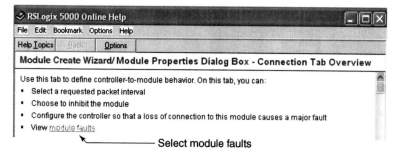

Figure 6-9 Select module faults from Help screen.
Used with permission Rockwell Automation, Inc.

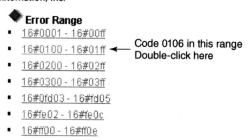

Figure 6-10 Select fault code from list.
Used with permission Rockwell Automation, Inc.

28. _____ Clicking the second range of numbers displays the screen as illustrated in Figure 6-11. Notice the error code, the string of text displayed in the Module Faults box and an explanation of possible solutions.

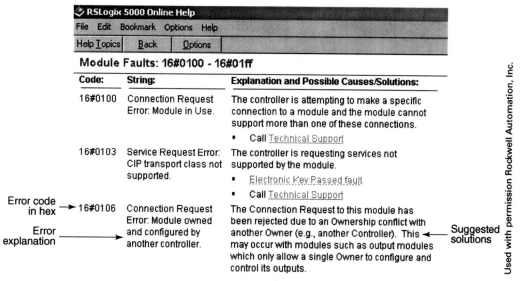

Figure 6-11 Error information from Help screen.

29. _____ If you are viewing the Module Fault Help screen, close out the screen and return to the Connection tab.

30. _____ Notice the module's status as displayed in the bottom-left corner of the Connection tab in Figure 6-8. The module is currently faulted. If you have one or more faulted modules, the I/O Not Responding LED will be blinking on the online toolbar, as in Figure 6-12. If you were viewing the actual toolbar, the I/O Not Responding LED

156 MODULAR CONTROLLOGIX I/O CONFIGURATION

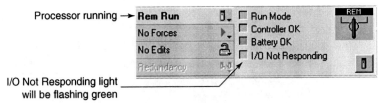

I/O Not Responding light will be flashing green

Figure 6-12 Online tool bar showing the I/O Not Responding light.
Used with permission Rockwell Automation, Inc.

should be blinking; the controller, however, is still in Run mode. Under normal conditions, a faulted I/O module does not cause a controller fault. Additional programming is required to identify and resolve the problem.

31. _____ Figure 6-13 shows the I/O Configuration folder and the Quick Pane view displaying information about the faulted module; a yellow cautionary triangle shows the user where the problem is.

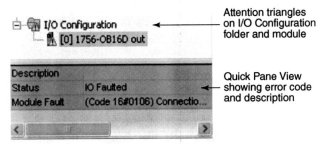

Figure 6-13 Faulted I/O module cautionary triangle displayed in I/O Configuration and Quick Pane View.
Used with permission Rockwell Automation, Inc.

32. _____ Remember the status in the bottom-left corner of the Connection tab window from Figure 6-8? It displays the current status of this module as faulted. We now move on to the next tab.

33. _____ Click the Module Info tab to see the read-only screen. When online with the controller, this screen is populated with module information, as illustrated in Figure 6-14. Note the module identification information on the left side of the screen. Status information is displayed on the right side.

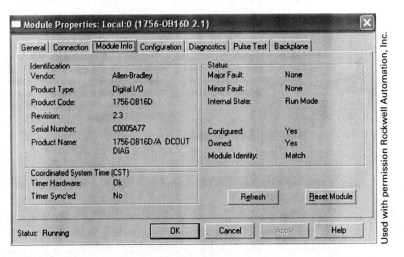

Figure 6-14 Module Info tab.

34. _____ Click the Configuration tab.
35. _____ This option sets up how each output point is to react when the controller goes into Program mode. Refer to Figure 6-15. If you want to modify the setting, click the drop-down arrow for each output point on the right-hand side of the Program Mode column. Select On, Off, or Hold. The default is Off. When the controller goes into Program mode, this output point will be turned on if it is not already on. Off means this output point will be turned off if it is not already off. Hold leaves the output point in its current state. The Program Mode selections define how the output point will behave in Program mode, Remote Test mode, and when the controller experiences a major recoverable fault. Figure 6-15 shows the Configuration tab for a 1756-OB16I module.

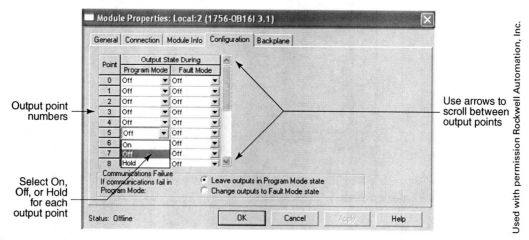

Figure 6-15 Select output point status when in Program mode.

36. _____ For each output, select what is to happen when the controller goes into Fault mode. Refer to Figure 6-16. Here Fault mode is defined as the controller has experienced a major nonrecoverable fault. To configure, click the drop-down arrow for each point on the right-hand side of the Fault Mode column. Select On, Off, or Hold. When the controller experiences a major nonrecoverable fault. On means this output point will be turned on if it is not already on. Off means this output point will be turned off if it is not already off. Hold leaves the output point in its current state. The default is Off.

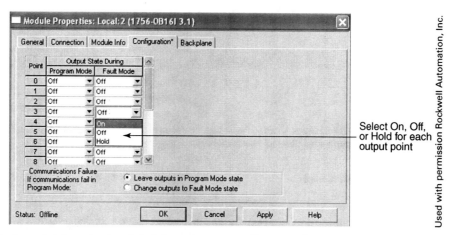

Figure 6-16 Select module's output point status when controller experiences a major nonrecoverable fault.

158 MODULAR CONTROLLOGIX I/O CONFIGURATION

37. _____ The module configuration screens in Figures 6-15 and 6-16 are examples of the Configuration tab features from a nondiagnostic output module. The module configuration screen is a 1756-OB16I, or isolated point output module. The module we are currently configuring is a 1756-OB16D or diagnostic output module. Diagnostic modules have additional features. See Figure 6-17.

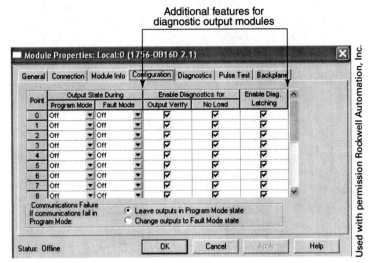

Figure 6-17 Set up diagnostic features.

Diagnostics for Output Verification

38. _____ Note the column heading "Enable Diagnostics for" and the subheading "Output Verify" in the center of the screen in Figure 6-17.

Diagnostic output modules have the ability to determine the actual state of the output point when the point is commanded to be on. Check the box for each output point you wish to test if the output is actually on when it has been commanded to be on. If, for some reason, the output was commanded to be on but found to be off, there would be a 1 in the associated OutputVerifyFault tag. The default condition is checked or enabled. Figure 6-18 shows the OutputVerifyFault tag from the controller scoped tags collection. Notice the tag name. Even though this is an output module, there is status information provided to the controller as input tags. The module referenced in the figure has an OutputVerifyFault on output point 10.

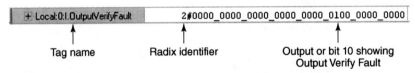

Figure 6-18 Output Verify Fault tag from controller scoped tags collection.
Used with permission Rockwell Automation, Inc.

Diagnostics for No Load

39. _____ Enable Diagnostics for No Load.

To the right of the Output Verify column in Figure 6-17 is the No Load column. Diagnostic output modules have the ability to detect an open field circuit or no load. When the output is commanded to be off, a small amount of current is sent through each enabled output circuit. If the module sees the current return, the circuit is deemed as good. If the signal does not return, the circuit is considered as open or no load. The signal sent out is not enough to turn on an output field device.

As part of the I/O configuration, a NoLoad tag is created. Each bit of the NoLoad tag associates to an output point. A no load fault for any output point results in a 1 being placed in the associated bit position within the NoLoad tag. The default state for each output point is checked or enabled. Figure 6-19 shows the NoLoad tag from the controller scoped tags collection. Even though this is an output module, this is status information provided to the controller as an input tag as shown. This module has a no load fault on output point 8.

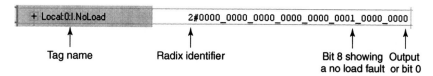

Figure 6-19 NoLoad tag in the Monitor Tags view of the controller scoped tags collection.
Used with permission Rockwell Automation, Inc.

A common problem encountered with diagnostic modules is that when a module is configured, the check boxes are left at default, which is checked, in the enable diagnostics for the No-Load section of the modules configuration tab. Any output point where there is no actual output field device connected always has a no load fault. As a result, the fault LEDs on the module will always be illuminated when there actually is no problem. Remember to uncheck any no load diagnostics not currently used. Also remember to re-enable diagnostic features when I/O points are wired.

Diagnostics for Electronic Fuse Blown

As discussed in Chapter 1, module part numbers with a suffix of "E" identify this module as an electronically fused output module. Each output point is electronically fused for modules with this feature. An example is a 1756-OA8E or 1756-OB16E. Diagnostic modules such as the 1756-OB16D are also electronically fused. Electronic fusing is used to protect the module output point, not the field device or load. There is a FuseBlown tag with a bit for each output point on the module. It is the designer's or installer's responsibility to determine how to protect the load. Figure 6-20 is that of a FuseBlown tag. Can you determine which output points have electronic fuses tripped?

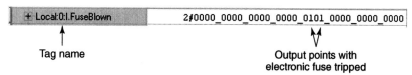

Figure 6-20 FuseBlown tag from the monitor tags controller scoped tags collection.
Used with permission Rockwell Automation, Inc.

Module Fault Tag

If a 1 is in any bit position in the Open Wire, NoLoad, or OutputVerifyFault tags, there will also be a 1 in the associated bit position in the Fault tag. The Electronic Fuse tag is a bit different because the fuses are in groups of eight. A single bit is used to represent whether a fuse has tripped in the group. If it has, then the first bit in the byte representing the group is the only bit set in the module fault word. There are also other modules that use groups to represent fault data. Refer to the module's user's manual for more information. The tags we looked at earlier as well as the Fault tag are shown in the module's input tag collection in Figure 6-21. A 1 in the Fault tag turns on the associated bit's fault light on the front of the module.

160 MODULAR CONTROLLOGIX I/O CONFIGURATION

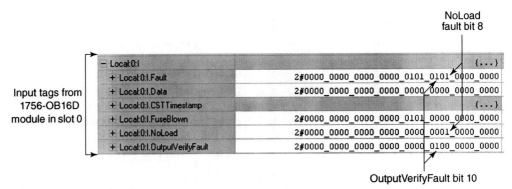

Figure 6-21 1756-OB16D output module's input tags.
Used with permission Rockwell Automation, Inc.

Communications Failure

40. _____ Near the bottom of the screen is the Communications Failure section (see Figure 6-22). If communications to this module fail when the controller is in Program mode, do you want to put the outputs in the selected Program mode state or Fault mode state? A communications fault will set all 32 bits in the fault tag regardless of the module density.

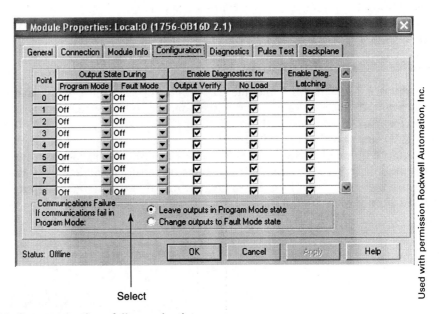

Figure 6-22 Communications failure selections.

Enable Diagnostic Latching

Diagnostic latching provides a mechanism for the controller to remember or latch onto a module fault condition and is selected by checking the Enable Diagnostic Latching box for an output point. In the default setting, all output point boxes are checked. In the case of the 1756-OD16D, a loose connection to an output point leads the no load bit to toggle between a 1 and a 0 as the connection is lost and regained. The Fault tag also toggles in a like manner with the associated FLT LED flickering on and off as the signal is made and lost. If the Enable Diagnostic Latching box for an output point is checked, on the other hand, then if the connection is lost and regained, the FLT LED for the module's output point will be latched onto until someone unlatches it.

Reset or Unlatch Diagnostic Latching or Electronic Fuse Fault

To reset latched diagnostics or an electronic fuse blown fault, click the appropriate reset button on the Diagnostics tab, as shown in Figure 6-23. Resetting these could also be accomplished by programming a message instruction on a ladder rung. Many of the tags, such as Electronic Fuse or Fault tag, cannot be directly written to using a clear instruction to zero out or clear their status using ladder logic. See Figure 6-23.

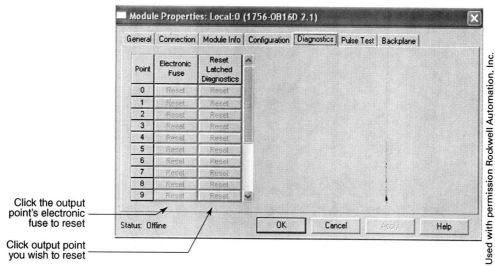

Figure 6-23 Resetting electronic fuse tag and latched diagnostics.

Pulse Test Screen

The Pulse Test feature provides the user the ability to verify that the output circuitry actually works without changing the state of the output field device. Even though the output does physically change as a result of this test, the change is too fast for the output field device to react. This feature is available on the 1756-OA8D and 1756-OB16D. Simply click the output point you wish to test. The results column registers as pass or fail as a result of the test. Figure 6-24 shows the Pulse Test screen.

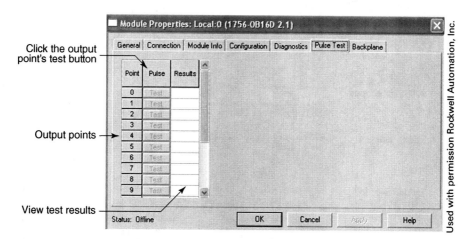

Figure 6-24 Pulse Test will determine whether module outputs are capable of switching.

Backplane Tab

As noted in Chapter 1, the ControlLogix backplane is a ControlNet network referred to as ControlBus. The Backplane tab is typically used for troubleshooting backplane problems.

Figure 6-25 illustrates the Backplane tab. This is the last tab that appears when completing the I/O configuration. There is nothing to configure on this tab, however.

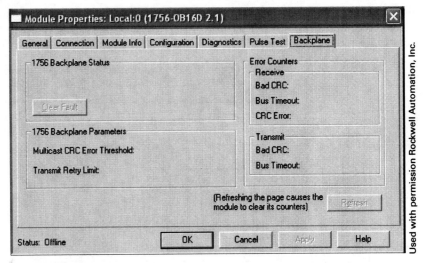

Figure 6-25 ControlBus status screen.

41. _____ Click OK when completed. Figure 6-26 illustrates how your I/O configuration should look.

Figure 6-26 Completed I/O configuration for 1756-OB16D in chassis slot 0.
Used with permission Rockwell Automation, Inc.

42. _____ Open your controller scoped tags collection and go to the Monitor Tags tab. You should see the tags created for this module, as in Figure 6-27.

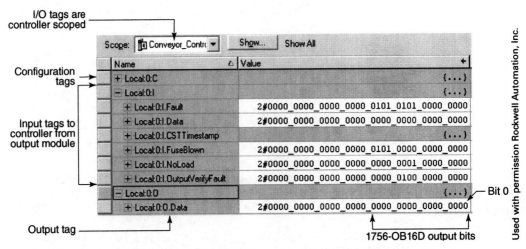

Figure 6-27 I/O tags created for the 1756-OB16D module in chassis slot 1.

43. _____ Save your work.

Before we leave this section, we will look at the tags generated with a listen only connection for an output module such as this 1756-OB16D. A listen only connection provides the programmer the option to listen to the input information or tags coming back to the controller from the module. Remember: The owner controller provides the connection and configuration data for the module. A controller with a listen only connection can only monitor or listen to the module only as long as the owner controller has communication with the module. If the owner loses the connection to the module, the listen only connection is also broken. Figure 6-28 shows the controller scoped tags we created for the output module in slot 0. There are configuration, input, and output tags for this module. Also shown are the listen only tags for another output module in slot 5. Notice that the configuration and output tags are not there. Because this controller is not the owner, there is no configuration or output tags. Because there are no output tags, this controller cannot control the outputs. This controller can listen to or use the input tags in their ladder routines.

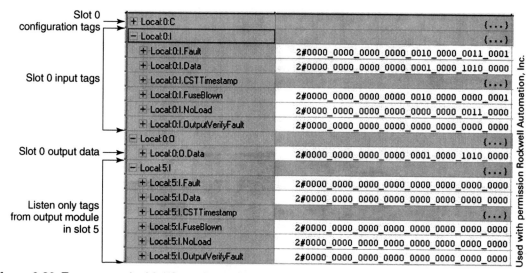

Figure 6-28 Tags created with I/O configuration.

44. _____ This completes the I/O configuration for the slot 0 1756-OB16D. Next, we configure a 1756-IB16D module in slot 2.

LAB EXERCISE 2: 1756-OB16D Configuration

This lab exercise steps you through the I/O configuration for a 1756-IB16D module in chassis slot 2. We continue working on the same project we used for configuring the output module in the last section.

1. _____ Near the bottom of the controller organizer is the I/O Configuration. Notice the output module we just configured is listed as well as the chassis size; the controller is shown in chassis slot 1.
2. _____ Right-click 1756 Backplane as illustrated in Figure 6-29 to start your configuration.
3. _____ Select New Module.
4. _____ The screen shown in Figure 6-30 should appear. There are four ways to continue from here. You can select the module from the By Category, By Vendor, or Favorites tabs, or use the Find tab to search. Keep in mind that RSLogix 5000 version 20 I/O configuration screens will look a bit different.
5. _____ Click the Find feature to search for our module.

164 MODULAR CONTROLLOGIX I/O CONFIGURATION

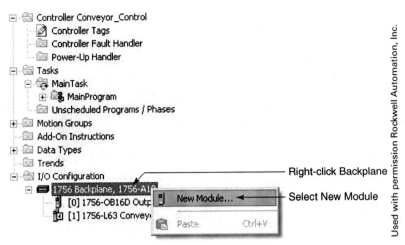

Figure 6-29 I/O configuration showing slot 0 module just configured and controller in slot 1.

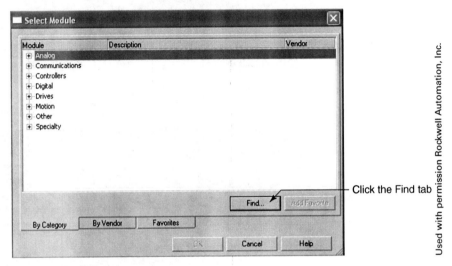

Figure 6-30 Select Module screen.

6. _____ Referring to Figure 6-31, enter the input module's part number as shown.

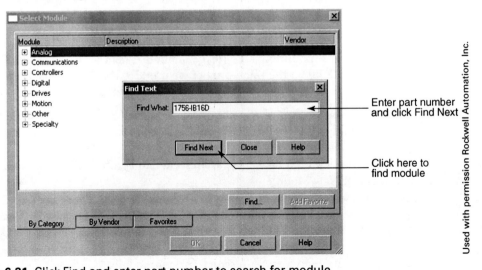

Figure 6-31 Click Find and enter part number to search for module.

7. _____ Click Find Next to locate the module.
8. _____ The module should be located as illustrated in Figure 6-32.

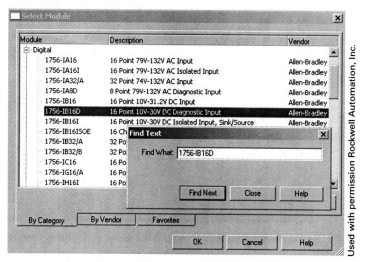

Figure 6-32 Specified module highlighted.

9. _____ Click Close on the Find Text window.
10. _____ With the module selected, click OK.
11. _____ The Select Major Revision screen should appear. One place to find the module's major and minor revision is on the label on the side of the module. Near the upper-right corner of the label is a dotted area. F/W REV is the module's firmware revision level. A 3.2 in that area signifies a major revision of 3 and minor revision 2. RSLinx could also be used to determine the module revision. After completing the I/O configuration, the Module Info tab when online lists the module firmware level on the left side of the display. Keep in mind that if the module's firmware had been upgraded in the field, the module label might not have been updated, so information might not be correct. Figure 6-33 shows selecting the module's major revision as 3. Select the correct level for your specific module.

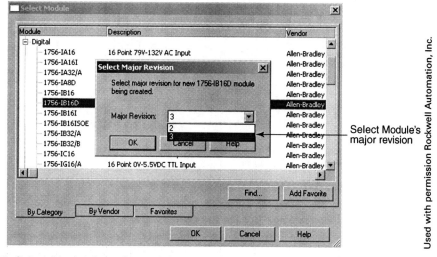

Figure 6-33 Select Module's major revision.

12. _____ After selecting your module's major revision, click OK.

MODULAR CONTROLLOGIX I/O CONFIGURATION

13. _____ The New Module window, as in Figure 6-34, should be displayed.

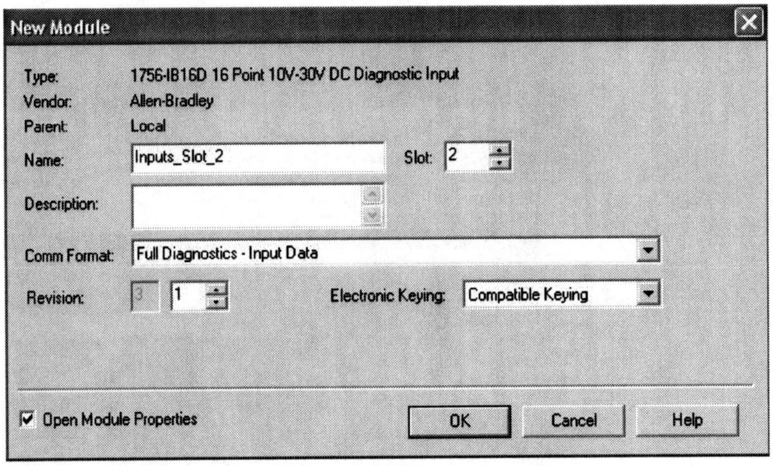

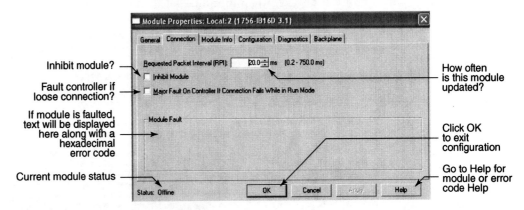

Figure 6-34 Select ownership or the module's communication format.

14. _____ Give the module a name. Remember that the name is not part of the I/O points tag.
15. _____ Select the module's slot number.
16. _____ Add a description if you wish.
17. _____ Select Full Diagnostics – Input Data as the communications format. This designates this controller as the module's owner.
18. _____ Select Electronic Keying. Compatible Keying is the most common.
19. _____ Make sure Open Module Properties is checked in the lower left-hand corner of the screen. This will display the remaining configuration screens.
20. _____ Click OK when completed with this screen.
21. _____ The module's configuration screen (Figure 6-35) should be displayed.

Figure 6-35 Select Module's properties.
Used with permission Rockwell Automation, Inc.

22. _____ Fill in this screen's information as you wish.
23. _____ Click the Module Info tab when completed. If you click OK, you will exit the configuration and not see the remaining screens.
24. _____ The Module Info screen should display. Because we are currently configuring this module, no information is displayed for it.

Figure 6-36 shows an example of a different module's Module Info screen so you can see how the screen would look for a running module. Keep in mind that this screen only has

MODULAR CONTROLLOGIX I/O CONFIGURATION   167

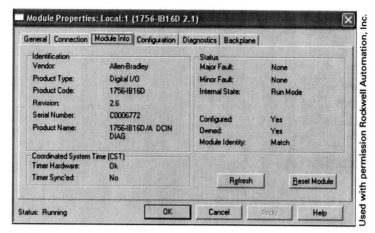

Figure 6-36 1756-IB16D read-only Module Info screen.

information when online with the Programmable Automation Controller (PAC). Note the following:

- The module's status bottom left: The module is running.
- The module revision is 2.6.
- The Status section on the right side shows that this module is in Run mode, configured, and has an owner. The module identity is a match.

25. _____ Click the Configuration screen tab.
26. _____ If you click OK, you will exit the configuration and not see the remaining screens.
27. _____ Figure 6-37 shows the configuration options for the 1756-IB16D diagnostic module, and Figure 6-38 shows the features for the 1756-IB16. The 1756-IB16 is not a diagnostic module, so it will not have the Open Wire or Enable Diagnostic Latching columns. Here we describe the features for the IB16D module. Figure 6-37 shows the default settings for all features. All boxes are checked.

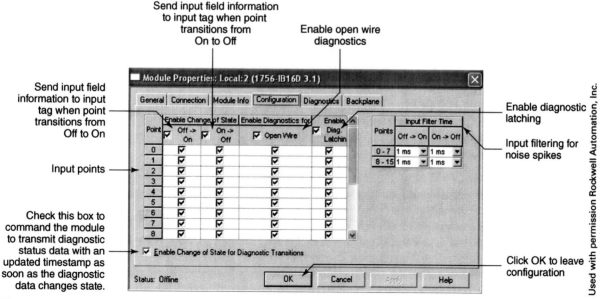

Figure 6-37 1756-IB16D Configuration screen.

168 MODULAR CONTROLLOGIX I/O CONFIGURATION

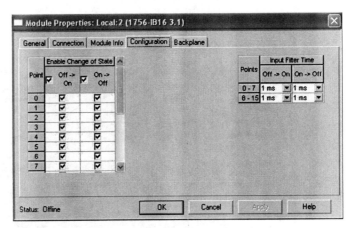

Figure 6-38 1756-IB16 has no diagnostic features.
Used with permission Rockwell Automation, Inc.

28. _____ Notice the input points are listed on the left side of the screen. The scroll bar is on the right.

ENABLE CHANGE OF STATE

29. _____ To immediately transfer input field device status to the input tag when a transition from off to on is sensed, check the Enable Change of State Off → On for that particular point. This sends immediate input data to the input tag when a field device transition is sensed, rather than waiting for the RPI to expire. This would be similar to an immediate input instruction used in other PLCs. This feature is also used to immediately trigger an event task as soon as the event occurs.

30. _____ To immediately transfer input field device status to the input tag when a transition from on to off is sensed, check the Enable Change of State On → Off. This is set up for each input. This will send immediate input data to the input tag when a field device transition is sensed rather than waiting for the RPI to expire.

31. _____ Input Filter Time is used to filter out noise spikes coming in on input signals.

The following are diagnostic features and are only available on diagnostic input modules.

Open-Wire Diagnostics

32. _____ Open Wire allows checking on a point-by-point basis for an open wire, loose connection, or broken wire. If the module detects an open circuit, a 1 is placed in the appropriate bit position in the OpenWire tag, which is shown in Figure 6-39.

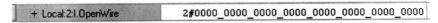

Figure 6-39 OpenWire tag for IB16D module in chassis slot 2.
Used with permission Rockwell Automation, Inc.

A 1 in the OpenWire tag also places a 1 in the Fault tag, as illustrated in Figure 6-40. Just like the output module configured earlier, a 1 in the fault tag turns on the associated fault LED on the front of the diagnostic input module.

A loose connection will result in the open-wire fault bit flickering between a 1 and a 0 as the connection is lost and then restored. To trap or latch an open-wire fault bit, check the Enable Diagnostic Latching box for the desired points. As an example, if there is a loose connection because of vibration, latching the fault bit on will ensure that the maintenance individual can verify that the problem actually occurred. We hope it is evident that unused inputs should have their associated open-wire check boxes unchecked so as to not have a continuous fault for

MODULAR CONTROLLOGIX I/O CONFIGURATION   169

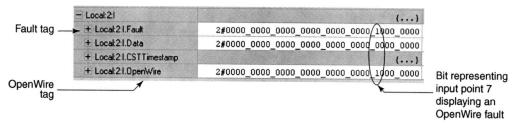

Figure 6-40 OpenWire and Fault tags for module.
Used with permission Rockwell Automation, Inc.

input points currently not used. On the other hand, when wiring additional inputs, their associated diagnostic features need to be configured because the input point is now going to be used.

33. _____ When done, click Apply to accept any changes.
34. _____ Click the Diagnostics tab.

RESET OR UNLATCH DIAGNOSTIC LATCHING

To reset latched diagnostics, click the appropriate reset button on the Diagnostics tab. Latch diagnostics cannot be directly cleared using a ladder logic clear instruction. Resetting could also be accomplished by programming a message instruction on a ladder rung.

Backplane Tab

As noted in Chapter 1, the ControlLogix backplane is a ControlNet network referred to as Control Bus. The Backplane tab is typically used for troubleshooting backplane problems. This is the last tab you see when completing the I/O configuration. There is nothing to configure on this tab.

35. _____ Click OK when completed. Figure 6-41 illustrates how your I/O configuration should look.

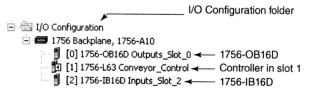

Figure 6-41 Completed I/O configuration for 1756-IB16D in chassis slot 2.
Used with permission Rockwell Automation, Inc.

36. _____ Open your controller scoped tags collection and go to the Monitor Tags tab; you should see the tags created for this module. The tags created are illustrated in Figure 6-42.

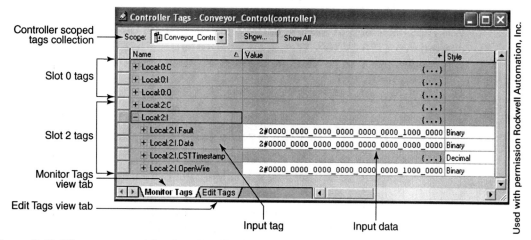

Figure 6-42 I/O tags created for the 1756-IB16D module in chassis slot 2.

37. _____ Save your work.
38. _____ This completes the I/O configuration for the slot 2 1756-IB16D.

MODULAR CONTROLLOGIX ANALOG MODULES I/O CONFIGURATION

An analog input module is an interface module that takes an analog input signal such as 0 to 10 Volts DC or 4 to 20 milliamps, converts it into a value, and stores it in a buffer on the module. When the RPI expires, the value is transferred from the buffer to the associated input tag in the controller. The controller takes the value stored in an output tag and, at the RPI, sends the value from the tag to the analog output module. The module converts the value to an analog signal such as 0 to 10 volts or 4 to 20 milliamps, to power a field device like a meter.

The conversion of the analog incoming signal to the initial value stored in the module is called the raw data. The raw value, which could be 0 to 32,767, is typically converted, or scaled, to a useable number. For example, a 0- to 10-volt input signal may be scaled to simply display 0 to 100 percent. Analog output data work in the same manner, but in reverse, from our output tag to the field device. The raw data are typically converted into meaningful application-specific information such as a flow rate (for example, gallons per minute [gal/min]), speed (for example, inches per minute), pressure (for example, pounds per square inch [psi]), or motor speed (revolutions per minute [rpm]). For instance, we receive a tag value of 0 to 1,750 rpm as data input from an operator interface terminal as a motor speed command for a variable frequency drive.

Analog scaling is part of the I/O configuration. As with the Rockwell Automation PLC-5, there are no scaling instructions included in the ControlLogix ladder logic language instruction set. Scaling for ControlLogix and PLC-5s is accomplished in the I/O configuration. The SLC 500 PLC handles analog differently as the scale with parameters (SCP) and scale (SCL) instruction included in the ladder logic instruction set.

If ControlLogix scaling is not accomplished in the I/O configuration because there are no ladder scaling instructions, there are three possible methods to use: Use math instructions to do your scaling; jump into a function block subroutine and use the scale instruction that is included with function block programming language; or build your own scaling instruction using the RSLogix 500 add-on instruction feature.

This lab exercise consists of configuring a 1756-IF6I analog input module and a 1756-OF6VI analog output module. The suggested analog modules for our modular ControlLogix are listed here:

- 1756-IF6I analog in chassis slot 4. This is a six-channel analog isolated input module. We assume a 0 to 10 VDC input signal from two or more potentiometers. We are monitoring channel 0 and channel 1 potentiometer data for the exercise.
- The analog output module in slot 5 is a six-channel isolated output module, part number 1756-OF6VI. We are going to assume a 0 to 10 VDC signal being output to two or more 0 to 10 VDC analog meters. We are controlling channel 0 and channel 1 meters for this lab.

LAB EXERCISE 3: Configuring the Analog Input Module

1. _____ Continue to use the same project used for the digital I/O configuration lab, completed in the last section, to configure the two analog modules.
2. _____ Right-click Backplane from the I/O configuration.
3. _____ Select the 1756-IF6I analog input module.
4. _____ The New Module window should display, as illustrated in Figure 6-43.

MODULAR CONTROLLOGIX I/O CONFIGURATION   171

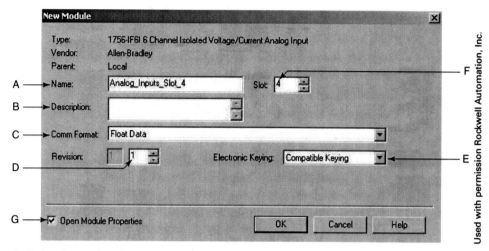

Figure 6-43 Begin configuring analog input module.

5. _____ Enter a unique name for the module in position A in Figure 6-43.
6. _____ A description can be entered in B if desired.
7. _____ The communication format (C) determines ownership data format of the module similar to the digital modules. There are two primary choices: Float Data or Integer Data. Available features are as follows:
   a. Float Data input features:
      - Independent channel configuration
      - Input range selection analog scaling configured in the I/O configuration
      - Results in a slower analog to digital conversion than integer data format
      - Latchable High–High, High, Low, and Low–Low process alarms
      - Dead band associated with process alarms
      - Chanel-by-channel configurable real-time sampling rate
      - Latchable rate alarm
      - Sensor offset
      - Notch filter
      - Digital filter
   b. Float Data output features:
      - Independent channel configuration
      - Sensor offset
      - Hold for initialization
      - Configurable output state in Program mode, Fault mode, and when communications fail
      - Latchable high and low clamping
      - Configurable ramp when transitioning to Run, Fault, and Program modes
   c. Integer Data input features:
      - The fastest analog-to-digital conversion
      - Input range selection
      - Notch filter
      - Single configurable real-time sampling rate for module

172 MODULAR CONTROLLOGIX I/O CONFIGURATION

   d. Integer Data output features:
      - Hold for initialization
      - Configurable output state in Program mode, Fault mode, and when communications fail
      - Latchable high and low clamping
      - Configurable ramp when transitioning to Run, Fault, and Program modes

For this lab, we select Float Data, so we configure the module features.

8. _____ Label D shows where to enter the minor revision level for this module. Remember this information can be found on the module label.
9. _____ E is where electronic keying is selected.
10. _____ Enter the module's slot number in position F.
11. _____ Click Open Module Properties (G), from Figure 6-43, so we can continue to configure the module. If this box is unchecked, clicking OK will accept all of the default settings for the remaining screens and complete the I/O configuration.
12. _____ Click OK when this is completed.
13. _____ The Connection tab should display, as illustrated in Figure 6-44.

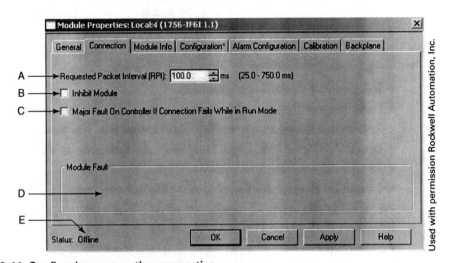

Figure 6-44 Configuring connection properties.

14. a. _____ Enter the desired RPI.
    b. _____ Do you wish to inhibit this module?
    c. _____ If the controller looses communication with this module should the controller fault?
    d. _____ If the module was faulted, a hexadecimal error code would be displayed here with some general information regarding the fault. The help screens can provide additional fault information.
    e. _____ Displays current status of the module.
15. _____ As you configure the module, the module status is offline. Click Apply, and then click the Module Info tab.
16. _____ The Module Info screen should display, as in Figure 6-45, showing the module information. Note Status in the bottom-left corner; the figure shows the module is currently in Run mode so you can see how the populated screen would look. This screen is read only and only displays information when online. When you are offline or configuring the module, the module data to the right of the categories listing are not displayed.
17. _____ Click the Configuration tab to continue.

MODULAR CONTROLLOGIX I/O CONFIGURATION   173

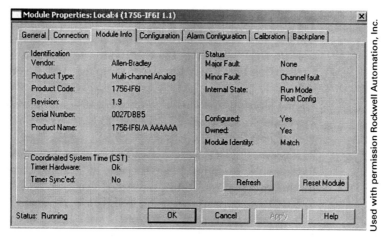

Figure 6-45 Module Info tab.

18. _____ The Configuration tab, as shown in Figure 6-46, should display so each channel's properties can be configured to the user's specifications. Each of the channel's features is outlined here.

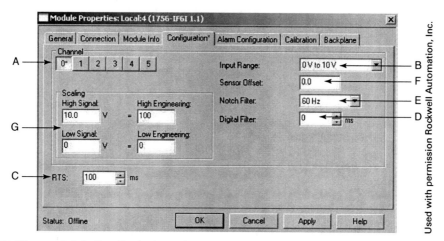

Figure 6-46 The module's Configuration tab.

a. _____ Click the channel button you wish to configure. In Figure 6-46, channel 0 has been selected.

b. _____ Click the drop-down menu, as illustrated in Figure 6-46, to select the module's analog input signal. As we are using this module and a 0- to 10-volt input signal, make the appropriate selection. The drop down in Figure 6-47 shows the input signal selections for this specific module. Refer to the ControlLogix Selection Guide for additional information on other module part numbers and their available signal selections.

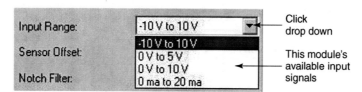

Figure 6-47 Click drop down and select the channel's input signal.
Used with permission Rockwell Automation, Inc.

174 MODULAR CONTROLLOGIX I/O CONFIGURATION

c. _____ The real-time sampling (RTS) is the time that the module samples the incoming data for each channel and stores that data in a buffer on the module. This data is sent from the module to the controller's input tags at the configured RPI on the Connection tab. Enter the signal sampling time if you wish to change from the default.

d. _____ Digital filtering smoothes input noise transients on a channel-by-channel basis.

e. _____ For example, the United States has 60 Hz electrical power, so the Notch filter is typically set for 60 Hz.

f. _____ The Sensor Offset is a correction factor where if the scaled maximum value were, for example, configured for 100 percent but only reached 98 percent when the maximum signal was input, the scaled value would be corrected. By adding the difference of approximately 2.0 in this case, the scaled value can be corrected to 100. Adjusting this number up and down a little may be required for fine-tuning to get the correct value. Keep in mind that when adjusting the maximum value up, as in the example, the bottom value is also adjusted up from 0 by about the same amount.

g. _____ The input signal's low and high values and the scaling of that signal is completed in the Scaling area. Figure 6-48 shows that channel 0 has been selected. The High and Low Signals represent the input voltage or current coming into the module. The figure shows the module's channel 0 input analog signal as 0 to 10 volts. The Low and High Engineering areas are where the scaled values are entered. The figure shows this input channel's 0- to 10-volt input signal being scaled to 0 to 100.

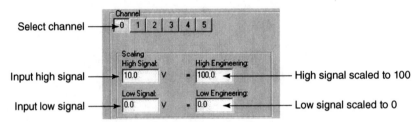

Figure 6-48 Channel 0 input high and low signals to be scaled to 0 to 100.
Used with permission Rockwell Automation, Inc.

19. _____ When you completed, click Apply and select the Alarm Configuration tab. An alarm configuration screen similar to that shown in Figure 6-49 should display.

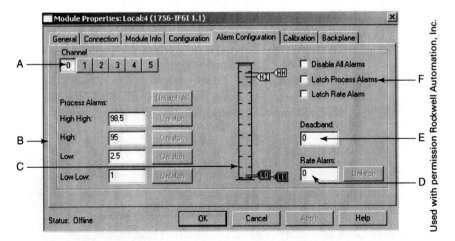

Figure 6-49 Analog alarm configuration screen.

20. _____ Configuration features for the alarm screen are listed here.
   a. _____ Select the channel you wish to configure.
   b. _____ You can configure up to four process alarms. There is a High–High alarm, a High alarm, a Low alarm, and a Low–Low alarm. The alarm values can be entered into their respective areas in the area identified as B. If the input signal goes outside, the boundaries of the set point, the associated tag become true. Figure 6-50 shows only the four-process alarm tags associated with channel 0. Currently, the input signal has exceeded the high alarm configured value, which is 95, as displayed in Figure 6-49. Remember, tags are created while performing an I/O configuration and stored in the controller scoped tags collection.

Local:4:I.Ch0HAlarm	1
Local:4:I.Ch0HHAlarm	0
Local:4:I.Ch0LAlarm	0
Local:4:I.Ch0LLAlarm	0

Figure 6-50 Process alarm tags.
Used with permission Rockwell Automation, Inc.

When the input signal exceeds any of the programmed process alarm values for channel 0, the tags in Figure 6-50 will become true. Logic has to be programmed to use this information to determine what action to take. Figure 6-51 shows a ladder rung monitoring the channel 0 high alarm tag from Figure 6-50.

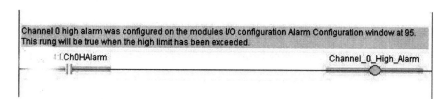

Figure 6-51 Channel 0 high alarm monitoring logic.
Used with permission Rockwell Automation, Inc.

   c. _____ The process alarm tags referred to in the last step can also be set and adjusted by simply dragging the desired pointer on the bar graph into position.
   d. _____ The Rate Alarm can be configured to alert you if the input signal changes too fast. Entering a value of 1in the rate alarm box, for example, means that if our 0- to 10-volt signal changes more than 1 volt per second, the Rate Alarm tag will become true, as on the ladder rung illustrated in Figure 6-52.

Figure 6-52 Channel 0 Rate Alarm tag monitored by ladder logic.
Used with permission Rockwell Automation, Inc.

With the Latch Rate Alarm check box checked, as in Figure 6-53, if the alarm occurs, the tag will contain a 1 and will be latched. Click Unlatch to the right of the Rate Alarm box to clear the alarm. A Message instruction can also be used to unlatch the alarm. If the Latch Rate Alarm box is not checked, the tag will be cleared when the condition clears.

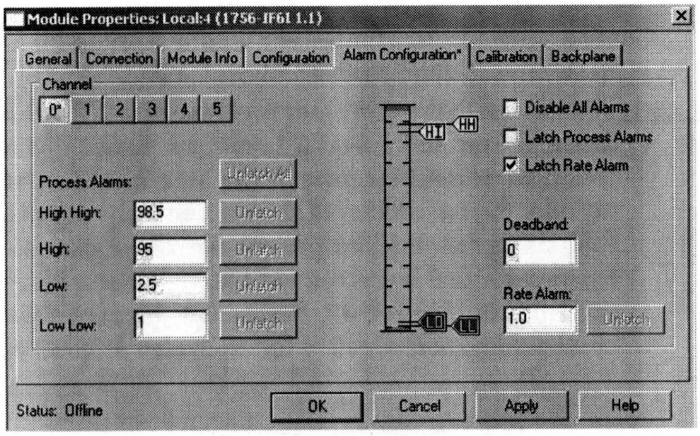

176 MODULAR CONTROLLOGIX I/O CONFIGURATION

Figure 6-53 Channel 0 Latch Rate Alarm box checked so as to latch bit on if condition exists.
Used with permission Rockwell Automation, Inc.

e. _____ Use the Dead band feature to trap when the input signal exceeds a process alarm limit but the alarm tag will not go false until the signal falls below a predefined low limit. The difference between the high and low limit is referred to as the dead band. Each of the process alarms uses the same band or value as entered into the dead-band parameter. Refer to Figure 6-54 and note the red area below the High–High (HH) and High Alarm (HI) set points, as well as the blue area below the Low (LO) and Low–Low (LL) process set point values. The colored area represents the dead band. In this example, if the input signal exceeds 80, the high process alarm tag will become true; however, the tag will not go false until the signal falls below 70, which is a dead band of 10 engineering units.

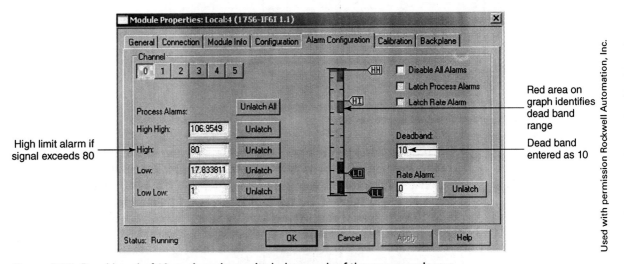

Figure 6-54 Dead band of 10 engineering units below each of the process alarms.

Refer to Figure 6-54 and note the following:

- Channel 0 high process alarm is set at 80.
- The dead band parameter is set at 10.
- Latch process alarms must not be checked or latched.
- The bar graph contains a red band of 10 units below the High–High and High alarm points. This is the dead band.
- The Local:4:I.Ch0HAlarm tag is true when the value exceeds 80, and it will not go false until the signal falls below 70.

MODULAR CONTROLLOGIX I/O CONFIGURATION

f. _____ Process or rate alarms, as discussed earlier, can be latched or unlatched by selecting or deselecting the check boxes. Selecting the top check box disables all alarms. Clicking the Unlatch button to the right of the numeric entry (B in the figure) will manually unlatch the associated process alarm. Clicking on Unlatch All will unlatch all process alarms.

21. _____ The Calibration tab is used to check the calibration status of the current module. In the event that the module needs recalibration, this tab is where calibration is carried out.

22. _____ The Backplane tab is used to monitor the status of the chassis backplane to see whether there are excessive communication errors.

23. _____ When you have finished, click OK to exit the configuration screens for this module.

24. _____ Open the controller tags collection and find the module's input tags. The list will be extensive. Look through the list and identify some of the tags we have seen in examples as we configured this module. Find the data tags as displayed in Figure 6-55. The tags may be displayed together, as in the figure, or combined with their respective channel, depending on how the column is sorted.

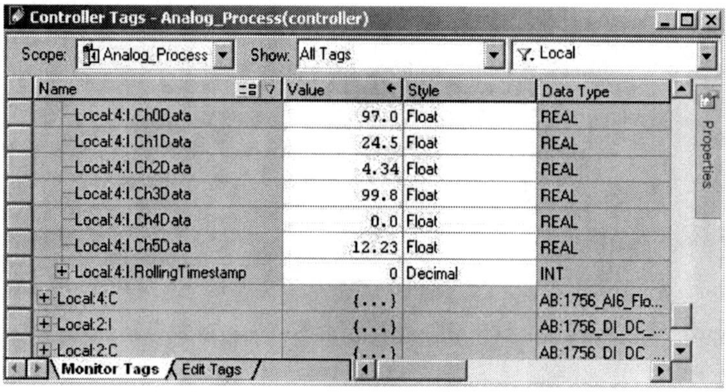

Figure 6-55 Controller scoped tags collection showing analog input tag data.
Used with permission Rockwell Automation, Inc.

1756-IF6I ANALOG INPUT MODULE CONFIGURED FOR INTEGER COMMUNICATIONS FORMAT

When the fastest analog-to-digital conversion of analog input data is desired, the integer communication format is selected when configuring the module. Figure 6-56 illustrates the General tab

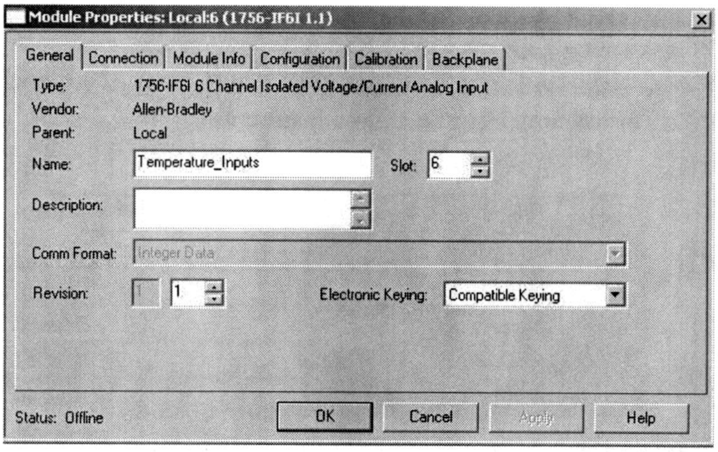

Figure 6-56 1756-IF6I General tab with integer communcations format selected.

with the communications format (Comm Format) as Integer Data. Note that the figure does not include the Alarm Configuration tab, as those features are not available in this communications format. Refer back to Figure 6-49 to view the features of the Alarm Configuration tab, which are available only for the floating point communications format.

Although the Configuration tab is available, shown in Figure 6-57, this tab also has limited feature selections. Refer back to Figure 6-46 to compare this tab with available features for the floating point configuration.

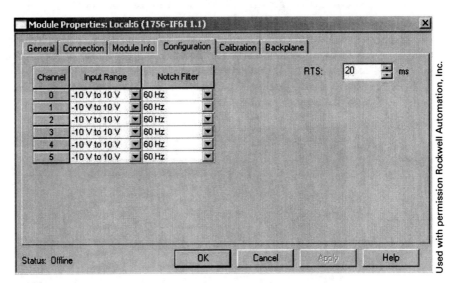

Figure 6-57 Channel configuration options for integer communcations format.

Figure 6-58 illustrates the input tags as integers. In this example, the raw data for this module are 0 to 32,767 for a 0 to 10 VDC incoming signal. If using Integer Data communications format, be sure to refer to the user manual for the analog module so you can verify the actual data range values for your specific module and input signal. The integer value transfers directly to the input tags as the scaling feature is not available for this communication format. The following options are available for scaling the analog raw input data:

- Program math instructions in ladder logic.
- Jump to a function block subroutine and use the scaling (SCL) instruction. (We work with this scaling option in Chapter 18, Introduction to the RSLogix 5000 Function Block Programming Language.)
- Create your own scaling instruction, such as the RSLogix 500 Scale with Parameters (SCP) instruction, using an add-on instruction.

⊞ Local:6:I.Ch0Data	32767	Decimal	INT
⊞ Local:6:I.Ch1Data	25000	Decimal	INT
⊞ Local:6:I.Ch2Data	3650	Decimal	INT
⊞ Local:6:I.Ch3Data	7500	Decimal	INT
⊞ Local:6:I.Ch4Data	0	Decimal	INT
⊞ Local:6:I.Ch5Data	0	Decimal	INT

Figure 6-58 Module's input tags are integers.
Used with permission Rockwell Automation, Inc.

LAB EXERCISE 4: I/O Configuration for the 1756-OF6VI

Next, we configure the analog output module. The suggested module for your lab trainer is a 1756-OF6VI, which is a six-channel isolated voltage output module.

1. _____ In the I/O configuration, right-click the Backplane.
2. _____ Select the module's part number.
3. _____ From the New Module window, as illustrated in Figure 6-59, enter a unique name for the module.

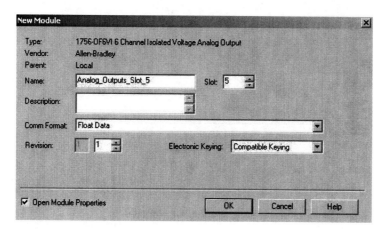

Figure 6-59 1756-OF6VI analog output module configuration.
Used with permission Rockwell Automation, Inc.

4. _____ Enter slot number.
5. _____ Select the Communications Format as Float Data. As you remember from the input module configuration, to retain all features Float is selected for the Communications Format.
6. _____ Select the desired electronic keying.
7. _____ Verify that the Open Module Properties box is checked.
8. _____ Click OK.
9. _____ The Connection tab, as shown in Figure 6-60, should open.
10. _____ Configure the options to suit your application.

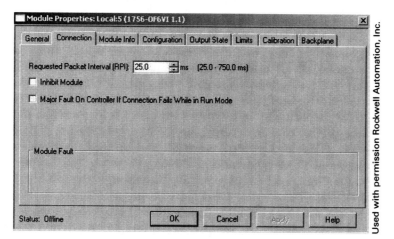

Figure 6-60 Configuring the Connection tab.

180 MODULAR CONTROLLOGIX I/O CONFIGURATION

11. _____ Click the Module Info tab. Module information is displayed in a manner similar to that in the other modules we have configured.
12. _____ Click the Configuration tab.
13. _____ Configure each channel on the Configuration tab, as shown in Figure 6-61, by clicking appropriate buttons.

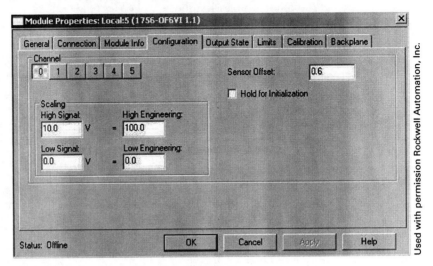

Figure 6-61 Configuring analog scaling for channel 0.

14. _____ The High Low Signals represent the analog output level. If you have a 0- to 10-volt meter, the values to be entered are 0 to 10.
15. _____ The High and Low Engineering parameters contain the scaling information. The figure shows 0 to 100. The output tag is in the range of 0 to 100 to match the scaling. A 0 to 100 value in the output tag deflects a 0- to 10-volt meter between 0 and 100 percent.
16. _____ Sensor Offset is the correction factor that can be applied similarly to the same feature on the input module.
17. _____ Hold for Initialization associates with PID loops. Checking this box causes the output to hold its present state until the value at the screw terminal is within 0.1 percent of the value commanded by the controller, thus providing a bumpless transfer.
18. _____ When this is complete, click Apply and go to the Output State tab.
    a. Refer to A in Figure 6-62 as you select the channel you wish to configure.
    b. Output State in Program Mode parameters (B): When the controller transitions to Program mode, has a major recoverable fault, or is in Remote Test mode, do you want the channel to hold its last state, or do you want it to ramp to the user-defined value entered in the User Defined Value box? Figure 6-62 illustrates the User Defined Value of 50 selected. The figure shows that when the controller transitions to Program or Remote Test mode or encounters a major recoverable fault, because the Ramp to User Defined Value check box has been selected, the analog output ramps to 50.0 at the ramp rate displayed at C. If the analog output were at a specific level and then was commanded to another level, the signal would ramp at the ramp rate too. This feature limits the rate at which an output can change, thus preventing fast changes that could cause machine damage.
    c. With the Ramp to User Defined Value selected, when the controller transitions into either of the listed operating modes, the channel output value ramps at 5.25 engineering units per second. The ramp rate is entered on the Limits tab. If the controller were in Run mode and was changed to Program mode, the current value of the output channel

MODULAR CONTROLLOGIX I/O CONFIGURATION   181

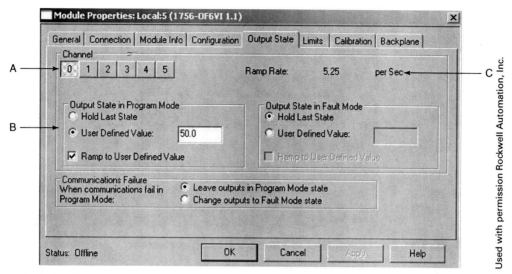

Figure 6-62 Configuring how each channel is to react when transitioning to Program mode or in Fault mode.

would ramp to 50.0 at a rate of 5.25 engineering units per second. When the controller transitioned back to Run mode, the channel output would ramp back to the original value at a rate of 5.25 engineering units per second.

d. Figure 6-63 shows the Output State in Fault Mode parameters: When the controller experiences a major nonrecoverable fault, is the channel to hold its last state or ramp to the user-defined value entered in the User Defined Value box? The Hold Last State and Ramp features operate the same as for the Output state in Program mode selections.

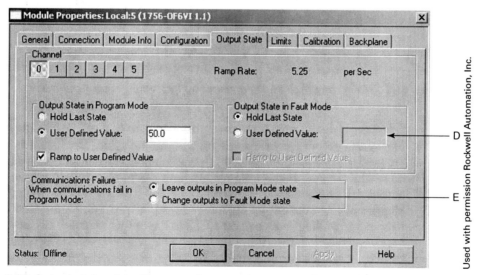

Figure 6-63 Output state when the controller experiences a major nonrecoverable fault or when communications fail.

e. Communications Failure parameters are shown in Figure 6-64: If communications were lost, would the channel transition to Program mode state or Fault mode state?

19. _____ When you have completed with the Output State tab, click Apply to accept changes and move to the Limits tab, as shown in Figure 6-65.

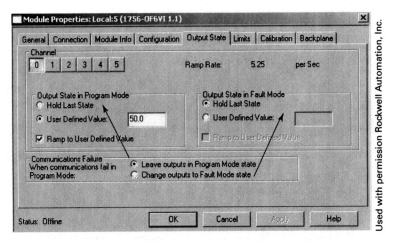

Figure 6-64 Communications failure state for output channel.

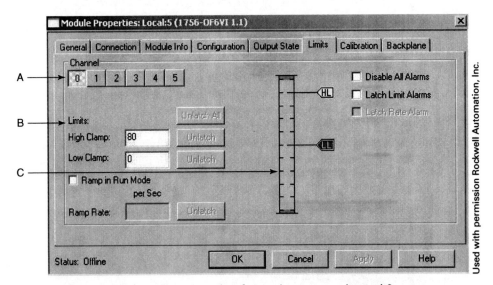

Figure 6-65 Configuring high and low clamping for analog output channel 0.

a. Select the channel you wish to configure. Each channel will be configured independently.
b. The analog output can be clamped at a maximum value. The high clamp value in the figure is 80, representing 80 percent. Being clamped, the output value will not be allowed to exceed 80 percent. In our example, we have a 0- to 10-volt output from this module. The output will be clamped at 80 percent, or 8 volts. As a result, the meter will not go above 80 percent when the output tag value exceeds 80 percent. Figure 6-66 illustrates a Move instruction sending a value of 82.10739 to the output channel. Because the output value has been exceeded, the High Limit Alarm tag will be true. As the result of the clamping, the output is not allowed to exceed 8.0 volts.

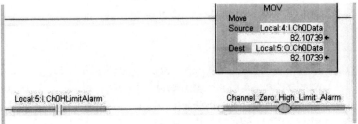

Figure 6-66 Channel 0 high limit alarm logic.
Used with permission Rockwell Automation, Inc.

c. This is a graphical representation of the values entered into the High Clamp and Low Clamp boxes. If desired, rather than entering a value in the boxes, use the mouse to position the HL or LL positions on the graphical representation picture. Notice that Figure 6-67 shows the Latch Limit Alarms check box is checked in the upper-right-hand portion of the window. With this box checked, if either clamp value is exceeded, the associated tag will be latched true. Being latched, if the value returns to within the acceptable limits, the alarm tag will not go false until the Unlatch button to the right of the High or Low Clamp value entry boxes is clicked. If the Latch Limit Alarms box is unchecked, the alarm tag will go false when the value returns to within limits. The Unlatch All button unlatches both alarms. Rather than having to open the software and use a mouse to click the Unlatch alarms buttons, a message instruction could be programmed on a ladder rung and configured to unlatch the alarms. A touchscreen object or other input device could trigger the message instruction and unlatch the alarms.

Figure 6-67 Configuring limits alarm latching.

If the output value ramps, or changes too rapidly, the Ramp in Run mode feature can trap the rapid change. Figure 6-68 shows the Ramp in Run Mode box checked and a 5.0 entered in the Ramp Rate box. Because this module was scaled as 0 to 100, if the output value for channel 0 changes faster than 5 engineering units per second, the Local:5:I.Ch0RampAlarm tag will become true. If the ramp rate changes too fast and the Latch Rate Alarms box is checked, as Figure 6-68 shows, the tag will be latched until cleared. Latching and unlatching the rate alarm works in the same manner as the High Limit Alarm described above.

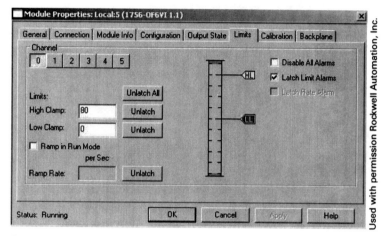

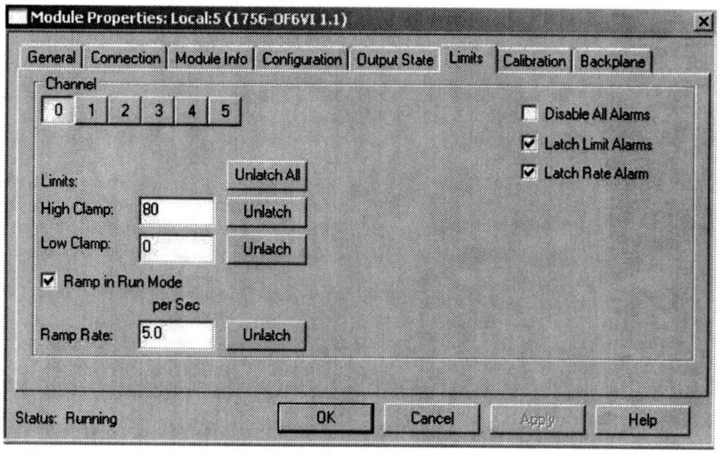

Figure 6-68 Configuring Ramp in Run mode and alarm latching.

Figure 6-69 shows a ladder rung using the Ramp Alarm tag to alert of a sitution where the ramp rate is changing too fast.

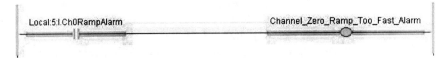

Figure 6-69 Above channel 0 ramp, too fast alarm logic.
Used with permission Rockwell Automation, Inc.

Click Apply to accept changes. When configuring the module has been completed, analog output data can be viewed in the Monitor Tags view of the controller scoped tags collection, as illustrated in Figure 6-70.

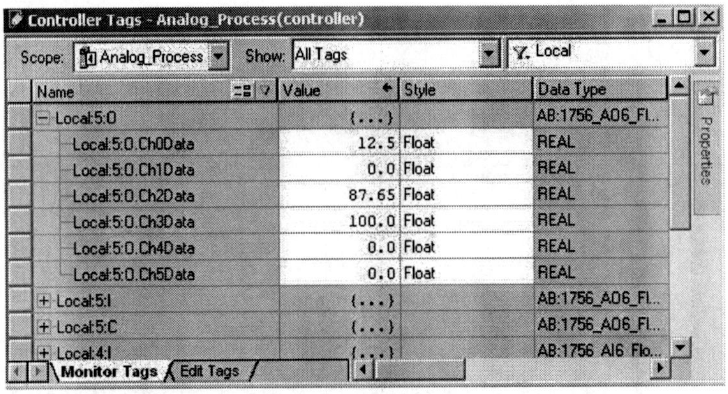

Figure 6-70 Slot 5 analog output module tags as displayed in the controller scoped tags collection.
Used with permission Rockwell Automation, Inc.

1756-OF6VI ANALOG OUTPUT MODULE CONFIGURED FOR INTEGER COMMUNICATIONS FORMAT

The Configuration tab associated with the floating point communications format is where channel scaling for this module was configured. When integer format is selected, scaling of the outputs is not configurable in the I/O configuration. The Configuration tab illustrated in Figure 6-71 combines the available features from the floating point configuration tab and output tab. Even though the format of this Configuration tab is a bit different, all of the output state tab features are present.

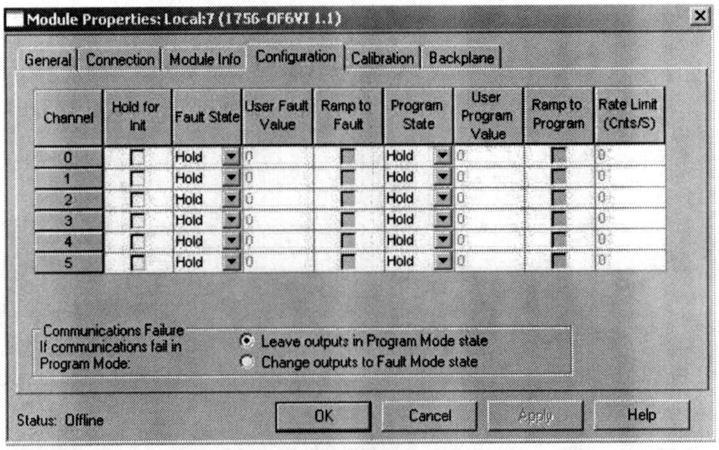

Figure 6-71 Configuration tab for 1756-OF6VI integer communications format.

Used with permission Rockwell Automation, Inc.

Figure 6-72 shows the output tags from the controller scoped tags collection; even though the tags are the same, the output tag data are in integer format. For this module, an output tag value of 0 to 32,767 outputs a 0- to 10-volt DC signal to the field device connected to that channel.

⊟ Local:7:O	{...}	AB:17...
⊞ Local:7:O.Ch0Data	12500 Decimal	INT
⊞ Local:7:O.Ch1Data	32767 Decimal	INT
⊞ Local:7:O.Ch2Data	22252 Decimal	INT
⊞ Local:7:O.Ch3Data	0 Decimal	INT
⊞ Local:7:O.Ch4Data	0 Decimal	INT
⊞ Local:7:O.Ch5Data	0 Decimal	INT

Figure 6-72 Output tags for 1756-OF6VI configured for integer communications.
Used with permission Rockwell Automation, Inc.

LAB EXERCISE 5: Modular ControlLogix Analog I/O Configuration

INTRODUCTION

This exercise provides practice opening a project and configuring analog inputs and outputs. After configuring communications, download the project and go into Run mode. While in Run mode, you will turn the potentiometers, monitor analog tags, and view the meters as they respond to different analog features.

EQUIPMENT NEEDED TO COMPLETE THIS LAB

We assume the minimum chassis configuration as we complete this lab exercise:

- 16-point output module in chassis slot 0
- 1756-IF6I analog input module in chassis slot 4
- 1756-OF6VI analog output module in chassis slot 5
- Two potentiometers sending a 0 to 10 VDC input signal into input channel 0 and channel 1 of the analog input module

We also assume the following conditions have been met:

- The slot 5 analog output module channels 0 and 1, providing a 0 to 10 VDC signal to each of two meters.
- You have connected the proper communications cables between your personal computer and PLC.
- Your instructor has demonstrated configuring personal computer to ControlLogix communications options for your classroom lab trainer unit, and you are able to configure communications, set the RSLogix 5000 communications path, download your project, and put the controller into Run mode. Refer to Section 2 of this book for lessons on communicating communications.

ANALOG MODULAR I/O CONFIGURATION

1. _____ Open the ControlLogix Analog I/O Configuration project.
2. _____ Add the analog input module to the I/O configuration.
3. _____ Go to the I/O configuration and begin the configuration for the 1756-IF6I analog input module.
   a. Give the module a unique name.
   b. Select the slot number.

186 MODULAR CONTROLLOGIX I/O CONFIGURATION

     c. Communications Format = Float Data.
     d. Select compatible keying.

Note: Make sure you click Apply before moving to another Configuration tab.

4. _____ Configure input channel 0 using the following specifications:
   a. Input range 0 to 10 volts
   b. High signal =10
   c. Low signal = 0
   d. Low engineering = 0
   e. High engineering = 100
   f. Alarm configuration:
      - High–High alarm = 95
      - High alarm = 80

5. _____ Configure input channel 1 with the following specifications:
   a. Input range 0 to 10 volts
   b. High signal =10
   c. Low signal = 0
   d. Low engineering = 0
   e. High engineering = 250
   f. Alarm configuration:
      - High–High alarm = 225
      - High alarm = 200
      - Check the box to latch process alarms.

6. _____ Click OK when you are done.
7. _____ Add the output module to the I/O Configuration.
   a. Give the module a unique name.
   b. Select the slot number.
   c. Communications Format = Float Data
   d. Select compatible keying.

Note: Make sure you click Apply before moving to another Configuration tab.

8. _____ Configure output channel 0 with the following specifications:
   a. Output range 0 to 10 volts
   b. Low engineering = 0
   c. High engineering = 100
   d. Verify limits are at the minimum and maximum.

9. _____ Configure output channel 1 with the following specifications:
   a. Output range 0 to 10 volts
   b. Low engineering = 0
   c. High engineering = 250
   d. Clamp output at 80 percent.
   e. Latch limit alarm.
10. _____ Verify the proper communications cables are connected for the desired communication method to download your project.
11. _____ Download your project.
12. _____ Put the controller into Run mode.

13. _____ Where do you go to monitor the I/O tags?
14. _____ Referring to the answer to the previous question, which tab should you select to monitor the tags created as a result of the I/O configuration you created in this lesson?
15. _____ Record the tag to monitor the potentiometer value for channel 0.
16. _____ Find and monitor that tag. As you turn the potentiometer from 0 to 100 percent, what values do you see as you monitor the tag?
17. _____ Monitoring the channel 0 meter, the needle should go from 0 to approximately 10, or 100 percent, as the potentiometer is turned.
18. _____ Why is the tag displaying those particular numbers?
19. _____ As you turn the channel 0 potentiometer, monitor the tag Local:4:I.Ch0Data. Output Light 2 on your lab trainer should go true when the tag value and the meter needle exceed 80, as the high process alarm has been exceeded.
20. _____ Continue turning the potentiometer until the tag value exceeds 95. Output Light 3 on your lab trainer should turn on when the analog tag value of 95 is exceeded.
21. _____ Slowly turn the potentiometer so that the tag value goes below 95 and 80. Why do the output lights go off?
22. _____ As you monitor the tag, turn the channel 1 potentiometer from 0 to 100 percent.
23. _____ Explain what tag values are being displayed and where the numbers come from.
24. _____ Output light Local:0:O.Data.4 should be on.
25. _____ Turn the potentiometer down from 100 percent back to 0.
26. _____ Did the light turn off? Explain your answer.
27. _____ Where do you go and how do you clear the process alarm so that the output light turns off when it is below the limit?
28. _____ Referring to question 27, there is another option for clearing the alarm. What is it?
29. _____ Turn both potentiometers to 0.
30. _____ Go to the analog output module's Limits tab for channel 0.
31. _____ Set the high clamp to 75.
32. _____ Click Apply.
33. _____ Because the controller is still in Run mode, a window displays that asks you to apply module configuration to the module.
34. _____ Turn the channel 0 potentiometer from 0 to 100 percent.
35. _____ As you monitor the meter, explain what you observe.
36. _____ Turn the channel 0 potentiometer to 0.
37. _____ What would happen if you modified the Limits tab for channel 0 and checked the Latch Limit Alarms check box?
38. _____ Make the modification and test your answer. Turn the potentiometer from 0 to 100 percent and then back to 0. The output light Local:0:O.Data.5 should turn on and latch when the clamp value is exceeded. Even though currently the potentiometer input tag value has fallen below the threshold, the tag is latched and output remains illuminated.
39. _____ Unlatch the High Clamp Alarm tag. Light 5 should turn off.
40. _____ Return clamping to the minimum and maximum values for both channels.
41. _____ Turn the potentiometer for channel 0 to 100 percent.
42. _____ Configure output channel 0 to ramp to the user-defined rate of 75 when the controller transitions to Program mode. Configure a ramp rate of 5 per second.
43. _____ Test that your configured ramping operates correctly when you change the controller to Program mode. As you monitor the meter, describe what you observe.
44. _____ While observing the meter, describe what happens when you put the controller back into Run mode.
45. _____ Turn channel potentiometer to 100 percent. The meter should go to 10.

46. _____ Go back and configure channel 0 output module's Limits tab to ramp in Run mode.
47. _____ Click Apply when you have finished, and click OK to accept the change.
48. _____ Quickly turn the potentiometer to 0 percent. Monitor the channel 0 meter and explain what you observe.
49. _____ What do you think would happen if you rapidly turned the potentiometer to a value like 60 percent?
50. _____ Turn the potentiometer to a value of your choosing to test your answer.
51. _____ Modify the channel 0 output properties to not ramp in Run mode.
52. _____ To configure an alarm if an analog input changes too quickly, go to the Alarm Configuration for input channel 0.
53. _____ While in Run mode, enter a value of 5 for the rate alarm.
54. _____ Check the Latch Rate Alarm box.
55. _____ Click Apply and then Yes to accept your changes.
56. _____ Turn the potentiometer for Channel 0 rapidly. Did output light 6 illuminate on your lab trainer?
57. _____ What tag would you monitor to view the bit that was set as a result of this alarm?
58. _____ How would you reset the alarm and clear the alarm bit?
59. _____ Reset the alarm condition and verify the bit was reset to 0.
60. _____ Save your project.

SUMMARY

I/O configuration in the ControlLogix systems is a manual process completed by a programmer or maintenance or electrical individual when adding to or modifying the current system hardware. Because there are many options when configuring I/O modules, the process is manual so as to provide programming individuals the opportunity to configure their systems to their exact specifications. As a result, there is no read I/O configuration feature with RSLogix 5000 software as would be found in Rockwell Automation's SLC 500 system and its associated RSLogix 500 software.

When creating a new project, the I/O modules that are used must be manually configured by performing an I/O configuration. Until the modules are configured, there are no I/O tags available in the Monitor or Edit tags views of the controller scoped tags collection. When adding a new I/O module to an existing RSLogix 5000 project, the module must be configured before its I/O tags are available for programming.

If you are using diagnostic modules and wire up additional inputs or outputs, the I/O configuration may need to be updated to turn on the diagnostic features for the newly installed I/O devices.

Modular ControlLogix provides the option to have multiple owners for input modules, whereas only one owner is allowed for output modules. With ControlLogix, one cannot assume the I/O modules in any chassis are necessarily controlled or owned by the controller in that chassis. Because ControlLogix is a multicontroller system, only one controller can own an output module.

REVIEW QUESTIONS

Note: For ease of handing in assignments, students are to answer using their own paper.

1. The _____ is the time you are requesting input information to be transferred from an I/O module and sent to the associated input tags.
2. Define Communications Fault mode. _____
3. Why would you want to check the box to have a major fault on the controller if communications fail while in Run mode?

MODULAR CONTROLLOGIX I/O CONFIGURATION   189

4. What is the RPI? _____
5. Until the modules are _____, there are no I/O tags available in the Monitor or Edit Tags views of the controller scoped tags collection.
6. The module _____ is not part of the I/O points tags associated with the module's input points.
7. Clicking _____ turns the module off as far as this controller is concerned.
8. Inhibiting a module is typically done during _____ or _____.
9. If the module is faulted, there will be a _____ error code and explanation in the module fault box. More information regarding the fault can be found by clicking the _____ button.
10. If there were an error code of 16# ff0b on the Connection tab for the 1756-OB16D module in slot 2, using your RSLogix 5000 software where would you go to find more information on the fault? _____
11. Using the answer from the previous question, use your software to find and write down the problem. _____
12. What is electronic keying? _____
13. What is a compatible module when setting up electronic keying? _____
14. When adding a new I/O module to an existing RSLogix 5000 project, the module has to be configured before its _____ are available for programming.
15. Explain Electronic Keying, Exact Match. _____
16. Interpret the tags in the 1756-OB16D module in Figure 6-73 and answer the following questions.

Name	Value
+ Local:0:C	(...)
− Local:0:I	(...)
+ Local:0:I.Fault	2#0000_0000_0000_0000_0010_0000_0011_0001
+ Local:0:I.Data	2#0000_0000_0000_0000_0001_0000_1010_0000
+ Local:0:I.CSTTimestamp	(...)
+ Local:0:I.FuseBlown	2#0000_0000_0000_0000_0010_0000_0000_0001
+ Local:0:I.NoLoad	2#0000_0000_0000_0000_0000_0000_0011_0000
+ Local:0:I.OutputVerifyFault	2#0000_0000_0000_0000_0000_0000_0000_0000
− Local:0:O	(...)
+ Local:0:O.Data	2#0000_0000_0000_0000_0001_0000_1010_0000

Figure 6-73 Question 16 tags interpretation.
Used with permission Rockwell Automation, Inc.

a. Looking at the tags, is this module in a remote or local chassis? _____
b. In what slot does this module reside? _____
c. Describe any fuse blown problems. _____
d. List any no load problems you see. _____
e. Identify any OutputVerifyFault problems. _____
f. The tag Local:0:I.Fault has several 1s in it. Explain what these set bits indicate. _____
g. The tag Local:0:I.Data has several set bits. Explain what this tag and these bits signify. _____
h. The tag Local:0:O.Data has several set bits. Explain what this tag and these bits signify. _____

17. Define Disable Keying. _____
18. When selecting the module to configure, modules can be selected by _____, _____, or from the Favorites tab.
19. What are the selections when configuring an output module's state when the controller transitions to Program mode? _____

20. When configuring an output module's state when the controller transitions to Program mode, is the configuration set up for the module or for each point separately? _____
21. If you are using diagnostic modules and wire additional inputs or outputs, the _____ may need to be updated so as to turn on the diagnostic features for the newly installed I/O devices.
22. Interpret the input tags for a 1756-OB16D module. Refer to Figure 6-74.

Name	Value
− Local:0:I	{...}
+ Local:0:I.Fault	2#0000_0000_0000_0000_1111_1111_1100_0000
+ Local:0:I.Data	2#0000_0000_0000_0000_0000_0000_0001_0010
+ Local:0:I.CSTTimestamp	{...}
+ Local:0:I.FuseBlown	2#0000_0000_0000_0000_0000_0000_0000_0000
+ Local:0:I.NoLoad	2#0000_0000_0000_0000_1111_1111_1100_0000
+ Local:0:I.OutputVerifyFault	2#0000_0000_0000_0000_0000_0000_0000_0000

Figure 6-74 1756-OB16D input tags.
Used with permission Rockwell Automation, Inc.

  a. Referring to the tags, do you see any problems? If so explian what a common cause to this problem might be. _____
  b. Looking at the front of this module, explain what you will see regarding this problem. _____
  c. What actions can you take to correct the problem? _____
23. As you interpret the I/O configuration in Figure 6-75, answer the following questions.

Name	Value
+ Local:0:C	{...}
− Local:0:I	{...}
+ Local:0:I.Fault	2#0000_0000_0000_0000_0000_1100_0000_1010
+ Local:0:I.Data	2#0000_0000_0000_0000_0011_0000_0100_0000
+ Local:0:I.CSTTimestamp	{...}
+ Local:0:I.FuseBlown	2#0000_0000_0000_0000_0000_0100_0000_0010
+ Local:0:I.NoLoad	2#0000_0000_0000_0000_0000_1000_0000_0000
+ Local:0:I.OutputVerifyFault	2#0000_0000_0000_0000_0000_0000_0000_1000
− Local:0:O	{...}
+ Local:0:O.Data	2#0000_0000_0000_0000_0011_0000_0100_0000
+ Local:2:C	{...}
− Local:2:I	{...}
+ Local:2:I.Fault	2#0000_0000_0000_0000_0001_0000_0000_0000
+ Local:2:I.Data	2#0000_0000_0000_0000_0000_0001_0000_0110
+ Local:2:I.CSTTimestamp	{...}
+ Local:2:I.OpenWire	2#0000_0000_0000_0000_0001_0000_0000_0000

Used with permission Rockwell Automation, Inc.

Figure 6-75 I/O configuration or 1756-OB16D in slot 0 and 1756-IB16D in slot 1.

  a. What is the output tag name? _____
  b. What output points are on? _____
  c. What inputs are currently energized? _____
  d. List any problems you see regarding our inputs. _____
  e. Identify any problems you see with the outputs. _____
  f. Looking at the output module in slot 0, what LEDs would you expect to see on? _____
  g. Specifically, what does the 1 in the Local:2:I.Fault tag cause to happen? _____
24. A controller with a listen only connection can only monitor or listen to the module only as long as the _____ has communication with the module.

25. An analog input module is an interface module that takes an analog input signal such as a 0- to 10-volt DC signal, converts it into a value, and stores it in a _____ on the module.
26. Explain the real-time sampling rate.
27. The analog output module takes the value stored in an output tag, which is either an integer or a real number, and at the_____ sends the value from the tag to the analog output module.
28. The analog output module converts the tag value to an analog value, such as 0 to 10 volts or 4 to 20 milliamps, and sends it to the _____.
29. If we check the box associated with Major fault on the controller, then if connection fails in the Run mode box and there is a loss of communications between the module and the _____, the controller will _____.
30. The conversion of the analog incoming signal to the initial value stored in the module is called the _____.
31. The analog input value from the field devices is stored in the module buffer and transferred to the associated input tag in the controller when the module's _____ expires.
32. The raw data are typically _____ into meaningful application-specific information such as a flow rate (for example, gallons per minute [gal/min], speed like inches per minute, pressure (for example, pounds per square inch [psi]), or motor speed (revolutions per minute [rpm]).
33. Where does one set up analog scaling?
34. Why would the scaling feature not be available in the analog module's configuration windows?
35. If the analog scaling feature was not available in the I/O configuration, or data in need of scaling were transferred across a network, what would the scaling options be?
36. To answer the following questions, refer to Figure 6-76.

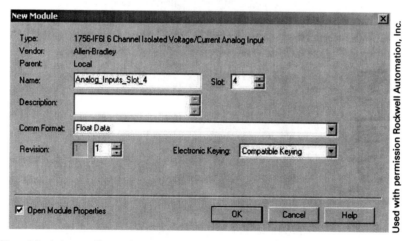

Figure 6-76 New Module configuration window for question 36.

    a. What module are we configuring?
    b. Explain the correlation between the name you give to the module and the I/O tag associated with this module, assuming the module resides in the local chassis.
    c. The module's major revision is displayed as "1," and is grayed out. If this module needed to be replaced with a newer module that was a major revision 2, would the newer module be considered a compatible module?
    d. In response to the situation referred to in question c, how would you fix the problem if using the newer module were the only option?
    e. The module revision is listed as major revision 1 and the minor revision as 1. If the module were to fail and an older spare with a revision of 1.0 was the only replacement, would this replacement module be considered compatible? Why or why not?

192 MODULAR CONTROLLOGIX I/O CONFIGURATION

37. To answer the following questions, refer to Figure 6-77.

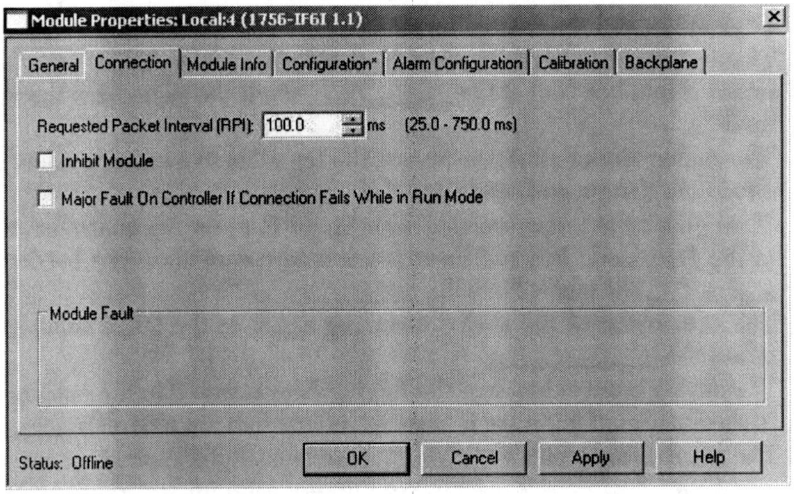

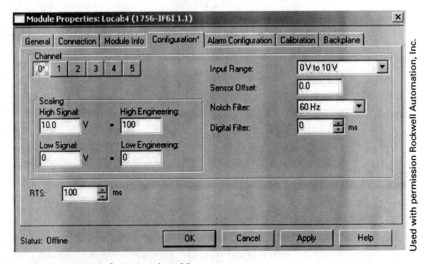

Figure 6-77 Configuration for question 37.

    a. Explain how the RPI works in association with the RTS for an analog module.
    b. What happens if we inhibit this module?
    c. If we check the box associated with Major fault on controller, then if connection fails in Run mode and there is a loss of communication between the module and the owner controller, after three RPIs the controller will fault. Explain specifically what will happen if the same communications loss occurs and the check box is unchecked.
    d. Where can one look to find additional information regarding fault information displayed?

38. Refer to Figure 6-78 as you answer the following questions.

Figure 6-78 Configuration tab for question 38.

    a. How do the High and Low signals in the scaling area correlate to the input range selection in the upper-right-hand section of this window?
    b. Also in the scaling portion of the window, explain the function of the High Engineering and Low Engineering entries.
    c. What is RTS? Explain your answer.

39. Refer to Figure 6-79 as you answer the following questions.

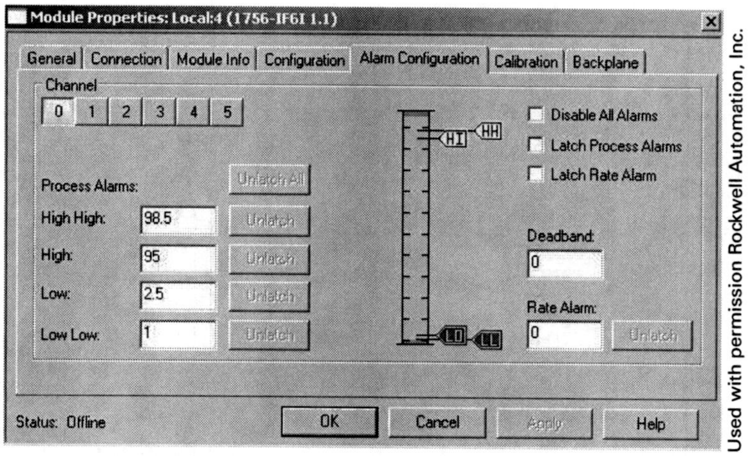

Figure 6-79 Analog alarm configuration screen.

   a. Explain the function of the process alarms.
   b. What would happen if the Disable All Alarms check box were not checked?
   c. Explain the function of checking the Latch Process Alarms box.
   d. How does the rate alarm work? What tag would you program to monitor this condition for channel 1?
40. Draw a rung below to monitor the channel 3 low alarm if the value were to fall below the programmed Low alarm value in the figure.
41. The following questions pertain to Figure 6-80.

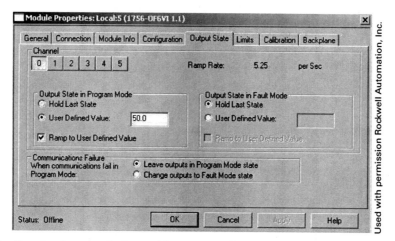

Figure 6-80 Configuring how each channel is to react when transitioning to Program mode or in Fault mode.

   a. What is the module part number for the module being configured from Figure 6-80?
   b. Explain the four conditions in which the Output State in Program Mode feature selections will be used.
   c. Explain the conditions in which the Output State in Fault Mode feature selections will be used.

42. Answer the following questions as you refer to Figure 6-81.

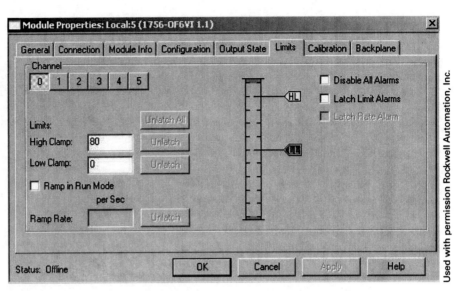

Figure 6-81 Configuring high and low clamping for analog output channel 0.

a. If the current channel were scaled as 0 to 100, explain how the high and low clamp features would operate in conjunction with the output signal.
b. Explain how the module will behave if the latch limit alarms check box is checked.
c. What does the ramp in Run mode feature do?
d. What unit is the ramp rate per second?
e. List below the tag that would be used to monitor channel 4 of the output module in chassis slot 8 if the ramp rate changed too rapidly.
f. List the options to unlatch the ramp rate alarm.

CHAPTER

7

CompactLogix I/O Configuration

OBJECTIVES

After completing this lesson, you should be able to

- Understand terminology associated with modules and their configuration.
- Perform digital I/O module configuration for either a 1768 or 1769 CompactLogix.
- Perform an analog I/O module configuration for either a 1768 or 1769 CompactLogix.

INTRODUCTION

As you remember from Chapter 1, the CompactLogix members of the ControlLogix family include the 1768 and the 1769 CompactLogix platforms. Both platforms use the same 1769 I/O modules.

When a user creates a new project, the I/O modules to be used must be manually configured by performing an I/O configuration. Until the modules are configured, no I/O tags are available in the Monitor or Edit Tags view of the controller scoped tags collection. If a new I/O module is being added to an existing RSLogix 5000 project, then the module must be configured before its I/O tags are available for programming.

LAB EXERCISE 1: 1769 CompactLogix Modular Digital I/O Configuration

Complete this section if you are using a 1769 CompactLogix. If you are using a 1768 CompactLogix, then skip to the section titled "1768 CompactLogix Modular Digital I/O Configuration."

EQUIPMENT NEEDED TO COMPLETE THIS LAB

This lab exercise guides you through completing an I/O configuration using a 1769 CompactLogix that uses RSLogix 5000 software. For this lab, we assume you are using the following CompactLogix hardware:

- 1769-L35E processor
- RSLogix 5000 software version 18 (used to create this lab, but if you have a different version, then the steps to complete the lab and the software screens, as you will see, will be similar)
- 1769-IQ16 input module in slot 1
- 1769-OB16 output module in slot 2
- 1769-IF4 analog module in slot 3
- 1769-OF2 analog output module in slot 4

See your instructor for specific instructions if you do not have this particular hardware.

THE LAB

1. _____ Open your RSLogix 5000 software.
2. _____ Create a new project. Refer to Figure 7-1.

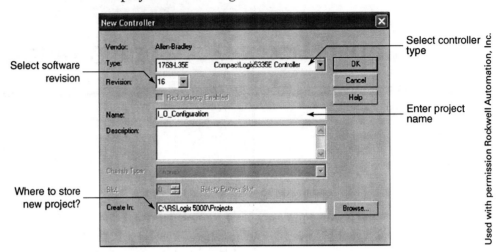

Figure 7-1 Starting a new 1769 CompactLogix project.

3. _____ Select your specific controller type.
4. _____ Select your software version.
5. _____ Give the project the name "Configuration_1769."
6. _____ Change the location to store your project if needed.
7. _____ Click OK when done.
8. _____ The bottom of the controller organizer is where the I/O configuration will be found. Refer to Figure 7-2 as you start your configuration.

Figure 7-2 Starting 1769 CompactLogix I/O configuration.

9. _____ Right-click CompactBus Local and select New Module.
10. _____ To continue the configuration, skip to "Selecting CompactLogix I/O Modules."

LAB EXERCISE 2: 1768 CompactLogix Modular Digital I/O Configuration

Complete this section if you are using a 1768 CompactLogix.

EQUIPMENT NEEDED TO COMPLETE THIS LAB

This lab exercise guides you through completing an I/O configuration using a 1768 CompactLogix using RSLogix 5000 software. For this lab, we assume you are using the following CompactLogix hardware:

- 1768-L43 controller
- RSLogix 5000 software version 18 (was used to create this lab, but if you have a different version, then the steps to complete the lab and the software screens you will see will be similar)
- 1769-IQ16 input module in slot 1
- 1769-OB16 output module in slot 2
- 1769-IF4 analog module in slot 3
- 1769-OF2 analog output module in slot 4

See your instructor for specific instructions if you do not have this particular hardware.

THE LAB

1. _____ Open your RSLogix 5000 software.
2. _____ Create a new project. Refer to Figure 7-3.
3. _____ Select your specific controller type.

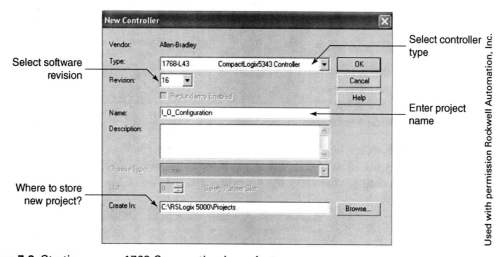

Figure 7-3 Starting a new 1768 CompactLogix project.

4. _____ Select your software version.
5. _____ Give the project the name "Configuration_1768."
6. _____ Change the location to store your project if needed.
7. _____ Click OK when done.
8. _____ The bottom of the controller organizer is where the I/O configuration can be found. Refer to Figure 7-4 as you start your configuration.
9. _____ Right-click 1769 Bus and select New Module.
10. _____ Continue to the next section.

198 COMPACTLOGIX I/O CONFIGURATION

Figure 7-4 Starting 1768 CompactLogix I/O configuration.
Used with permission Rockwell Automation, Inc.

SELECTING COMPACTLOGIX I/O MODULES

1. _____ A screen like that in Figure 7-5 appears. Figure 7-5 shows the Select Module for later software revisions through version 19. Starting with RSLogix 5000 version 20, the window will be different. Carefully studying the window, you will see the same information in a slightly different format. There are three ways to continue from here. First, you can select the module by category from the list, as illustrated in the figure. This is the select By Category tab. Second, you can select the By Vendor tab to select by manufacturer. Third, you can select the Favorites tab.

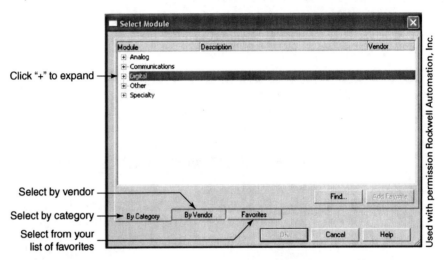

Figure 7-5 Module selection options.

2. _____ For now, select the By Category tab and click + in front of Digital, as shown in Figure 7-5.
3. _____ A list of available digital modules as shown in Figure 7-6 should appear. We will configure the 1769-IQ16 module in slot 1. Select the module, as shown in Figure 7-6.

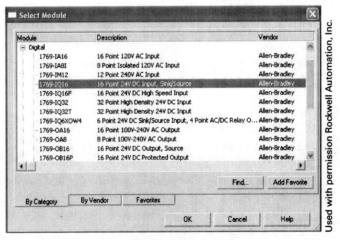

Figure 7-6 Select the 1769-IQ16 module to configure.

Another way to find the module is to search for it. To do such a search, click the Find button and enter the module part number, as illustrated in Figure 7-7.

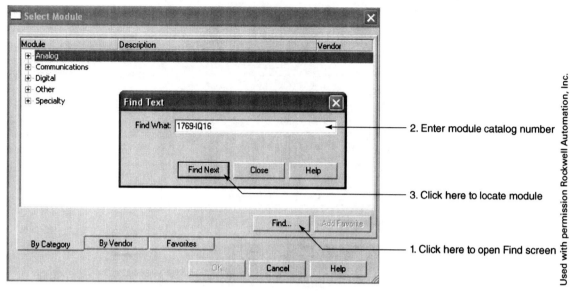

Figure 7-7 Select 1769-IQ16 module using the Find Text feature.

When you find the module, select it by double-clicking it, as shown in Figure 7-8.

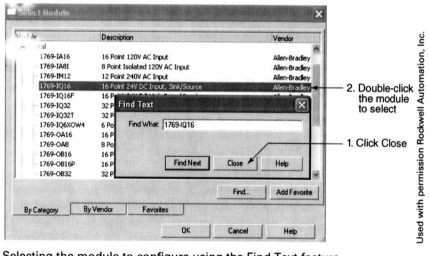

Figure 7-8 Selecting the module to configure using the Find Text feature.

4. _____ You should see the screen illustrated in Figure 7-9.
5. _____ Verify that the correct module part number is being displayed.
6. _____ For this exercise, give the module the name "Slot_1_Inputs." Remember: The module name is not part of the I/O points tags associated with the module's input points.
7. _____ Select the slot number where the module resides. For this exercise, the module is in slot 1.
8. _____ Add a description if you wish.
9. _____ In the Module Definition section, click the Change button.
10. _____ After clicking Change, you should see the screen illustrated in Figure 7-10.
11. _____ Enter the series of the module.
12. _____ Enter the major and minor revisions.
13. _____ Select the desired form of electronic keying.

200 COMPACTLOGIX I/O CONFIGURATION

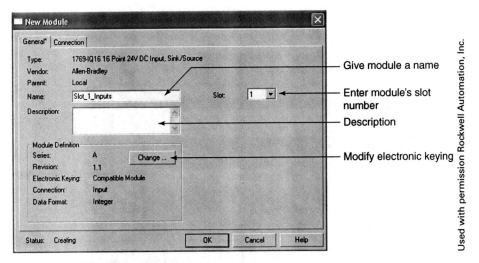

Figure 7-9 Enter module information.

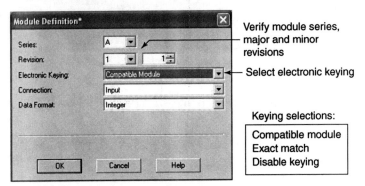

Figure 7-10 Configuring the module definition.
Used with permission Rockwell Automation, Inc.

When selecting electronic keying, consider this question: If the module needs to be replaced, how close of a match to the original does the replacement need to be? Your choices are Exact Match, Compatible Module, or Disable Keying.

Exact Match An exact match requires the same part number, manufacturer, and major and minor revision. An example of exact match could be a manufacturing area in a pharmaceutical company under jurisdiction of the federal Food and Drug Administration. If the module fails and is replaced with an exact match, then typically only a notation that the failed module was replaced with an exact match is required. If the replacement is not an exact match, then that part of the system or the entire system may need to be revalidated. This means the system could be down for revalidation for days, weeks, or months.

Compatible Module A compatible module must have the same part number, manufacturer, and major revision. The minor revision must be equal to or greater than the value entered into the revision box. A compatible module is typically the most common selection, but it is up to the programmer to decide whether this is appropriate.

Disable Keying When keying is disabled, the module type of a newer module must match the module to be replaced—but a module with any firmware level can be placed in the slot. If keying is disabled, then the controller will attempt to communicate with the module; if the module is not the same type, the connection will be rejected. If one attempted to put an output module in a slot configured for an input module, the connection would be rejected by the controller because the module type would not match.

- Disable keying is not used under normal conditions. Typically, it is used by project developers in the early stages of a project when the specific modules have not been installed.

- The end user should not use disable keying as a shortcut in lieu of deleting the original module from the I/O configuration and redoing the I/O configuration when a newer replacement module would have a newer major firmware revision. In many cases, a replacement module with a higher major revision level could have new features that the older module did not have. As a result, the amount and format of the data sent by the newer module to the controller and the amount of and format of the data received from the controller would probably not be compatible, and the module would probably not work properly. When keying is disabled, the controller uses the major and minor revision information contained in the I/O configuration as its reference as to what to send to and receive from the module regardless of what the module actually sends and is expecting to receive. In other words, if one were to replace a module that was configured as a version 2.1 with a version 8.0 and keying were disabled, the controller would treat the module as if it was a 2.1 because it was configured for that. In such a case, the data format and amount would not match, and there would probably be problems with the operation of the module.

14. _____ When you are finished with the Module Definition screen, click OK.
15. _____ Click the Connection tab. A screen like that in Figure 7-11 should be displayed.

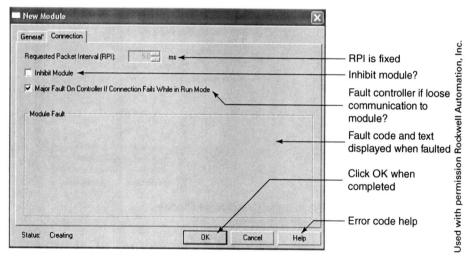

Figure 7-11 Module properties selections.

16. _____ If you want a major fault on the controller in the event that communication (the connection) fails to this module in Run mode, then click the check box.
17. _____ Click Inhibit Module if you wish to turn the module off as far as this controller is concerned. Inhibiting a module is typically used during startups and debugging.
18. _____ Notice that the Requested Packet Interval (RPI) is fixed and not changeable. Because this is an input module, the interval is the time in which you are

requesting input information from this input module be sent to the associated input tag.

19. _____ If the module is faulted, then there will be a hexadecimal error code and explanation in the Module Fault box. More information regarding the fault can be found by clicking the Help button.

20. _____ Notice the word Status in the bottom-left corner of the window. This displays the module's current status.

21. _____ Click OK when done.

22. _____ Open the controller scoped tags collection and go to the Monitor Tags tab; you should see the tags created for this module (also see Figure 7-12). The tag: Local:1:I. Data is your input tag. Input bits can be viewed in the value column. Bit 0 is on the far right of the bit string. The 2# signifies the radix of the displayed information is binary.

23. _____ This input module is configured. Next, we configure an output module.

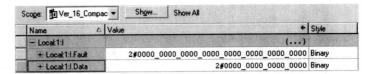

Figure 7-12 Resulting tags from I/O configuration.
Used with permission Rockwell Automation, Inc.

CONFIGURING A 1769-OB16

For this portion of the lab, we configure a 1769-OB16 output module. The input module configured in the last section should be displayed in the I/O configuration. You can see part of its part number in Figure 7-13.

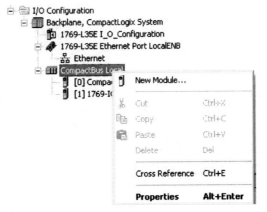

Figure 7-13 1769 CompactLogix: Select New Module.
Used with permission Rockwell Automation, Inc.

1. _____ Right-click CompactBus Local if configuring a 1769 CompactLogix (see Figure 7-13), or if using a 1768 CompactLogix, click 1769 Bus (see Figure 7-14) and select New Module as illustrated.

Figure 7-14 1768 CompactLogix: Select New Module.
Used with permission Rockwell Automation, Inc.

2. _____ Select the 1769-OB16 module, as illustrated in Figure 7-15. Figure 7-15 illustrates the Select Module window up through software version 19. Version 20 software will look somewhat different.
3. _____ If you wish to add a module to your favorites list, then select the module and click the Add Favorite button. See Figure 7-15. Next time you configure modules, select the module from the list of favorites. Click OK when completed. You should see a window similar to that shown in Figure 7-16.
4. _____ Fill in the correct information for the module's General tab. See Figure 7-16.
5. _____ The bottom left area of the General tab is the Module Definition area. Click the Change button to go to the Module Definition window. Refer to Figure 7-17.
6. _____ Enter the correct module series, major revision (left box), minor revision (right box), and desired electronic keying.

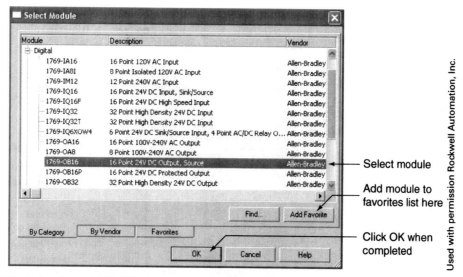

Figure 7-15 Selecting a 1769-OB16 module for configuration.

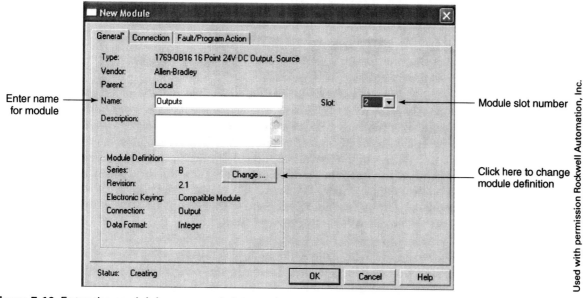

Figure 7-16 Enter the module's name and slot number.

204 COMPACTLOGIX I/O CONFIGURATION

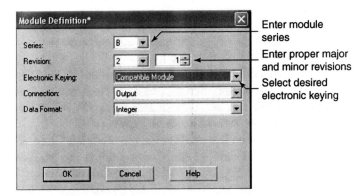

Figure 7-17 Configure Module Definition.
Used with permission Rockwell Automation, Inc.

7. _____ Click OK when done.
8. _____ Click the Connection tab.
9. _____ Set up this screen's properties as desired. See Figure 7-18 for a reference to the screen's parameters.

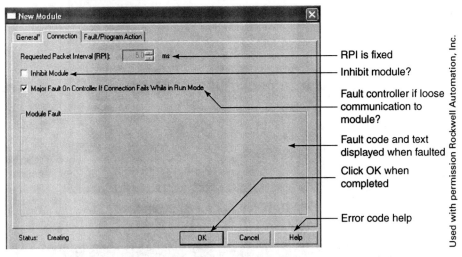

Figure 7-18 Configure the 1769-OB16 Connection tab.

10. _____ Click OK when done.
11. _____ Open the Fault/Program Action tab.
12. _____ For each output, select what is to happen when the controller goes into Program mode. Program mode is defined as when the controller goes into Program mode, Remote Test mode, experiences a major recoverable fault or experiences a communications failure. Communications failure behavior is configured below. Refer to Figure 7-19. To change output points, click the drop-down arrow for each point on the right-hand side of the Program Mode column. Select On, Off, or Hold. When the controller goes into Program mode, On means this output point will be turned on if it is not already on. Off means this output point will be turned off if it is not already off. Hold leaves the output point in its current state.
13. _____ For each output, select what is to happen when the controller goes into Fault mode. Refer to Figure 7-19. Here Fault mode is defined as when the controller experiences a major nonrecoverable fault. To configure, click the drop-down arrow for each point on the right-hand side of the Fault Mode column. Select On, Off, or Hold. When the controller goes into Communication Fault mode, On means this output

COMPACTLOGIX I/O CONFIGURATION   205

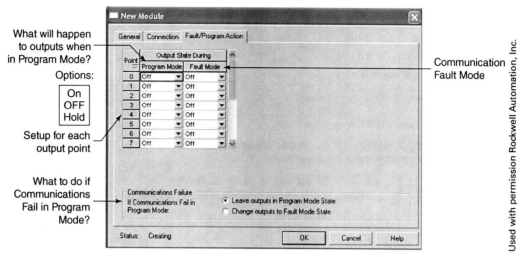

Figure 7-19 Configure Program mode and Fault mode actions.

point will be turned on if it is not already on. Off means this output point will be turned off if it is not already off. Hold leaves the output point in its current state.

14. _____ Near the bottom of the screen is the Communications Failure section. If communications to this module fail when the controller is in Program mode, do you want to put the outputs in the selected Program mode state or Fault mode state as configured above?

15. _____ Click OK when done. Figure 7-20 illustrates how your 1769 CompactLogix I/O configuration should look; Figure 7-21 illustrates how your 1768 CompactLogix I/O configuration should look.

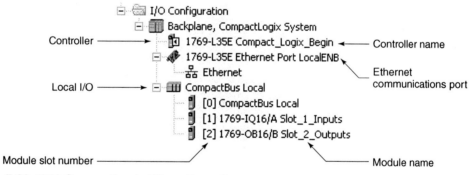

Figure 7-20 1769 CompactLogix I/O configuration.
Used with permission Rockwell Automation, Inc.

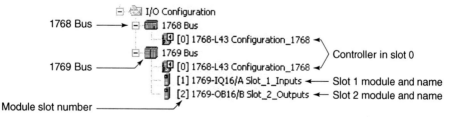

Figure 7-21 1768 CompactLogix I/O configuration.
Used with permission Rockwell Automation, Inc.

16. _____ Figure 7-22 shows the input and output tags created in the controller scoped tags collection Monitor Tags screen for both configured modules.

206 COMPACTLOGIX I/O CONFIGURATION

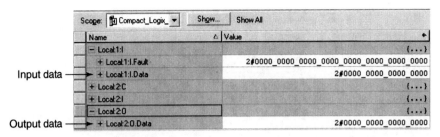

Figure 7-22 Resulting I/O controller scoped tags.
Used with permission Rockwell Automation, Inc.

17. _____ Figure 7-23 illustrates a rung of ladder logic and two tags and their associated instructions. Each piece of the tag is identified for your reference.

Figure 7-23 CompactLogix tag formats.
Used with permission Rockwell Automation, Inc.

18. _____ This completes this section of the lab.
19. _____ Remember to save your project.

LAB EXERCISE 3: CompactLogix Analog Input Module Configuration

This portion of the lab configures a 1769-IF4 four-channel analog input module. The I/O modules configured in the last two sections should be displayed in the I/O configuration.

1. _____ Right-click CompactBus Local or 1769 Base (see Figure 7-24).

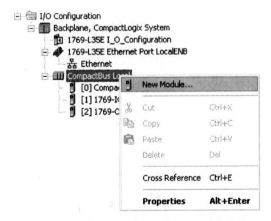

Figure 7-24 Select New Module to configure.
Used with permission Rockwell Automation, Inc.

2. _____ Select the 1769-IF4 module.
3. _____ Fill in the correct information for the module's General tab, as illustrated in Figure 7-25.
4. _____ Open the Change Module Definition window.
5. _____ Enter the correct module series, major and minor revisions, and desired electronic keying.
6. _____ Click OK when completed.

COMPACTLOGIX I/O CONFIGURATION **207**

7. _____ Click the Connection tab.
8. _____ Set up this screen's properties as desired. Figure 7-26 shows the screen parameters.
9. _____ Click OK when completed.

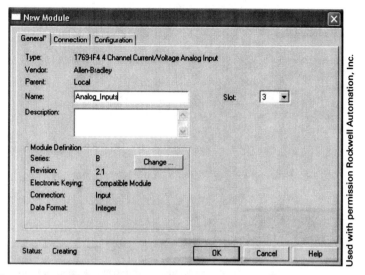

Figure 7-25 Configure the module's General tab.

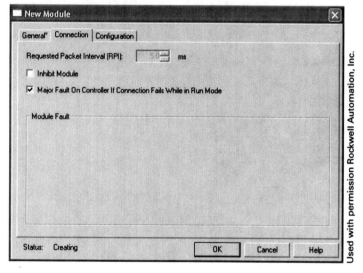

Figure 7-26 The module's Connection tab.

10. _____ Click the Configuration tab.
11. _____ In Figure 7-27, four channels are listed on the left side of the screen. Check the boxes in the first column to enable the channels you wish to use. Unchecked channels will be turned off or not enabled.
12. _____ Click the drop-down arrow in the Output Range column and select each channel's output range.
13. _____ Select filtering from the drop-down arrow in the Filter column.
14. _____ Click the drop-down arrow in the Data Format column and select the format for each channel, as in Figure 7-28.
15. _____ Click OK when completed.
16. _____ Figure 7-29 shows the controller scoped tag collection's Monitor Tags tab for the modules configured so far.

208 COMPACTLOGIX I/O CONFIGURATION

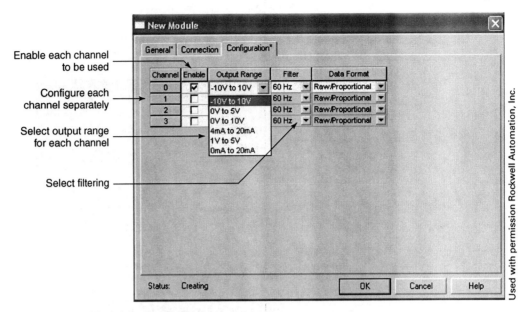

Figure 7-27 New Module's Configuration tab.

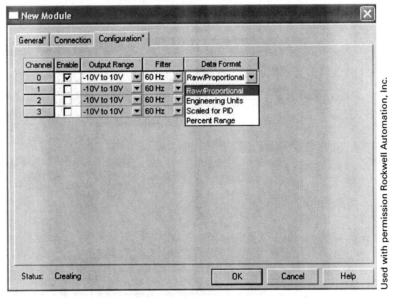

Figure 7-28 Select data format for each channel.

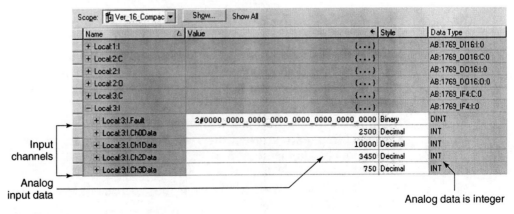

Figure 7-29 Resulting tags created.
Used with permission Rockwell Automation, Inc.

17. _____ Note the analog channel tag under the Name column.
18. _____ Note the analog input values under the Value column.
19. _____ Note the analog channel data type in the Data Type column.
20. _____ This completes this section of the lab.
21. _____ Remember to save your project.

COMPACTLOGIX ANALOG OUTPUT MODULE CONFIGURATION

For this portion of the lab, we configure a 1769-OF2 four-channel analog output module. The I/O modules configured in the last three sections should be displayed in the I/O configuration.

1. _____ Right-click CompactBus Local or 1769 Base and select New Module.
2. _____ Select the 1769-OF module. Refer to Figure 7-30. Figure 7-30 illustrates the Select Module window up through software version 19. Version 20 software will look somewhat different.

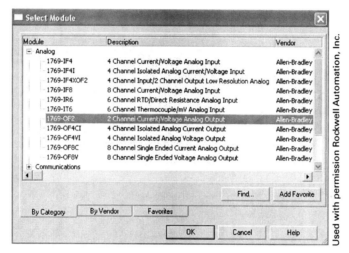

Figure 7-30 Select the module to configure.

3. _____ Click OK.
4. _____ Fill in the correct information for the module's General tab (see Figure 7-31).

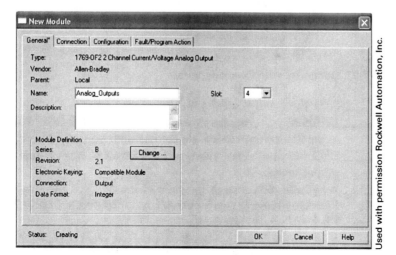

Figure 7-31 Analog module's General tab for configuration.

210 COMPACTLOGIX I/O CONFIGURATION

5. _____ Open the Change Module Definition window.
6. _____ Enter the correct module series, major and minor revisions, and desired electronic keying (see Figure 7-32).

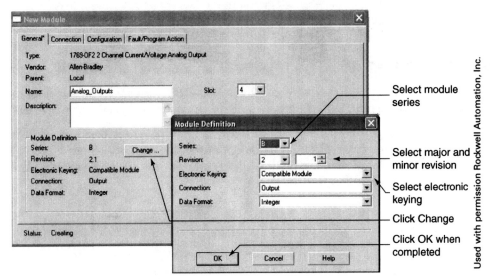

Figure 7-32 Configure Module Definition.

7. _____ Click OK when completed.
8. _____ Click the Connection tab.
9. _____ Set up this screen's properties as desired. Figure 7-33 shows the screen's parameters.

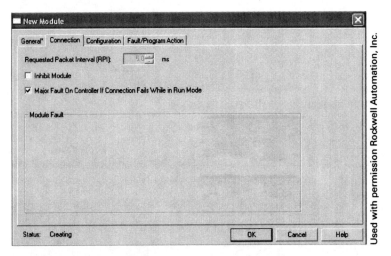

Figure 7-33 Configure your Connection tab.

10. _____ Click the Configuration tab.
11. _____ In Figure 7-34, two channels are listed on the left side of the screen. Check the boxes in the first column to enable the channels you wish to use.
12. _____ Click the drop-down arrow in the Output Range column and select each channel's output range.
13. _____ Select the data format for each channel. Click the drop-down arrow in the Data Format column for each channel, as illustrated in Figure 7-34.
14. _____ Click OK when completed.
15. _____ Next, we configure how each channel will react to transitioning from Run to Program mode or when there is a fault.
16. _____ Open the Fault/Program Action tab.

COMPACTLOGIX I/O CONFIGURATION 211

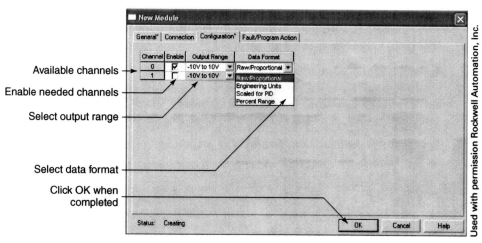

Figure 7-34 Configure channel properties.

17. _____ For each output, select what is to happen when the controller goes into Fault mode (refer to Figure 7-35). Here Fault mode is defined as when the controller experiences a major nonrecoverable fault. To configure, click the drop-down arrow for each point on the right-hand side of the Fault Mode column. Select On, Off, or Hold. When the controller goes into Communication Fault mode, On means this output point will be turned on if is not already on. Off means this output point will be turned off if it is not already off. Hold leaves the output point in its current state.

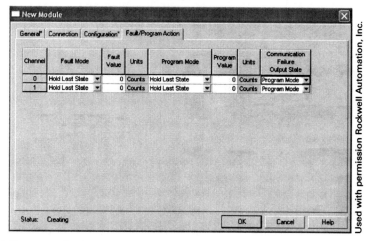

Figure 7-35 Fault/Program Action.

Fault Mode:
How will the channel react when the controller experiences a major nonrecoverable fault? Options are Hold Last State or Use Fault Value. If communications with the controller fail, then outputs from the controller are ignored, and the Fault mode selected will be used. Hold Last State is the default.

Fault Value:
When Use Fault Value is selected, the value entered in this parameter will be output. A 0 outputs the lowest value for the output range set up on the channel's Configuration tab. If your output range is 0 to 10 volts, a 0 will output 0 volts.

Program Mode:
How will the channel respond when the controller goes into Program mode? Program mode is defined as when the controller goes into Program mode, Remote Test mode, experiences a major recoverable fault or experiences a communications failure. Communications failure behavior is configured below. Options are Hold Last State or Use Program Value. If the controller

transitions into Program mode outputs from the controller are ignored, then the mode selected will be used. If Use Program Value is selected, the value in the program value parameter will be used. Hold Last State is the default.

Communication Failure:

When there is a communication failure, do you wish to change the outputs to Program Mode value or change outputs to the Fault Mode value? The default is Program Mode.

18. _____ Click OK when completed. Figures 7-36 and 7-37 illustrate how your 1769 and 1768 I/O configurations should look, respectively.

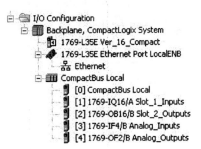

Figure 7-36 Completed 1769 I/O configuration.
Used with permission Rockwell Automation, Inc.

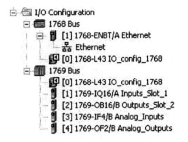

Figure 7-37 Completed 1768 I/O configuration.
Used with permission Rockwell Automation, Inc.

19. _____ Figure 7-38 shows the Monitor Tags tab for the controller scoped tags collection. The modules we used for this example are all illustrated. The analog output module output tags from the module we just configured are displayed.

Name	Value	Style	Data Type
+ Local:1:I	{...}		AB:1769_DI16:I:0
+ Local:2:C	{...}		AB:1769_DO16:C:0
+ Local:2:I	{...}		AB:1769_DO16:I:0
+ Local:2:O	{...}		AB:1769_DO16:O:0
+ Local:3:C	{...}		AB:1769_IF4:C:0
+ Local:3:I	{...}		AB:1769_IF4:I:0
+ Local:4:C	{...}		AB:1769_OF2:C:0
+ Local:4:I	{...}		AB:1769_OF2:I:0
− Local:4:O	{...}		AB:1769_OF2:O:0
+ Local:4:O.Ch0Data	10000	Decimal	INT
+ Local:4:O.Ch1Data	7500	Decimal	INT

Figure 7-38 Resulting controller scoped tags as a result of the I/O configuration.
Used with permission Rockwell Automation, Inc.

20. _____ Figure 7-39 shows a move instruction using the slot 3 input module, channel 1 input data, and slot 4 analog output module channel 1 tags.

Figure 7-39 Sample ladder rungs using analog tags.
Used with permission Rockwell Automation, Inc.

SUMMARY

I/O configuration in ControlLogix systems is a manual process completed by a programmer or maintenance or electrical individual when adding to or modifying the current system hardware. Because there are many options when configuring I/O modules, the process is manual so that programming individuals have the opportunity to configure their system to their exact specifications. As a result, there is no read I/O configuration feature with RSLogix 5000 software such as you find in Rockwell Automation's SLC 500 system and its associated RSLogix 500 software. When creating a new project, the I/O modules that will be used must be manually configured by performing an I/O configuration. Until the modules are configured, no I/O tags are available in the Monitor or Edit Tags views of the controller scoped tags collection. When adding a new I/O module to an existing RSLogix 5000 project, the module must be configured before its I/O tags are available for programming.

REVIEW QUESTIONS

Note: For ease of handing in assignments, students are to answer using their own paper.

1. Explain what inhibiting a module does. _____
2. Define Communication Fault mode. _____
3. Why would you want to check the box to have a major fault on the controller if communications fail while in Run mode? _____
4. Define RPI. _____
5. Until the modules are configured, there are no I/O tags available in the _____ or _____ Tags views of the controller scoped tags collection.
6. The module _____ is not part of the I/O points tags associated with the module's input points.
7. Clicking _____ turns the module off as far as this controller is concerned.
8. Inhibiting a module is typically used during or _____.
9. If the module is faulted, there will be an error code and explanation in the Module Fault box. More information regarding the fault can be found by clicking the _____ button.
10. What is electronic keying? _____
11. What is a compatible module when setting up electronic keying? _____
12. When adding a new I/O module to an existing RSLogix 5000 project, the module must be _____ before its I/O tags are available for programming.
13. The _____ is the time you are requesting input information from this I/O module to be sent to the associated tags.
14. When configuring analog channels, checking the boxes to enable the channels turns each channel on. Unchecked channels are _____ or not enabled.
15. Explain Electronic Keying, Exact Match. _____
16. Define Disable Keying. _____
17. When selecting the module to configure, modules can be selected by category, by _____, or from the _____ tab.
18. When we configure an output module's state when the controller transitions to Program Mode, what are the selections? _____

CHAPTER

8

Communicating between Your Personal Computer and Your ControlLogix

OBJECTIVES

After completing these lab exercises you should be able to

- Define common communication terms.
- Determine personal computer to PAC communication options.
- Configure the proper communications path for your specific ControlLogix or CompactLogix hardware.
- Download an RSLogix 5000 project.
- Put the controller in Run mode.
- Check a project for errors.
- Monitor tags.

INTRODUCTION

RSLogix 5000 software is used to develop and edit RSLogix 5000 projects. A second software package, RSLinx Classic software, is used to set up communications between the PAC controller and a personal computer.

As RSLogix software is upgraded from one version to the next, RSLinx Classic software must also be updated in order to use new features available with the newer software. Refer to the RSLogix 5000 software Release Notes for your specific release of RSLogix software, and review the minimum personal computer, Windows, and RSLinx Classic specifications required to use the RSLogix 5000 software release in question. The Release Notes can be found under the RSLogix 5000 Help tab in the software, or on the Rockwell Automation Web site in the Publications Library.

In order to monitor PAC activity, upload a PAC project into your personal computer, or download a project from your personal computer to your PAC, a communications link must be established between the personal computer and the programmable controller. RSLinx Classic software, usually referred to simply as RSLinx or Linx software, is used to configure the communications driver. Those who have purchased and installed a printer at home or at work have worked with setting up a printer driver. The driver is required to set up communications between a word processor application on a personal computer and a printer. The principle is the same for PACs. A communication driver is required to be configured using RSLinx so that we can communicate. Once the proper communications driver has been configured, a communications path is configured, and then communications can take place.

Because there are numerous ways to connect between a personal computer and a specific PAC, this lab steps you through the download process. Part 2 of this book is entitled "ControlLogix and RSLinx Communications." Numerous labs therein step you through configuring the needed RSLinx communication driver for your specific hardware. After your instructor shows you the proper driver configurations for your specific lab hardware, select the appropriate lab from this chapter to download your project to your PAC. If using an L 7 series controller and the USB port to download, go to Chapter 22. Labs 1 through 2C list the steps to configure communications and how to download for your future reference.

Before we begin setting up communications for downloading, we need to review some terms.

Communication Terms

The following terms are important to know in conjunction with setting up communications and transferring a copy of a project between a personal computer and a PAC controller:

Offline Working on a project that resides in computer memory is working offline. When working offline, a computer is not communicating with a PAC controller across a communications cable.

Upload Transferring a copy of the PAC controller's project to a personal computer's volatile memory is called *uploading*. To keep a copy on the personal computer's hard drive, it must be saved. Because rung documentation does not reside in the controller, uploading a project does not provide any documentation. The only exception to this is that RSLogix 5000 alias tags are stored in the controller and will be uploaded. If a copy of the entire project on a computer matches the project file in the controller, the two projects will be compared during uploading; if they match, the documentation will be displayed.

Download Transferring a copy of a project from a personal computer to a controller is *downloading*. Before downloading begins, current controller memory is cleared before the project is transferred. Once the original project in the controller is replaced with the downloaded version, the older project cannot be recovered. One important thing regarding downloading to a controller: Anytime an edit is made to an offline project file, the project must be redownloaded. Downloading transfers the updated project file to the controller. Sometimes people modify their ladder rungs and then go online rather than download. If one goes online in this case, the preedited project that was residing in the controller will be displayed on the computer screen. Until the modified project is downloaded, the controller does not have the new information.

Go online Going online means viewing a project in the PAC controller through a personal computer via a communication link. Online mode falls into two categories:

1. After completing a download and putting the controller in Run mode, live data can be viewed on the computer screen as you are online.

2. If a controller has faulted, connect to the controller and go online. In this situation, we want to monitor what is going on inside the controller. We probably would not wish to upload or download, just monitor. Monitoring can be accomplished by going online.

Driver A driver is an independent software application that is used to configure communication between a personal or industrial computer and a hardware device such as a PAC, variable frequency drive, or operator interface terminal. To program, edit, or monitor a ControlLogix project, RSLogix 5000 software is used. Communication drivers are set up using a separate piece of software called RSLinx.

RSLinx RSLinx Lite is the communication driver software bundled with RSLogix software. An RSLinx driver is configured to allow communications between the personal computer and the PAC controller.

LAB EXERCISE 1: Downloading to a 1768 or 1769 CompactLogix or Modular ControlLogix Using Serial Communications

Complete this lab if you are using a 1768 or 1769 CompactLogix or a modular ControlLogix and wish to download your project using serial communications. Many of the screen shots are for the 1769-L35E controller. If you are using the 1768-L43 or modular ControlLogix controller, follow the same steps. The screens you see will be very similar.

1. _____ Have your RSLogix 5000 software open.
2. _____ Make sure you have the correct communications cable.
   Plug the communications cable into your personal computer's serial port and the other end into your PLC serial port.
3. _____ Have the project to be downloaded verified and open.
4. _____ Refer to Figure 8-1 as you click Communications on the menu bar.
5. _____ Click Who Active, as illustrated in the figure.

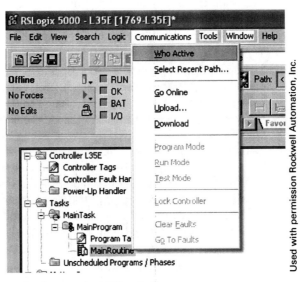

Figure 8-1 From Communications, select Who Active.

6. _____ From the Who Active screen, Figure 8-2 for a 1769-L35E, Figure 8-3 for a 1768-L43, or Figure 8-4 for a modular ControlLogix, follow the next steps to set up your communications path.
   a. _____ Select the RSLinx driver (A) and click + to expand the list. The default name for a serial driver is "AB_DF1-1." There may be more than one driver configured in your Who Active screen. For this lab exercise, we are only using the serial driver.
   b. _____ Click your controller once to select it. See B in the figure. Note that the path is being created in the lower-left portion of the window as we make our selections.

COMMUNICATING BETWEEN YOUR PERSONAL COMPUTER AND YOUR CONTROLLOGIX **217**

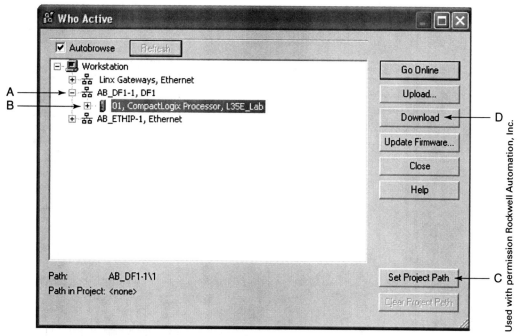

Figure 8-2 Who Active Screen for a 1769-L35E controller.

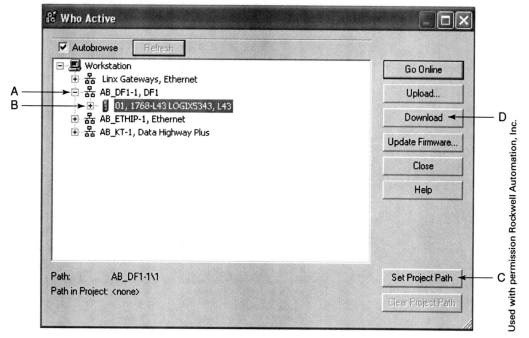

Figure 8-3 Who Active screen for a 1768-L43 controller.

    c. _____ Once there is a valid path, click Set Project Path (C) to replace the current path in the Path toolbar with this updated, or current, path. Figure 8-2 shows the RSWho screen for a 1769-L35E controller, whereas Figure 8-3 is the same RSWho screen for a 1768-L43 controller and Figure 8-4 shows the same screen when using a modular ControlLogix.

    d. _____ Click D to download.

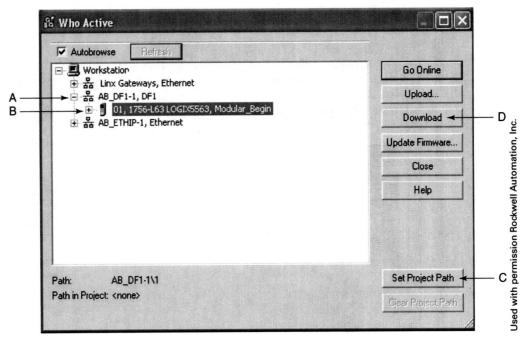

Figure 8-4 Who Active screen for a modular ControlLogix.

7. _____ You should see a screen similar to that shown in Figure 8-5. Click Download to continue.

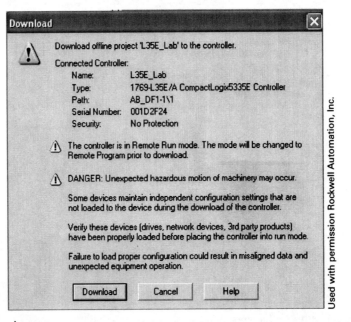

Figure 8-5 Download screen.

8. _____ You will probably see two screens as the project is being downloaded. When the download is complete, the screen similar to that shown in Figure 8-6 could be displayed. Click Yes to put the controller into Run mode. The window shown in Figure 8-6 assumes the controller was in Run mode before the offline project was downloaded. If the controller was in Remote Program mode before the download began, then the controller will be returned to Remote Program mode when the download is complete. The controller must be manually placed in Run mode. One method is shown in Figure 8-7.

COMMUNICATING BETWEEN YOUR PERSONAL COMPUTER AND YOUR CONTROLLOGIX 219

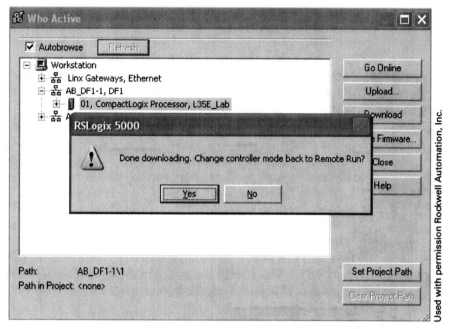

Figure 8-6 Put controller into Run mode.

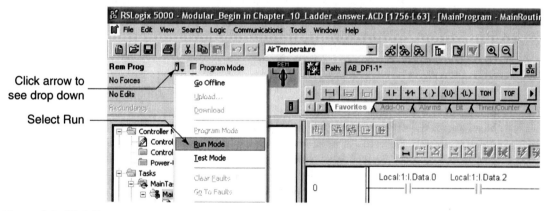

Figure 8-7 Click Run to put controller in Run mode from Remote Program mode.
Used with permission Rockwell Automation, Inc.

9. _____ With the download complete, refer to your RSLogix 5000 software and Figure 8-8 and note the following:

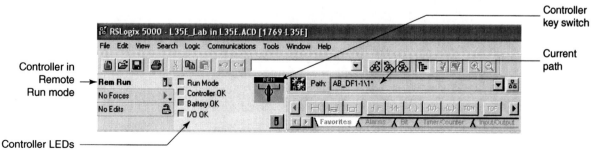

Figure 8-8 Monitor controller status.
Used with permission Rockwell Automation, Inc.

a. Controller is in Remote Run mode.
b. The controller status LED information should be reflected in the LEDs on your controller.
c. The controller key switch position displayed in the software should reflect the controller key switch position. As you physically change the controller key switch, note that the key in your software responds.
- Move the key from Remote (REM) to Run.
- Move the key from Run to Program (PRG) mode. The controller is now in Program mode.
- Switch the key from PRG to REM. What operating mode is the controller now in?
- Do not be confused by moving the key switch from PRG to REM mode and assuming that the controller is going back into Run mode. Many people assume going from Run mode or REM mode to Program mode and back to REM will return the controller to Run mode; in actuality, the controller is changing from Program to Remote Program mode.

10. _____ You have now successfully downloaded your project and changed the controller into Run mode. The gears should be turning to the right of the key switch. This signifies you are online. This completes this lab exercise.

ETHERNET COMMUNICATIONS

When we want to communicate with our controller and we determine that the current communications path is not correct, we will need to configure a new path. This lab guides you through setting up an Ethernet communications path. One should always verify whether the communications path is correct before attempting communications with any device across an Ethernet network. If the path is not correct, after the new path has been configured, we will replace the current path with the newly configured path and then download our project into our controller. After downloading, we will put our controller into Run mode.

Before we can configure our path three things need to be completed. First, if your personal computer does not have a compatible IP address, or you are not sure if your personal computer has a compatible IP address, refer to chapter 23. Your Ethernet communication module or CompactLogix controller must also have a compatible IP address. Refer to Part two of this book and review the chapter associated with your specific communications module or CompactLogix controller. Last, your RSLinx Ethernet driver must be configured and operating properly. Refer to Chapter 27 for guidance on setting up your RSLinx driver.

Figure 8-9 shows the Path toolbar in our RSLogix 5000 software. Before we can download our project, we must verify that the current path is correct. Always remember that the path displayed in the Path toolbar is the one that was used by the last individual who worked with this project. It is the responsibility of the current user to verify that the path is correct for the intended work.

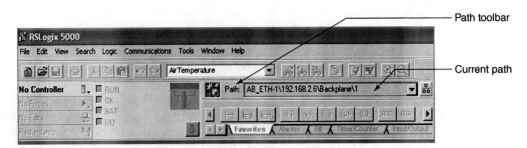

Figure 8-9 RSLogix 5000 Path toolbar.
Used with permission Rockwell Automation, Inc.

UNDERSTANDING THE PATH

Understanding the path and setting up the communications path is critical for communications to take place. Here we investigate how a path is put together. When using Ethernet, ControlNet, or Data Highway Plus, there are four pieces to the path:

1. The name of the RSLinx driver to be used for communications
2. The address of the communication module
3. Moving from the communications module to the chassis backplane
4. The controller slot number

Figure 8-10 illustrates a possible Ethernet path. The following list correlates to the items in the previous list.

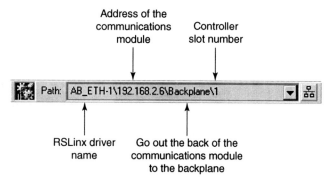

Figure 8-10 Ethernet path configuration.
Used with permission Rockwell Automation, Inc.

1. The RSLinx driver name is "AB_ETH-1."
2. The IP address of the Ethernet communications module is 192.168.2.6.
3. Go out the back of the Ethernet communications module to the chassis backplane.
4. Go to the controller in chassis slot 1.

Downloading an Offline RSLogix 5000 Project to the Controller Using Ethernet/IP Communications

This lab guides you through downloading your newly created offline RSLogix 5000 project using Ethernet/IP communications. When the project has been downloaded, the controller will be put into Remote Run mode. You then progress to Lab Exercise 3: Monitoring Data.
　　If you are using a 1769-L35E CompactLogix, complete lab 2A.
　　If you are using a 1769-L43 CompactLogix, complete lab 2B.
　　If you are using a modular ControlLogix, complete lab 2C.

LAB EXERCISE 2A: Downloading to a 1769-L35E CompactLogix Using Ethernet/IP Communications

Complete the following lab to download and run your project if you are using the 1769-L35E CompactLogix. If the current path on the Path toolbar is not correct, we have to configure a path. To do so, refer to Figure 8-11 as we begin to set up our path. Follow the steps below as we create a path and get ready to download the I/O configuration project completed in the last chapter.

1. _____ Open the project to be downloaded and select Communications from the RSLogix 5000 menu bar.
2. _____ Refer to Figure 8-11 as you select Who Active.

222 COMMUNICATING BETWEEN YOUR PERSONAL COMPUTER AND YOUR CONTROLLOGIX

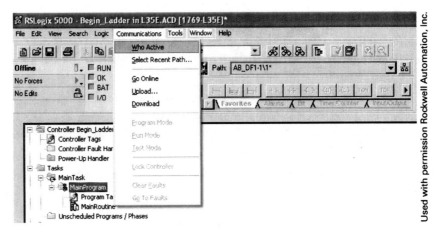

Figure 8-11 From the Communications drop-down menu, select Who Active.

Refer to Figure 8-12 as you complete the following steps.

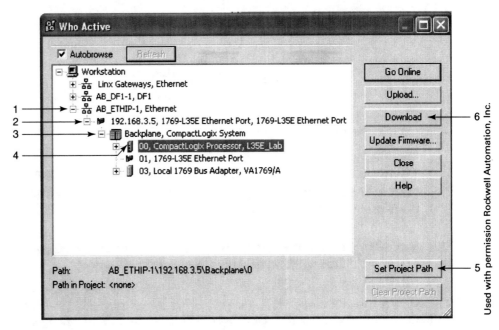

Figure 8-12 Building the communications path.

1. _____ Click + in front of Ethernet driver.
2. _____ Click + in front of IP address.
3. _____ Click + in front of Backplane to expand.
4. _____ Select the CompactLogix Controller by single-clicking it.
5. _____ To set current path in the project, click Set Project Path.
6. _____ Click Download.

3. _____ Click Download. Refer to Figure 8-13.
4. _____ You should see a few screens display as the download progresses. When the download process is complete, Figure 8-14 should display, assuming the controller was in Remote Run mode when we started our download. Click Yes to change the controller back to Remote Run mode.
5. _____ The window, as shown in Figure 8-14, will disappear and the controller will be placed into Remote Run mode. Monitor the identified parts shown in Figure 8-15 to see how our RSLogix 5000 software will display the current status and current path for this project.

COMMUNICATING BETWEEN YOUR PERSONAL COMPUTER AND YOUR CONTROLLOGIX

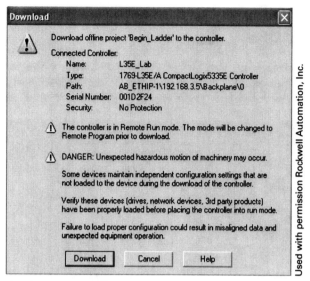

Figure 8-13 Click Download to continue.

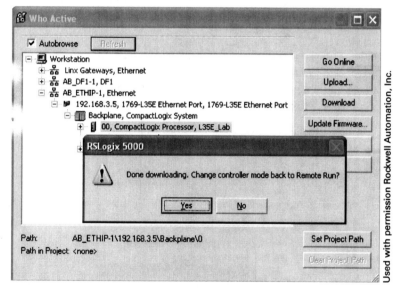

Figure 8-14 Click Yes to change controller to Remote Run mode.

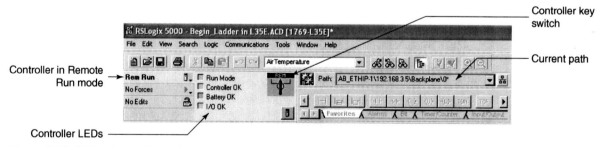

Figure 8-15 RSLogix toolbars showing controller status.
Used with permission Rockwell Automation, Inc.

6. _____ Your ControlLogix project has been downloaded, and the controller is running your project. Go to Lab Exercise 3 (Monitoring Data) and continue.

LAB EXERCISE 2B: Downloading to a 1768 CompactLogix Using Ethernet/IP Communications

Complete the following lab to download the I/O configuration project completed in the last chapter if you are using the 1768-L43 CompactLogix. If the current path on the Path toolbar is not correct, then we must configure a path. To configure a path, refer to Figure 8-16 as we begin. Follow the steps listed here as we create a path and get ready to download our project.

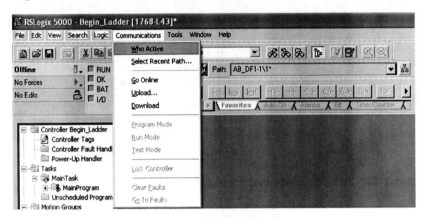

Figure 8-16 From Communications, select Who Active.
Used with permission Rockwell Automation, Inc.

1. _____ Open the project to be downloaded and select Communications from the RSLogix 5000 menu bar.
2. _____ Refer to Figure 8-16 as you select Who Active.
3. _____ Refer to Figure 8-17 as you build the communications path.

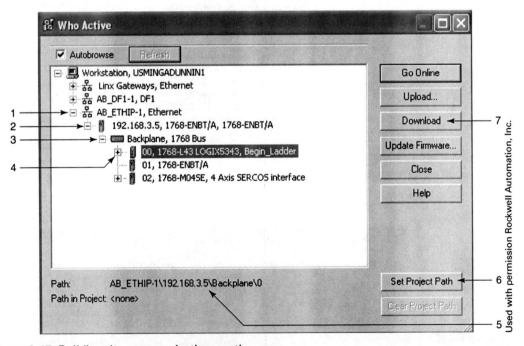

Figure 8-17 Building the communications path.

1. _____ Click + in front of Ethernet driver.
2. _____ Click + in front of IP address.
3. _____ Click + in front of Backplane to expand.
4. _____ Select the CompactLogix Controller by single-clicking it.

COMMUNICATING BETWEEN YOUR PERSONAL COMPUTER AND YOUR CONTROLLOGIX 225

    5. _____ Note the path is built here as each selection is made.
    6. _____ To set the current path in the project, click Set Project Path.
    7. _____ Click Download.
4. _____ Refer to Figure 8-18 and click Download.

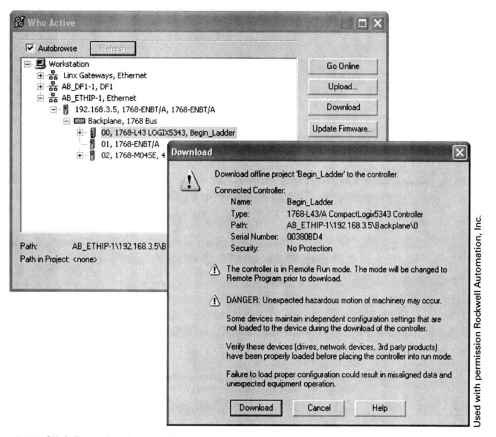

Figure 8-18 Click Download to continue.

5. _____ You should see a few screens display as the download progresses. When the download process is complete, Figure 8-19 should display.

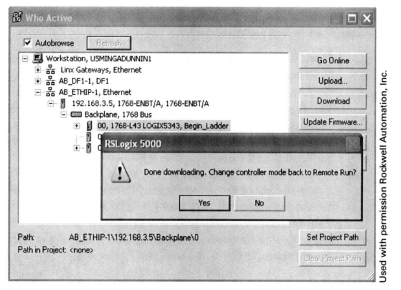

Figure 8-19 Download complete: Go to Remote Run mode?

6. ____ Click Yes to change the controller back to Remote Run mode.
7. ____ When the download process is completed, view the RSLogix 5000 toolbars for controller status and the current path. See Figure 8-20.

Figure 8-20 RSLogix tool bars showing controller status.
Used with permission Rockwell Automation, Inc.

8. ____ Your ControlLogix project has been downloaded, and the controller is running your project. Go to Lab Exercise 3 (Monitoring Data) and continue.

LAB EXERCISE 2C: Downloading to a Modular ControlLogix Using Ethernet/IP Communications

Complete the following lab to download and run your project if you are using a modular ControlLogix. If the current path on the Path toolbar is not correct, then we must configure a path. To do so, refer to Figure 8-21 as we begin to set up our path. Follow the steps listed here to create the path and get ready to download the I/O configuration project completed in the last chapter.

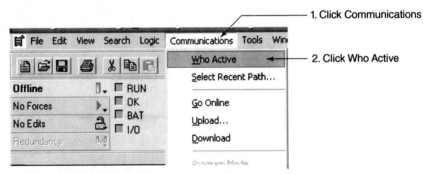

Figure 8-21 Starting to build a communications path.
Used with permission Rockwell Automation, Inc.

1. ____ Open the project to be downloaded and select Communications from the RSLogix 5000 menu bar.
2. ____ Click Who Active.
3. ____ You should see a Who Active screen, as shown in Figure 8-22.
4. ____ The first step in configuring a path is to determine what method of communications to use and then to select the appropriate driver. There may be multiple drivers configured and running in the RSWho window. For our lab, we use the Ethernet driver. From the RSWho window, select the AB_ETHIP-1 driver by clicking the + sign to the left of the driver name. Note that the driver name is displayed in the Path area in the lower-left corner of the RSWho window.
5. ____ The Ethernet communication module's IP address should be displayed directly below the driver you selected. Click the + sign to the left of the IP address to select it and expand to the next level (see Figure 8-23).

COMMUNICATING BETWEEN YOUR PERSONAL COMPUTER AND YOUR CONTROLLOGIX 227

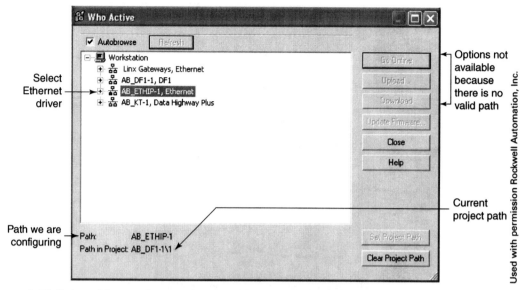

Figure 8-22 Select driver and expand to next level by clicking the + sign.

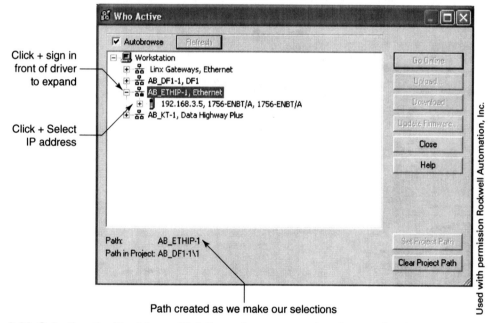

Figure 8-23 Selecting the IP address. Click the + sign to expand to the next level.

6. _____ The backplane should be displayed. Remember that by selecting the IP address, we have gotten only as far as the Communications module. The next step is to go out the back of the Ethernet Communications module and on to the chassis backplane.
7. _____ Refer to Figure 8-24 as you click the + sign to left of the Backplane.
8. _____ The Backplane should have been added to the display as well as to the path in the lower-left part of the window. See Figure 8-25.
9. _____ Refer to Figure 8-26 as you select the controller into which to download the project. Keep in mind that the modular ControlLogix system can have the controller in any chassis slot. There can also be multiple controllers in the chassis. Note that the controller slot number has been added to the path.

228 COMMUNICATING BETWEEN YOUR PERSONAL COMPUTER AND YOUR CONTROLLOGIX

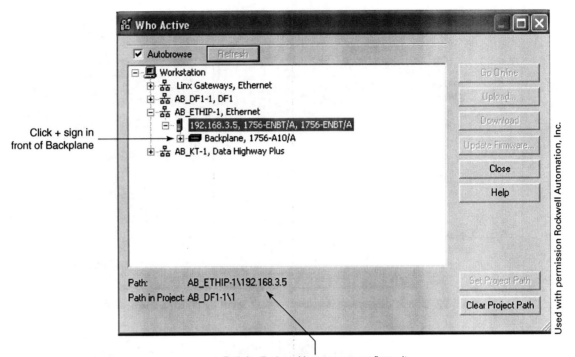

Figure 8-24 Expanding the path to contain the Backplane.

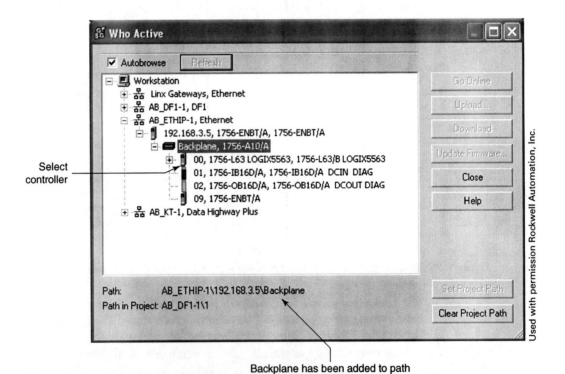

Figure 8-25 Controller and modules displayed in the ControlLogix chassis.

10. _____ Now that we have a valid path, the buttons in the upper right of the RSWho window are available. At this point, you can go online, upload, or download.

11. _____ If you click Set Project path, as shown in Figure 8-26, the project path toolbar will have the newly created path displayed, as shown in Figure 8-27.

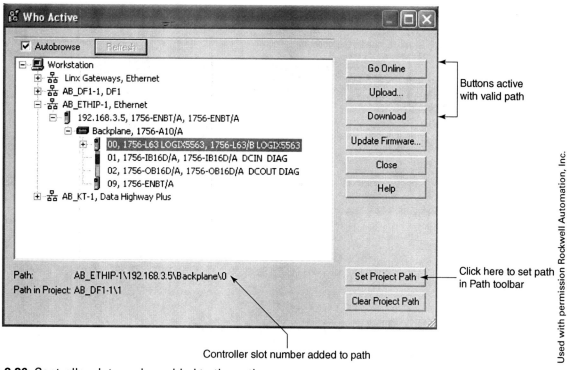

Controller slot number added to path

Figure 8-26 Controller slot number added to the path.

Figure 8-27 RSLogix 5000 project path tool bar updated with the newly created path.
Used with permission Rockwell Automation, Inc.

12. _____ With the correct path configured, we are now ready to download. Click Download from the Who Active window (Figure 8-26) to start the download process.
13. _____ The Download window, similar to that shown in Figure 8-28, should display. Before continuing, note the cautionary messages.
14. _____ Click Download to continue the download process.
15. _____ When the download process is completed, the controller may return to Remote Program mode or ask whether you want to return to Run mode. The message you see is determined by the state the controller was in at the beginning of the download process.

If the controller was in Remote Run mode before the download, then select Yes from the window asking whether you wish to return to Remote Run mode. If the controller was in Remote Program mode before the download, then refer to Figure 8-30 and put the controller in Remote Run mode. Figure 8-29 shows that the controller is now in Remote Program mode. As you refer to the figure, note the following:
- The current operating mode is Remote Program.
- LEDs reflect the current state of the LEDs on the controller because we are online.

230 COMMUNICATING BETWEEN YOUR PERSONAL COMPUTER AND YOUR CONTROLLOGIX

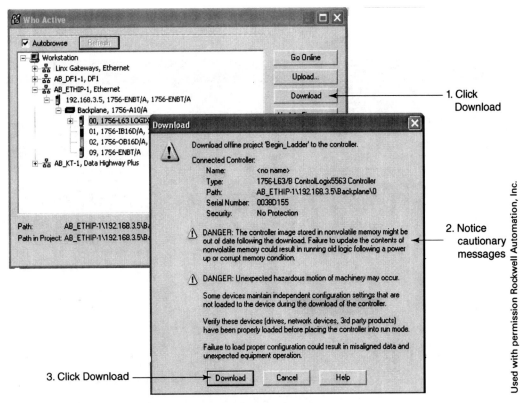

Figure 8-28 Download window with cautionary messages.

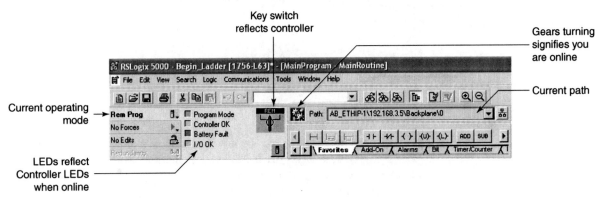

Figure 8-29 RSLogix 5000 toolbars showing current state of communications.
Used with permission Rockwell Automation, Inc.

- Key switch also reflects the current position of the key switch on the controller.
- Because the gears are turning, we are online with the controller displayed in the Path toolbar.
- The Path toolbar is displaying the current path to the controller we are online with.

16. _____ Referring to Figure 8-30, because we are currently in Remote Program mode, click the icon, as illustrated in Figure 8-30, to display the drop-down list.
17. _____ To go into Run mode, click Run Mode.
18. _____ The message in the lower-right corner of Figure 8-31 asks whether you wish to go into Run mode. Select Yes.
19. _____ After the controller transitions into Run mode, note the changes in the toolbar in Figure 8-31. Go to Lab Exercise 3 (Monitoring Data).

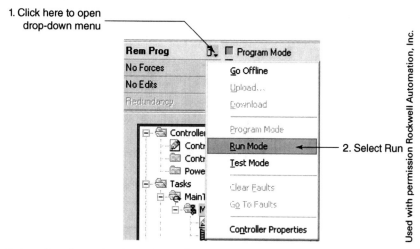

Figure 8-30 Select Run from the drop-down list to change the controller operating mode from Remote Program to Remote Run mode.

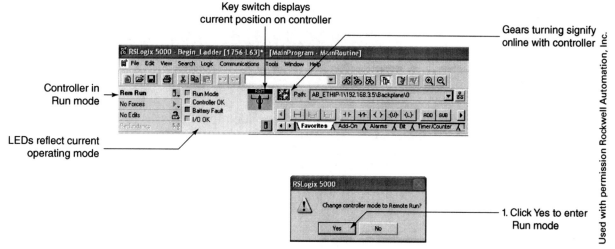

Figure 8-31 Click Yes to change mode to Remote Run.

LAB EXERCISE 3: Monitoring Data

There are many ways to communicate between a personal computer and a PLC. The specific ControlLogix or CompactLogix available hardware dictates your communications options, some of which are listed here:

- Direct serial connection
- USB to serial converter
- Ethernet communications at 10 M
- Ethernet/IP communications at 100 M
- USB to a Series 7 controller or CompactLogix 5370 series PLC
- ControlNet
- Data Highway Plus

Examine your hardware and determine your options. Your instructor will guide you through RSLinx driver configuration for your classroom hardware. If you are using this text as a self-study program, after determining your communication options, refer to the chapters in Part 2 to learn how to configure the necessary communications drivers.

CHECKING THE PROJECT FOR ERRORS: VERIFICATION

Before a project can be transferred to the controller, the project must be checked for errors. If you are confident there are no errors, the project can simply be downloaded to the controller. RSLogix 5000 software will check for errors in the background. If there are no errors, the project will be downloaded. If there are errors, the project will not be downloaded and all errors will have to be removed. A programmer can check for errors anytime by clicking the verification icons near the top of the RSLogix 5000 software's standard Windows toolbar, which is illustrated in Figure 8-32.

Figure 8-32 Verification icons.
Used with permission Rockwell Automation, Inc.

The Verify Routine icon will check the current routine for errors, whereas the Verify Controller icon will check the entire project for errors. If after clicking either of these icons errors are found, they will be listed in the Results window, as shown in Figure 8-33. Notice the tab called Errors—this is where any errors will be listed. You may have to scroll up the list to find the errors. Once an error is located in the list, clicking that error takes you right to it on the ladder, as noted in the figure. With the error displayed, it can now be easily fixed.

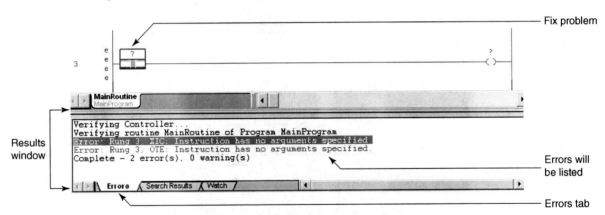

Figure 8-33 Results window displayed as a result of errors.
Used with permission Rockwell Automation, Inc.

THE LAB

We are going to download and monitor the I/O configuration project you completed for Chapter 6 or Chapter 7 and monitor the tags from the Controller Tags monitor window.

1. _____ Open the RSLogix 500 project completed for Chapter 6 or Chapter 7.
2. _____ Verify your project.
3. _____ Select and follow the desired download procedures 1 through 2C in this chapter, and download the project.
4. _____ Put the PAC into Run mode.
5. _____ Open the controller tags collection.
6. _____ Click the Monitor tab.
7. _____ As you turn each input switch on and off, correlate the switch being closed to the LED illuminating on the respective input module.
8. _____ Energize the input buttons or switches on your demo unit, and correlate each input device with the associated input bit in the input tag. You should see the bit change to a 1 when the input device is closed.
9. _____ Examine the label on the inside of each of the module's doors. You should be able to correlate each input device wiring to its respective screw terminal and the module's LED, and the tags bit status. If your ControlLogix has a removable terminal block on the module, note that the numbers stamped on the terminal block are screw terminal wires only and do not directly correlate to the specific input bit in the input tag.
10. _____ If you have a diagnostic input and output module, examine the diagnostic tags introduced in Chapter 6, "Modular ControlLogix Digital I/O Configuration."
11. _____ If you have an electronically fused output module examine the electronic fuse tag, also introduced in Chapter 6.
12. _____ Turn on each of your analog potentiometers (pots), and find and monitor the analog input tags.
13. _____ You should see the analog numeric data display as the scaled data configures in the I/O configuration exercise.
14. _____ Even though there is no logic created to turn on any outputs at this time, examine the output tags and correlate the tags to the output module LEDs, screw terminals, and—if a 1756 output module—the screw terminal numbers stamped on the removable terminal block. In a later lesson, we will monitor live logic, including live input and output status.
15. _____ Go offline.
16. _____ Save your project.

CHAPTER

9

Creating and Monitoring RSLogix 5000 Tags

OBJECTIVES

After completing this lesson, you should be able to

- Create base tags.
- Create alias tags.
- Assign a tag a data type.
- Configure a tag's style.
- Enter a tag description.
- Create controller scoped tags and programmed scoped tags.
- Identify controller scoped and programmed scoped tags.
- Introduce ControlLogix arrays.

INTRODUCTION

Traditional PLCs allocate a portion of the controller's physical memory for the storage of program-related information. The memory is segmented into data files so it is easier for the programmer to work with. As an example, the PLC-5 and SLC 500 PLCs have the following data files automatically created when a new project file is created:

- output status table
- input status table
- status file
- binary file
- timer file
- counter file
- control file
- integer file
- floating point file

ControlLogix is a tag-or name-based PAC. ControlLogix and its associated RSLogix 5000 software have no predefined data files as found in traditional PLCs. When creating a tag in RSLogix 5000 software, assign a unique name—a *tag*—and then select the type of data this particular tag represents; this is called the tag's *data type*. This lesson introduces creating tags and alias tags in the Edit Tags window.

An alias tag can be attached to a base I/O tag as a way to assign a name that is easier to understand and work with. Here we introduce the Monitor Tags window and show you how to monitor data. Online data monitoring will take place in a later lesson. Finally, we introduce grouping data together in a block called an *array*. An array can be multidimensional: one, two, or three dimensions. Arrays can be used as a way to become more efficient as well as to increase organization of tags in a RSLogix 5000 project.

TAGS

ControlLogix is a name-based or tag-based PAC. When creating a tag, give it a unique name and then identify what that tag is—its data type. If we create a tag and call it "start machine" to represent a push button, it is a single-bit, or a *Boolean* (BOOL) tag. To create a tag that represents how long to mix a solution, we could create a tag called "mixing timer." Because a timer is used to track the mixing time, the data type for this tag is *timer*. Figure 9-1 reviews information from Chapter 3 that identifies the basic data types for the ControlLogix.

ControlLogix Data Type		Number of Bits	Data Range
BOOL	Boolean	1	1 or 0
	Nibble	4	
SINT	Short integer	8	−128 to +127
INT	Integer	16	−32,768 to +32,767
DINT	Double integer	32	−2,147,483,648 to +2,147,483,647
REAL	Floating point	32	Extremely small or large numbers with decimal point

Figure 9-1 ControlLogix data types.
© Cengage Learning 2014

ControlLogix's basic allocation of memory for any tag is 32 bits, or a double word. As an example, if we created a BOOL tag, 32 bits of memory would be consumed. A 16-bit tag, similar to an integer in a PLC-5 or SLC 500, also would consume 32 bits. Figure 9-2 illustrates the basic ControlLogix data types and the memory required for each tag created with that data type.

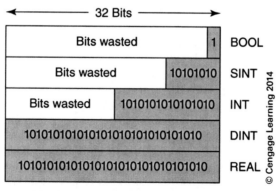

Figure 9-2 Memory consumed for basic data types.

Figure 9-3 illustrates a portion of the RSLogix 5000 controller scoped tags collection's Monitor Tags view. Notice the tag named "Total": Its data type is *integer* (INT) with a value of 22,316. Selecting INT as the data type results in the data being a 16-bit signed integer. The tag named "Production Total" is a *double integer* (DINT), and its current value is 75,499,876. Selecting a DINT as the data type results in the data being a 32-bit signed integer. Temperature is a REAL number data type with the value of 234.8. Start and stop are single bits, or BOOL data.

Tag name →	Name	Value	Data Type	← Tag data type
	Start	0	BOOL	
	Stop	1	BOOL	
	+ Production_Total	75499876	DINT	
	Temperature	234.8	REAL	
	+ Total	22316	INT	
	+ Value_1	43	DINT	← Tag value

Figure 9-3 Portion of an RSLogix 5000 controller scoped tags collection.
Used with permission Rockwell Automation, Inc.

CONTROLLER SCOPED AND PROGRAM SCOPED TAGS

There are two collections of tags within a ControlLogix project: controller scoped and program scoped. Controller scoped tags are global tags—that is, they can be used anywhere within the current RSLogix 5000 project. Program scoped tags are local to a specific program. Even though there is only one controller scoped collection of tags, each program has its own program scoped tags collection. Figure 9-4 shows an RSLogix 5000 controller organizer and points out the controller scoped tags collection as well as each individual program's program scoped tags collection.

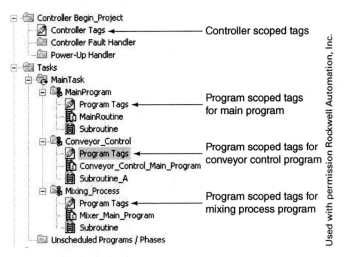

Figure 9-4 Controller and program scoped tags.

When we performed an I/O configuration in Chapters 6 and 7, we saw that I/O tags are automatically stored in the controller scoped tags collection. I/O tags are global; they can be used in any program and routine. Figure 9-5 shows the controller organizer on the left and the controller scoped collection of tags for our example on the right. Notice that the I/O configuration tags are listed at the top of the list of tags. The tags display is the Monitor Tags view in the RSLogix 5000 software. The top left of the tags window identifies the collection or scope of tags currently being displayed. Note a controller-like icon with a left-pointing arrow in the left portion of the scope display; this icon signifies that this is the controller scoped tags collection. Keep in mind that the controller icon was first used in RSLogix 5000 software version 15. Also displayed are the listed tag names, current value, and data type.

Program scoped tags are local to their specific program. Tags stored in the Conveyor_Control program scoped tags collection can only be used in that program. Program scoped tags can be used as a way to organize project tags and also easily reuse tags. By default, tags are arranged alphabetically, A through Z. A project that has 5,000 controller scoped tags arranged A through Z is quite a long list of tags to try and manage. Program scoped tags can be used to group tags with their associated program. The program scoped tags can be stored in the Conveyor_Control program scoped tags collection as a way to organize them and provide an easy way to work with the tags.

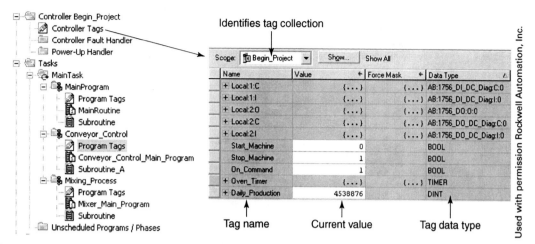

Figure 9-5 Controller scoped tags collection.

It is much easier to work with a program scoped collection of tags with, say, 200 tags that pertain to that specific program and its associated routines rather than attempt to find them in the list of 5,000 controller scoped tags. When creating the initial project, a programmer could enter the tags for one program and copy and paste those tags into a second program that could use the similar tags. Figure 9-6 shows the Conveyor_Control program scoped tags collection. In the scope view, note the folder to the left of the name of the tags collection; it identifies the tag collection as program scoped.

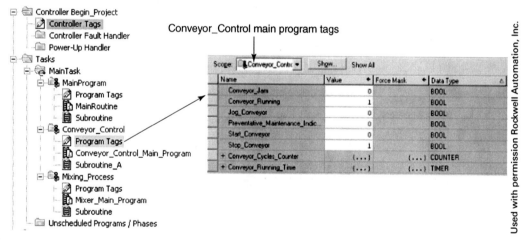

Figure 9-6 The Conveyor_Control main program scoped tags collection.

Figure 9-7 shows the Mixing_Process program's collection of program scoped tags. In the scope view, note the folder to the left of the name of the tags collection. Entering the tags associated with their specific program provides an easy method to organize the project tags.

To select a tags collection from either the Monitor Tags or Edit Tags view, click the down arrow to display the tag collections available. Refer to Figure 9-8 and select the desired collection to display.

Edit Tags and Monitor Tags Windows

RSLogix 5000 has two windows for working with tags. The Edit Tags window is for creating and editing tags, whereas the Monitor Tags window is used for monitoring the project's tags. As illustrated in Figure 9-8, one can navigate between tag collections by clicking the drop-down arrow

238 CREATING AND MONITORING RSLOGIX 5000 TAGS

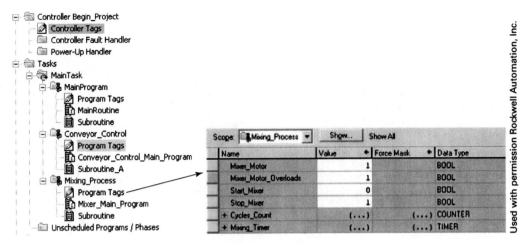

Figure 9-7 Mixing_Process program scoped collection of tags.

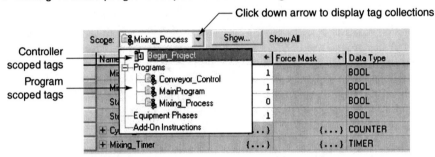

Figure 9-8 Selecting a tags collection from the Monitor or Edit Tags window.
Used with permission Rockwell Automation, Inc.

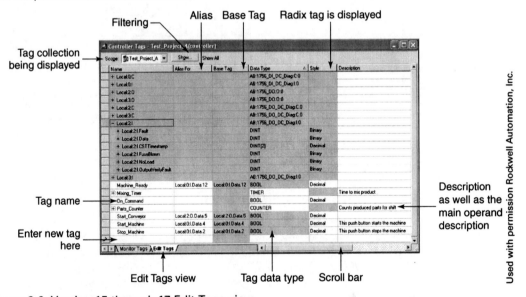

Figure 9-9 Version 15 through 17 Edit Tags view.

and selecting the desired tag collection. Figure 9-9 shows the Edit Tags window. Starting top left, the columns are as follows:

Name This is the tag name to select from the drop-down menu when assigning a tag to an instruction.

Alias For The tag in the Name column is an alias for the tag name displayed here. We introduce and work with alias tags in the next section of this chapter.

CREATING AND MONITORING RSLOGIX 5000 TAGS 239

Base Tag The Base Tag is the base, or starting, tag name. With ControlLogix, there can be multiple levels of aliasing. The Base Tag is the starting tag. The Alias For column could have the first alias, and the Name column could have the second alias.

Data Type This column displays each tag's data type.

Style Style defines the radix in which the tag's information is displayed in an instruction. The default style is decimal. Binary style is selected to display the tag's information in binary when viewing an instruction on a ladder.

Description Text that describes the tag name's use can be entered in the Description column. This documentation is for better understanding of what this particular tag does or is used for. The text entered in this column is also the main operand description, or the text displayed above the base tag and its alias on a ladder rung. An example of a main operand description can be seen in Figure 2-2. We will work on entering text documentation, including main operand descriptions, in Chapter 13.

Show The Show button near the top center is used to filter which tags are being displayed in the tags window. Currently, the text to the right of the Show button shows all that is displayed. This means that all tags will be displayed. The user has many options regarding which type of tags will be displayed. As an example, only I/O tags, or timers, counters, or BOOL tags could be displayed.

Note the empty row on the bottom of the tags list in Figure 9-9. This is where a new tag is created. The minimum entry would be the tag name and data type. The style, description, and alias can be entered or modified if desired.

The Monitor Tags window is displayed in Figure 9-10. This window is for monitoring the tags in a system. If a user were online with the controller and wanted to monitor an I/O point, a timer, a counter, production data, and so on, this is where the user would go to view or monitor current values. Most of the column heads have the same name and perform the same function as listed in the Edit Tags view. The Monitor column is where a tag's data can be viewed. The Force Mask column is used to force inputs and outputs when troubleshooting.

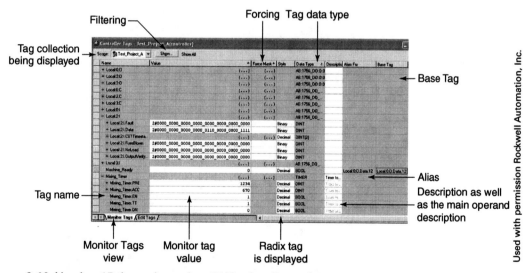

Figure 9-10 Version 15 through version 17 Monitor Tags view.

Figures 9-9 and 9-10 show Edit and Monitor Tags windows for RSLogix version 15 through version 17. Figure 9-11 illustrates a tag collection window beginning with software version 18. New features are identified in Figure 9-11. The Show area performs the same function even though it looks a bit different. The user can type in a partial tag name in the area to the right of the show area for help in finding a tag where the entire name is not known. Assuming the user wanted to find tags that had the word air, start typing air in this area. After entering a few letters only tags

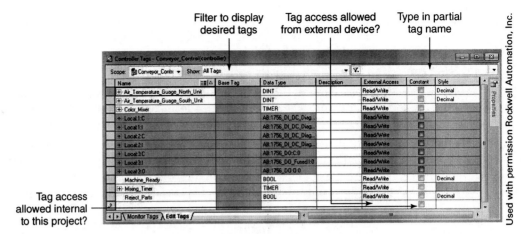

Figure 9-11 Edit and Monitor Tags view, beginning software version 18.

with the word air would be displayed. The External Access option is used to grant or deny access to the tag from an external device. Options from the drop down include Read/Write, Read Only, or No Access. The Constant option grants or denies access to the tag internal to this project's logic.

LAB EXERCISE 1: Creating Tags in an RSLogix Project

This lab steps you through creating basic tags in the Edit Tags window that we just examined. We add new tags to the I/O Configuration project created in Chapter 6 or 7. Follow the steps below to complete the lab.

1. _____ Open the RSLogix 5000 software.
2. _____ Open the I/O Configuration you created in either Chapter 6 or 7.
3. _____ Double-click Controller Tags to open the controller scoped tags collection.
4. _____ Click the Edit Tags tab to open the Edit Tags window. Figure 9-12 lists the tags to create.
5. _____ Refer to Figure 9-13 and follow the steps below to create the following tags.

Tag Name	Data Type	Description
Start–Machine	BOOL	This push button starts the machine
Stop–Machine	BOOL	This push button stops the machine
On–Command	BOOL	Machine is on
Oven–Timer	Timer	This is how much time the product is in the oven
Daily–Production	DINT	Daily production total

Figure 9-12 Tags to create.
© Cengage Learning 2014

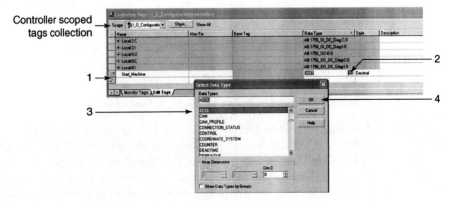

Figure 9-13 Creating new tags.
Used with permission Rockwell Automation, Inc.

6. _____ Figure 9-13 shows the controller scoped tags collection. As you begin to create our first tag, the last row in the Edit Tags collection is blank. As you enter the name and press the space bar, the underscore is inserted. Spaces are not allowed in tag names. Refer to the number 1 with the arrow in the figure as you click the blank row and enter "Start_Machine."

7. _____ The default data type for a tag is a DINT. Refer to the number 2 in the figure as you click the right side of the Data Type column. You should see the small button with the three periods. Click that spot; you should see the Select Data Type window, as illustrated.

8. _____ Probably the easiest way to select the data type from the list is to enter the first letter of the data type on your computer keyboard. Because this is a BOOL tag, press your keyboard's B key. This takes you to the data types that start with the letter B. Because there is only one data type starting with B, BOOL is selected. See the number 3 in the figure. When there are multiple tags with the same starting letter, you can scroll to the desired tag, or keep entering letters until the desired data type is selected.

9. _____ Click OK (item number 4) to accept the data type.

10. _____ Click in the Description box to display the box in which you enter the tag description. Refer to Figure 9-14 as you enter the description.

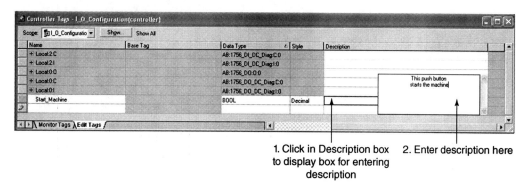

1. Click in Description box to display box for entering description
2. Enter description here

Figure 9-14 Entering tag description.
Used with permission Rockwell Automation, Inc.

11. _____ When finished entering the description, click off the cell to update your entry.

12. _____ On your own, enter the remaining tags. Next, we look at alias tags and then create them in the lab exercise.

ALIAS TAGS

Alias is another name for a tag. A base tag of Local:2:I.Data.2 could have an alias of Start_Machine. The tag Local:2:I.Data.4 could have an alias of Stop_Machine. When working with the software, it is easier to work with and search for components with names such as Start_Machine than to use the base tag. Once established, the alias tag and its base tag are married together in the current project. One important point to remember: Once an alias tag is established, always select it when programming, to be consistent. Figure 9-15 illustrates the basic rung components. Note the rung number, base tags, and alias tags. The main operand description is also displayed as a result of the description entered when creating the tag.

When programming, if Enable Look Ahead is checked in the Workstation Options window (Figure 9-16), left-click once on the question mark (?) above an instruction and start entering the alias name and the software will find tags that are close to what was entered as you enter the beginning letters. If the desired tag is not displayed, then it can be selected out of the drop-down list. The Enable Look Ahead feature is identified in Figure 9-16. To get to Workstation Options, click Tools and then Options.

242 CREATING AND MONITORING RSLOGIX 5000 TAGS

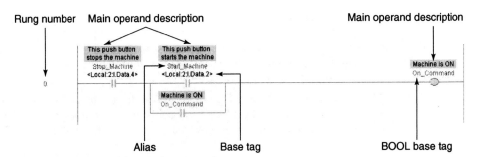

Figure 9-15 Ladder rung showing components.
Used with permission Rockwell Automation, Inc.

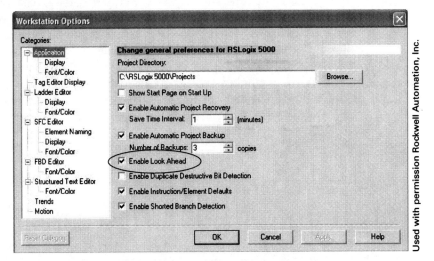

Figure 9-16 Enable Look Ahead feature for ease of programming alias tags.

LAB EXERCISE 2: Creating Alias Tags

This lab exercise assigns alias tags to the base tags we created in Lab Exercise 1. Figure 9-17 lists the base tags and their alias tags to assign. Refer to Figure 9-18 and follow the steps as you assign alias tags.

1. _____ Here, we modify the tags created in the last lab exercise.
2. _____ Open the controller scoped tags collection.
3. _____ Refer to the number 1 in Figure 9-18 as you select the Start_Machine tag and click in the right portion of the cell, under the Alias For column.
4. _____ Scroll down to find Local:2:I. See the number 2 in the figure.
5. _____ Click the right portion of the line when the arrow appears, as shown in the number 3 item.
6. _____ Refer to item number 4 as you click the bit number from the table and complete your address.
7. _____ Note item 5 in the figure. The Windows Tool Tips show whether the selected bit has already been used in your project. Later versions of the software will also display color and indent bit number areas for easy identification of unused bits.

Base Tag	Alias Tag
Local:2:1.Data.2	Start–Machine
Local:2:1.Data.4	Stop–Machine

Figure 9-17 Alias tags to assign.
© Cengage Learning 2014

CREATING AND MONITORING RSLOGIX 5000 TAGS   243

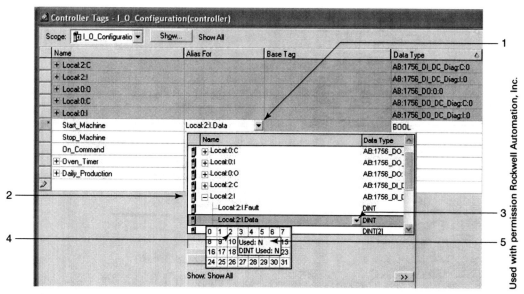

Figure 9-18 Assign alias tag.

8. _____ If there are columns that are not displayed in your Edit Tags window, they can be turned on or off by clicking View and selecting Toggle Column. Verify that the columns you want to be displayed are checked. Refer to Figure 9-19.

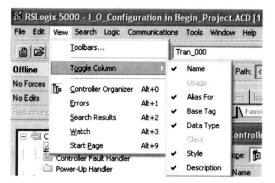

Figure 9-19 Turn columns on or off in the Edit or Monitor Tags window.
Used with permission Rockwell Automation, Inc.

9. _____ On your own, assign the other alias.
10. _____ Save your completed project.
11. _____ Ask your instructor whether he or she needs to check off your completion of the exercises.

LAB EXERCISE 3: Create Program Scoped Tags

Lab Exercise 3 provides hands-on experience in modifying your current project and adding program scoped tags. We add more project components and tags to the project completed for Lab Exercise 2.

1. _____ Open the completed project from Lab Exercise 2.
2. _____ Modify your project controller organizer to look like that shown in Figure 9-20.
3. _____ Refer to Figure 9-21 as you create the program scoped tags for the Conveyor_Control program. Refer back to Figure 9-6 for reference. Do not modify the Value column in your project.

244 CREATING AND MONITORING RSLOGIX 5000 TAGS

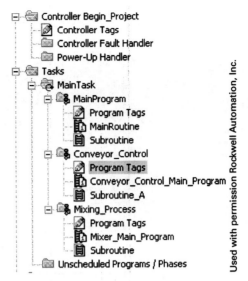

Figure 9-20 MainTask modification for Lab Exercise 3.

Scope: Conveyor_Control			
Name	Value	Force Mask	Data Type
Conveyor_Jam	0		BOOL
Conveyor_Running	1		BOOL
Jog_Conveyor	0		BOOL
Preventative_Maintenance_Indic...	0		BOOL
Start_Conveyor	0		BOOL
Stop_Conveyor	1		BOOL
+ Conveyor_Cycles_Counter	{...}	{...}	COUNTER
+ Conveyor_Running_Time	{...}	{...}	TIMER

Figure 9-21 Program scoped tags for the Conveyor–Control main program.
Used with permission Rockwell Automation, Inc.

4. _____ Refer to Figure 9-22 as you create the program scoped tags for the Mixing_Process program. Refer back to Figure 9-7 for reference. Do not modify the Value column in your project.

Scope: Mixing_Process			
Name	Value	Force Mask	Data Type
Mixer_Motor	1		BOOL
Mixer_Motor_Overloads	1		BOOL
Start_Mixer	0		BOOL
Stop_Mixer	1		BOOL
+ Cycles_Count	{...}	{...}	COUNTER
+ Mixing_Timer	{...}	{...}	TIMER

Figure 9-22 Program scoped tags for the Mixing–Process program.
Used with permission Rockwell Automation, Inc.

5. _____ Next, assign the alias tags as listed in Figure 9-23.

Base Tag	Alias Tag
Local:0:O.Data.11	Mixer–Motor
Local:2:I.Data.5	Mixer–Motor–Overloads
Local:2:I.Data.6	Start–Mixer
Local:2:I.Data.7	Stop–Mixer
Local:0:O.Data.6	Conveyor–Jam
Local:0:O.Data.7	Conveyor–Running
Local:2:I.Data.8	Jog–Conveyor
Local:0:O.Data.12	Preventative–Maintenance–Indicator
Local:2:I.Data.9	Start–Conveyor
Local:2:I.Data.10	Stop–Conveyor

Figure 9-23 Alias tags to assign.
© Cengage Learning 2014

INTRODUCTION TO ARRAYS

Figure 9-2 illustrated the basic ControlLogix data types and the memory required for each tag created with that data type. Remembering the rule introduced earlier in this chapter, the minimum memory allocation for any tag is a double integer. Figure 9-24 is the same as Figure 9-2. The figure illustrates the amount of memory consumed for each of the basic ControlLogix data types. As shown in the figure, assigning a BOOL, SINT, or INT as a data type can be inefficient for memory allocation. One way to become more efficient is to group data of the same data type together in an array. Any array must be entirely of the same data type. We can create an array of any of the basic data types: an array of counters, an array of timers, or an array of integers. A counter array can only contain counters. An array of timers can only contain timers. An array of integers can only contain integers. In other words, we cannot create an array that contains timers, counters, and integers. To group dissimilar data types, a user-defined data can be created. We do not work with user-defined data types in this lesson.

Figure 9-24 Memory consumed for basic data types.

Say we had four tags: tag A with the value 1,234, tag B with the value 2,345, tag C with the value 3,456, and tag D with the value 23,456, all with the data type of INT. Notice in Figure 9-25 that one double integer was consumed in memory for each of the integer tags. In this situation, 50 percent of the memory was wasted as the result of assigning the data type of integer to each of these four tags. We could create a group or array of the same integers.

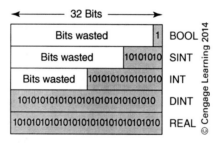

Figure 9-25 Memory used for tags defined as Integer data type.

Figure 9-26 illustrates memory allocation through the creation of an array of integer tags.

32 Bits	
2,345	1,234
23,456	3,456

Figure 9-26 Efficient memory arrangement of tags within an array.

Note how each integer has been stored as one-half of a double integer. One integer is in the upper word; the other integer is in the lower word. Here no memory was wasted. This concept also can be carried forward with BOOL and SINTs. Keep in mind that tags are arranged alphabetically in the Edit and Monitor Tags windows, so one cannot see how the information is actually stored in memory. To allocate tags and memory efficiently when working with tags, remember the basic concept of how data are stored, as represented in Figures 9-25 and 9-26.

Array Dimensions

An array can be created as one, two, or three dimensions. A single-dimension array resembles a list. The list of integers shown in Figure 9-27 could represent production data for the week and could be represented as a one-dimensional array.

Work Shift	Total Parts Produced
Sunday day	23,465
Sunday night	22,567
Monday day	31,432
Monday night	29,456
Tuesday day	28,654
Tuesday night	32,675
Wednesday day	15,790
Wednesday night	32,454

Figure 9-27 Representing production totals with a one-dimensional array.
© Cengage Learning 2014

A two-dimensional array could contain a table of data. The table has rows and columns, or two dimensions. The data type of the array depends on the largest number to be stored. An array of integers could store a value as large as 32,767 in each element. Larger numbers could require the array to be DINT or real numbers. If a decimal point is required, the array would need to be an array of real numbers. Remember: Data types cannot be mixed in an array. The table in Figure 9-28 is storing some type of data. The table has rows and columns, or two dimensions. The maximum array size currently allowed with RSLogix 5000 software is 2 megabytes. Keep in mind there is no place to go and view the table illustrated in Figure 9-28. Tags are listed alphabetically in the tags collection as we studied earlier.

2,982	2,237	234	3,346	7,657
4,567	4,434	9,989	43	567
23	980	2,345	11,123	34
34,789	2,234	12	89	44,564
23,457	9	2,346	6	1,234

Figure 9-28 A two-dimensional array representing a table of data.
© Cengage Learning 2014

An example of a three-dimensional array could be a two-dimensional array with multiple layers or pages—sort of an electronic book. Arrays can be one, two, or three dimensions as long as they are arrays of the basic data types. A timer or counter, for example, has multiple parts. Each part is considered a structure because it comprises three DINTs—status bits, preset value, and accumulated value—each one being a separate DINT. Multidimensional arrays can also be created when working with a structure.

Figure 9-29 shows the Monitor Tags view of a project with examples of arrays. Note the following:

- For the tag Integers, the data type is INT[10]. This is a one-dimensional array of integers with the name Integers. The square brackets [10] in the Data Type column identify this as a single-dimension array. The array contains 10 integers.
- For the tag Production_Data, the data type is DINT[10,10]. This is a two-dimensional array of double integers with the name Production_Data. The square brackets [10,10] in the Data Type column identify this as a two-dimensional array. This array is a table that is 10 double integers wide by 10 double integers deep.
- For the tag Counter_Array, the data type is COUNTER[100]. This is a one-dimensional array of counters. This array contains 100 counters.
- For the tag Product_Information, the data type is DINT[10,10,10]. This is a three-dimensional array of double integers. The square brackets [10,10,10] in the Data Type column identify this as a three-dimensional array.

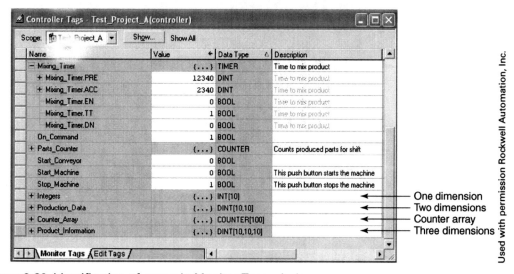

Figure 9-29 Identification of arrays in Monitor Tags window.

Figure 9-30 shows an expanded array to show the elements. Figure 9-29 has this array with a plus sign in front of the tag. Click the + to expand the array. Refer back to Figure 9-26 and see how the data are stored in the array in Figure 9-30. The element names can be aliased using the same procedures as in Lab Exercise 3. Assigning aliases for these elements will be done in the next lab exercise.

Figure 9-31 shows a two-dimensional array that has 10 by 10 elements. This array could be used to store the data in the table illustrated in Figure 9-28.

Figure 9-32 is an array of 100 counters. Early versions of the software only allowed structures to be single dimensional. The first counter element, Counter_Array[0], has been expanded by clicking + to show its elements.

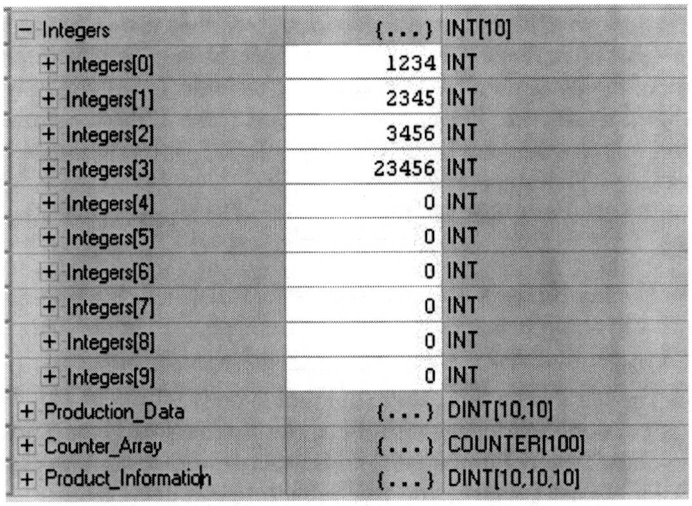

− Integers	{...}	INT[10]
+ Integers[0]	1234	INT
+ Integers[1]	2345	INT
+ Integers[2]	3456	INT
+ Integers[3]	23456	INT
+ Integers[4]	0	INT
+ Integers[5]	0	INT
+ Integers[6]	0	INT
+ Integers[7]	0	INT
+ Integers[8]	0	INT
+ Integers[9]	0	INT
+ Production_Data	{...}	DINT[10,10]
+ Counter_Array	{...}	COUNTER[100]
+ Product_Information	{...}	DINT[10,10,10]

Figure 9-30 Expanded one-dimensional array of integers showing its elements.

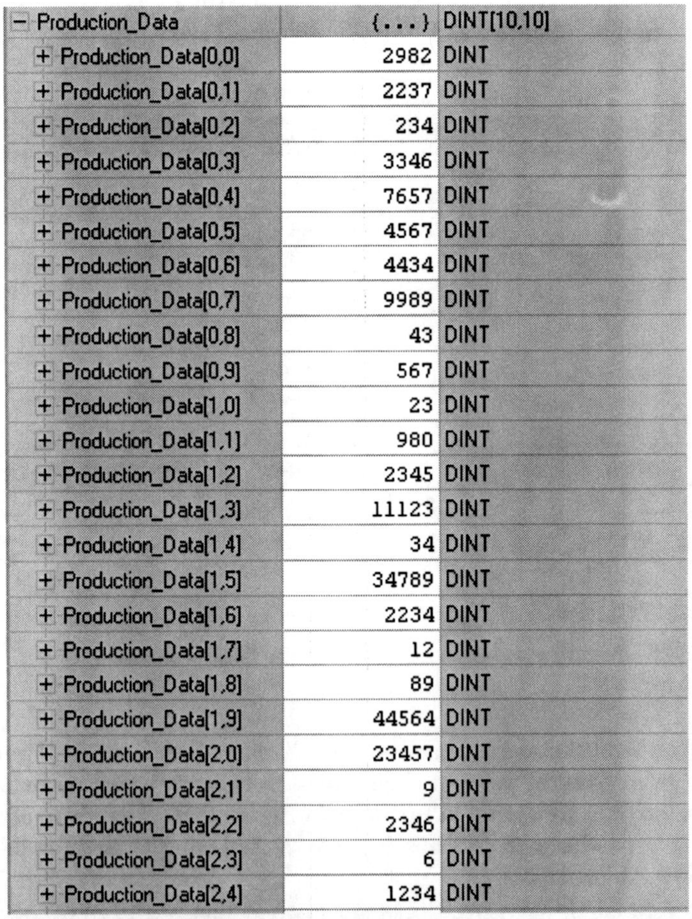

− Production_Data	{...}	DINT[10,10]
+ Production_Data[0,0]	2982	DINT
+ Production_Data[0,1]	2237	DINT
+ Production_Data[0,2]	234	DINT
+ Production_Data[0,3]	3346	DINT
+ Production_Data[0,4]	7657	DINT
+ Production_Data[0,5]	4567	DINT
+ Production_Data[0,6]	4434	DINT
+ Production_Data[0,7]	9989	DINT
+ Production_Data[0,8]	43	DINT
+ Production_Data[0,9]	567	DINT
+ Production_Data[1,0]	23	DINT
+ Production_Data[1,1]	980	DINT
+ Production_Data[1,2]	2345	DINT
+ Production_Data[1,3]	11123	DINT
+ Production_Data[1,4]	34	DINT
+ Production_Data[1,5]	34789	DINT
+ Production_Data[1,6]	2234	DINT
+ Production_Data[1,7]	12	DINT
+ Production_Data[1,8]	89	DINT
+ Production_Data[1,9]	44564	DINT
+ Production_Data[2,0]	23457	DINT
+ Production_Data[2,1]	9	DINT
+ Production_Data[2,2]	2346	DINT
+ Production_Data[2,3]	6	DINT
+ Production_Data[2,4]	1234	DINT

Figure 9-31 Two-dimensional array storing data from Figure 9-28.

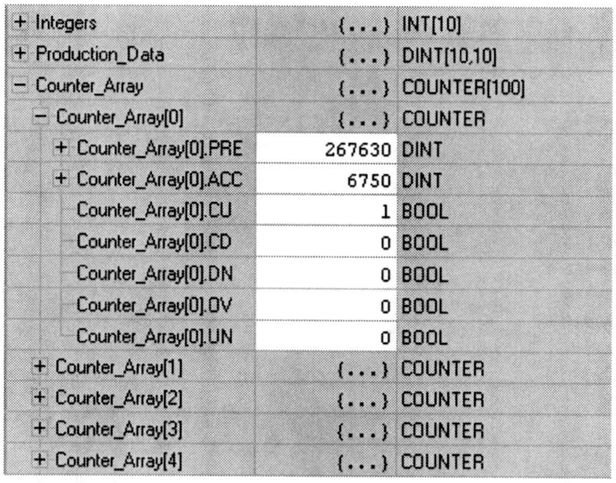

CREATING AND MONITORING RSLOGIX 5000 TAGS 249

+ Integers	{...}	INT[10]
+ Production_Data	{...}	DINT[10,10]
− Counter_Array	{...}	COUNTER[100]
− Counter_Array[0]	{...}	COUNTER
+ Counter_Array[0].PRE	267630	DINT
+ Counter_Array[0].ACC	6750	DINT
Counter_Array[0].CU	1	BOOL
Counter_Array[0].CD	0	BOOL
Counter_Array[0].DN	0	BOOL
Counter_Array[0].OV	0	BOOL
Counter_Array[0].UN	0	BOOL
+ Counter_Array[1]	{...}	COUNTER
+ Counter_Array[2]	{...}	COUNTER
+ Counter_Array[3]	{...}	COUNTER
+ Counter_Array[4]	{...}	COUNTER

Figure 9-32 Array of 100 counters.

Figure 9-33 is the three-dimensional array of product information from Figure 9-29.

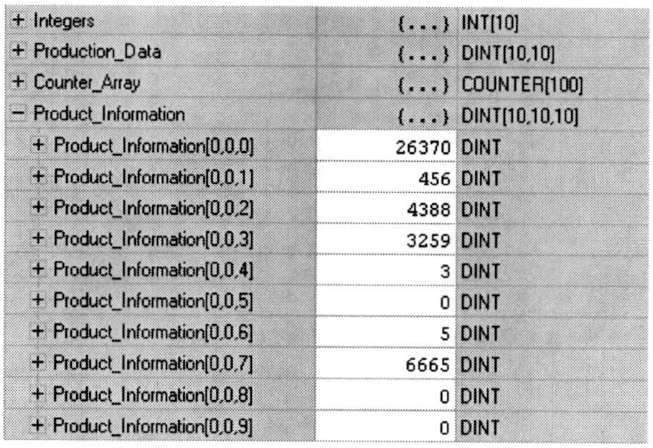

+ Integers	{...}	INT[10]
+ Production_Data	{...}	DINT[10,10]
+ Counter_Array	{...}	COUNTER[100]
− Product_Information	{...}	DINT[10,10,10]
+ Product_Information[0,0,0]	26370	DINT
+ Product_Information[0,0,1]	456	DINT
+ Product_Information[0,0,2]	4388	DINT
+ Product_Information[0,0,3]	3259	DINT
+ Product_Information[0,0,4]	3	DINT
+ Product_Information[0,0,5]	0	DINT
+ Product_Information[0,0,6]	5	DINT
+ Product_Information[0,0,7]	6665	DINT
+ Product_Information[0,0,8]	0	DINT
+ Product_Information[0,0,9]	0	DINT

Figure 9-33 Product_Information three-dimensional array.
Used with permission Rockwell Automation, Inc.

LAB EXERCISE 4: Creating a One-Dimensional Array

This lab steps you through creating the single-dimensional array we just looked at in the preceding figures. We add these arrays to the completed Lab Exercise 3. After creating an array, we alias off of the array elements.

1. _____ Open the project you created for Lab Exercise 3.
2. _____ Open the controller scoped tags collection.
3. _____ Open the Edit Tags window. We will create the one-dimensional array in Figure 9-30.
4. _____ Click the bottom row of the tags list and enter the tag Integers.
5. _____ Refer to Figure 9-34 as you create your array.
6. _____ Click item labeled number 1 in the figure to get to the Select Data Type screen.
7. _____ Select Integer as the data type. See the item shown by the number 2.
8. _____ Refer to item number 3—dimension 0 (Dim 0)—and either use the up arrow or enter in 10 for the number of elements.
9. _____ Select OK when done.
10. _____ You might remember the table storing production data for each day. The figure is included below. We need to modify our array to be able to store one week's production data. Because we have two shifts per day, we need 14 elements to store a week's data.

250 CREATING AND MONITORING RSLOGIX 5000 TAGS

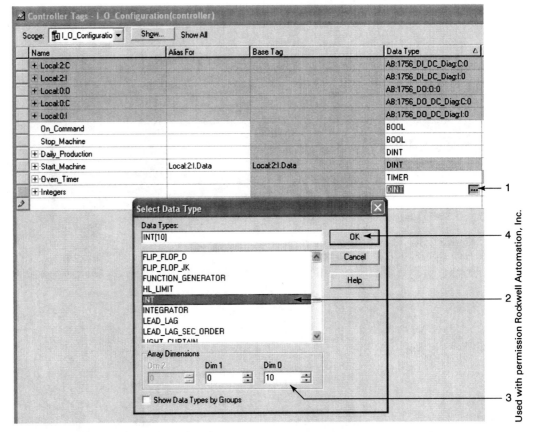

Figure 9-34 Creating a one-dimensional array.

11. _____ Return to the way you entered the array's dimensional value and change it to 14 elements.
12. _____ While still in the Edit Tags window, create aliases for each array element to the proper day of the week and shift. Enter the tag "Sunday_Day_Shift" in the bottom row of the Edit Tags window, just like you did when you created aliases earlier. Now make this tag an alias off Integers[0].
13. _____ Create the aliases for each day of the week, just like in Figure 9-35. When completed, you should have aliased all 14 elements to the proper day and shift. Your tags collection should look like those in Figure 9-36.

Work Shift	Total Parts Produced
Sunday–Day–Shift	23,465
Sunday–Night–Shift	22,567
Monday–Day–Shift	31,432
Monday–Night–Shift	29,456
Tuesday–Day–Shift	28,654
Tuesday–Night–Shift	32,675
Wednesday–Day–Shift	15,790
Wednesday–Night–Shift	32,454

Figure 9-35 Alias off the array elements to store production shift data.
© Cengage Learning 2014

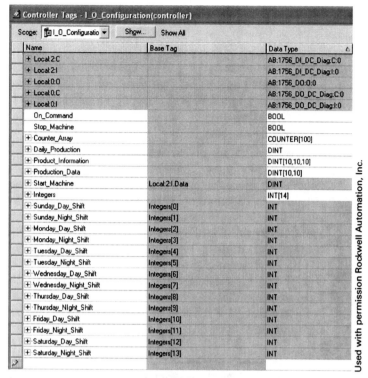

Figure 9-36 Days and shifts aliases to elements of our array.

LAB EXERCISE 5: Creating a Two-Dimensional Array

This lab guides you through the creation of the two-dimensional array from Figure 9-31. We add these arrays to the completed Lab Exercise 4.

1. _____ Open the project you created for Lab Exercise 4.
2. _____ Open the controller scoped tags collection.
3. _____ Open the Edit Tags window. We will create the two-dimensional array from Figure 9-31.
4. _____ Click the bottom row of the tags list and enter the tag Production Data.
5. _____ Refer to Figure 9-34 as you create your array.
6. _____ Click the area noted by the number 1 in the figure to get to the Select Data Type screen.
7. _____ Select DINT as the data type. Use number 2 from the figure as a guide.
8. _____ Referring to the item numbered 3, select dimension 0 (Dim 0) and either use the up arrow or enter 10 for the number of elements.
9. _____ From Figure 9-34, note the Dim 1 box to the left of Dim 0. Enter the second dimension—10—in this box.
10. _____ You do not need to enter the values. Select OK when done.
11. _____ If you expand your array, it should look similar to that shown in Figure 9-31.

LAB EXERCISE 6: Creating a Counter Array

This lab steps you through creating a single-dimensional array of counters similar to that in Figure 9-32. We add this array to Lab Exercise 5.

1. _____ If it is not already open, open the project you created for Lab Exercise 5.
2. _____ Open the controller scoped tags collection.

3. _____ Open the Edit Tags window.
4. _____ Click the bottom row of the tags list and enter the tag Counter_Array.
5. _____ Refer to Figure 9-34 as you create your array.
6. _____ Click the area numbered 1 in the figure to get to the Select Data Type screen.
7. _____ Select Counter as the data type. Use number 2 from the figure as a guide.
8. _____ Referring to item number 3 (Dim 0), either use the up arrow or enter 100 for the number of elements.
9. _____ Select OK when done.
10. _____ Expand the array so you can see all of the counters.
11. _____ Expand array element Counter_Array[0]. Your array should look similar to that in Figure 9-32. Because you do not have any counters programmed and in use in your project, all of the numbers in the value column should be 0. Counter_Array[0].PRE is the preset value of the counter. The preset is how many counts the counter will count before it is considered done counting. The tag Counter_Array[0].ACC is the accumulated value. The accumulated value tracks the number of seen or accumulated counts. The CU, CD, DN, OV, and UN tags are the counter's status bits. The tag Counter_Array[0].DN is the done bit. The done bit is true when the accumulated count is greater than or equal to the preset tag value. If you go to Counter_Array[0].PRE and enter a value, you have assigned the preset value to that counter.

LAB EXERCISE 7: Creating a Three-Dimensional Array

This lab guides you through the creation of the three-dimensional array in Figure 9-33. We add these arrays to Lab Exercise 6.

1. _____ From the Edit Tags window, create the three-dimensional array from Figure 9-33.
2. _____ In Figure 9-34, note the Dim 2 box to the left of Dim 1. Enter the third dimension—10—in this box.
3. _____ Select OK when done.
4. _____ If you expand your array, it should look similar to that in Figure 9-33.

LAB EXERCISE 8: Creating Arrays

This lab gives you practice creating arrays on your own. Add the following arrays to your existing project.

1. _____ Create a single dimensional array of 25 timers.
2. _____ Alias off the first four timers. Create the following aliases:
   a. Fruit punch mixing timer
   b. Grape punch mixing timer
   c. Cherry punch mixing timer
   d. Orange punch mixing timer
3. _____ Create a single-dimensional array of 100 BOOL.
   a. When you click off the line on which the tag was created, what happens to the size of the array?
   b. Explain why this happened.

4. _____ Create a three-dimensional array of DINTs to store part rejection data for production line 1 in your plant. We want to store data for three shifts, Sunday through Saturday. There are three shifts per day. Let's assume Sunday is day 1. The first shift is shift 1.
   a. Design your array.
   b. Create your array adding it to your existing project.
   c. Alias off the array to provide human-readable tags.

SUMMARY

Traditional PLCs like the Allen-Bradley PLC-5 and SLC 500 allocate a portion of the controller's physical memory for storage of program-related information. To make it easier for the programmer to work with, the memory is segmented into data files. ControlLogix is a tag- or name-based PLC. ControlLogix and its associated RSLogix 5000 software have no predefined data files as found in traditional PLCs. When creating a tag in RSLogix 5000 software, assign a unique name called a *tag* and then select the data type. This lesson introduced creating tags and alias tags. The lab exercises provided hands-on experience in creating arrays as a way to group data together in a block. An array can be multidimensional: one, two, or three dimensions. Arrays can be used as a way to become more efficient as well as increase tag organization in your RSLogix 5000 project. Remember: Any array must be entirely of the same data type. When using RSLogix 5000 software, dissimilar data types can be grouped together into a structure called a *user-defined data type*.

REVIEW QUESTIONS

Note: For ease of handing in assignments, students are to answer using their own paper.

1. If you create a tag that represents a push button, it is a single-bit, or a _____ tag.
2. To create a tag representing how long to mix a solution, we could create a tag called *mixer*. The data type for this tag is _____.
3. If we wanted to create a tag with fractional accuracy to store values such as 1,654.5940, the data type is a _____.
4. A ControlLogix tag that is of the integer data type has a data range of _____ to positive _____.
5. ControlLogix basic allocation of memory for any tag is _____ bits, or a _____.
6. There are two collections of tags within a ControlLogix project: controller scoped and _____ scoped.
7. Fill in the missing pieces in the table in Figure 9-37. You do not need to fill in the Floating Point-Data range entry.

Common Term	ControlLogix Data Type	Definition	Data Range
Bit			
		Short integer	
			−32,768 to 32,767
Double Word			
Floating Point	REAL		

Figure 9-37 Data Type terminology.
© Cengage Learning 2014

8. A DINT comprises _____ words and _____ bytes.
9. _____ scoped tags are global tags—that is, they can be used anywhere within the current RSLogix 5000 project.

10. ControlLogix and its associated RSLogix 5000 software have no predefined _____ files as found in traditional PLCs.
11. When creating a tag in RSLogix 5000 software, we first assign a unique name that is called a _____.
12. After a tag name has been assigned, the kind of information or _____ is selected.
13. As an example, if you created a BOOL tag, _____ bits of memory would be consumed.
14. A 16-bit tag, similar to an integer in a PLC-5 or SLC 500, would consume _____ bits.
15. Creating an array is a way of _____ data together.
16. An array can be multidimensional. Arrays can be created as _____, _____, or _____ dimensions.
17. Describe what the tag Part_Serial_Number[3] represents. _____
18. Program scoped tags are local to a specific _____.
19. RSLogix 5000 has two windows for working with tags. The _____ Tags window is for creating and editing tags, and the _____ Tags window is used for monitoring a project's tags.
20. Arrays can be used as a way to become more _____ and provide better organization of tags in your RSLogix 5000 project.
21. Any array must be entirely of the same _____.
22. RSLogix 5000 software allows grouping dissimilar data types together into a structure called a _____.
23. One tag in an array is called an _____.
24. Even though there is only one controller scoped collection of tags, each _____ has its own program scoped tags collection.
25. Describe what the tag Production_Data[2,3] represents. _____
26. No _____ are allowed in a tag name.
27. Looking in the Monitor Tags window, we see the following:
    Tag = Counters
    Data Type is Counters[500]
    a. How many counters are included?
    b. The counter _____ would be numbered from 0 to _____.
28. When creating arrays, the dimensions are assigned in the _____ window.
29. With ControlLogix, there can be _____ levels of aliasing. The _____ tag would be the starting tag. The Alias For column could have the _____ alias, and the _____ column could have the second alias.
30. If _____ is checked in the Workstation Options window and you start entering the alias name above the instruction base tag, the software will find the tag as you enter the beginning letters.
31. What does the Style parameter set up in the Edit or Monitor Tags window? _____
32. When adding two numbers together where the answer could possibly exceed 32,767, a _____ data type or _____ data type would be selected.
33. Describe the uses for the Description column in the Edit or Monitor Tags window. _____
34. If you created a BOOL tag, _____ bits of memory would be consumed.
35. A 16-bit tag, similar to an integer in a PLC-5 or SLC 500, consumes _____ bits of memory.

CHAPTER 10

Introduction to Logic

OBJECTIVES

After completing this chapter, you should be able to

- Explain AND logic.
- Explain OR logic.
- Explain NOT logic.
- Develop and fill in a truth table for specified logic.
- Determine whether a PAC ladder rung is true or false under specified conditions.
- Identify ControlLogix Boolean function blocks.

INTRODUCTION

This chapter introduces the concepts of how logic functions are executed when solving a PAC program. We introduce how a PAC solves a user program using AND, OR, and NOT logic.

Conventional hardwired relay ladder diagrams represent actual hardwired control circuits. In a hardwired circuit, there must be electrical continuity before the load will energize. Even though PAC ladder logic was modeled after the conventional relay ladder, there is no electrical continuity in PAC ladder logic. PAC ladder rungs must have logical continuity before the output will be directed to energize. The general rule for solving ladder logic is that if there is a path of true instructions to an output instruction, then the output instruction is true. Evaluating ladder rungs while thinking about current flow leads to problems in understanding how the rung will execute.

Function block diagram programming is a relatively new programming language for American PACs. Rockwell Automation's ControlLogix family of PACs is the first Rockwell Automation/Allen-Bradley PAC that supports function block diagram programming. *Function block*, as it is called, is an additional programming language option where boxes called function blocks replace traditional relay ladder instructions. This chapter introduces Boolean function blocks. There are no ladder rungs in function block. The workspace is called a *sheet*—like a sheet of paper. When creating function block routines, the function blocks are placed on sheets and joined together with lines called *wires*. Boolean function blocks replace the traditional normally open and normally closed instructions. We look closer at the Function Block programming language in Chapter 18.

CONVENTIONAL LADDERS VERSUS PAC LADDER LOGIC

The familiar electrical ladder diagram is the traditional method of representing an electrical sequence of operation in hardwired relay circuits. Ladder diagrams are used to represent the interconnection of field devices. Each rung of the ladder clearly illustrates the relationship of turning on one field device and shows how it interacts with the next field device. Due to industry-wide acceptance, ladder diagrams became the standard method of providing control information to users and designers of electrical equipment. With the advent of the programmable controller, one of the specifications for this new control device was that it had to be easily programmed. As a result, programming fundamentals were developed directly from the old familiar ladder diagram format, with which electrical maintenance personnel were already familiar. Even though there are multiple programming languages available for modern PACs, ladder logic is still the most popular. As a reference point, some numbers have been suggested regarding the use of current PAC programming languages. These numbers suggest that ladder logic is used approximately 80 to 85 percent of the time, with function block used approximately 15 percent. Structured text and sequential function chart comprise 2 to 3 percent each.

The difference between a PAC ladder program and relay ladder rungs involves continuity. An electrical schematic rung has electrical continuity when current flows uninterrupted from the left power rail to the right power rail. Electrical continuity, as illustrated in Figure 10-1, is required to energize the load. An electrical current flows from L-1 through SW1 and onto the load, returning by way of L-2.

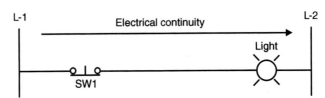

Figure 10-1 Hardwired relay circuit.
© Cengage Learning 2014

Even though a PAC ladder program closely resembles an electrical schematic, PLC ladder symbols represent ladder program instructions. A PAC program is a set of instructions that is stored in memory. These instructions tell the PAC what to do with input signals and then, as a result of following the instructions, where to send the output signals.

A PAC relay ladder program uses an electrician's ladder schematic as a model. Even though a PAC ladder program employs familiar terms like *rungs*, *normally open*, and *normally closed* contacts, relay ladder logic has no electrical continuity between an input and the controlled output. There is no physical conductor that carries the input signal through to the output. As illustrated in Figure 10-2, a PAC input signal follows these six steps:

1. The input signal is seen by the input module.
2. The input module isolates and converts the input signal to a low-voltage signal with which the PAC can work.
3. The ON or OFF signal from the input section is sent via the backplane at the RPI to the input tag, where it is stored.
4. The controller will look at each input's ON or OFF level as it solves the associated instruction.
5. The resulting ON or OFF action, as a result of solving each rung, is sent to the output tag for storage.
6. When the output module's RPI expires the ON or OFF signals from each output tag are sent to the associated output module by way of the backplane.

The PAC follows, or executes, the instructions stored in its memory the same way you might follow instructions to make, say, packaged grape drink. The package instructions for grape drink instruct you to do the following:

1. Measure 1 cup of sugar.
2. Put 2 quarts of water and the sugar into a container.
3. Add the contents of the package.
4. Mix until uniform.
5. Grape drink is ready to serve.

INTRODUCTION TO LOGIC 257

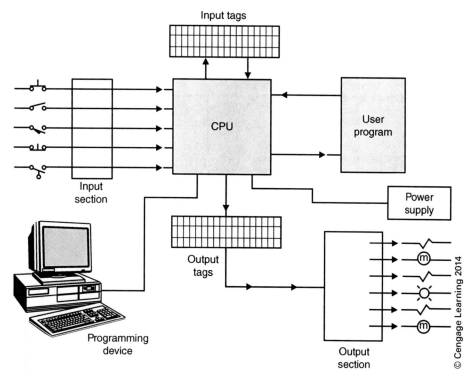

Figure 10-2 Signal flow into and out of a PAC. Notice that there is no electrical continuity between the inputs and the controlled output.

Following this procedure provides an end product (grape drink). Likewise, the PAC follows the instructions programmed on the ladder rungs to achieve an end product.

WHAT IS LOGIC?

Devices in an electrical schematic diagram are described as being open or closed. PAC ladder instructions are typically referred to as either true or false. When a PAC solves the user program, it is said to be solving the ladder logic.

See Figure 10-3 for a look at RSLogix ladder rungs.

Each rung is a program statement. A program statement consists of a condition, or conditions, along with some type of action. Inputs are the conditions, and the action, or output, is the result of the conditions. Each PAC ladder rung can be looked at as a problem the controller has to solve. The PAC combines ladder program instructions similar to those for physically wiring hardware devices in series or parallel. However, rather than working in series or parallel, the PAC combines instructions logically, using logical operators. Logical operations performed by a PAC are based on the fundamental logic operators: AND, OR, and NOT. These operators are used to combine the instructions on a PAC rung so as to make the outcome of each rung either true or false. The symbol that represents the result of solving the input logic on a particular rung is an output.

OVERVIEW OF LOGIC FUNCTIONS

To understand and program programmable controllers, we must understand basic logic. Three logic functions are introduced here.

258 INTRODUCTION TO LOGIC

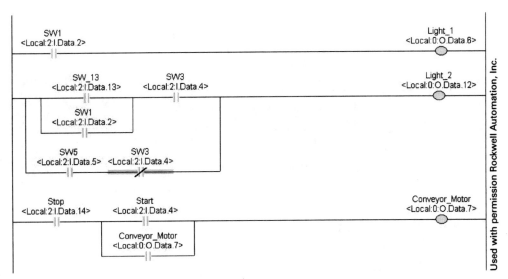

Figure 10-3 Ladder rungs for ControlLogix using Rockwell Automation's RSLogix 5000 software.

One Instruction Combined in Series with Another Is "AND"

You performed a logical operation when you mixed the grape drink. Although you might not have realized it, you performed AND logic. Let's rewrite our drink-mixing task to see how it relates to AND logic:

1. Measure 1 cup of sugar.
2. AND put 2 quarts of water AND the sugar into a container.
3. AND add contents of package.
4. AND mix until uniform.
5. Grape drink is ready to serve.

AND logic is similar to placement in series, as all series devices must pass continuity before the outcome is allowed to happen. Likewise, when mixing the grape drink, all steps must be performed before a satisfactory beverage is produced.

One Instruction Combined in Parallel with Another Is OR

Our drink-mixing example can be made into OR logic (a parallel operation). When mixing our drink, there may be a choice whereby grape flavoring OR orange flavoring can be used. By following the same instructions but then choosing either grape flavor or orange flavor, OR logic is carried out.

The Opposite of a Normally Open Instruction Is a Normally Closed Instruction

If a set of normally open contacts is represented by a normally open instruction and a set of normally closed contacts is represented by a normally closed instruction, the normally open instruction must be energized to become true, whereas the normally closed instruction must not be energized to be true. In PAC logic, the instructions do not really become energized. The normally open instruction actually examines the state of the input bit stored in the input tag being referenced by the instruction. If the normally open instruction finds a 1 in that bit, then the instruction is true. Rockwell Automation PACs use the term *examine* if closed. The normally open instruction examines the bit to see whether it is closed. As a result, the instruction will be referred to as an *examine* if closed, or XIC, instruction. If the bit being examined contains a 1, then the XIC instruction is considered true. The normally closed instruction also examines the state of the input bit stored in the input tag being referenced by the instruction. If the normally closed instruction finds a 0 in that bit, then the instruction is true. Rockwell Automation PACs

INTRODUCTION TO LOGIC 259

use the term *examine* if open. The normally closed instruction examines the bit to see whether it is open, or a 0. As a result, the instruction is referred to as an *examine* if open, or XIO, instruction. If the bit being examined contains a 0, then the XIO instruction is considered true. Because there are only two states in digital logic, the normally closed instruction represents the opposite of the normally open instruction.

Let's explore these logic functions and see how they relate to PAC ladder programs.

SERIES—THE *AND* LOGIC FUNCTION

The old familiar series circuit can also be referred to as an AND logic function. In the series circuit (Figures 10-4 and 10-5), switch 1 AND switch 2 must be closed to have electrical continuity. When there is electrical continuity, output (Light 1) will energize. The key word here is AND.

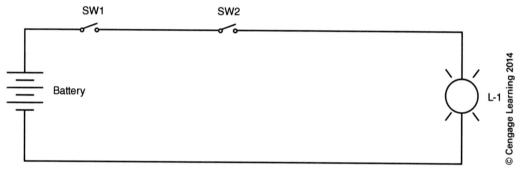

Figure 10-4 Conventional series circuit.

The circuit in Figure 10-4 is represented as a schematic diagram ladder rung in Figure 10-5.

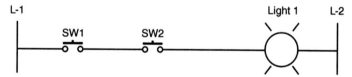

Figure 10-5 Series circuit represented as a conventional ladder rung.
© Cengage Learning 2014

Closing switch 1 and switch 2 provides power, or electrical continuity, to Light 1. This is illustrated in Figure 10-6.

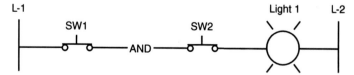

Figure 10-6 Switch 1 AND Switch 2 need to be closed to energize Light 1.
© Cengage Learning 2014

Let us look at all the possible combinations that switch 1 (SW1) and switch 2 (SW2) can have, and the resulting output signals (Figure 10-7).

POSSIBLE SWITCH CONDITIONS AND RESULTING OUTPUT		
SW1	SW2	Light 1
OFF	OFF	OFF
OFF	ON	OFF
ON	OFF	OFF
ON	ON	ON

Figure 10-7 Truth table for AND logic.
© Cengage Learning 2014

Figure 10-7 is called a truth table. All possible input configurations for switches 1 and 2 are listed in the two left-hand columns. The expected output signal for Light 1 is listed in the right-hand column. From the truth table, you can see that only when switch 1 AND switch 2 are ON will the output (Light 1) energize.

Figure 10-8 is an example of the ladder program instructions that would be entered using RSLogix 5000 software. The ladder rung in Figure 10-8 is identical to that in Figure 10-6 except that the symbols have been changed to PAC ladder format.

Figure 10-8 RSLogix 5000 representation of the rung represented in Figure 10-6.
Used with permission Rockwell Automation, Inc.

The instructions tell the controller to load input Local:2:I.Data.2 into memory, AND it with Local:2:I.Data.3, and then send the result to the output tag Local:0:O.Data.6. The resulting output is determined by the truth table (Figure 10-7). The truth table represents the rules for ANDing two inputs together.

Before we look at the truth table in Figure 10-9, let's update our input status from OFF/ON to a more commonly accepted form. OFF is the same as no power, or power not ON, and can be represented by the symbol 0. ON is the same as power ON, which can be represented by the symbol 1. We use the commonly accepted symbol 1 to represent the presence of a valid signal and the commonly accepted symbol 0 to represent the absence of a valid signal.

Our truth table for the previous example would look as follows (Figure 10-9):

TWO-INPUT *AND* TRUTH TABLE		
SW1	SW2	Light 1
0	0	0
0	1	0
1	0	0
1	1	1

Figure 10-9 Truth table for two-input AND logic. This truth table represents the same situation as the one in Figure 10-7.
© Cengage Learning 2014

The truth table in Figure 10-9 can also be represented as shown in Figure 10-10.

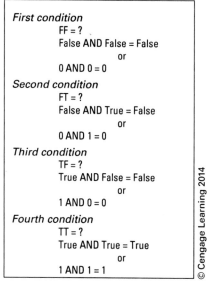

Figure 10-10 Explanation of Figure 10-9 truth table.

THREE-INPUT *AND* LOGIC

Figure 10-11 has three switches in series controlling the load L-1. The conventional series circuit is shown in Figure 10-12. Figures 10-11 and 10-12 state that switch 1 AND switch 2 AND switch 3 must be energized before output L-1 will occur.

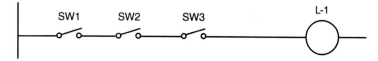

Figure 10-11 Three-input series circuit.
© Cengage Learning 2014

Figure 10-12 Three-input AND circuit.
© Cengage Learning 2014

Figure 10-13 illustrates the converted PLC ladder rungs for the SLC 500 PLC as the top rung. The center rung is for a PLC-5, and the bottom rung is for the ControlLogix. Each rung has its respective input and output addresses.

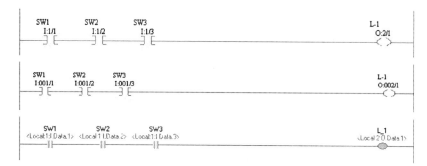

Figure 10-13 PLC three-input AND logic.
Used with permission Rockwell Automation, Inc.

Figure 10-14 is a truth table illustrating the expected outputs for three-input AND logic.

THREE-INPUT *AND* LOGIC			
Switch 1	Switch 2	Switch 3	Light 1
0	0	0	0
0	0	1	0
0	1	0	0
0	1	1	0
1	0	0	0
1	0	1	0
1	1	0	0
1	1	1	1

Figure 10-14 Three-input AND logic truth table.
© Cengage Learning 2014

FUNCTION BLOCK DIAGRAM *AND* LOGIC

Rockwell Automation's ControlLogix family can be programmed in function block diagram language in addition to standard ladder logic. The PLC-5 and SLC 500 families do not support function block.

The principles of combining function block inputs are basically the same as ladder logic. Instead of normally open ladder logic symbols, boxes are used and referred to as function blocks. Figure 10-15 illustrates three-input AND ladder logic. A function block diagram represents the same logic using a Boolean AND (BAND) function block. Figure 10-16 illustrates a BAND function block. Notice the three blocks to the left of the BAND function block, which include Data.1, Data.2, and Data.3 as part of the information inside. These are function block input references, which represent the input tags where the information is coming from. In this example, we are going to AND the three input references together. Here we are looking to see whether Data.1 AND Data.2 AND Data.3 are all true. That is the job of the BAND block.

Figure 10-15 Three-input AND ladder logic.
Used with permission Rockwell Automation, Inc.

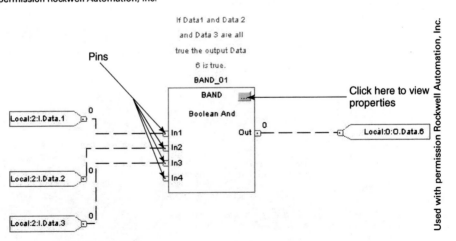

Figure 10-16 Function block BAND.

On the left side of the function block are the input points to the block, called *pins*. In1 through In4 are shown. This function block can contain up to eight input pins. In this example, because there are only three input references, only three input pins are used. When the inputs represented by Data.1 AND Data.2 AND Data.3 are true, the BAND function block will be true. Being true, the output (out pin) of the block will be true. The dotted line or wire connected between the output pin and the output reference symbol identified as Data.6 represents the output data as a bit. Notice the 0 just to the right of the input reference and out pin. This identifies the logical state of the input reference or output pin for the instruction. A 0 identifies the reference tag as false, whereas a 1 signifies the reference tag is true. In this example, each input reference has a 0 representing its input state as false. Because we do not have logical continuity, when executed, the BAND block will mark its output as false, a 0. As a result, the associated output reference and its tag will also be false. This function block diagram is equivalent to the ControlLogix rung from Figure 10-15.

Clicking the View Properties box in the upper-right-hand corner of the BAND function block reveals the BAND Properties view as illustrated in Figure 10-17. Notice that there are eight input pins for this function block, labeled In1 through In8. Checking the boxes in the visibility (Vis) column turns on the four input pins and displays them on the function block. Because In5 through In8 are not checked, these pins will not be displayed on the function block. This illustrates how the programmer can select the function block options that are specifically needed for this application. Notice that the Out pin is also checked. Other properties and their associated pins that are not checked will not be displayed on the function block.

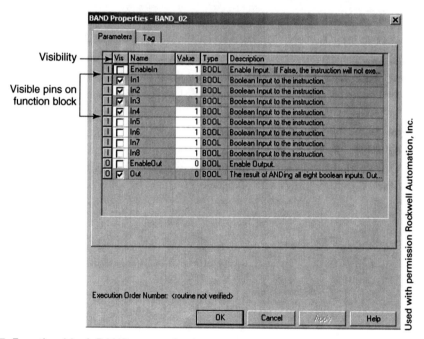

Figure 10-17 Function block BAND properties box.

PARALLEL CIRCUITS—THE *OR* LOGIC FUNCTION

The familiar parallel circuit can also be referred to as the OR logic function. The rule of OR logic is that if any input is true, the output will also be true. OR logic also states that if all inputs are true, the output will be true. In Figure 10-18, if switch 1 OR switch 2 is energized, Light 1 will energize. If both SW1 and SW2 are true, the output will also energize.

Figure 10-19 illustrates Figure 10-18 converted to a ControlLogix ladder rung. Remember, when drawing programmable controller ladder diagrams, do not use the conventional switch symbols such as we employed in the previous examples. A PAC rung of logic has normally open or normally closed contacts instead of normally open or closed switch symbols.

264 INTRODUCTION TO LOGIC

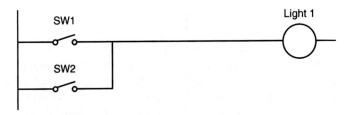

Figure 10-18 Conventional parallel circuit where switch 1 or switch 2 can energize the load Light 1.
© Cengage Learning 2014

Figure 10-19 RSLogix 5000 ladder rung for the ControlLogix.
Used with permission Rockwell Automation, Inc.

Figures 10-18 and 10-19 illustrate the same logic with different symbols.
A two-input OR truth table representing Figures 10-17 and 10-18 is illustrated in Figure 10-20.

TWO-INPUT *OR* TRUTH TABLE		
SW1	**SW2**	**Light 1**
0	0	0
0	1	1
1	0	1
1	1	1

Figure 10-20 Two-input OR truth table.
© Cengage Learning 2014

Notice that in the case of the OR circuit, if either switch is ON, the output will be true. In addition, if both switches are ON, the output will be ON.

Figure 10-21 shows a three-input parallel circuit, using three-input OR logic.

Figure 10-21 Three-input OR logic.
Used with permission Rockwell Automation, Inc.

What are the expected outputs from three-input OR logic for Figure 10-22? Remember that if one or more inputs are true, the output will be true.

The function block Boolean OR (BOR) is illustrated in Figure 10-23. This function block has the same functional components as the BAND. For this example if Data.1 OR Data.2 OR Data.3 are true, the output of the BOR function block will be true, or a 1. As with ladder OR logic, if any combination of input references is true the function block will be true. With the function block true, the output reference tag Data.6 will be true. Even though the input reference tags are different, this BOR function block logic is the same as the ladder rung illustrated in Figure 10-21.

INTRODUCTION TO LOGIC

THREE-INPUT OR LOGIC			
Switch 1	Switch 2	Switch 3	Light 1
0	0	0	0
0	0	1	1
0	1	0	1
0	1	1	1
1	0	0	1
1	0	1	1
1	1	0	1
1	1	1	1

Figure 10-22 Three-input OR logic truth table.
© Cengage Learning 2014

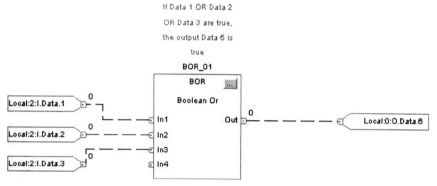

Figure 10-23 Function block BOR.
Used with permission Rockwell Automation, Inc.

The NOT logic operator works in a manner opposite from the AND and OR logic with which we have been working. The next section introduces the normally closed instruction and NOT logic.

NOT LOGIC

A normally closed hardwire relay contact passes power any time the relay coil is not energized. Likewise, the normally closed PAC ladder logic instruction is true any time the input tag bit is not a 1. This means that the physical hardware input is not sending an input signal into the PAC's input module. NOT logic is the opposite of a normally open PAC instruction. It can be used in conjunction with AND or OR logic when a logical 0 in the status file is expected to activate some output device. The NOT logic function is used when an input must not be energized for an output to be energized. Likewise, the NOT logic function is used when a logical 1, or true input, is necessary to make the instruction false or deactivate an output device.

The truth table in Figure 10-24 simply states that a normally closed instruction on a PAC ladder rung is the inverse, or opposite, of the input tag associated with the specific instruction. If the input tag bit is a 1, or true, the normally closed instruction will be false. In comparison, when the input tag bit is false, or a 0, the associated normally closed instruction will be true.

TWO-STATE LOGIC FUNDAMENTALS		
Input Signal to Input Module	Normally Open PAC Instruction	Normally Closed PAC Instruction
ON	TRUE	FALSE
OFF	FALSE	TRUE

Figure 10-24 Truth table for NOT logic.
© Cengage Learning 2014

The NOT logic function is somewhat difficult to grasp. Let's look a little closer at the relationship between the normally open contact and how it controls the output in comparison to the normally closed contact. Figure 10-25 illustrates two rungs, the first with a normally open instruction and the second with a normally closed instruction. Rungs are displayed in their off line state.

Figure 10-25 ControlLogix ladder logic with XIC and XIO instructions.
Used with permission Rockwell Automation, Inc.

ANALYSIS OF RUNG #1

Instruction SW1 energizes the output only when there is a logical 1 in its associated input tag bit. A 1 in this bit position causes the normally open instruction (XIC) to become true and change state. In changing state, the instruction allows logical continuity to pass on to the output instruction and make it true. SW1 is considered true when it passes logical continuity. If there is no valid input signal from the field device attached to SW1's screw terminal on the input module, a logical 0 will be placed in the input tag. A logical 0 in the input tag results in the normally open (XIC) input instruction becoming false. Being false, the instruction will not pass logical continuity.

ANALYSIS OF RUNG #2

The normally closed instruction works much like the normally closed contacts on a hardware relay. Being normally closed, instruction SW2 will energize the output only when there is a logical 0 in its associated input tag. Even though there is a logical 0, or false input signal, in the tag, the normally closed (XIO) instruction is true and passes logical continuity on to the output instruction. If there is a valid ON input signal from the field device attached to SW2's input -module screw terminal, a logical 1 will be placed in the input module tag. A logical 1 in the input tag causes the normally closed (XIO) instruction to change state. The normally closed (XIO) instruction changes from true to false. Being false, the normally closed (XIO) instruction will not pass logical continuity to the output instruction. Without logical continuity, the output instruction becomes false.

The function block Boolean NOT (BNOT) is illustrated in Figure 10-26. If the input reference representing input Data.1 is true, the output pin of the BNOT function block will be false, or a 0. See Figure 10-26. The output is the opposite of the input.

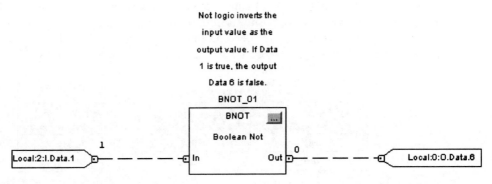

Figure 10-26 Function block NOT.
Used with permission Rockwell Automation, Inc.

PARALLEL *NOT* LOGIC

Figure 10-27 is that of a parallel ladder rung with a normally open (XIC) instruction in parallel or ORed with a normally closed (XIO) instruction. This is called OR NOT logic. This rung will be true under the conditions shown in Figure 10-28. SW1 must be true OR SW2 must NOT be true to make this rung true and make the Light_1 output true. Rungs are displayed in their off line state.

Figure 10-27 Parallel NOT logic.
Used with permission Rockwell Automation, Inc.

INPUTS		OUTPUTS
SW1	SW2	Light_1
0	0	1
0	1	0
1	0	1
1	1	1

Figure 10-28 NOT logic truth table.
© Cengage Learning 2014

Not all PLC manufacturers use the same terminology to identify the normally open and normally closed contact instruction. Figure 10-29 is a sample of terminology used with different PLCs. When using Rockwell Automation's PLC-5, SLC 500, MicroLogix, or ControlLogix PACs, the normally open instruction is called an Examine if Closed, or XIC, instruction. The normally closed instruction is the Examine if Open, or XIO.

DIFFERENT TERMINOLOGY USED FOR NORMALLY OPEN AND NORMALLY CLOSED INPUT INSTRUCTIONS	
—||—	Normally open Examine if closed AND
—|/|—	Normally closed Examine if open AND invert OR invert AND NOT

Figure 10-29 Normally open and normally closed instruction identification.
© Cengage Learning 2014

OR logic states that when any or all inputs are true, the associated output will be true. Can we develop logic to give a true output if one or the other parallel inputs is true, but not both? To solve this problem, we look at exclusive OR logic in the next section.

EXCLUSIVE *OR* LOGIC

Reviewing a truth table for a two-input OR logic function, we see that there are three input conditions that will give us an output signal (see Figures 10-30 and 10-31):

1. If SW1 input is false and input SW2 is true, then Light_1 will be true.
2. If SW1 input is true and input SW2 is false, then Light_1 will be true.
3. If SW1 input is true and input SW2 is true, then Light_1 will be true.

Figure 10-30 Two-input OR logic.
Used with permission Rockwell Automation, Inc.

TWO-INPUT *OR* TRUTH TABLE		
SW1	SW2	Light 1
0	0	0
0	1	1
1	0	1
1	1	1

Figure 10-31 Two-input OR truth table.
© Cengage Learning 2014

The exclusive OR logic function allows either input 01 OR input 02, but not both together, to control the output (see Figure 10-32).

TRUTH TABLE FOR EXCLUSIVE *OR* LOGIC IN FIGURE 10-33		
SW1	SW2	Light_1
0	0	0
0	1	1
1	0	1
1	1	0

Figure 10-32 Exclusive OR logic truth table.
© Cengage Learning 2014

The logic for exclusive OR (sometimes referred to as XOR) looks as follows (see Figure 10-33).

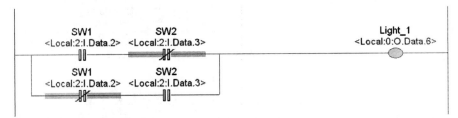

Figure 10-33 Exclusive OR logic.
Used with permission Rockwell Automation, Inc.

ANALYSIS OF EXCLUSIVE *OR* LOGIC

If XIC input SW1 is true and XIO input SW2 is left as is, the logic on the main rung will become true, thus enabling the output. As for the state of the XIO SW1 instruction on the parallel branch, with the XIC input instruction true, the XIO for input SW1 on the parallel branch will be false. With the XIC from input SW1 false on the parallel branch, input SW2 cannot control the output.

Input SW2's logic operates in the same manner. If SW2's XIC instruction becomes true while SW1's instruction remains in its normal state, the parallel branch will become true. With the parallel branch true, the rung will be true. The rung will be true as there is logical continuity in the branch. With the XIC SW2 true on the parallel branch, the XIO SW2 on the main rung will go false and prevent SW1 from controlling the output.

If by chance both SW1 and SW2 are energized (and therefore change from their normal state), their XIO counterparts will both become false. With a false on the main rung and the parallel branch, there is no way for the rung to become true.

Figure 10-34 illustrates the function block Boolean exclusive OR (BXOR). If the input reference representing Data.1 OR the input reference representing Data.2 is true, but not both, the BXOR output will be true. The figure illustrates that both input references are true. As a result, the output of the BXOR will be false. With the output false, output reference Data.6 will also be false.

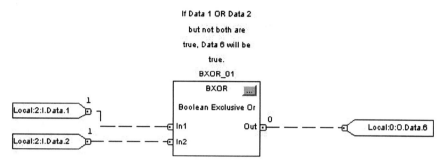

Figure 10-34 Function block XOR.
Used with permission Rockwell Automation, Inc.

COMBINATIONAL LOGIC

Most ladder rungs include some combination of AND, OR, and NOT logic. No matter what the logic combination or how many logic elements or instructions are on a rung, there must be at least one path of true instructions before the output can be made true.

Figure 10-35 illustrates an example of combinational input as well as combinational output logic.

If SW1 AND SW2 OR SW3 AND SW4 are all true:

- Light_1 will be true.
- AND if SW5 is true, Light_2 will be true.
- Light_3 will be true.

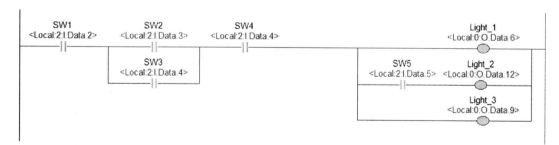

Figure 10-35 Combinational AND-OR logic.
Used with permission Rockwell Automation, Inc.

In Figure 10-36, the ladder rung has four logical paths by which it can be true:

1. If SW2 AND SW4 AND SW5 are all true, then output Light_1 will be true.
2. If SW2 AND SW4 AND SW1 are all true, then output Light_1 will be true.
3. If SW3 AND SW4 AND SW5 are all true, then output Light_1 will be true.
4. If SW3 AND SW4 AND SW1 are all true, then output Light_1 will be true.
5. If all input switches are true, then output Light_1 will be true.

270 INTRODUCTION TO LOGIC

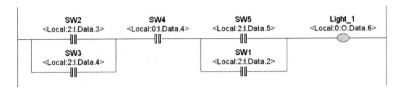

Figure 10-36 Combinational AND-OR logic. Four paths can make this rung logically true.
Used with permission Rockwell Automation, Inc.

In Figure 10-37, the ladder rung has many logical paths by which it can be true. Study the figure and see how many paths you can find.

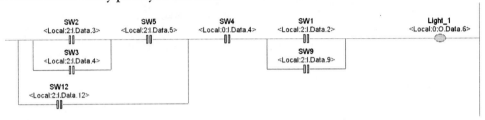

Figure 10-37 Combinational input logic.
Used with permission Rockwell Automation, Inc.

The ladder rungs we have been looking at in this chapter were programmed with input instructions on the left side of the ladder rung and feeding into a single output instruction on the right side of the rung. Even though Figure 10-35 illustrated a rung in which outputs could be put in parallel, outputs cannot be programmed in series. This has been the rule with earlier PLCs such as the PLC-5 and SLC 500 families. ControlLogix with its RSLogix 5000 software offers a new feature in which inputs and outputs can be interlaced on the same rung.

INTERLACING INPUTS AND OUTPUTS WITH RSLOGIX 5000 AND CONTROLLOGIX

A new programming feature with RSLogix 5000 software is the ability to interlace inputs and outputs on a ladder rung as long as the last instruction on a rung is some type of output instruction. When evaluating the rung in Figure 10-38, remember the rule that if there is a path of true instructions to an output, then that output is true. Thinking about solving rungs while thinking of current flow through the rungs will lead to major difficulties.

Figure 10-38 RSLogix 5000 interlaced inputs and outputs on same rung.
Used with permission Rockwell Automation, Inc.

Remember the rule: If SW2 AND SW5 is true, then Light_4 will be true. At this point, do not worry about the rest of the rung. If we have a path of true instructions to the output, then Light_4, the output, will be true. If we continue evaluating the rung, if SW1 is also true, then we have a path of true instructions to the outputs Light_1 and Light_2. Light_1 and Light_2 both will be true. Now assume SW2 was true and SW5 was false. Output Light_4 would be false because there is no path of true instructions to output Light_4. Likewise, if SW1 were also true, then Light_1 and Light_2 both would be false because currently there is no path of true instructions to the outputs if SW5 were currently false. Let's look at another example in Figure 10-39.

Look at the ladder rung and consider the following statements:

1. If there is a path of true instructions to Light_4, then the output Light_4 will be true.
2. If there is a path of true instructions to Light_1, then the output Light_1 will be true.

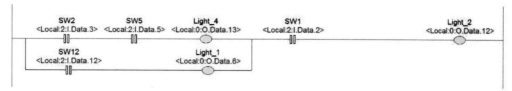

Figure 10-39 RSLogix 5000 interlaced inputs and outputs with OR logic.
Used with permission Rockwell Automation, Inc.

3. If Light_4 is true OR if Light_1 is true AND SW1 is also true, then Light_2 will be true.
4. If Light_4 is false OR if Light_1 is true AND SW1 is also true, then Light_2 will be true.
5. If Light_4 is true OR if Light_1 is false AND SW1 is also true, then Light_2 will be true.
6. If both Light_4 and Light _1 are false, even though SW1 might be true, Light_2 cannot be true because there is no path of true instruction to Light 2's output instruction.

Let's examine another possible ControlLogix ladder rung by looking at Figure 10-40.

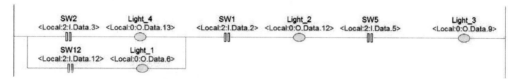

Figure 10-40 RSLogix 5000 interlaced inputs and outputs with combinational AND with OR logic.
Used with permission Rockwell Automation, Inc.

The table in Figure 10-41 shows the different combinations of inputs and the associated output states.

SW2	Light_4	SW1	Light_2	SW5	Light_3
T	T	T	T	T	T
F	F	T	F	F	F
F	F	T	F	T	F
F	F	F	F	T	F
T	T	T	T	F	F
T	T	F	F	T	F

Figure 10-41 AND logic truth table.
© Cengage Learning 2014

The table in Figure 10-42 shows the different combinations of inputs for the parallel branch and the associated output states. Remember that when using OR logic, if all inputs are true, then the output, Light_3, will be true.

SW12	Light_1	SW1	Light_2	SW5	Light_3
T	T	T	T	T	T
F	F	T	F	F	F
F	F	T	F	T	F
F	F	F	F	T	F
T	T	T	T	F	F
T	T	F	F	T	F

Figure 10-42 OR logic truth table.
© Cengage Learning 2014

Figure 10-43 is another example of interlacing inputs and outputs possible with ControlLogix and RSLogix 5000 software. If SW9 is true, then Light_4 will be true. If SW1 is also true, then the add instruction will execute and then the subtract instruction will execute. Because this rung has three outputs, Light_4, the add instruction along with the subtract instruction, the rung would not pass verification, error checking with the PLC 5 and its associated RSLogix 5 software, or the SLC500 and MicroLogix PLCs and their associated RSLogix 500 software. Currently, the only Rockwell Automation PAC that supports ladder rungs with inputs and outputs interlaced on the same rung is the ControlLogix family of PACs and its RSLogix 5000 software.

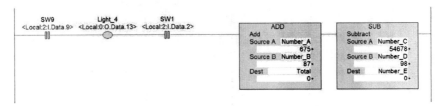

Figure 10-43 Interlaced math output instructions on a ControlLogix ladder rung.
Used with permission Rockwell Automation, Inc.

SUMMARY

The main focus of this chapter was to introduce the concepts of how logic functions are executed when solving a PAC program. We looked at how a PAC solves its user program using AND, OR, and NOT logic. PAC ladder logic differs when compared to conventional relay logic in one important aspect: Relay logic has electrical continuity, whereas PAC logic has only logical continuity. Where there is actual current flow in a hardwired relay circuit, there is only logical continuity on a PAC rung.

PAC instructions are considered either true or false. The normally open instruction is called Examine If Closed (XIC) instruction in Rockwell Automation PLCs or PACs. This instruction examines the bit in the associated tag for a 1. If there is a 1 in that tag, then the instruction is true. If the tag contains a 0, then the instruction is false. The normally closed instruction is called Examine If Open (XIO) instruction. This instruction examines the bit in the associated tag for a 0. If there is a 0 in that tag, then the instruction is true. If the tag contains a 1, then the instruction is false. Keep in mind that because the PAC examines the tag, the actual input field device cannot be seen by the controller, and only the tag bit value is considered when determining whether a specific instruction is to be evaluated as true or false. When evaluating ladder logic, the rule is this: If there is a path of true instructions to an output, then the output will be true. The last instruction on any rung must be some type of output instruction. If the last instruction on a rung is not an output, then the rung will not pass verification. All rungs must pass verification, or the PAC project cannot be downloaded to the controller. Currently, the only Rockwell Automation PAC that supports ladder rungs with inputs and outputs interlaced on the same rung is the ControlLogix family of PACs and its RSLogix 5000 software.

REVIEW QUESTIONS

Note: For ease of handing in assignments, students are to answer using their own paper.

1. PLCs use which of the following logic operators?
   a. AND
   b. OR
   c. NOT
   d. all of the above
2. The normally open instruction is called _____ or _____ instruction in Rockwell Automation PLCs and PACs.
3. The last instruction on any ladder rung must be some type of _____.

4. When evaluating ladder logic, if there is a path of true instructions to a(n) _____, then the output will be true.
5. The normally closed instruction is called _____ or _____ instruction.
6. Fill in the following truth table for NOT logic (Figure 10-44).

NOT LOGIC TRUTH TABLE		
Input Signal to Input Module	Normally Open PLC Instruction	Normally Closed PLC Instruction
ON		
OFF		

Figure 10-44 NOT logic truth table.
© Cengage Learning 2014

7. The logical AND function is similar to
   a. in parallel
   b. in series
   c. inverted logic
   d. both A and C
   e. depends on the application
8. The logical OR function is similar to
   a. in parallel
   b. in series
   c. inverted logic
   d. both A and C
   e. depends on the application
9. The logical OR NOT function is similar to:
   a. in parallel
   b. in series
   c. inverted logic
   d. both A and C
   e. depends on the application
10. Illustrate a two-input AND logic PLC ladder rung.
11. Develop a truth table for the answer in question 10.
12. Will the following rung in Figure 10-45 be true or false?

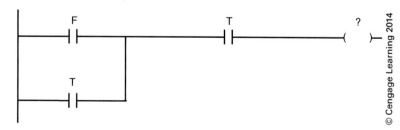

Figure 10-45 Rung for question 12.

13. The BAND function block is similar to
    a. Series ladder logic.
    b. Parallel ladder logic.
    c. NOT ladder logic.
    d. Boolean ladder logic.

14. The BAND function block
    a. ANDs up to twelve inputs.
    b. ANDs up to eight input references.
    c. logically combines up to eight input references.
    d. performs a logical AND/OR function.
15. A BXOR function block
    a. is true when all inputs are true.
    b. is true when neither input is true.
    c. is true only when one or the other input pin is true.
    d. is false when only one input is true.
    e. is false when both input references are true.
    f. C and E are correct.
    g. B and D are true.
16. A BNOT function block is true
    a. when both inputs are true.
    b. when the function block is true.
    c. when the order of execution is true.
    d. when the input is false.
    e. both B and C.
17. Currently, the only Rockwell Automation PLC or PAC that supports ladder rungs with inputs and outputs interlaced on the same rung is the _____ family.
18. This instruction examines the bit in the associated tag for a 1. If there is a 1 in that tag, then the instruction is true. What instruction is this? _____
19. All rungs must pass _____, or the PAC project cannot be downloaded to the controller.
20. The _____ instruction examines the bit in the associated tag for a 0. If there is a 0 in that tag, then the instruction is true.

LAB EXERCISE 1: Developing Rungs from Functional Specifications

Note: For ease of handing in assignments, students are to answer using their own paper.

Before you can develop successful ladder programs you must be able to translate functional specifications into PAC ladder rungs. As an example, take the following specification:

When push button PB2 is pressed, pilot light PL2A will turn on.

Now develop a rung of PAC ladder logic from this specification. Figure 10-46 illustrates a rung of logic where the push button is represented as a normally open contact (XIC) instruction, and the output pilot light is represented by an output coil (OTE) instruction.

Figure 10-46 PLC ladder rung developed from functional specification.
Used with permission Rockwell Automation, Inc.

For each of the following functional specifications, develop the correct PAC ladder rung or rungs in the space provided. For this exercise, do not worry about the ControlLogix base tags. Use only the alias tags listed in the specifications. We will create ladder rungs using the RSLogix 5000 software in a future lesson.

1. When push button PB2 is pressed and switch SW2 is closed, pilot light PL2A will turn on.

2. When push button PB2 is pressed or limit switch LS1 is closed, pilot light PL2A will turn on.

3. If inductive proximity switches SW1, SW2, and SW3 all sense a target, motor M1 will start.

4. If any of the four doors of an automobile are open, the dome light will come on.

5. If limit switch SW1, and limit switch SW2 or limit switch SW3, are true, the full case is in position; energize glue gun to apply glue to case flaps.

6. If limit switches SW1 or SW2, and SW3, are true, energize solenoid SOL3A to move product into position.

7. If product is in position from the movement of SOL3A from question 6 above, turn on Part in Position Pilot Light.

8. Input Sensor is a sensor that determines whether there are box blanks in the feeder. If the boxes are not replenished by the operator and the feeder runs out of box blanks, the conveyor will be shut down and an alarm bell will sound.

9. A conveyor line is used to label and fill cans with a product. A bar-code reader is used to read the bar code on the can's label to determine that the proper label has been placed on the can.

A photoelectric sensor is used to trigger the bar-code reader when there is a can in position. The bar-code reader then reads the can's bar code. If the bar-code reader fails to see a label with a bar code or sees a bad or damaged bar code, a no-read discrete signal will be sent to the PAC. Develop two rungs of PAC logic: one rung for the bar-code read trigger and a second rung to alert the PAC of a no-read condition.

10. A variable frequency drive has four preset speeds it can run at, depending on the conditions of three inputs from a four-position selector switch into our PAC. The drive is an Allen-Bradley PowerFlex Variable Frequency Drive with an interface card to accept 120-volt AC control signals. Input signal patterns into terminals 26, 27, and 28 determine at which preset speed the drive runs. The table in Figure 10-47 lists the conditions terminals 26, 27, and 28 need to be in to select a specific preset speed. Create logic to control VFD speed in response to the 4 position switch input signals.

Preset Speed	DRIVE OPTION CARD INPUT SIGNALS FOR SPEED SELECTION		
	26	27	28
Preset Speed 1	False	False	False
Preset Speed 2	False	False	True
Preset Speed 3	False	True	False
Preset Speed 4	True	True	False

Figure 10-47 Variable-speed drive preset speed input truth table.
© Cengage Learning 2014

Figure 10-48 illustrates the target table for the four-position selector switch.

FOUR-POSITION SELECTOR SWITCH TARGET TABLE				
Position and Preset #1	Position and Preset #2	Position and Preset #3	Position and Preset #4	Switch Circuit Input to PAC
0	0	0	X	26
0	0	X	X	27
0	X	0	0	28

Figure 10-48 Four-position selector switch target table.
© Cengage Learning 2014

LAB EXERCISE 2: Determining Rung Continuity

Note: For ease of handing in assignments, students are to answer using their own paper. In this lab we determine when ladder rungs are true or false. Evaluate the following and explain when the outputs on the following rungs are true.

1. Refer to Figure 10-49. When will Local:0:O.Data.10 be true?

Figure 10-49 Ladder rung for question 1.
Used with permission Rockwell Automation, Inc.

2. When will Local:0:O.Data.6 be true?
3. Looking at the rung in Figure 10-50, list the different combinations of true instructions to make each output true.

Figure 10-50 Ladder rung for question 3 evaluation.
Used with permission Rockwell Automation, Inc.

4. Looking at the rung in Figure 10-51, list the different combinations of true instructions to make each output true.

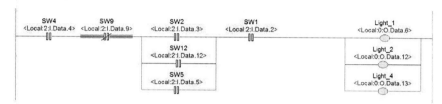

Figure 10-51 Evaluating ladder logic, question 4.
Used with permission Rockwell Automation, Inc.

5. Evaluate the rung in Figure 10-52, and list the different combinations of true instructions to make each output true.

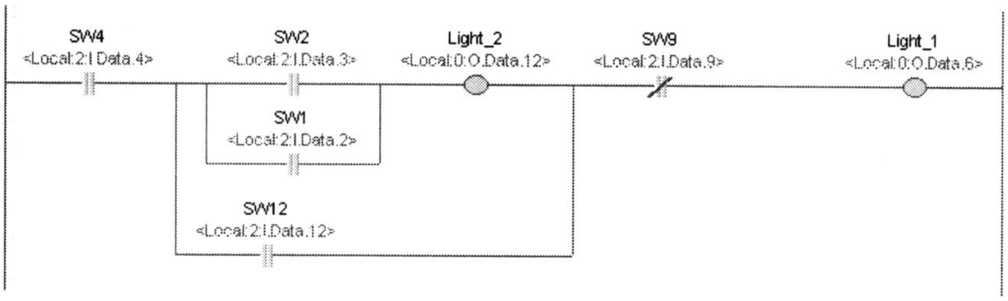

Figure 10-52 Ladder rung for question 5.
Used with permission Rockwell Automation, Inc.

6. What has to happen to make the rung in Figure 10-53 true?

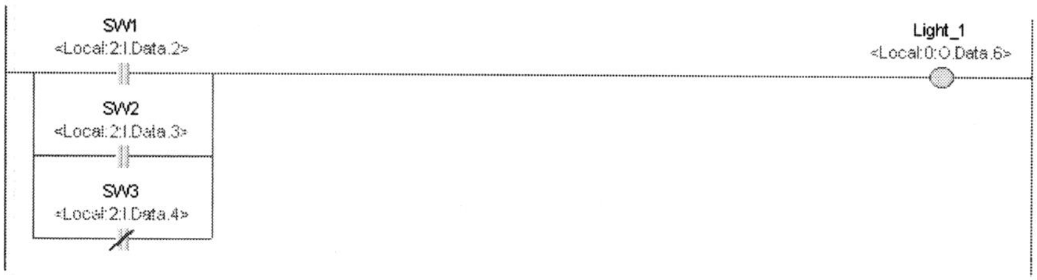

Figure 10-53 Ladder rung for question 6.
Used with permission Rockwell Automation, Inc.

EVALUATING RUNGS FROM AN RSLOGIX 5000 PROJECT

To this point, we have been looking at RSLogix 5000 ladder rungs and determining the possible combinations of when the rung would be true or false. In this lab exercise, we look at rungs in an RSLogix 5000 project and how instructions are represented on a computer screen when they are true or false. Notice the two intensified instructions on the rung in Figure 10-54. The first and third instructions are intensified, signifying they are currently true. The second instruction is not intensified, meaning that instruction is currently false. Because there is not a path of true instructions, the output is false, or not intensified. Also note the rung number to the left of the left vertical line called the *power rail*.

Figure 10-54 Interpreting true and false instructions on a ladder rung.
Used with permission Rockwell Automation, Inc.

The same rung is displayed in Figure 10-55. Notice that all three input instructions are currently highlighted. This means they are all true. Because all inputs are true, the output is true, or highlighted. Keep in mind that the highlights on either side of the instruction signify that the instruction is true. Sometimes individuals assume the normally open instruction symbol will change to normally closed when the instruction becomes true. This is not correct. Highlights that signify the instruction is true can only be seen on relay-type instructions. Other instructions that are represented as a box, such as timers, counters, data manipulation, and math instructions, will not intensify. You need to know how the instruction works because there will be no intensification or blinking to identify its true or false status.

Figure 10-55 Highlighted instructions signify a path of true instructions to the output.
Used with permission Rockwell Automation, Inc.

Figure 10-56 shows a rung with interlaced inputs and outputs. Because Local:2:I.Data.12 and Local:2:I.Data.4 are both true, there is a path of true instructions to the output. Output Local:0:O.Data.15 is true. Local:3:I.Data.11 is not highlighted, so the instruction is false. Being false, there is not a path of true instructions to output Local:0:O.Data.6. As a result, that output is false.

Figure 10-56 Interpretation of highlighted instructions on a rung with interlaced inputs and outputs.
Used with permission Rockwell Automation, Inc.

LAB EXERCISE 3: Interpreting RSLogix 5000 Ladder Rungs

Note: For ease of handing in assignments, students are to answer using their own paper.

The following rungs are also from RSLogix 5000 software. For this lab, we interpret the status of the rung by identifying which instructions are true or false by whether they are highlighted. This lab is exactly what you will see using the software on a personal computer. Refer to Figure 10-57 as you answer the questions below.

1. What is the rung number for the top rung?
2. Is rung 3 currently true or false? Explain your answer.
3. What is the current status of rung 4?
4. Looking at rung 5, explain the status of output Light_3.
5. Because Local:2:I.Data.2 is true, will the math instructions execute? Explain your answer.

LAB EXERCISE 4: Function Block Diagram Logical Continuity

Note: For ease of handing in assignments, students are to answer using their own paper.

In this lab, we determine when function block logic is true or false. Evaluate the following function blocks to determine whether the output devices are true or false.

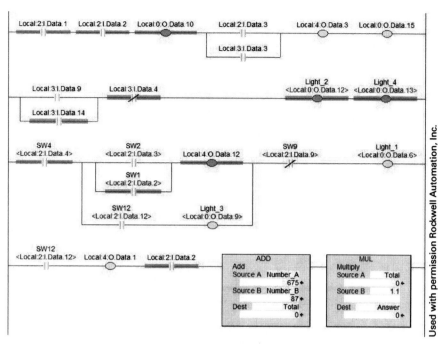

Figure 10-57 Interpreting ladder rungs for Lab Exercise 3.

1. Referring to Figure 10-58, explain the function of the function block illustrated.

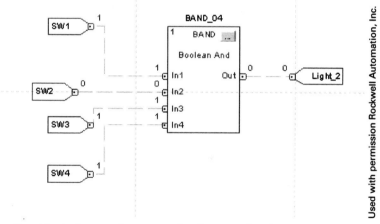

Figure 10-58 Function block diagram interpretation for question 1.

2. Evaluate the logic and determine the status of Light_2. Explain your answer.
3. Explain the function of the BOR function block from Figure 10-59.
4. List what would have to happen to make Light_2 become true.
5. If the function block diagram in Figure 10-60 were running, what would be the current status of Light_2?
6. How does the BNOT function block work?
7. List the states of SW1, SW2, and SW9, along with SW3, to make Light_2 turn on.
8. Interpret the function block diagram in Figure 10-61, and explain how the BXOR works.
9. What has to happen for Light_2 to become true?
10. Is there more than one way to make Light_2 true?

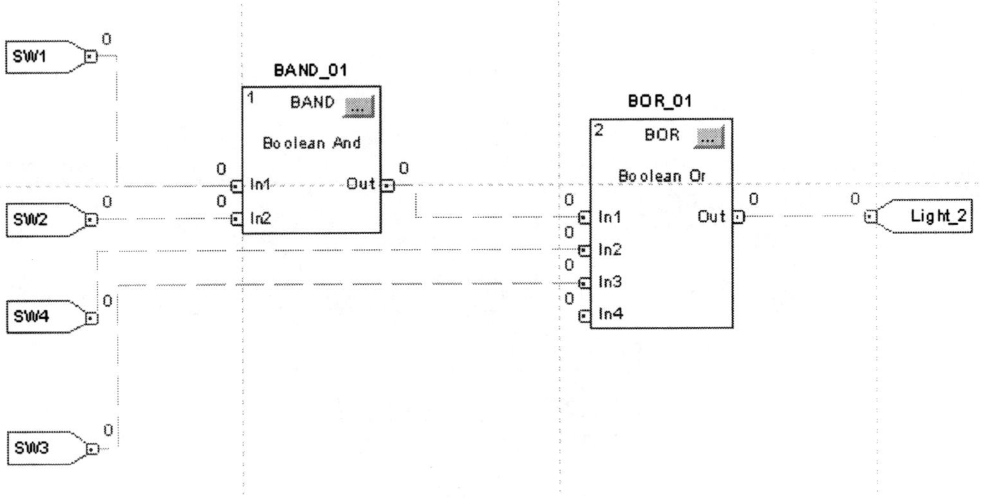

280 INTRODUCTION TO LOGIC

Figure 10-59 Function block diagram interpretation for question 3.

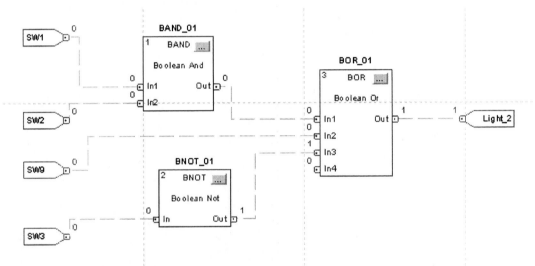

Figure 10-60 Function block diagram interpretation for question 5.

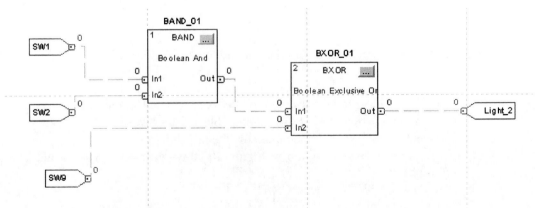

Figure 10-61 Function block diagram interpretation for question 8.

INTRODUCTION TO LOGIC

LAB EXERCISE 5: Logic and Truth Tables

Note: For ease of handing in assignments, students are to answer using their own paper.

The following questions provide practice in creating truth tables and help you understand how the logic operators work.

11. Fill in the following truth table in Figure 10-62 as to what the expected output will be for the two-input AND logic.

Two-Input *AND* Truth Table		
In1	In2	Out
0	0	
0	1	
1	0	
1	1	

Figure 10-62 Fill in truth table for question 11.
© Cengage Learning 2014

12. Fill in the truth table in Figure 10-63 for XOR logic.

Two-Input *XOR* Truth Table		
In1	In2	Out

Figure 10-63 XOR truth table for question 12.
© Cengage Learning 2014

13. Fill in the truth table in Figure 10-64 for NOT logic.

NOT Truth Table	
In1	Out

Figure 10-64 NOT logic truth table for question 13.
© Cengage Learning 2014

14. Fill in the truth table in Figure 10-65 for three-input AND logic.

Three-Input *AND* Truth Table			
			Out

Figure 10-65 Three-input AND logic truth table for question 14.
© Cengage Learning 2014

Challenge Lab

In this lab, you have an opportunity to evaluate a more complex function block diagram. Currently, all inputs are false.

1. As you evaluate Figure 10-66, what has to happen to make Light_4 turn on?

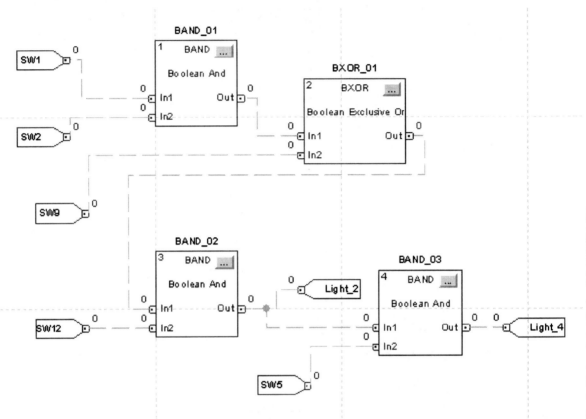

Figure 10-66 Challenge lab function block diagram.

2. List what has to happen to make Light_2 true.

CHAPTER

11

The Basic Relay Instructions

OBJECTIVES

After completing this lesson, you should be able to

- Describe the function of a normally open or examine if closed instruction.
- Explain the function of a normally closed or examine if open instruction.
- Describe the function of the one-shot instruction.
- Explain the function and programming of the latch and unlatch.
- Describe the function of the output energize instruction.
- Create ladder rungs on paper from functional specification.
- Create ladder rungs using the RSLogix 5000 software from specification.

INTRODUCTION

Each manufacturer's programmable automation controllers (PACs) have their own vocabulary of instructions. A programmable automation controllers repertoire of instructions is called its *instruction set*. Although different controllers have different instruction sets, basic instructions are shared by all programmable automation controllers. This chapter introduces the basic relay instructions. Each instruction is illustrated by looking at a rung of actual ladder logic from Rockwell Automation's RSLogix 5000 software. After looking at the sample rung, we explain how each instruction functions in a program. Even though we illustrate the instructions with RSLogix 5000 software, the instructions are similar when using Rockwell Automation's ControlLogix family, SLC 500, and MicroLogix, as well as the PLC-5. In other words, the instructions introduced in this and future lessons in this book can be applied to PLC-5 and SLC 500 applications with little need to learn new information.

LADDER RUNG EXECUTION SEQUENCE

Rungs are executed from left to right, or from inputs to outputs, starting at rung 0 and moving to the highest rung number. Figure 11-1 illustrates a number of ladder rungs from RSLogix 5000 software. Note the rung numbers to the far left. Rung numbering always starts with rung 0 and increments to the last rung in the routine. The last rung in any routine contains the End instruction. The End instruction alerts the controller that there are no additional rungs to execute here. The End instruction cannot be deleted, and nothing else can be programmed on the End rung.

LADDER COMPONENTS

The main components of ladder logic are rungs, branches, and instructions. Figure 11-2 illustrates ladder rungs and the major components of ladder logic.

284 THE BASIC RELAY INSTRUCTIONS

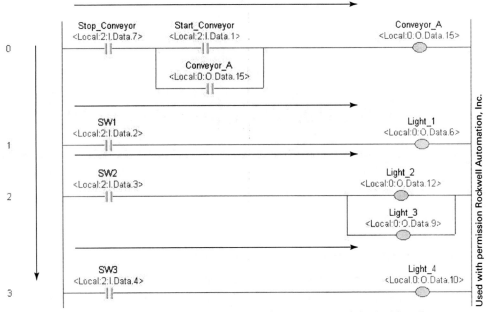

Figure 11-1 Rungs are executed from left to right and from the top of the ladder diagram to the bottom.

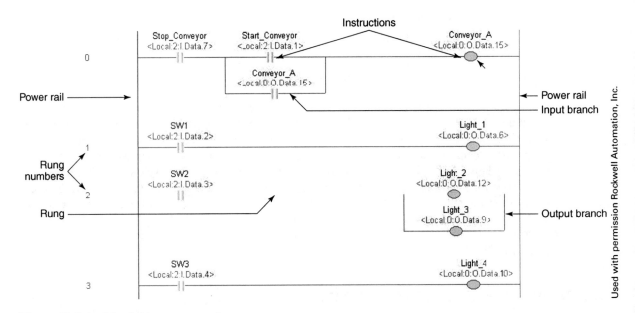

Figure 11-2 Ladder logic components.

A typical ladder rung starts with conditional input instructions on the left side that feed logical continuity to the output instruction on the right side of the rung. In all cases, the last instruction on any rung must contain some type of output instruction. Figure 11-3 shows rungs 3 and 4 from an RSLogix 5000 ladder routine. Notice the conditional input instructions on the left. The output instruction is the last instruction on the rung before the right power rail. Rung 4 illustrates an add instruction. The add instruction is an output instruction. Now that we have a basic understanding of the mechanics of ladder logic, we can start looking at the basic relay-type instructions.

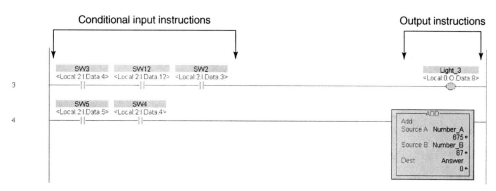

Figure 11-3 Conditional input instructions start on the left side of a ladder rung feeding logical continuity to the output instruction on the right.
Used with permission Rockwell Automation, Inc.

BIT OR RELAY INSTRUCTIONS

Contacts and coils are the basic symbols found on a ladder diagram. Normally open or normally closed contact symbols are programmed on a given rung to represent input conditions that are to be evaluated by the controller as it solves the user ladder. Rung contacts are evaluated to determine how output instructions are to be controlled by the PAC. Each output is represented by a coil symbol. Contacts and coils are also referred to as bit or relay instructions. Each real-world input or output is represented by a separate bit in an input or output tag. This information is used to represent actual inputs coming in from, and going out to, the outside world. Figure 11-4 presents a generalized overview of the basic instructions available.

BIT INSTRUCTIONS		
Instruction	**Symbol**	**Use This Instruction**
Normally Open, or Examine ON	—| |—	As a normally open, or examine if ON, input instruction on your ladder rung
Normally Closed, or Examine OFF	—|/|—	As a normally closed, or examine if OFF, input instruction on your ladder rung
One-Shot	—(ONS)—	To input a single digital pulse from an input signal
Latch Output Coil	—(L)—	To latch an output ON. Output stays ON until the unlatch instruction becomes true
Unlatch Output Coil	—(U)—	To unlatch a latched ON instruction with the same tag
Output Coil	—( )—	As an output instruction that becomes true when all inputs on the rung are true

Figure 11-4 Overview of bit instructions.
© Cengage Learning 2014

Instructions direct the PAC as to how to respond to bits found in its memory. Bits in PLC memory are typically input tags representing ON or OFF signals from an input module. Figure 11-5 provides an overview of how the normally open instruction is evaluated by the PAC controller.

THE NORMALLY OPEN OR XIC INSTRUCTION

The normally open instruction is used by all PLCs, although each manufacturer may have its own name for the instruction. Rockwell Automation PLCs, such as the PLC-5, SLC 500, and ControlLogix family members, use the term *examine if closed* (XIC) to represent the normally open instruction. The examine if closed instruction tells the controller to test for an ON condition from the referenced tag bit. The referenced bit could be from an input device, an examine-an-output-tags bit, an internal BOOL bit, or a status bit from another instruction. The abbreviation XIC is called the *instructions mnemonic*. You will see the mnemonic used to represent the instruction in the software. The table in Figure 11-5 illustrates the rules for the execution of the XIC instruction.

EXAMINE ON, OR XIC, INSTRUCTION	
Description: Input instruction that examines 1 bit for an ON condition	
If the Input Is	Then the XIC Instruction Is
ON or true represented by a 1 in the tag	true
OFF or false represented by a 0 in the tag	false

Figure 11-5 XIC instruction rules for interpretation.
© Cengage Learning 2014

Refer to Figure 11-6 and the following XIC signal flow explanations:

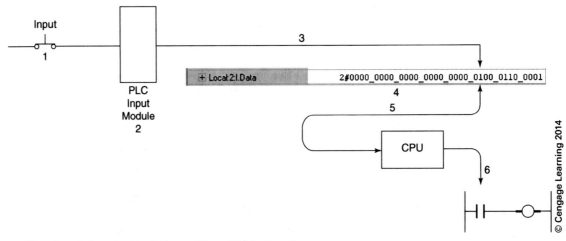

Figure 11-6 Input signal association with an XIC instruction.

1. With the input switch closed, an input signal is seen by the input module's input point.
2. Assume the input signal was 120 volts AC. The input module changes the 120 V into low-voltage signals the PAC can work with. Internally, an ON signal could be represented by +5 V DC, whereas an OFF signal is represented by 0 V DC.
3. The ON or OFF signals are sent by way of the chassis backplane to the input tag at the time specified by the RPI.
4. Callout 4 is an example of an input tag from the controller scoped tags collection. In our example, the input switch is physically closed. The input signal is represented in the input tag as a 1.
5. As the controller solves the ladder rungs, it examines the bit stored in the input tag.
6. Because an XIC instruction is programmed on the ladder rung, the CPU examines the input tag bit position for a closed condition, or a 1 in the bit position. In this case, the CPU finds a 1 in the associated bit position and makes the instruction true.

THE OUTPUT INSTRUCTION OUTPUT ENERGIZE

The output instruction is typically represented as an output coil. Rockwell Automation refers to the output instruction as the *output energize* (OTE) instruction. Even though a rung is not required to have any input instructions, there must be at least one output. Figure 11-7 illustrates a rung with an XIC input instruction and one OTE instruction.

THE BASIC RELAY INSTRUCTIONS **287**

Figure 11-7 XIC instruction as input and OTE as output instruction.
Used with permission Rockwell Automation, Inc.

The output instruction is the last instruction before the right power rail. A traditional PLC, such as the PLC-5 or SLC 500, is allowed to have only one OTE instruction per rung. The only exception to this rule is if the OTE instructions are in parallel. As noted in Chapter 10, ControlLogix allows input and output instructions to be interlaced as long as the last instruction on any rung is some type of output.

Figure 11-8 shows a few of many possible configurations of ControlLogix OTE instructions.

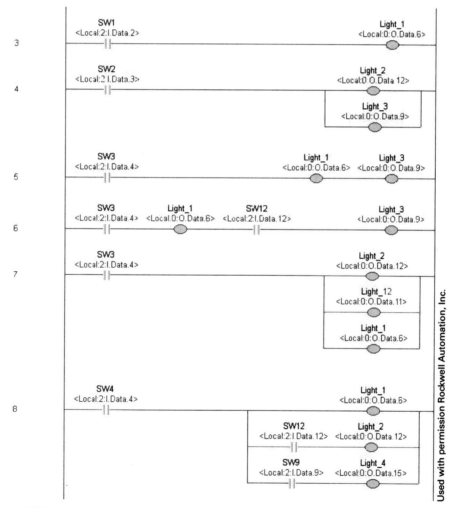

Figure 11-8 OTE instruction variations.

Figure 11-9 illustrates output energize instruction programming options referenced in Figure 11-8 and if the rungs would be verifiable for ControlLogix, SLC 500 and PLC-5 platforms.

An output instruction represents the action that is to be taken when the solved logic results in a logically true rung. When either programming or trying to understand someone else's ladder logic, remember this rule: If there is a path of true instructions to an output instruction, then the output is true. It is difficult to try evaluating ladder logic while thinking of current flow. Figure 11-10 is from Chapter 10. If XIC SW2, and XIC SW5 are true, then the OTE Light_4 will be true. If XIC SW1 is also true, then OTE instructions Light_1 and Light_2 will be true. Again, if there is a path of true instructions from the left power rail to an OTE instruction, then that instruction will be true.

OTE INSTRUCTION VARIATIONS FOR ROCKWELL AUTOMATION CURRENT PLATFORMS			
Rung Number	ControlLogix	PLC-5	SLC 500/MicroLogix
Rung 3	Yes	Yes	Yes
Rung 4	Yes	Yes	Yes
Rung 5	Yes	No	No
Rung 6	Yes	No	No
Rung 7	Yes	Yes	Yes
Rung 8	Yes	Yes	Yes

Figure 11-9 Examples of possible output instruction configurations from Figure 11-8.
© Cengage Learning 2014

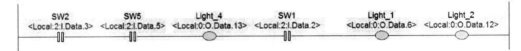

Figure 11-10 Interlaced inputs and outputs.
Used with permission Rockwell Automation, Inc.

THE NORMALLY CLOSED, OR EXAMINE IF OPEN, INSTRUCTION

The normally closed instruction is used by all PLCs; however, each manufacturer may have its own name for the instruction. Rockwell Automation PLCs such as the PLC-5, SLC 500, and ControlLogix family members use the term *examine if open* (XIO) to represent the normally closed instruction. The XIO instruction tells the controller to test for an OFF condition from the referenced tag bit. If the controller sees an OFF condition, then the instruction is true. The referenced bit could be from an input device, examination of an output tags bit, an internal BOOL bit, or a status bit from another instruction. The abbreviation XIO is called the *instructions mnemonic*. The table in Figure 11-11 illustrates the rules for the execution of the XIO instruction.

EXAMINE OFF, OR XIO, INSTRUCTION	
Description: Input instruction that examines 1 bit for an OFF condition	
If the Input Is	Then the XIO Instruction Is
OFF (represented by a 0 in the tag)	true
ON (represented by a 1 in the tag)	false

Figure 11-11 XIC instruction rules for interpretation of the XIO instruction.
© Cengage Learning 2014

Figure 11-12 is an actual rung from RSLogix 5000 software. The XIO instruction is highlighted, or true. Because we have a path of true instructions to the OTE instruction, the output is also true, as we would see it in the software. The highlights on either side of the OTE instruction signify the instruction is true. It should be evident from the table in Figure 11-11, that the SW3 tag currently has the value of 0.

Figure 11-12 XIO instruction as input instruction.
Used with permission Rockwell Automation, Inc.

THE ONE-SHOT INSTRUCTION

The *one-shot rising instruction* (ONS) is an input instruction that allows an event to occur only once per trigger. Figure 11-13 shows a rung with SW_1 as input logic to the ONS instruction. The ONS instruction has a BOOL tag named "One_Shot" associated with it. The ONS tag must be a unique BOOL tag. The ONS instruction controls the one-shot output pulse to the ADD output instruction. The one-shot instruction is typically used with an instruction that only has to be triggered once as a result of an input signal. Many output instructions, such as the ADD instruction in the picture, execute every time the controller scans the rung as long as the input conditions remain true. When programmers want the instruction to execute only once as the result of an input trigger, then they use the one-shot instruction.

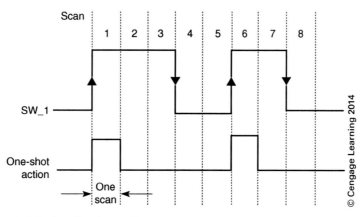

Figure 11-13 One-shot instruction controlling the action of an ADD instruction.
Used with permission Rockwell Automation, Inc.

The "rising" portion of the instruction name means that the instruction is looking for the rising edge of the input pulse, or false-to-true, transition of the input logic in front of the one-shot instruction. When the ONS sees that transition, the instruction sends a single pulse—that is, a "one-shot" signal—to the output instruction. The output instruction executes only once. Figure 11-14 illustrates how the one-shot instruction works in conjunction with the rung's input conditions.

Figure 11-14 One-shot instruction operation.

1. SW_1 transitions from OFF to ON.
2. The one-shot instruction sees the OFF-to-ON transition from the input instruction.
3. The ONS outputs one pulse to the rung's output instructions.
4. Even though SW_1 stays true for three scans of the ladder logic, the ONS outputs only one pulse.
5. On scan 6, the SW_1 goes true again and the ONS outputs only one pulse.

The Rockwell Automation platform being worked with dictates which one-shot instructions will be included in the instruction set. One-shot instructions come as either input or output instructions. Because the ControlLogix has the PLC-5 instruction set, ControlLogix has the same three one-shot instructions found in the PLC-5. The ONS is a one-shot input instruction, whereas the OSR is a *one-shot rising* output instruction. ControlLogix and the PLC 5 also have the *one-shot falling* (OSF) output instruction. Figure 11-15 illustrates how the OSF instruction works. The top two pieces represent the one-shot rising instruction action we looked at a moment ago. The bottom trace illustrates how the one-shot falling instruction works. SW_1 transitions from ON to OFF during scan 4. This is known as the *falling edge* of the input pulse. The falling edge of the input pulse triggers the one-shot instruction to send one pulse to the associated output instructions.

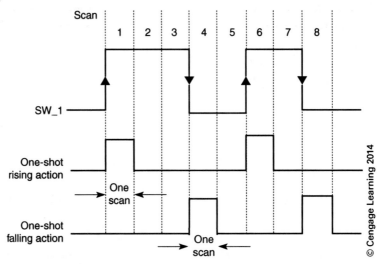

Figure 11-15 One-shot falling operation.

WHEN TO USE THE ONE-SHOT INSTRUCTION

You could use this instruction to start an event triggered by a push-button input where you want the event to happen only once per actuation of the push button, no matter how long it is held in. As an example:

1. A one-shot can be used to reset desired conditions in a single scan.
2. You can use a push button to read the current value of a thumbwheel switch one time.
3. A momentary push-button actuation could be used to increment speed on a motor. Speed would increment one step for each push of the button.
4. A single push button could be used to start and stop a motor. The first closure of the button starts the motor and the second closure stops it.
5. Use the one-shot instruction with a math instruction to perform a calculation once per trigger.
6. A one-shot instruction can be used to bring in changing analog input data, which can be sampled at a predetermined rate.
7. Use a push button or an internal bit and the one-shot instruction to send data to output display devices. A typical output device could be an LED numerical display. The one-shot instruction allows rapidly changing data to be "frozen" and output to an LED display in a timely manner. Timely updates ensure a readable, stable display. If data were allowed to be output continuously and were changing rapidly, the display could be hard to read.

An example of an LED display where data changes rapidly is a filling station gasoline pump displaying total gallons as they are pumped. Another example is the display on the pump showing the total dollars spent on gasoline. As you fill your tank, both displays flash numbers rapidly. These displays are instantaneously updated. This is a situation in which receiving instantaneous updates is not a problem. As the individual pumping the gasoline, you want to know exactly how much gasoline has been dispensed and how much money has been spent.

Output instructions are either retentive or nonretentive. The OTE instruction introduced earlier in this chapter is a nonretentive instruction. Being nonretentive, the instruction does not retain its state when the rung goes false for any reason. An example of programming the OTE is a conveyor application. When hardwiring a conveyor control circuit, a start and stop push-button station and a mechanical motor starter could be used. This circuit is typically wired as three-wire control. If power were lost as the result of an electrical storm, then for safety reasons the conveyor would not automatically restart when power was resumed. On the other hand, there could be a ventilation fan in the ceiling of the manufacturing area. If power were lost as the result of an electrical storm, then one might wish the fan to come back on when power resumes. This fan could be wired as a two-wire control. Two-wire control could be programmed using the latch and unlatch instruction. We investigate latching instructions next.

THE OUTPUT-LATCH INSTRUCTION

An output-latching instruction is an output instruction used to maintain, or latch, an output ON even if the status of the input logic that caused the output to energize changes.

When any logical path on the ladder rung containing the latching instruction has continuity, the output referenced to the latching instruction is turned ON and remains ON, even if the rung's logical continuity or PAC system power is lost. Because the latch instruction retains its state through a system power loss, the latching instruction is called a *retentive instruction*. Remember, the controller's battery must be in good condition for the latching status to be remembered (or retained by the controller) in case of a power failure.

The latched instruction will remain in a latched ON condition until an unlatch instruction with the same reference tag is energized. Latch and unlatch instructions are directly used in pairs. Each instruction is typically located on a separate rung.

The table in Figure 11-16 summarizes the *output-latch* (OTL) and *output-unlatch* (OTU) instructions.

LATCH AND UNLATCH INSTRUCTIONS		
Operation	Instruction Name	Mnemonic
If the OTL instruction sees a false-to-true transition for a minimum of one scan of the program, the OTL instruction will be true, or latch the output on.	Output Latch	OTL
If the OTU instruction sees a false-to-true transition for a minimum of one scan of the program, the OTU instruction will be true and unlatch, or turn the output off.	Output Unlatch	OTU

Figure 11-16 Overview of the latch and unlatch instructions.
© Cengage Learning 2014

Figure 11-17 illustrates two ladder rungs. The first rung contains the latching instruction, and the second rung contains the unlatching instruction.

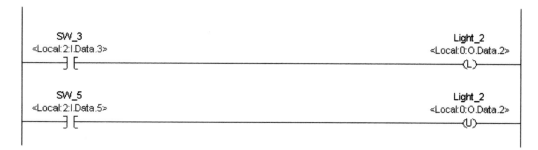

Figure 11-17 Latching and unlatching ladder logic.
Used with permission Rockwell Automation, Inc.

The first rung, containing the OTL instruction, functions similarly to the OTE instruction, although once the output bit is set, it is latched ON and retains its state even if the rung goes false. The bit is reset with the OTU.

Rung Operation

When SW_3 input on the first rung is energized, the OTL instruction Light_2 is energized. The output-latch instruction remains latched ON and is unaffected no matter how SW_3 changes. The output-unlatch instruction is used to turn off, or unlatch, the output Light_2 that was turned on by the latching instruction. The unlatch instruction with the same tag must be energized to unlatch the output tag Light_2. These particular example rungs have only one input each. Any valid input logic instructions may be used as inputs to the latch and unlatch instructions. The following rules pertain to most latch and unlatch instructions.

1. Latch and unlatch instructions are typically used in pairs.
2. Latch and unlatch pairs of instructions must have the same reference address.
3. The latch and unlatch ladder rungs do not need to be grouped together in the ladder program.
4. Latching and unlatching instructions are retentive, provided your controller battery is installed and in good condition.
5. Use an unlatch instruction to unlatch, or clear, status bits.
6. If an unlatch instruction is left energized, the associated latching output cannot be latched.
7. Output latch instructions are retentive. This means that if the controller loses power, is switched to Program or Test mode, or detects a major fault, the output-latch instruction will retain the state of the latched bit in controller memory. Even though all outputs will be turned OFF during these controller conditions, retentive outputs return to their previous states when the controller returns to Run mode.

Programming Considerations

The placement of the latch and unlatch instruction rungs within your ladder program can affect the behavior of these instructions. Figure 11-17 illustrates a latching instruction programmed before the unlatch instruction. If both instructions are true at the end of the scan, the last instruction programmed on the rung takes precedence over the other instruction. In this example, the output instruction is always unlatched.

Figure 11-18, on the other hand, has the unlatch instruction programmed after the latch instruction. In this case, the last instruction—the latch instruction—takes precedence and keeps the output latched, provided both the latch and unlatch rungs are true at the end of the scan.

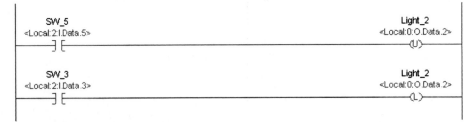

Figure 11-18 Latching instruction programmed after the unlatch instruction.
Used with permission Rockwell Automation, Inc.

INTERNAL BIT–TYPE INSTRUCTIONS

There are instances when you have to control instructions other than an output on a rung or rungs other than the current rung. An instruction is needed that is easily programmed on the current rung and that does not represent a real-world field device. An internal bit–type instruction that could be programmed as either normally open or normally closed really helps in program development.

Most PLCs have some method of incorporating internal bits into the user program when other than real-world field devices are needed as input or output reference instructions. Different manufacturers name these internal bit–type instructions differently. Some use the terms *internal bits*, *internal coils*, or *internal relays* to identify internal bits programmed as non–real-world field devices.

An internal bit used as an output is sometimes referred to as an *internal relay*, *internal coil*, or *internal output*. An internal bit is used as a rung output when a real output is not desired. An internal bit as an output (used like a control relay, as an example) is used when the logical resultant of a rung is used to control other internal logic.

ControlLogix and RSLogix 5000 software identify internal bits as the BOOL data type. Figure 11-19 illustrates an example of using a BOOL bit as an output. The tag On_Command is a BOOL tag. When someone pushes the start button, the On_Command output becomes true.

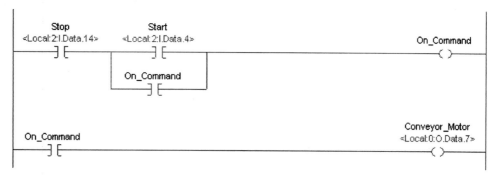

Figure 11-19 Internal BOOL bit On_Command used as an output.
Used with permission Rockwell Automation, Inc.

The XIC instruction ORed with the start examines the tag On_Command to see whether there is a 1 or 0 stored in the tag. Because the output is true, the tag contains a 1, which makes the XIC instruction with the tag On_Command also true. Now the individual pushing the start button can release the button, and the rung will remain true, or "sealed in." The next rung also uses the On_Command tag with an XIC instruction to control the output Conveyor_Motor. Because the BOOL bit referenced to On_Command is true, this rung's XIC instruction is also true and makes the OTE Conveyor_Motor true. This is a simple example of using an internal, or BOOL, bit to control other rungs of logic. Figure 11-20 shows a similar logic to that in Figure 11-19. Hovering the cursor over the OTE instruction causes the Tool Tips to display. Note the box of text just below the On_Command OTE instruction. The Tool Tips identify the tag name, data type as BOOL, and tag as controller scoped. The current value of the tag is 0.

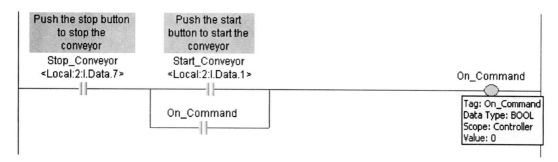

Figure 11-20 Internal BOOL bit On_Command.
Used with permission Rockwell Automation, Inc.

Figure 11-21 shows a portion of the tag collection where the On_Command tag is stored. Note that the tag is BOOL data type. The current value is 1, and the style in which the information is displayed is decimal.

294 THE BASIC RELAY INSTRUCTIONS

On_Command	1	Decimal	BOOL
One_Shot	0	Decimal	BOOL
+ Shift_One	35467	Decimal	DINT
+ Shift_Two	67543	Decimal	DINT
Start	0	Decimal	BOOL
Stop	0	Decimal	BOOL
SW_1	0	Decimal	BOOL

Figure 11-21 Portion of the tags collection showing the On_Command tag.
Used with permission Rockwell Automation, Inc.

CONVERTING FROM RELAY LADDER TO PLC LADDER LOGIC

The following examples illustrate the conversion from a standard relay ladder diagram to PAC ladder logic. The ladder rungs in the figures are actual rungs from the RSLogix 500 software.

Example 1:
   Two series limit switches controlling solenoid 1 are shown in Figures 11-22 and 11-23.

```
      LS1   LS2        SOL1
    ──┤ ├──┤ ├────────⌇/⌇──
```

Figure 11-22 Relay ladder diagram for example 1.
© Cengage Learning 2014

```
   Limit_Switch_1   Limit_Switch_2                          Solenoid_1
   <Local:2:I.Data.16>  <Local:2:I.Data.0>              <Local:0:O.Data.22>
 ──┤ ├──────────────┤ ├──────────────────────────────────────( )──
```

Figure 11-23 Converted RSLogix 5000 ladder rung for example 1.
Used with permission Rockwell Automation, Inc.

Example 2:
   Conversion of combinational input logic controlling pilot light 1 to PAC ladder format is shown in Figures 11-24 and 11-25.

```
        LS5    CR-1    PL-1
      ──┤ ├─┬──┤ ├────(G)──
       LS6  │
      ──┤ ├─┘
```

Figure 11-24 Relay ladder diagram for example 2.
© Cengage Learning 2014

Figure 11-25 Converted RSLogix 5000 ladder rung for example 2.
Used with permission Rockwell Automation, Inc.

Example 3:
   Conversion of parallel input logic controlling pilot light 2 to PAC ladder logic is shown in Figures 11-26 and 11-27.

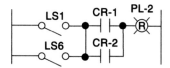

Figure 11-26 Converted parallel input logic controlling pilot light 2 to PAC ladder logic.
© Cengage Learning 2014

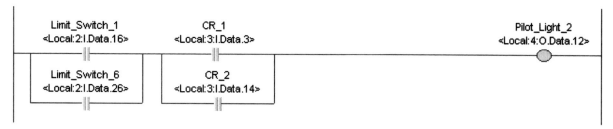

Figure 11-27 Converted RSLogix 5000 ladder rung for example 3.
Used with permission Rockwell Automation, Inc.

Example 4: Converting a Standard Hardwired Start–Stop Circuit to Ladder Logic

For this example, we convert a conventional two-button start–stop push-button station for interface to a PAC. Figure 11-28 illustrates a standard start–stop schematic for a standard start–stop push-button station. Note that the start push button is momentary normally open, whereas the stop push button is momentary normally closed.

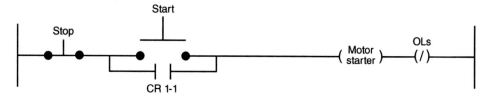

Figure 11-28 Conventional schematic start–stop circuit.
© Cengage Learning 2014

Figure 11-29 is a conceptual drawing of a standard start–stop push-button station.

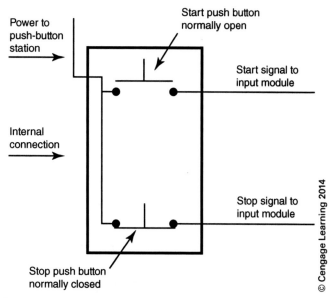

Figure 11-29 Typical start–stop push-button wiring.

One of the first things we need to do as we start our conversion is to determine the inputs and outputs. Figure 11-30 lists the inputs and their normal state, the normal signal to the input point, and the normal bit found in the input tag.

PLC INPUTS			
Device	Normal State	Normal Signal to Input Point	Normal Bit in Input Tag
Start push button	Momentary normally open	OFF	0
Stop push button	Momentary normally closed	ON	1
AUX contacts for seal in	Normally open	OFF	0
Overload contacts	Normally closed	ON	1

Figure 11-30 PAC inputs and their states for our start–stop circuit conversion.
© Cengage Learning 2014

The table in Figure 11-31 lists our output and its normal ON energized state.

PLC OUTPUTS			
Device	Normal State	Normal Signal from Output Point	Normal Bit in Output Tag
Starter coil	OFF	OFF	0

Figure 11-31 Motor starter output coil.
© Cengage Learning 2014

Figure 11-32 illustrates the physical input module wiring. Notice that each input device is wired into a separate input point.

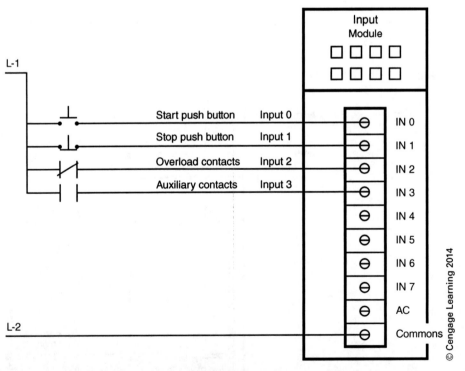

Figure 11-32 Typical input wiring of a motor starter to an input module.

Figure 11-33 shows typical wiring of mechanical motor starter to generic PAC input and output modules.

THE BASIC RELAY INSTRUCTIONS **297**

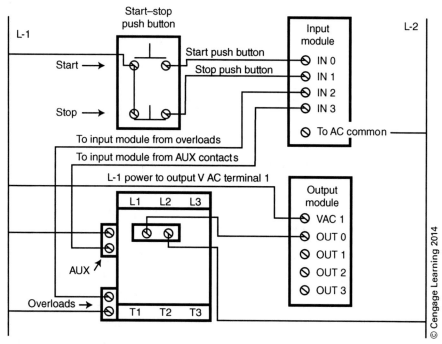

Figure 11-33 Typical motor starter interface to a PAC.

As we begin the conversion process, we need to remember that input instructions examine the bit in the tag associated with the instruction to determine whether the instruction is to become true or false. Because the controller cannot see the field devices, their physical states are of no interest to the controller as it solves the logic. As humans, we need to not only look at the input tag bit value but also consider the field devices' physical state and the associated input signals being sent into the input module and on to the input tags.

1. Let us determine what instruction to program to represent our start push button. Because the start push button is normally open, there is a 0 in the input tag whenever the button is not pressed. When the start push button is pressed, a 1 is sent to the input tag by way of the input module. For our ladder rung, we want the instruction to be true when the start push button is pressed or sending a 1 to the input tag. Which instruction is true when there is a 1 in the input tag? The normally open, or XIC, instruction is true when there is a 1 in the input tag. The table in Figure 11-34 illustrates the rules for the execution of the XIC instruction.

EXAMINE ON, OR XIC, INSTRUCTION	
Description: Input instruction that examines 1 bit for an ON condition	
If the Input Is	Then the XIC Instruction Is
ON or true represented by a 1 in the tag	true
OFF or false represented by a 0 in the tag	false

Figure 11-34 XIC instruction operation.
© Cengage Learning 2014

2. Next we look at programming the normally closed push button. Because the stop push button is physically normally closed, a signal is always sent into the input module's screw terminal. That will result in a constant 1 in the input tag until someone presses the stop push button. The instruction programmed on our rung will need to be true all the time until

someone presses the stop button. With a 1 in the input tag until someone presses the stop button, which instruction needs to be programmed when there is normally a 1 in the input tag? The normally open or XIC instruction is true when there is a 1 in the input tag. On the surface, this may not seem to be the correct way to program the ladder rung. Keep in mind that the controller looks at the input tag and not the field device when determining how to evaluate the instruction. If we look at the field device as the primary determining point for either programming or evaluating an instruction, in some cases we will be incorrect.

3. The AUX contact is normally open. The instruction programmed must be true when the AUX contact closes and sends a 1 into the input tag. Referring to the table in Figure 11-34, we see that the XIC instruction is true when the input tag contains a 1.
4. Because the overload contacts are physically normally closed, there is a signal input to the module under normal conditions. With a signal input to the module, normally a 1 is in the input tag. Again, the XIC instruction is used on our ladder rung. Keep in mind the rule that the last instruction on a rung must be an output. Because of this, we need to move the overload instruction on the input side of the rung.
5. The output instruction on this rung is an output energize instruction. As we recall, the OTE instruction is a nonretentive instruction. In many situations, a motor starter is wired as a three-wire control so that the motor will not automatically restart when power is resumed following a power loss.
6. If we were going to program this rung, once the instruction type had been determined, we would need to determine and assign the input tag address. Assign the alias tag if desired.
7. Assign the output tag and alias.
8. Create tags.
9. Create ladder rung.
10. Program inputs and outputs.

Figure 11-35 illustrates the converted schematic rung to ladder logic. The base tags and alias have been assigned. A main operand description is the text directly above the tag and its alias.

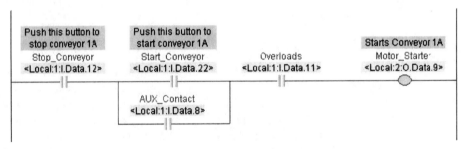

Figure 11-35 Converted stop–start to RSLogix 5000 ladder logic.
Used with permission Rockwell Automation, Inc.

TYPICAL SITUATIONS FOR CREATING LADDER RUNGS

PLC programming falls into one of four general categories:

1. The program to be written is being converted from standard conventional schematic to PAC control.
2. The programmer is starting completely from scratch for a new application.
3. The programmer is upgrading an older existing PLC system to a newer system.
4. The programmer is modifying a current project.

In many cases, we upgrade to a more powerful PLC because the current PLC platform cannot provide some or all of the following:

- Number of I/O points
- Programming instructions

- Adequate controller memory
- Newer programming languages
- Ability to integrate multiple pieces of the system into one platform
- Communication capabilities

Conversion utilities are available that can convert from an older PLC platform like the Rockwell Automation SLC 500 or PLC-5 platform to ControlLogix, but keep in mind that the utilities will only do 70 to 80 percent of the conversion. The remaining cleanup has to be manually done by the programmer.

The following steps are typically included in the creation of a ControlLogix ladder project.

1. Determine which controller or controllers are required.
2. Create a new project.
3. Perform an I/O configuration.
4. Determine real-world I/O and allocate tags.
5. When converting a conventional ladder diagram schematic to one in which we can develop our user program, we first determine input and output devices that are connected to real-world hardware devices. After the real-world I/O has been determined, allocate each I/O point a valid input or output tag and alias.
6. Determine whether we will be using controller and programmed scoped tags.
7. Separate logic into the three different available tasks. Which rungs go into the continuous task, periodic task, and event tasks?
8. If using periodic tasks, determine interrupt period.
9. If using event-driven tasks, determine interrupt triggers.
10. Determine the priority of the tasks.
11. Set up the watchdog timers for each task.
12. Determine whether programs are required and in which order they must be executed.
13. Create the main routine and assign it as the main routine.
14. Create any required subroutines.
15. Determine whether a fault routine is needed.
16. Find out whether any programs remain unscheduled.
17. Allocate internal references.

Once the real-world I/Os have been determined, internal coil instruction references need to be allocated. These references are analogous to relay system mechanical relays that do not drive real-world devices but rather interact with other internal system relays. Internal coils are ladder logic elements that interact with other non–real-world input or output logic internally in the controller.

18. Develop user programs.
19. Provide documentation of user programs.

A table should be developed that defines each input, output, and internal coil reference. This table should list each I/O's function. Drawings that indicate the wiring and its operation should be included for future reference. The tables should list every point available for use, even if it is only for future use. A table should be prepared for internal data storage tags. This table should be filled in during the development of the program. We will add some basic documentation to our ladder rungs shortly.

LAB EXERCISE 1: Creating Rungs in RSLogix 5000 Software
OBJECTIVES

- Create a new RSLogix 5000 project.
- Perform an I/O configuration.

300 THE BASIC RELAY INSTRUCTIONS

- Create tags.
- Create ladder rungs using RSLogix 5000 software.
- Verify ladder rungs and check for programming errors.

INTRODUCTION

This lab introduces creating ladder rungs using the RSLogix software. You can complete this lab if you are using the modular ControlLogix or CompactLogix.

THE LAB

1. _____ Create a new RSLogix 5000 project.
2. _____ Name the project "Begin Ladder."
3. _____ Select your controller.
4. _____ Select the correct software version.
5. _____ If you are using the modular ControlLogix, select the slot number of the controller.
6. _____ When the new project has been created, perform an I/O configuration. For this lab exercise, we set up our hardware as follows:
   - Controller in slot 0
   - 16-point input module in slot 1
   - 16-point output module in slot 2
7. _____ For this lab, we only use the continuous task, main program, and main routine.
8. _____ Find the Main Routine folder.
9. _____ Double-click Main Routine to open it.
10. _____ Follow the steps below as we program rung 0 as illustrated in Figure 11-36.

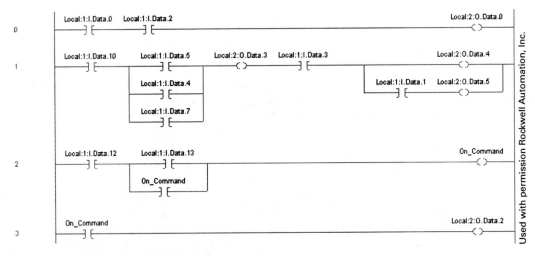

Figure 11-36 Rungs for this lab exercise.

11. _____ Refer to Figure 11-37 as we begin. Click rung number 0 to select the rung.
12. _____ On the Language Element toolbar, click the Bit tab.
13. _____ Click the normally open, or XIC, instruction.
14. _____ The instruction should display on the rung as illustrated by item number 3 in Figure 11-37. The lower case "e's" in blue on the left side signify that this rung is being edited. Refer to Figure 11-37.

THE BASIC RELAY INSTRUCTIONS 301

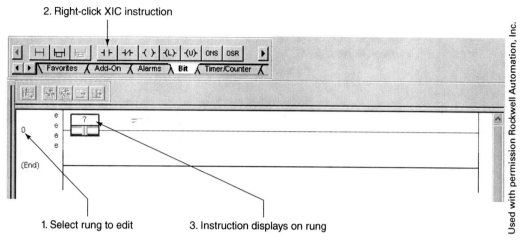

Figure 11-37 Creating rung 0.

15. _____ To add the instructions tag, double-click the question mark above the XIC instruction.
16. _____ Click the displayed down arrow. See item number 1 in Figure 11-38.

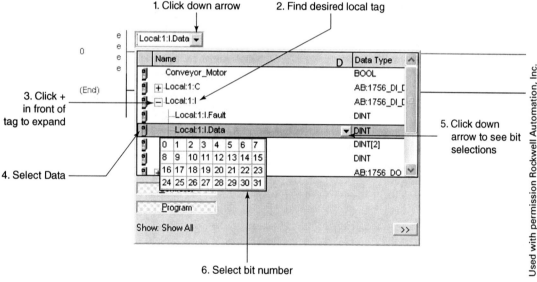

Figure 11-38 Selecting tag address.

17. _____ From the list displayed, find Local:1:I, as shown in item number 2 in the figure.
18. _____ Click the + in front of Local:1:I to expand it.
19. _____ Select Local:1:I.Data from the available list of tags. Refer to item number 4.
20. _____ Refer to item 5 in the figure and click the down arrow to display the bit selections.
21. _____ Click the bit number 0 from the table, as shown in item number 6.
22. _____ When your rung is displayed, click the desktop to complete the tag selection. Your rung should look like that in Figure 11-39.

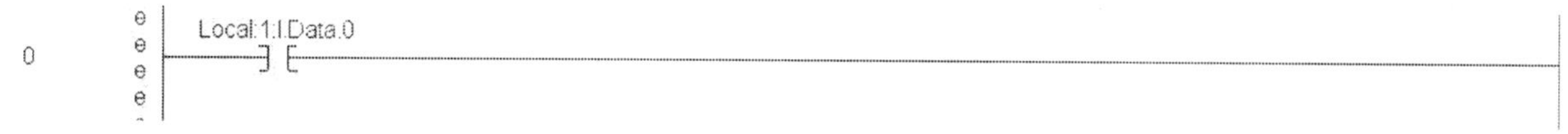

Figure 11-39 XIC instruction programmed on rung.
Used with permission Rockwell Automation, Inc.

23. _____ Before we add the next instruction, we will add three new rungs to our project. Click the new rung icon, as shown in Figure 11-40. The project should now have four rungs, 0 through 4.

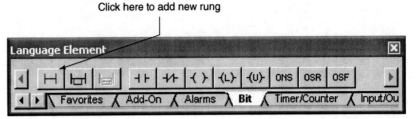

Figure 11-40 Add new rung icon.
Used with permission Rockwell Automation, Inc.

24. _____ To program the next XIC instruction, we will practice dragging and dropping the instruction to the proper position on the rung. As you refer to Figure 11-41, click the XIC instruction and drag it into position. When dragging the instruction, select the proper position of the instruction on the rung by determining which target will accept the instruction. The proper position for the instruction is next to the first instruction we programmed. Note that as you get close to the target, the shape and color of the target change. See item number 2 in the figure. When you have selected a target for the instruction and the target has changed state and color, drop the instruction by releasing your left mouse button. The instruction should be displayed as illustrated in Figure 11-42.

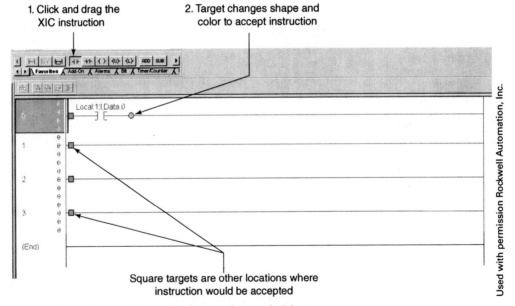

Figure 11-41 Dragging and dropping instruction on ladder rung.

Figure 11-42 Instruction 2 programmed on ladder rung.
Used with permission Rockwell Automation, Inc.

25. _____ Double-click the question mark above the XIC instruction and add the required tag. Refer to Figure 11-36.
26. _____ Add the OTE instruction and tag by clicking the instruction in the toolbar or by dragging and dropping.

Programming Rung 1

27. _____ Use the same programming procedures to program rung 1, instructions, and tags as shown in Figure 11-43.

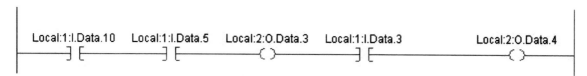

Figure 11-43 Rung 1.
Used with permission Rockwell Automation, Inc.

28. _____ Next we add the left branch to our rung. Click branch instruction, as illustrated in Figure 11-44. While holding down the left mouse button, move toward the target. When the target changes its shape and color, release the mouse button. Figure 11-45 shows our newly created branch.

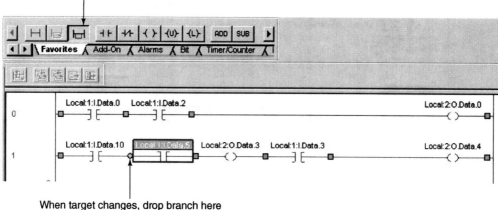

Figure 11-44 Adding branch to rung 1.
Used with permission Rockwell Automation, Inc.

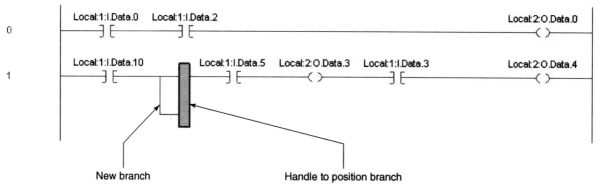

Figure 11-45 New branch on rung 1.
Used with permission Rockwell Automation, Inc.

29. _____ Click the branch handle and drag to target. Notice there are five possible targets where this branch could go. Drag the branch handle to the desired target. See Figure 11-46.
30. _____ Drag the XIC instruction to the branch and add the instructions tag.
31. _____ To add the next branch, right-click the lower-left corner of the current branch illustrated as item number 1 in Figure 11-47.

304 THE BASIC RELAY INSTRUCTIONS

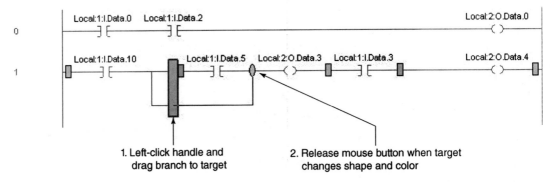

Figure 11-46 Click handle and drag to target.
Used with permission Rockwell Automation, Inc.

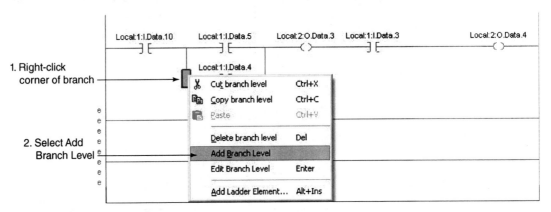

Figure 11-47 Add next branch level.
Used with permission Rockwell Automation, Inc.

32. _____ Select Add Branch Level, as shown in item number 2 in Figure 11-47.
33. _____ Program the remaining input branch levels, instructions, and tags.
34. _____ Create the output branch instruction by clicking the branch icon and dragging it down between Local:1:I.Data.3 and the OTE instruction.
35. _____ Program the XIC and OTE instructions on the branch. Add their respective tags.

Programming Rung 2

36. _____ Start creating rung 2 by programming the logic, as shown in Figure 11-48.

Figure 11-48 Beginning of rung 2.
Used with permission Rockwell Automation, Inc.

37. _____ Right-click the question mark above the OTE instruction.
38. _____ Select New Tag.
39. _____ Refer to Figure 11-49 as you enter "On_Command" for the name.
40. _____ Leave the tag as a base tag.

THE BASIC RELAY INSTRUCTIONS   305

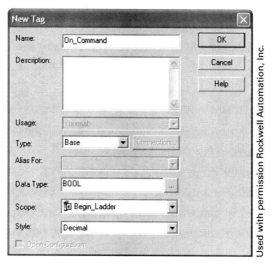

Figure 11-49 Creating a new tag from the ladder.

41. _____ Leave the data type as BOOL. This is an internal or BOOL bit. We use this bit to control another rung. This bit is not a real-world output.
42. _____ Click OK when you are completed.
43. _____ The On_Command tag is also used on the parallel branch below Local:1:I.Data.13. An easy way to associate the XIC instruction on the parallel branch is to copy the tag by simply dragging it from the OTE instruction.
44. _____ Refer to Figure 11-50 as you left-click and drag a copy of the On_Command tag to the parallel branch.

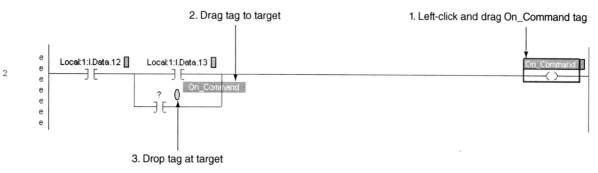

Figure 11-50 Dragging tag for reuse for another instruction.
Used with permission Rockwell Automation, Inc.

45. _____ As you drag the tag toward the parallel branch and its XIC instruction, you will see the target change size and color when you get close. When you see the target change state and color, release the left mouse button to drop the tag into position. See item number 3 in Figure 11-50. Rung 2 should be complete.

Programming Rung 3

46. _____ After you program the XIC instruction on the rung, drag the On_Command tag into position like you did for the last rung.
47. _____ Program the OTE instruction on the rung. This rung is an example of using an internal BOOL bit from rung 2 to control internal logic on rung 3 rather than real-world output.
48. _____ Save your completed project as we will use it again in Chapter 12.

LAB EXERCISE 2: Creating Ladder Rungs from Functional Specifications

Note: For ease of handing in assignments, students are to answer using their own paper.

OBJECTIVES

On completion of this lab exercise, you should be able to

- Develop ladder rungs from functional specifications.
- Develop ladder rungs with ANDed inputs.
- Develop ladder rungs with ORed inputs.
- Develop ladder rungs using AND along with OR logic.
- Develop ladder rungs with interlaced inputs and outputs.
- Develop ladder rungs with ANDed along with ORed outputs.

INTRODUCTION

This lab exercise provides practice creating ladder rungs using the RSLogix 5000 software.

Historically, the ladder diagram has been the traditional method for representing electrical sequences and operations for controlling machinery or equipment. The ladder diagram is accepted as the industry standard for providing control information from the designers to the users charged with equipment installation, modification, and maintenance. When the programmable controller was developed and introduced to industry, one of the requested features was that it also represent circuit control with the traditional ladder diagram format. With traditional ladder logic built into the PLC, maintenance and electrical personnel could easily adapt to the new technology because they were familiar with ladder logic. Even though there are new higher-level languages available today, most PLCs are still programmed with the old familiar ladder logic.

DEVELOPING LADDER RUNGS FROM FUNCTIONAL SPECIFICATIONS

Before you can develop successful ladder programs you must be able to translate functional specifications into PLC ladder rungs. As an example, take the following specification:

When push button PB2 is pressed, pilot light PL2A turns on.

Now develop a rung of PLC ladder logic from this specification. Figure 11-51 illustrates a rung of PLC logic where the push button is represented as a normally open contact (XIC) instruction and the output pilot light is represented by an output coil (OTE) instruction.

Figure 11-51 PLC ladder rung developed from functional specification.
Used with permission Rockwell Automation, Inc.

Create a new RSLogix 5000 project for the specific ControlLogix you are using. Perform an I/O configuration. We will introduce adding the symbolic names to our I/O points in Chapter 13. For each of the following functional specifications, create the PLC ladder rungs using RSLogix 5000 software. For questions with no tags assigned, select your own unused tags.

1. When push button PB2 (Local:1:I.Data.1) is pressed and switch SW2 (Local:1:I.Data.12) is closed, pilot light PL2A (Local:2:O.Data.1) turns on.
2. When push button PB2 (Local:1:I.Data.1) is pressed or limit switch LS1 (Local:1:I.Data.5) is closed, (Local:2:O.Data.13) turns on.
3. If inductive proximity switches SW1 (Local:1:I.Data.4), SW2 (Local:1:I.Data.8), and SW3 (Local:1:I.Data.9) all sense a target, motor M1 (Local:2:O.Data.3) will start.

4. If any of the four doors of an automobile are open, the dome light will come on.
5. If limit switches SW1 (Local:1:I.Data.7), and SW2 (Local:1:I.Data.11) or SW3 (Local:1:I.Data.13), are true, the full case is in position; energize glue gun to apply glue to case flaps (Local:2:O.Data.5).
6. If (Local:1:I.Data.3) or (Local:1:I.Data.10), and (Local:1:I.Data.14), are true, energize solenoid (Local:2:O.Data.6) to move product into position.
7. If product is in position from the movement of SOL3 from question 6 above, energize outputs Local:2:O.Data.2 and Local:2:O.Data.7.
8. Input Local:1:I.Data.6 is a sensor that determines whether there are box blanks in the feeder. If the boxes are not replenished by the operator and the feeder runs out of box blanks, the conveyor will be shut down and an alarm bell will sound.
9. A conveyor line is used to label and fill cans with a product. A bar-code reader is used to read the bar code on the can's label to determine that the proper label has been placed on the can.

    A photoelectric sensor is used to trigger the bar-code reader when there is a can in position. The bar-code reader then reads the can's bar code. If the bar-code reader fails to see a label with a bar code or sees a bad or damaged bar code, a no-read discrete signal is sent to the PAC. Develop two rungs of logic: one rung for the bar-code read trigger and a second rung to alert of a no-read condition.

SUMMARY

This chapter introduced the basic relay-type instructions that are used when programming PAC ladder programs.

Figure 11-52 summarizes XIC and XIO instruction use.

PROGRAMMING XIC AND XIO INSTRUCTIONS		
Instruction	Input's Physical State	To Make Instruction TRUE
XIC	Open	Close input device or set bit
XIC	Closed	Instruction is true
XIO	Open	Instruction is true
XIO	Closed	Open input device or reset bit

Figure 11-52 RSLogix XIC and XIO instructions.
© Cengage Learning 2014

Keep in mind that the PAC ladder program is only a logical representation of how real-world input devices are to interact with real-world outputs. The ladder program is only a graphical representation of what the processor is to do in response to programmed input conditions. As you look at a ladder rung, remember that the rung will be true if there is logical continuity from the left power rail, through a minimum of one continuous path of instructions, and onto an output instruction. The actual logical status of all input instructions is found in each instruction's associated tag. After the processor determines whether the rung is logically true or false, it sends a 1 or 0 bit to the tag assigned to the output instruction. Remember that not all input and output instructions represent real-world field devices. Internal bits are used in many situations where only internal program control is desired.

REVIEW QUESTIONS

Note: For ease of handing in assignments, students are to answer using their own paper.

1. A controller's repertoire of instructions is called its _____.
2. The main components of ladder logic are _____, _____, and _____.
3. Rungs are executed from left to right or from _____, starting at rung 0 and moving to the highest rung number.
4. Typical ladder rungs start with _____ input instructions on the left side of the rung feeding logical continuity to a(n) _____ instruction.

5. In all cases, the last instruction on any rung must contain some type of _____ instruction.
6. Rung numbering always starts with rung _____ and increments to the last rung in the routine.
7. ControlLogix family members use the term _____, or XIC, to represent the normally open instruction.
8. The examine if closed instruction tells the controller to test for a(n) _____ condition from the referenced tag bit.
9. The output instruction is typically represented as an output _____.
10. Rockwell Automation PACs refer to the output instruction as the output energize or the _____ instruction.
11. The OTE instruction is a non_____ instruction.
12. The abbreviation XIC is called the instruction's _____.
13. Output instructions are either _____ or _____.
14. The last rung in any routine contains the _____ instruction.
15. Even though a rung is not required to have any input instructions, there must be at least one _____.
16. ControlLogix family members use the term _____, or XIO, to represent the normally closed instruction.
17. A _____ instruction will not retain its state when the rung goes false for any reason.
18. ControlLogix allows input and output instructions to be _____ as long as the last instruction on any rung is some type of output.
19. Complete the table in Figure 11-53.

XIO AND XIC INSTRUCTIONS		
If the input Is	The XIO Instruction Is	The XIC Instruction Is
OFF represented by a 0 in the tag		
ON represented by a 1 in the tag		

Figure 11-53 XIC and XIO instructions properties.
© Cengage Learning 2014

20. The examine if open instruction tells the controller to test for a(n) _____ condition from the referenced tag bit.
21. The one-shot rising instruction, or _____, is an input instruction that allows an event to occur only once per trigger.
22. The ONS instruction has a _____ BOOL tag associated with it.
23. The "rising" portion of the ONS instruction means that the instruction is looking for a _____, or transition of the input logic in front of the one-shot instruction.
24. When the ONS sees a false-to-true transition, the instruction sends a "one-shot" signal to the output instruction, which will execute only _____.
25. As a review of the instructions, fill in the missing parts of the table in Figure 11-54.

RSLOGIX BIT INSTRUCTIONS			
Instruction	Symbol	Mnemonic	Explanation
Normally open, or examine ON			
Normally closed, or examine OFF			
One-shot			
Output latch			
Output unlatch			
Output coil			

Figure 11-54 Overview of bit instructions.
© Cengage Learning 2014

CHAPTER 12

ControlLogix Timer Instructions

OBJECTIVES

After completing this lesson, you should be able to

- Identify the TON, TOF, and RTO timers.
- Program the TON timer and understand the associated status bits.
- Configure a communications path for download.
- Download a project into the controller.
- Change controller mode to Run, Go online, and Monitor data.

INTRODUCTION

The ControlLogix family of programmable automation controllers (PACs) has three timer instructions. This section introduces the timer instructions. The lab exercise for this chapter steps you through programming the timer on-delay instruction. ControlLogix timers are very similar to the SLC 500 and PLC 5 timer and counter instructions.

INSTRUCTOR DEMONSTRATION

Follow along as the instructor demonstrates an RSLogix 5000 project containing the timer instructions.

TIMER INSTRUCTIONS

The three types of timers are on delay, off delay, and retentive. Figure 12-1 describes the RSLogix 5000 timer instructions.

The timer instruction comprises the following components (refer to Figure 11-2 and identify the components of the timers introduced):

A. *Timer mnemonic*—a three-letter abbreviation that is used to identify the timer instruction
B. *Timer instruction identification*
C. The *timer parameter*, which identifies the tag where the timer and its associated status information are stored
D. The *preset*, which is a DINT between 0 and 2,147,483,647 that identifies how many time base increments the timer is to time before the timer is considered done (the ControlLogix timer time base is fixed at 1 millisecond)

310 CONTROLLOGIX TIMER INSTRUCTIONS

TIMER INSTRUCTIONS			
Instruction	Mnemonic	Use Instruction to	Functional Description
On Delay	TON	Program a time delay before an instruction becomes true.	Use an on-delay timer when an action is to begin at a specified time after the input becomes true. As an example, a certain step in the manufacturing process is to begin 30 seconds after a signal is received from a limit switch. The 30-second delay is the on-delay timer's preset value.
Off Delay	TOF	Program a time delay to begin after rung inputs go false.	An external cooling fan on a motor is to run all the time the motor is running and for 5 minutes after the motor is turned off. This is a 5-minute off-delay timer. The 5-minute timing cycle begins when the motor is turned off.
Retentive	RTO	Retain accumulated value through power loss, controller mode change, or rung state going from true to false.	Use a retentive timer to track the running time of a motor for maintenance purposes. Each time the motor is turned off, the timer remembers the motor's elapsed running time. The next time the motor is turned on, time will increase from there. To reset this timer, use a reset instruction.
Reset	RES	Resets the accumulated value of a timer or counter.	A reset is typically used to reset a retentive timer's accumulated value to 0.

Figure 12-1 RSLogix timer instructions.
© Cengage Learning 2014

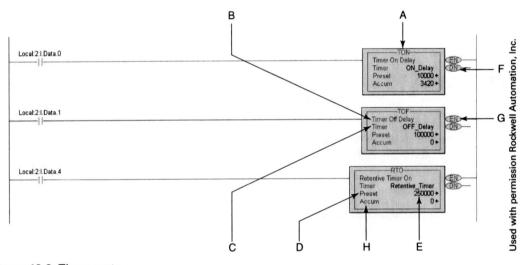

Figure 12-2 Timer parts.

    E. *Timer preset value* in milliseconds

    F. *DN* (done), which is a status bit indicator that displays when the timer is done timing (the DN bit is one of the timer's status bits)

    G. *EN* (enable), which is a status indicator that displays whether the timer instruction is true or enabled (the EN is one of the timer's status bits)

    H. The *accumulated value*, also a DINT, or our current position in the timing cycle (the accumulated value contains the current elapsed time)

TIMER STATUS BITS

In order for the PAC and human operator to monitor the current timer status, *status bits* are used. Timers have three status bits, as listed in Figure 12-3.

Status Bit	Identified as	Example Tag	Functional Description
Enable	EN	Mixing_Timer.EN	Bit is true when timer instruction is true or enabled.
Done	DN	Mixing_Timer.DN	The done bit is true when the accumulated value and preset value are equal.
Timer Timing	TT	Mixing_Timer.TT	The timer timing bit is true when the timer is actually timing.

Figure 12-3 RSLogix 5000 timer status bits.
© Cengage Learning 2014

INTERPRETING TIMER LADDER RUNGS

Now that we have introduced the three timers and their components, we look next at interpreting the timers and their associated ladder rungs.

TON Timer Operation

The first timer we look at is the TON instruction. Refer to Figure 12-4 as we illustrate how these rungs operate. The TON is a nonretentive instruction. If the rung on which the TON is programmed were to go false for any reason, then the timer would not retain its accumulated value. The accumulated value would be reset to 0. It is the nature of the instruction to be nonretentive. A nonretentive instruction's operation has nothing to do with the controller battery. The TON instruction is reset simply by making the rung go false.

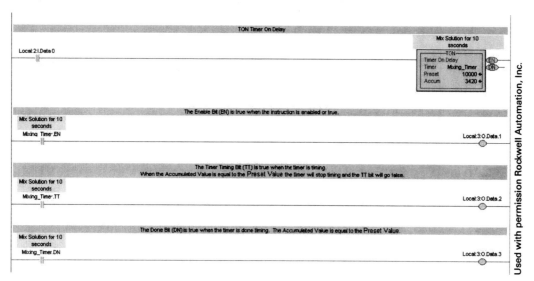

Figure 12-4 RSLogix 5000 TON ladder rungs.

1. When Local:2:I.Data.0 on the first rung is true, the TON instruction with the tag Mixing_Timer is true, or enabled.
2. With the timer instruction true, the tag Mixing_Timer.EN on rung 2 becomes true. With the enable bit true, the EN indicator on the upper right of the TON instruction is highlighted.
3. The TON preset value is 10,000. Because the timer time base is in 1-millisecond intervals, the timer times for 10 seconds.
4. When the timer becomes enabled, the timer starts timing. The timer timing (TT) bit, Mixing_Timer.TT, also becomes true on the third rung.
5. The accumulated value increments in millisecond intervals until it is equal to the preset.
6. When the accumulated and the preset values are equal, the timer stops timing and the timer timing bit goes false.
7. Now that the timer is done timing, the done bit is set on rung 4. The done bit tag is Mixing_Timer.DN.

8. If the timer rung stays true, the done bit remains set until the TON instruction goes false, which resets the accumulated value to 0.
9. Next time the TON rung goes true again, the instruction will start timing.

There are two ways the status bits can be used to control the outputs. The done bit turns on the output when the time has expired and maintains the output as true until the TON is reset by the rung going false. The timer timing bit, on the other hand, makes its associated output go true as soon as the timer starts timing. When the time has expired and the timing bit goes false, the output goes false. Do we want the output to come on immediately and turn off when the time has expired, or do we want it to come on after the time has expired and stay on until the instruction is reset?

TOF Timer Operation

The TOF is a timer off delay. The off-delay timing cycle starts after the input controlling the timer goes false. An example of an off-delay timer is a projector used to show a PowerPoint presentation. When the instructor pushes the button to start the projector for the presentation, the projector lightbulb and the fan that cools the bulb turn on immediately. Both are on for the duration of the presentation. When the presentation is completed, the standby button is pressed to turn the projector off. The light is turned off, but the fan that cools the bulb continues to run for a predetermined time. The bulb-cooling fan is controlled by an off-delay timer. The time delay starts after the projector is turned off. Figure 12-5 illustrates an example of an off-delay timer. The TOF is also a nonretentive instruction. If the rung on which the TOF is programmed were to go false for any reason, then the timer would start its off-delay timing cycle.

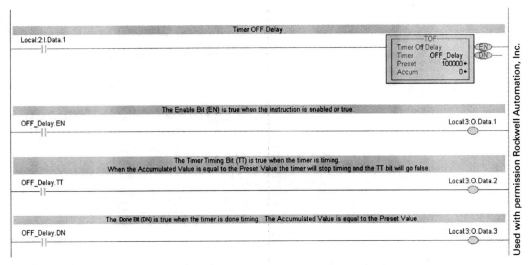

Figure 12-5 Timer_OFF Delay instruction.

1. When Local:2:I.Data.1 on the first rung is true, the TOF instruction with the tag OFF_Delay is true, or enabled.
2. With the timer instruction true, the tag OFF_Delay.EN becomes true. With the enable bit true, the second rung is true, and the indicator on the upper right of the TOF instruction is highlighted. The done bit indicator, as well as rung 4, also becomes true at this time. The timer stays in this state until the rung goes false.
3. The TOF preset value is 100,000. Because the timer time base is in 1-millisecond intervals, the timer has an off-delay timing cycle time of 100 seconds.
4. At such time as the timer rung goes false, the timer starts its off-delay timing cycle. The timer timing bit, OFF_Delay.TT becomes true on the third rung.

5. The accumulated value increments in millisecond intervals until equal to the preset.
6. When the accumulated and the preset values are equal, the timer stops timing and the timer timing bit goes false.
7. Now that the timer is done timing, the done bit will be reset. The done bit tag is OFF_ Delay.DN.
8. The timer is ready to go again.

RTO Timer Operation

The RTO is a retentive timer on delay. This timer works the same as the TON except this instruction is retentive. Being retentive, the instruction remembers its accumulated time value if the rung goes false. Also, if for some reason power is lost to the PLC and the battery is good, then the instruction remembers or retains its accumulated value. If the battery is dead or missing, after a short time the entire project will be lost (see Figure 12-6).

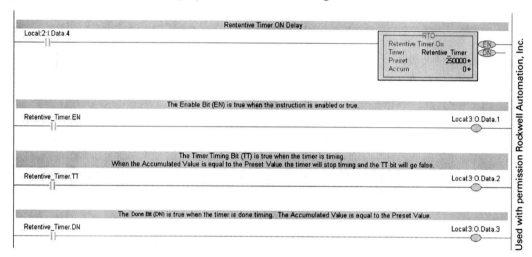

Figure 12-6 Retentive Timer_ON delay.

The RTO timer status bits are the same as the TON instruction's status bits. Because this timer is a retentive instruction, we cannot simply make the rung go false as with the TON to reset the instruction's accumulated value to 0. The reset, or RES, instruction is used to reset the timer's accumulated value to 0. Figure 12-7 shows a rung with a RES instruction. When the reset rung goes true, the accumulated value of Retentive_Timer is reset to 0. Anytime the RES instruction goes true, the accumulated value of the associated timer is reset to 0. If someone accidentally pushed the button Local:2:I.Data.4, the timer would reset.

Figure 12-7 Reset instruction programmed to reset the RTO with the tag Retentive_Timer accumulated value back to 0.
Used with permission Rockwell Automation, Inc.

ADDRESSING CONTROLLOGIX TIMERS

A ControlLogix timer is a three-DINT structure. A timer structure contains three DINTS that work together as a unit to give us a timer instruction. DINT 0 of the structure contains the timer's status bits. DINT 1 contains the preset value, and DINT 2 contains the accumulated value.

Timer instruction data are in one of two formats. Status bit data are single-bit data, whereas preset and accumulated value data are represented as a 32-bit signed integer. Figure 12-8 illustrates the format of the timer structure.

DINT 0	Status Bits
DINT 1	Timer preset value (.PRE) 0 -2,147,483,647
DINT 2	Accumulated value (.ACC) 0 -2,147,483,647

Figure 12-8 Timer structure data format.
© Cengage Learning 2014

Each timer structure has two DINTs that contain data that represent the preset and accumulated value. These two DINTs are subelements of the timer structure. A *subelement* is simply a separate piece of the structure or instruction that can be addressed as a stand-alone element or tag. The preset value and accumulated value can be addressed individually.

Timer Address Format

Figure 12-9 illustrates part of a controller scoped tags collection. There are three timers included in the collection: the OFF_Delay, ON_Delay, and Retentive_Timer tags.

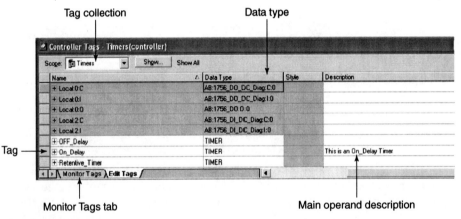

Figure 12-9 RSLogix 5000 Monitor Tags view.
Used with permission Rockwell Automation, Inc.

By clicking the + in front of the timer tag name, the timer can be expanded to show the preset, accumulated value, and status bits. As illustrated in Figure 12-10, current data or status can

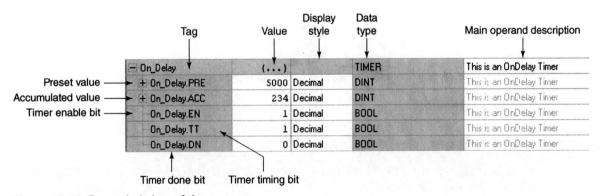

Figure 12-10 Expanded view of timer structure.
Used with permission Rockwell Automation, Inc.

be viewed in the value column of the Monitor Tags window when online with the controller. The display style identifies the base or radix in which the instruction data are displayed on the ladder rung. Data Type identifies they type of tag. The main operand description is text associated with the main tag and the instruction. We work with programming the other timer instructions and adding documentation in Chapter 13.

Examples of ControlLogix timer tag formats are as shown in Figure 12-11.

SAMPLE TIMER ADDRESSES			
Structure Member	Sample Tag	Data Type	Explanation
Preset value	On_Delay.PRE	DINT	How long is the timer to time? This is in milliseconds
Accumulated value	On_Delay.ACC	DINT	Tracks how much time has passed or accumulated
Timer timing bit	On_Delay.TT	BOOL	True only when the timer is timing
Enable bit	On_Delay.EN	BOOL	True when the timer instruction is true
Done bit	ON_Delay.DN	BOOL	True when the timer is done timing

Figure 12-11 Sample timer addresses for the timer with the tag On_Delay.
© Cengage Learning 2014

SELECTING TIMER INSTRUCTIONS FROM THE LANGUAGE ELEMENT TOOLBAR

Timer and counter instructions can be selected from the Timer/Counter tab on the Language Element toolbar. The desired instruction can be selected by clicking it or dragging and dropping it into the desired position on the ladder rung. The three timer instructions are identified in Figure 12-12.

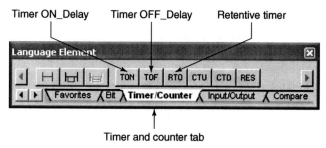

Figure 12-12 Language Element toolbar Timer/Counter tab.
Used with permission Rockwell Automation, Inc.

INTERPRETING TIMER INSTRUCTIONS

In this lab exercise we download a project containing each of the timer instructions and monitor each timer and its status bits as the instructions execute.

Timer On-Delay Instruction Operation

The first timer in your project is the timer on-delay or TON instruction. The timer and associated rungs containing the timer's status are shown in Figure 12-13. As you evaluate the rungs answer the following questions.

1. _____ Open the Timer Interpretation project.
2. _____ Download the project.
3. _____ Put the controller in Run mode.
4. _____ Explain the function of the timer on delay.
5. _____ Can the timers time base be modified?

316 CONTROLLOGIX TIMER INSTRUCTIONS

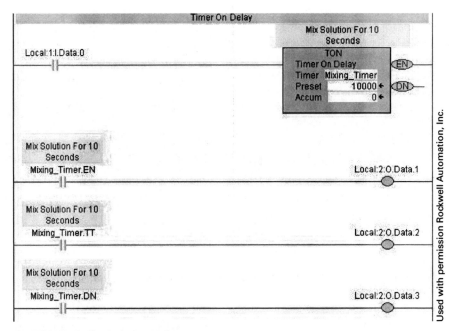

Figure 12-13 TON logic for interpretation.

6. _____ When is the enable bit true?
7. _____ Define when the done bit is true.
8. _____ Explain when the timer timing bit is true.
9. _____ What is the maximum value that can be entered into the timer's preset value?
10. _____ Where will you go to monitor the timer tag and its status bits?
11. _____ Go to that location and click the + to the left of the Mixing Timer tag to expand the structure.
12. _____ Make the input go true and monitor the timer accumulated value, timer timing bit, enable bit and done bit. Do they work as you expected?
13. _____ How do you reset the TON timer?
14. _____ Reset the timer.
15. _____ Start the timer and after a few seconds, make the rung go false. What happens to the timer?
16. _____ As the result of what you just observed, is the TON an retentive or nonretentive instruction.

Interpreting the Timer Off Delay

The second timer in your project is the timer off-delay or TOF instruction as shown in Figure 12-14. As you monitor the rungs answer the following questions.

1. _____ Explain how the TOF timer operates.
2. _____ Make the input instruction go true.
3. _____ What happens to the enable (EN) bit?
4. _____ When the instruction goes true what happens to the done (DN) bit?
5. _____ When does the timer timing (TT) bit go true?
6. _____ Make the instruction go false and test your answer to the previous question.
7. _____ Does the instruction behave as expected?
8. _____ Go to the Monitor Tags window and monitor the timer parameters as you execute the timer again.

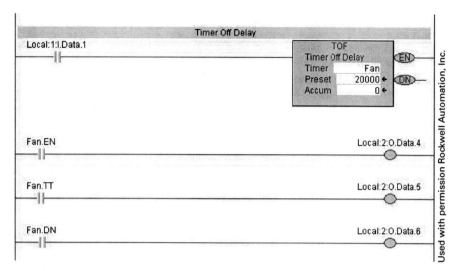

Figure 12-14 TOF timer logic.

Retentive Timer On Delay

Answer the following questions as you monitor the Retentive Timer On Delay in Figure 12-15.

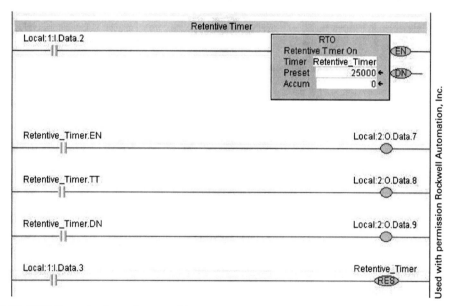

Figure 12-15 RTO timer logic and status bits.

1. _____ Execute the timer instruction and answer the questions below.
2. _____ As the timer runs but before the timer is done, make the timer instruction go false. Explain what you observed regarding the timer's accumulated value.
3. _____ Thinking about what you observed, is the RTO timer instruction retentive or nonretentive?
4. _____ Since the instruction is retentive, how does the timer get reset?
5. _____ Start the timer timing again. After the timer has timed for a while, but before the preset is reached, press the reset button. Explain what happens.
6. _____ If the reset button was held as true, will the timer time?
7. _____ This completes this lab exercise.

LAB EXERCISE 1: Programming the TON or ON_Delay Timer

We now modify the Lab Exercise 1 project we were working on for Chapter 11. We program the timer instruction ON_Delay timer on rung 4 to track the time the machine was on. The abbreviation or mnemonic is TON. When selecting the instruction, the TON is selected from the Language Element toolbar's Timer/Counter tab. Probably the best way to understand how the timer works is to program it and then download and run the project in order to observe the timer's operation. Next we program the timer on this rung (see Figure 12-16). Then we set up communications, download and run the project, and monitor the timer as it runs.

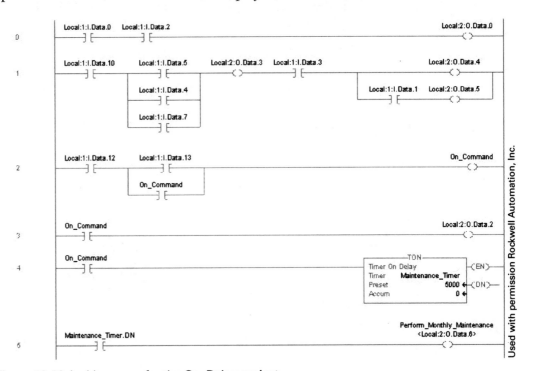

Figure 12-16 Ladder rungs for the On_Delay project.

1. _____ Open the Lab Exercise 1 project you were working on in Chapter 11.
2. _____ Add rung 4 to your project.
3. _____ Program the XIC instruction on rung 4. Drag the On_Command tag to the instruction.
4. _____ Click the Timer/Counter tab, #1 in Figure 12-17, on the Language Element toolbar.
5. _____ Left-click the TON button (#2 in the figure) and drag the instruction to the target, item number 3 in Figure 12-17. When the target accepts the instruction, drop the instruction by releasing the left mouse button. The TON automatically positions itself.
6. _____ With the instruction in position, as shown in Figure 12-18, right-click the question mark next to the word Timer (item A in the figure). Select New Tag from the list displayed.
7. _____ Enter the name "Maintenance_Timer" for the new tag, as shown in Figure 12-19. Note that the data type has already been filled in.
8. _____ Click OK when completed.
9. _____ Click the question mark for the preset (refer to item B in Figure 12-18).
10. _____ Enter the value 5,000 and press Enter on your computer keyboard.

The preset value (PRE) for a timer is the amount of time the timer is to time before it is done timing and the done bit is set. The time base for ControlLogix timers is fixed at 1 millisecond. A preset of 5,000 is equal to 5 seconds. The preset value is a DINT.

CONTROLLOGIX TIMER INSTRUCTIONS 319

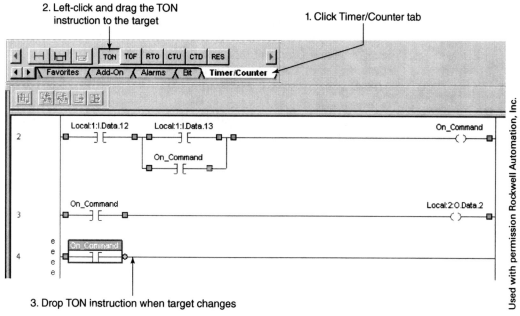

Figure 12-17 Drag TON instruction into position.

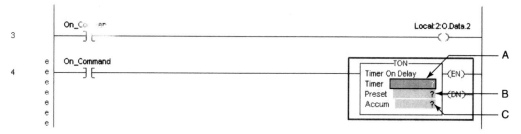

Figure 12-18 Programming the TON instruction.
Used with permission Rockwell Automation, Inc.

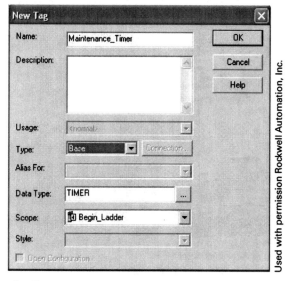

Figure 12-19 Enter name for timer.

11. _____ The accumulated value (ACC) is the amount of time that has accumulated or transpired since the timer instruction became true. When programming a timer,

the accumulated value is typically set to 0. When the timer is true, we are going to say the instruction is enabled. Now that the timer is enabled, the timer starts timing and the TT bit is true. The timer increments the accumulated value until it is equal to the preset. When the accumulated and the preset values are equal, the timer is done timing and the timer will stop. The timer timing bit goes false as the done bit becomes true.

The TON instruction is a nonretentive instruction. If the rung on which the timer is programmed goes false, then the timer forgets where it was in the timing cycle and resets the accumulated value to 0. Likewise, after the timer has completed timing, resetting the timer is accomplished by making the rung go false. Your timer instruction should look like that shown in Figure 12-20. The (EN) is an indicator that is illuminated when the instruction is true or enabled. The (DN) is an indicator showing that the timer is done timing.

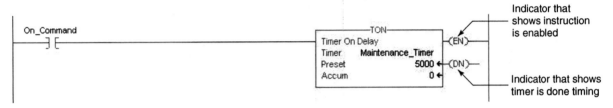

Figure 12-20 Completed timer instruction.
Used with permission Rockwell Automation, Inc.

On the next rung, we program the done status bit to control the output when the timer is done timing. Refer to Figure 12-16.

12. _____ Create a new rung.
13. _____ As you program the XIC instruction, refer to Figure 12-16.
14. _____ Double-click the question mark above the XIC instruction so we can select the timer done bit tag.
15. _____ Click the down arrow in the box that appears.
16. _____ Scroll through the list to find the timer tag. Tags should be alphabetical.
17. _____ When the time tag has been located, click + to expand the timer structure and display the elements. Refer to item number 1 in Figure 12-21.

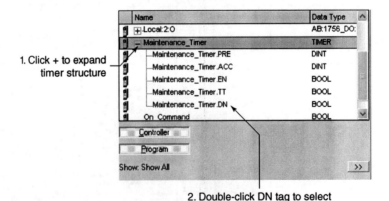

Figure 12-21 Selecting the timer's done bit as your tag.
Used with permission Rockwell Automation, Inc.

18. _____ Refer to item number 2 in Figure 12-21 as you double-click the timer's done bit from the list to select it as the tag for this instruction.
19. _____ Your XIC instruction and tag should look like that shown in Figure 11-22.

CONTROLLOGIX TIMER INSTRUCTIONS   321

Figure 12-22 Completed rung 5.
Used with permission Rockwell Automation, Inc.

20. _____ Program the OTE instruction on rung 5.

We will add an alias tag for our output on this rung. The tag for the OTE instruction will be "Local:2:O.Data.6." This is called the *base tag*. The text above the base tag Perform Monthly Maintenance is known as the *alias tag*. Follow the steps to create this alias tag.

21. _____ Right-click the question mark above the OTE instruction so we can create the alias tag.
22. _____ Select New Tag. The New Tag dialog box should appear.
23. _____ Refer to Figure 12-23 as you complete the following steps:

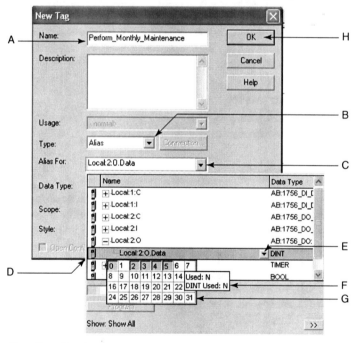

Figure 12-23 Creating the alias tag.
Used with permission Rockwell Automation, Inc.

    A. Enter alias tag name.
    B. Click the down arrow and select Alias.
    C. Click the down arrow to find Output Module and click + to expand the list.
    D. Find Local:2:O.Data.
    E. Click the down arrow.
    F. A new feature that appeared with version 15 provided a quick method to see whether a specific bit has been used. Note that the used bits are identified by their box being indented. Also, if you hover the cursor over the bit, the Tool Tips will determine whether the bit is used.
    G. Select bit 6 for this tag.
    H. Select OK when completed. Your completed rung should look like that shown in Figure 12-24.

322 CONTROLLOGIX TIMER INSTRUCTIONS

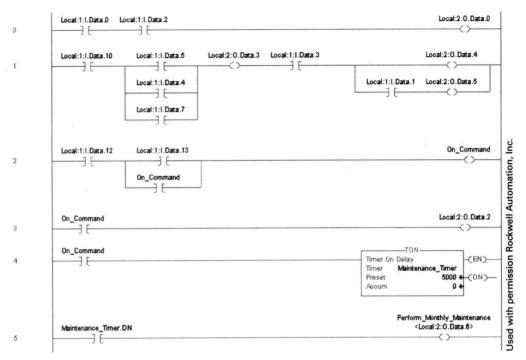

Figure 12-24 Completed rungs.

24. _____ Save your project.
25. _____ Download your project.
26. _____ Put the controller in Run mode.

MONITORING THE TON TIMER

With our project now running in our controller and our computer online with the controller, we can monitor the rungs we just completed programming. We assume you have 16 maintained push buttons or toggle switches wired to the input module in slot 1 and indicator lights wired to the output module in slot 2. If you do not have indicator lights connected to the output module, then use the output module's output LEDs.

1. _____ Push buttons Local:1:I.Data.10 and Local:1:I.Data.5. Notice that the input instructions are true as represented by the highlights to the right and left of the instruction. See Figure 12-25.
2. _____ The output Local:2:O.Data.3 is also true because it is highlighted. Write the rule regarding logical continuity and when a rung's outputs will be true.

_____

_____

_____

3. _____ What would happen if input Local:1:I Data.3 were also made true?
4. _____ Push button Local:1:I Data.3 to test your answer.
5. _____ Push button Local:1:I.Data.1. What happens?
6. _____ If you also push button Local:1:I.Data.7 and make the instruction go true, does anything change regarding the outputs? What is the term used to refer to this type of logic? What happens if you push button Local:1:I.Data.5 and make the input go false?
7. _____ Make input Local:1:I.Data.10 go false. Explain what you observe regarding the run's outputs.

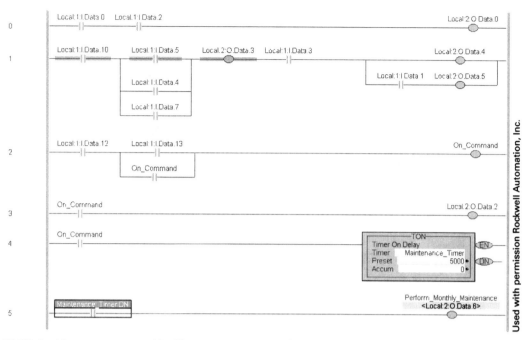

Figure 12-25 Ladder rungs created in Chapter 11 and modified in Chapter 12 downloaded for monitoring.

8. _____ To simulate a start–stop push-button station, let us assume Local:1:I.Data.12 is the normally closed stop push button. Local:1:I.Data.13 is the normally open start push button. Because the stop button on a start–stop push-button station is normally closed, push the simulated stop push button. What happens to the input instruction?

9. _____ Explain what happens if you push the simulated start push button and make the input true. To simulate a momentary start push button, push the button again to make the input go false. Explain what you observe.

10. _____ If you hover the cursor over the On_Command tag, what data type does the Tool Tips state that the tag is?

11. _____ The tag On_Command is a(n) _____ bit used to control other rungs rather than a real-world output.

12. _____ Watch the TON instruction as it executes, and answer the following questions.

13. _____ What type of timer is the TON instruction on rung 4?

14. _____ List and define the status bits associated with the TON instruction.

15. _____ What is the tag for the preset value?

16. _____ What is the tag for the accumulated value?

17. _____ List the tags for the TON status bits.

18. _____ What happens to the status bits when the timer becomes true?

19. _____ What happens to the TON status bits when the accumulated value is equal to the preset value?

20. _____ Explain what happens if the input on the rung goes false.

21. _____ Identify the parts of the window shown in Figure 12-26.

22. _____ On ladder rung 4, right-click the TON tag and select Monitor "Maintenance_Timer," as illustrated in Figure 12-27.

23. _____ How can you expand the Maintenance_Timer structure to show all of the associated tags so you can monitor them, as shown in Figure 12-26?

24. _____ Click + to expand the structure.

25. _____ Monitor the timer in the monitor window as you push the appropriate buttons to run the timer. Do the status bits operate as expected?

324 CONTROLLOGIX TIMER INSTRUCTIONS

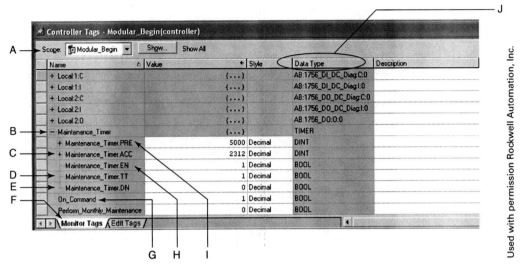

Figure 12-26 RSLogix display for question 17.

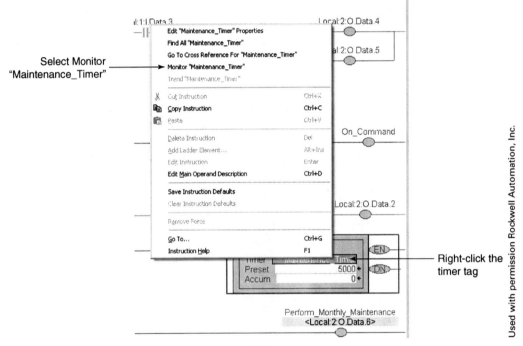

Figure 12-27 Right-click timer tag and select Monitor "Maintenance_Timer."

26. _____ Run the timer for approximately 3 seconds.
27. _____ Now reset the timer.
28. _____ What did you do to reset the timer?
29. _____ Did the accumulated value and status bits behave as expected?
30. _____ When the timer goes false, the accumulated value resets itself to 0 because the TON is a _____ instruction.
31. _____ Click + in front of your input and output tags to display the associated tags. If you have diagnostic modules, your Monitor Tags window should look similar to that shown in Figure 12-28.

   The figure is displaying a modular ControlLogix with a 1756-IB16D and 1756-OB16D modules. Remember, only the modular ControlLogix has diagnostic I/O modules.

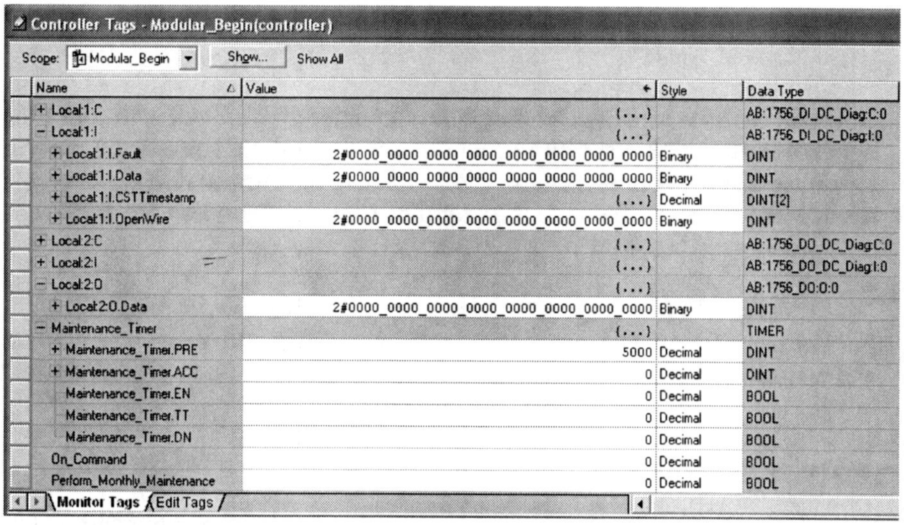

Figure 12-28 Monitor Tags view.

32. _____ Experiment with making your rungs become true and false by pressing the appropriate buttons. Monitor your ladder rungs and the Monitor Tags window to understand the correlation between the input push buttons and their associated tag bits. Also monitor the output rungs and the tag in the Monitor Tags view as they become true and false. Lastly, monitor the timer rungs and the associated tags in the Monitor Tags view.

33. _____ Now that you have a basic understanding of how ladder rungs are represented as they execute in the Ladder Logic window and how to monitor the instructions as they execute, interpret the RSLogix 5000 Monitor Tags window in Figure 12-29 for our project. Answer the following questions:

    A. Which input instructions are currently true?
    B. Which digital outputs are currently true?
    C. What is the status of the TON timer? Explain the displayed condition of the timer's status bits.

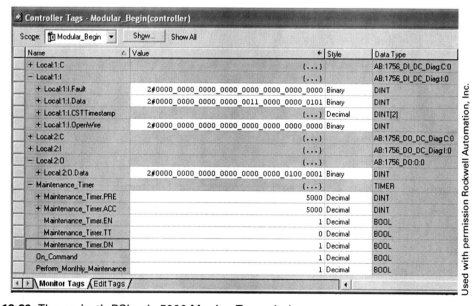

Figure 12-29 The project's RSLogix 5000 Monitor Tags window.

LAB EXERCISE 3: Traffic Light Application

INTRODUCTION

Now that we have experienced programming timers, we are going to start programming a traffic light application. We start with programming a couple of basic timer programming sequences for our traffic lights in this chapter. As we move into future chapters, we will add some of the newly learned programming features and incorporate them into our traffic control application.

THE LAB

For this programming lab, we have a basic intersection with north and south lanes as well as east and west lanes.

APPLICATION SPECIFICATIONS

- Start–stop buttons in the control cabinet control an on-command bit output.
- Program the on command to be retentive so the project will resume after a power loss.
- Program the PAC so that if it goes out of Run mode or faults, an output will signal a mechanical relay to energize a mechanical timer to cause all directions to flash red.
- Program rungs so that when the PAC starts, the system starts with the north and south lights green and the east and west lights red.
- After power is resumed from a power failure, the sequence automatically resumes.
- After sequence is started, the light sequence automatically repeats itself until the stop button is pressed.
- North–south traffic lights:
  - Green on for 20 seconds
  - Yellow on for 5 seconds
  - Red on for 25 seconds
- East–west traffic lights:
  - Green on for 20 seconds
  - Yellow on for 5 seconds
  - Red on for 25 seconds

Figure 12-30 provides a view of the intersection for our application.

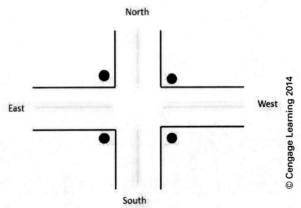

Figure 12-30 Intersection for the traffic light application.

PLANNING

Before beginning programming, we need to do some planning. Use tables 12-31 through 12-35 to determine and assign inputs, outputs, BOOL bits, and timer instructions. You may wish to copy the tables, as we will be adding additional logic to our project in future labs and lessons. The tables then can be used as reference as we expand our project's features. Ladder logic should be written out on paper before you start programming. The counter assignment sheet will be used in Chapter 14.

INPUT TAG ASSIGNMENT SHEET		
I/O Base Tag	Alias Tag	Description
Local:1:I.Data.0		
Local:1:I.Data.1		
Local:1:I.Data.2		
Local:1:I.Data.3		
Local:1:I.Data.4		
Local:1:I.Data.5		
Local:1:I.Data.6		
Local:1:I.Data.7		
Local:1:I.Data.8		
Local:1:I.Data.9		

Figure 12-31 Input tag assignment sheet.
© Cengage Learning 2014

OUTPUT TAG ASSIGNMENT SHEET		
I/O Base Tag	Alias Tag	Description
Local:2:O.Data.0		
Local:2:O.Data.1		
Local:2:O.Data.2		
Local:2:O.Data.3		
Local:2:O.Data.4		
Local:2:O.Data.5		
Local:2:O.Data.6		
Local:2:O.Data.7		
Local:2:O.Data.8		
Local:2:O.Data.9		

Figure 12-32 Output tag assignment sheet.
© Cengage Learning 2014

BOOL TAG ASSIGNMENT SHEET	
BOOL Tag	Description

Figure 12-33 BOOL tag assignment sheet.
© Cengage Learning 2014

TIMER TAG ASSIGNMENT SHEET		
Timer Base Tag	Timer Preset	Description

Figure 12-34 Timer tag assignment sheet.
© Cengage Learning 2014

COUNTER TAG ASSIGNMENT SHEET		
Counter Base Tag	Counter Preset	Description

Figure 12-35 Counter tag assignment sheet.
© Cengage Learning 2014

THE LAB

1. _____ Open a new project for your specific PAC.
2. _____ Perform an I/O configuration for the PAC you are using.
3. _____ Complete programming the needed logic.
4. _____ Save your project, as you will be making modifications in future lessons.
5. _____ Set up communications and download the project.
6. _____ Run your project and verify proper operation.

LAB EXERCISE 4: Modification of Traffic Light Application

INTRODUCTION

Traffic studies have shown that many times during the day there are few vehicles traveling on the east and west roadways. This causes the north and south traffic to stop for a red light when there may be no vehicles going east or west. As a result of this study, you have been directed to modify the project as listed here.

PROJECT MODIFICATIONS

- Rather than have a continuous sequence between the two traffic directions, modify the application so that unless there is a vehicle waiting to go through the intersection either east or west, the north–south light will maintain its green state.
- A sensor has been installed in the pavement for both the east and west traffic lanes. If either of these sensors detects a waiting vehicle, the PAC will start the sequence for the north and south lights to turn red, and the east and west lights will become green.
- East and west lights are green for 20 seconds.
- A sign has been placed on the shoulder of the north and south lanes, reading, "Prepare to stop when flashing." There are two lights directly above the sign.
  - The lights alternately flash, warning the north and south drivers that their green light is about to change.
  - These lights start flashing 20 seconds after a vehicle is sensed waiting in either the east or west traffic lanes.
  - The lights flash for 20 seconds before the north and south green-to-red transition begins.
  - The flashing lights continue to flash until the north and south lights return to green. At this point, the north and south lights remain green until another vehicle is detected on either the east or west sensors.
  - The times for other lights remain the same as in the last exercise.

Figure 12-36 shows the sign for the flashing-lights feature. Figure 12-37 illustrates the modified traffic-signaling application.

Figure 12-36 Flashing lights alerting north and south drivers their light is about to change.

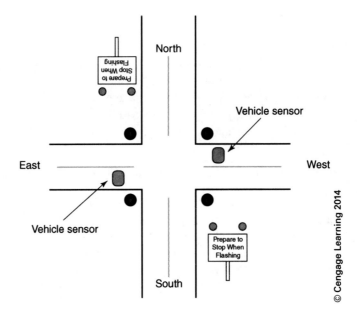

Figure 12-37 Modified view of our traffic control application.

PLANNING

Before you start creating ladder rungs, it is a good idea to do some preliminary planning. Use the previous tables to assign the new tags required. It is usually a good idea to write out your logic on paper before starting programming.

THE LAB

1. _____ Open your traffic light project completed in the last lab.
2. _____ Modify the project to incorporate the new specifications.
3. _____ Save your project with a new name of your choosing, as you will make modifications in future lessons.
4. _____ Set up communications and download the project.
5. _____ Run your project and verify proper operation.

SUMMARY

This lesson covered a lot of ground. We learned about the three timer instructions and how to edit the project started in Chapter 11 by adding a couple rungs and the TON timer. After we modified our original ladder routine, we verified that the project was free of errors. With the project verified, we were clear to set up a communications path and download our offline project into the ControlLogix controller, change the controller's operating mode to Remote Run, and go online. Once we were online, the screen on our personal computer displayed live information regarding the ladder rungs and their instructions, so we could monitor the rungs as they responded to our pushing input buttons and made the rungs true or false. Once we had a basic understanding of how the rungs look as they execute, we opened the controller scoped tags collection and monitored our ladder instructions from the Monitor Tags window. The next lesson introduces the other timers and adding text documentation to our project's ladder rungs. Once we modify our project, we will redownload it and monitor the new timers we have programmed. Because we will be reusing this same ladder project, be sure to save it.

REVIEW QUESTIONS

1. The abbreviation or mnemonic for the timer on delay is _____.
2. The three types of timers are _____, _____, and _____.
3. Fill in the missing information in the table in Figure 12-38.

Instruction	Mnemonic	Use This Instruction to	Functional Description
Timer ON_Delay			
Timer OFF_Delay			
Retentive Timer ON_Delay			
Reset			

Figure 12-38 Timer instructions.
© Cengage Learning 2014

4. After the timer has completed timing, resetting the TON timer is accomplished by making the rung go _____.
5. The _____ is the amount of time that has accumulated or transpired since the timer instruction became true.
6. When the timer is true, the instruction is _____.
7. The timer times until the _____ value is equal to the preset.
8. When the _____ and _____ are equal, the timer is done timing.
9. The timer stops timing when accumulated and preset values are _____.
10. Is the TON instruction a retentive or nonretentive instruction?
11. If the rung that the timer is programmed on goes false, the _____ timer forgets where it was in the timing cycle and resets the accumulated value to 0.
12. Fill in the missing information in the table in Figure 12-39.

Status Bit	Identified as	Example Tag	Functional Description
	Enable		
TT			
		Heating_Timer.DN	

Figure 12-39 RSLogix 5000 timer status bits.
© Cengage Learning 2014

13. The _____ is a Timer OFF_Delay.
14. Is the TOF a retentive or a nonretentive instruction?
15. If the rung that the TOF instruction is programmed on were to go _____, the timer would start its off-delay timing cycle.
16. Identify parts A–G shown in Figure 12-40.
17. The abbreviation or mnemonic for the Timer OFF_Delay is _____.
18. The _____ instruction is used to reset the accumulated value of the RTO back to 0.
19. The abbreviation or mnemonic for the retentive timer on delay is _____.
20. Identify parts A–J shown in Figure 12-41.

332 CONTROLLOGIX TIMER INSTRUCTIONS

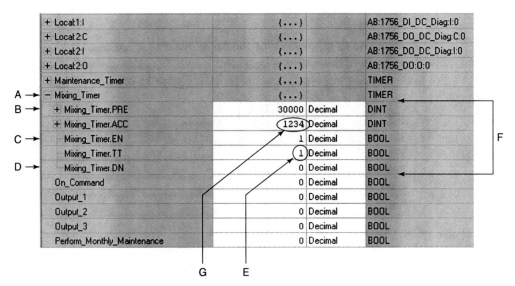

Figure 12-40 Identification of timer structure.
Used with permission Rockwell Automation, Inc.

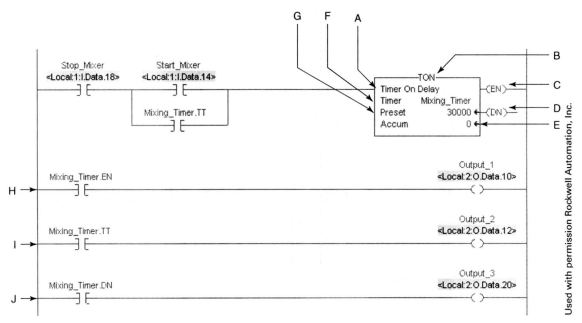

Figure 12-41 Interpreting timer ladder rungs.

CHAPTER

13

Adding Ladder Rung Documentation

OBJECTIVES

After completing this lesson, you should be able to

- Add alias tags.
- Add program documentation for main operand descriptions.
- Add program documentation for rung comment descriptions.
- Create a new project program and edit ladder rungs.
- Document, set up communications, download, and monitor a TOF timer.
- Modify our project to include a RTO timer to track running time of a motor.
- Add documentation to the project.
- Monitor the project as it runs.

INTRODUCTION

This lab introduces the timer off-delay (TOF) and *retentive timer on* (RTO) timers along with adding text documentation using the RSLogix 5000 software. RSLogix text documentation includes alias tags, main operand description, and rung comments. Refer to Figure 13-1 to view each type of documentation.

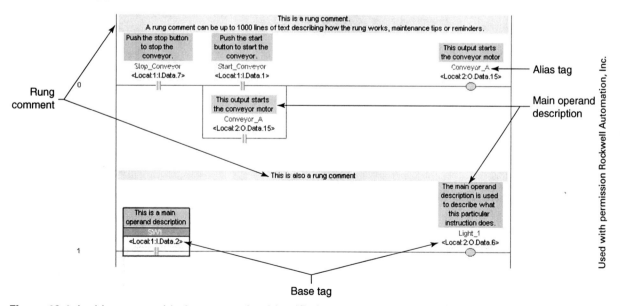

Figure 13-1 Ladder rungs with documentation identified.

Rung comment Each rung can have a rung comment, and each comment can display as many as 1,000 lines of text that explain the function of the ladder rung. Rung comments can be turned on or off under Workstation Options.

Main operand description A main operand description is text that can be added by the programmer to describe the function of the main tag—that is, the main operand of the instruction. The main operand is the destination tag for any instruction that has source and destination parameters. A single main operand description can be as many as 128 characters and 20 lines. The main operand description is displayed directly above the base tag and its associated alias. Remember that documentation can be turned on or off under Workstation Options.

Alias tag *Alias* is another name for a tag. It is easier to remember an alias tag such as Light_1 rather than the base tag like Local:0:O.Data.6. Refer to Figure 13-1 to see this alias and base tag. The alias tag and its base tag can be displayed or only the alias. This will also be set up in Workstation Options.

LAB EXERCISE 1: Adding Ladder Rung Documentation

This lab exercise uses the ladder rungs created in Chapter 12's lab exercise 2 and adds documentation to them.

1. _____ Open the project you created for lab exercise 2 of Chapter 12. Your current ladder rungs should look like that shown in Figure 13-2.

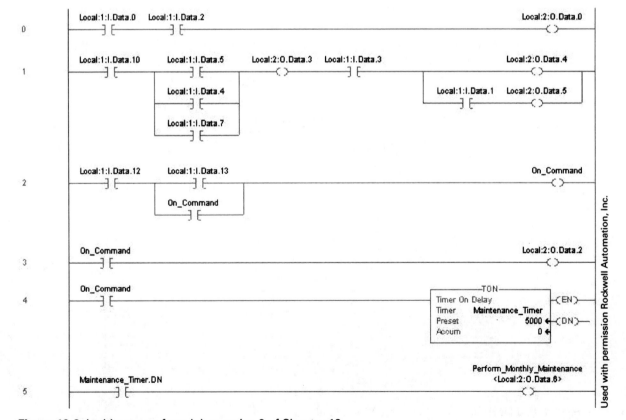

Figure 13-2 Ladder rungs from lab exercise 2 of Chapter 12.

2. _____ We will use the alias tag information in the table in Figure 13-3 and add the aliases to our rungs. Note the underscores between the words. Spaces are not allowed. The underscore will be inserted as you enter the tag name and press the space bar. Before adding the aliases, we will look at two different ways to create alias tags.

Local:1:I.Data.0	Limit_Switch_1
Local:1:I.Data.2	Limit_Switch_2
Local:1:I.Data.12	Stop_Machine
Local:1:I.Data.13	Start_Motor
Local:2:O.Data.0	Close_Gate
Local:2:O.Data.2	Machine_Running

© Cengage Learning 2014

Figure 13-3 Alias tags to assign to ladder rungs from the lab exercise of Chapter 12.

ASSIGNING ALIAS TAGS THROUGH THE CONTROLLER SCOPED EDIT TAGS TAB

Figure 13-4 guides you through assigning an alias tag through the Edit Tags view of the controller scoped tags collection. As an example, let us assign the Start_Motor as the alias for Local:1:I.Data.13. Refer to the figure as you follow the steps given here.

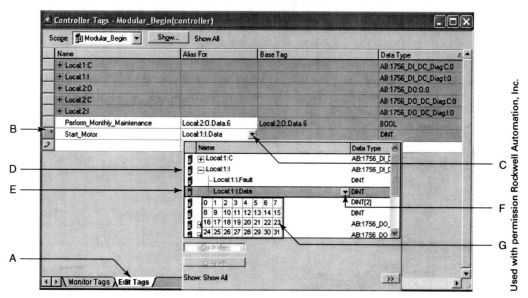

Figure 13-4 Assigning an alias tag.

A. Open the controller scoped tags collection and select Edit Tags.
B. Enter the tag name.
C. Move your mouse to the right side of the Alias For cell. Click the down arrow when it appears.
D. Select the module from the list. Click + to expand and see the tags.
E. Select Local:1:I.Data.
F. Click the down arrow.
G. Select the proper bit number. Click off row when done to update the tag.

Figure 13-5 shows the controller scoped tags collection and our newly created alias tag. Tags can have multiple alias tags. Because this tag only has one alias, the information in the Alias For and Base Tag columns is the same. Note that the Perform Monthly Maintenance alias tag from the lab exercises of Chapter 11 is also displayed.

ADDING LADDER RUNG DOCUMENTATION

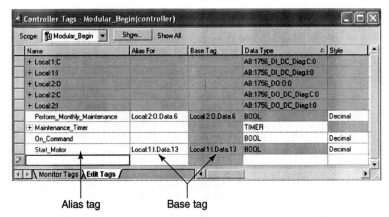

Figure 13-5 Controller scoped tags collection showing newly created alias tag.
Used with permission Rockwell Automation, Inc.

CREATING ALIAS TAGS FROM THE LADDER RIGHT-CLICK MENU

Alias tags can also be created from the ladder view as the instructions are programmed on the ladder rungs. To simulate this, we will delete the current tag and create a new tag.

1. _____ Double-click rung 2 and the Local:1:I.Data.12 tag.
2. _____ Delete the current tag.
3. _____ Click a blank spot on your ladder view to exit.
4. _____ The original tag should now be gone. Currently, there is no tag assigned to this instruction. This is like programming a new instruction and assigning a tag as you program it on the ladder rung.
5. _____ Right-click the question mark above the instruction as in Figure 13-6.

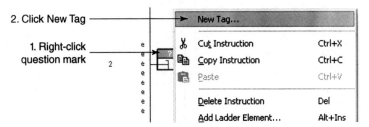

Figure 13-6 Creating an alias tag using the right-click menu from the Ladder Logic window.
Used with permission Rockwell Automation, Inc.

6. _____ Click New Tag.
7. _____ Refer to Figure 13-7 and follow the steps given here to create our new alias tag.
   a. Enter the tag's alias name.
   b. Click the drop-down menu and select Alias.
   c. Click the Alias For drop-down arrow.
   d. Scroll to Local:1:I and click the + sign.
   e. Scroll to Local:1:I.Data and click the down arrow.
   f. Select bit 13. If using RSLogix 5000 version 19 or eralier click OK. If using RSLogix 5000 20 or newer, click Create.
8. _____ Now that you know how to create alias tags, complete adding the needed alias tags listed in Figure 13-3. Try using both methods for assigning the alias tags.

ADDING LADDER RUNG DOCUMENTATION **337**

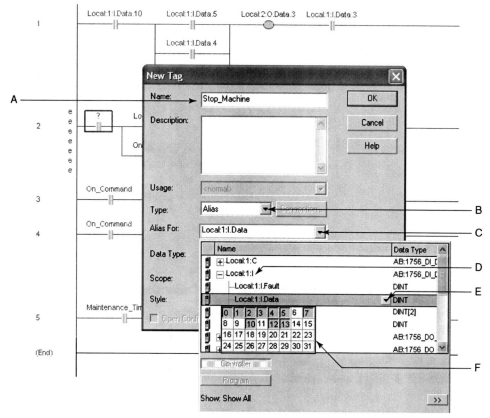

Figure 13-7 Creating a new alias tag.
Used with permission Rockwell Automation, Inc.

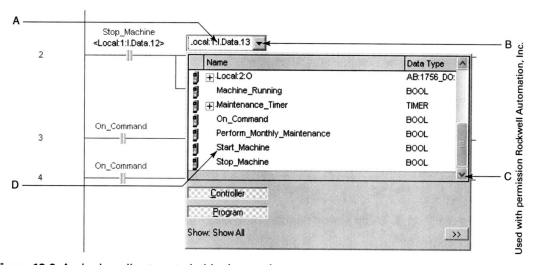

Figure 13-8 Assigning alias tags to ladder instruction.

9. _____ Because we had already assigned tags to our inputs and outputs in the lab exercises in Chapters 11 and 12, we will have to reassign the alias tags now that they are created. For an example, refer to Figure 13-8 and the following steps:
   a. Double-click the existing tag.
   b. Click the down arrow.
   c. Tags should be alphabetical. Scroll down to find the tag.
   d. Double-click the tag to select.
10. _____ Reassign any tags for which you created alias tags.

ADDING A MAIN OPERAND DESCRIPTION

A main operand description is additional text documentation that you can add to ladder instructions. A main operand description can have as many as 20 lines and 128 characters. Figure 13-9 shows an instructions base tag, alias tag, and main operand description. There are multiple ways to add a main operand description. We introduce two methods before we complete this portion of the lab.

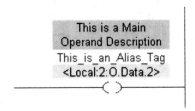

Figure 13-9 Instruction documentation.
Used with permission Rockwell Automation, Inc.

Adding a Main Operand Description from the Tags Collection

Main operand descriptions can be added anytime the programmer likes. In many cases, a program's documentation is added when the programmer initially creates the project. Documentation can also be added much later by the end user. The main operand description can be added to either the Edit or Monitor Tags view in either controller scoped or program scoped tags. Figure 13-10 illustrates adding a main operand description to a tag in the Edit Tags view.

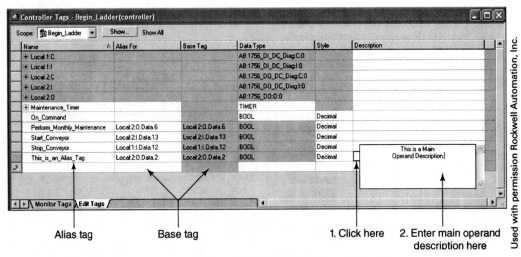

Figure 13-10 Creating a main operand description from the tags collection.

Adding a Main Operand Description from the Ladder

The main operand description can be added any time from the ladder instruction.

1. _____ As you refer to Figure 13-11, right-click the Machine Running OTE instruction.
2. _____ From the list displayed, click Edit Main Operand Description.
3. _____ Enter your text as shown in B in Figure 13-12. The main operand description is for the instruction identified as A in the figure.
4. _____ Click "C," the check mark to complete and accept the description.
5. _____ Your OTE instruction should like that shown in Figure 13-13.

ADDING LADDER RUNG DOCUMENTATION 339

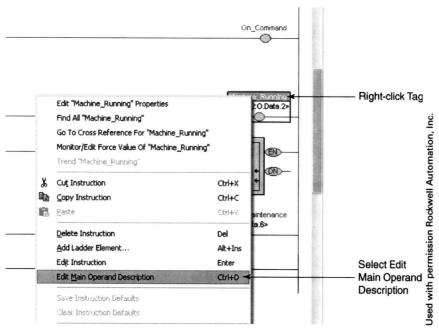

Figure 13-11 Creating an alias tag from the ladder rung.

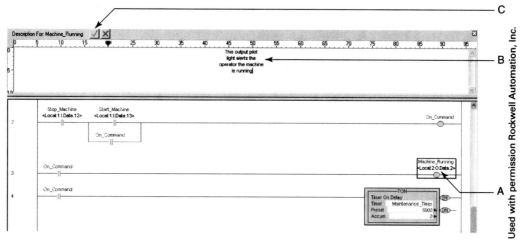

Figure 13-12 Creating an alias tag from the ladder rung.

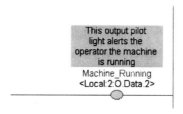

Figure 13-13 OTE instruction with alias tag and main operand description.
Used with permission Rockwell Automation, Inc.

6. _____ Now it is your turn to complete adding the main operand descriptions as listed in the table in Figure 13-14. Notice what happens when you enter the main operand description for the tag On_Command.

7. _____ Your documented rungs should look similar to those shown in Figure 13-15.

340 ADDING LADDER RUNG DOCUMENTATION

Local:1:I.Data.12	Push this button to stop the machine.
Local:1:I.Data.13	Push this button to start the machine.
On_Command	Machine has been turned on.
Local:2:O.Data.0	When this output is on, the diverter gate A is closed.
Local:2:O.Data.2	This output pilot light alerts the operator that the machine is running.
Local:2:O.Data.6	This output pilot light alerts the operator to remind the maintenance team to perform monthly maintenance.

Figure 13-14 Main operand descriptions to add to the ladder rungs from the lab exercise of Chapter 12.
© Cengage Learning 2014

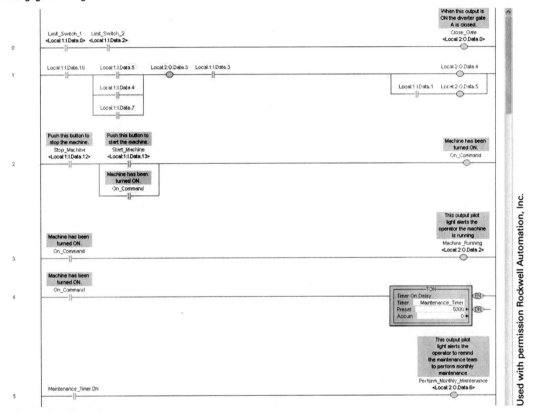

Figure 13-15 Completed rungs showing alias tags and main operand description.

PROGRAMMING RUNG COMMENTS

As many as 1,000 lines of text that explain what the rung does, how it operates, or maintenance and troubleshooting hints can be entered as a rung comment. The rung comment is an invaluable tool to assist anyone who has not worked with this application to understand the function of the rung. A sample rung is displayed in Figure 13-16 with the available documentation with RSLogix 5000 software.

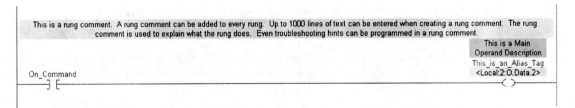

Figure 13-16 Ladder rung showing base tag, alias tag, main operand description, and the rung comment.
Used with permission Rockwell Automation, Inc.

ADDING LADDER RUNG DOCUMENTATION 341

1. _____ To add a rung comment for rung 2, right-click the rung number (A) in Figure 13-17.
2. _____ Click Edit Rung Comment (item B in the figure).

Figure 13-17 Select rung and right-click to start to enter a rung comment.
Used with permission Rockwell Automation, Inc.

3. _____ Refer to Figure 13-18 as you enter the rung comment.

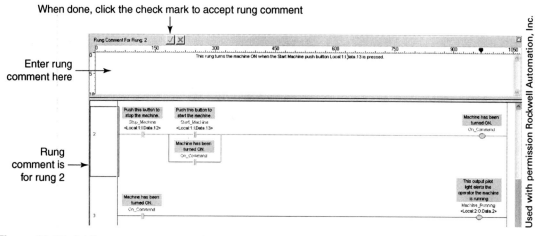

Figure 13-18 Adding a rung comment.

4. _____ When done, click the check mark. If you do not wish to keep the rung comment you just entered, click the X.
5. _____ Your completed rung should look similar to that shown in Figure 13-19.

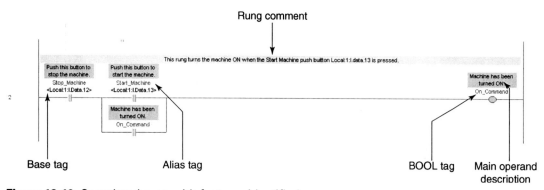

Figure 13-19 Completed rung with features identified.
Used with permission Rockwell Automation, Inc.

342 ADDING LADDER RUNG DOCUMENTATION

6. _____ Add the rung comments to the rungs as illustrated in Figure 13-20.

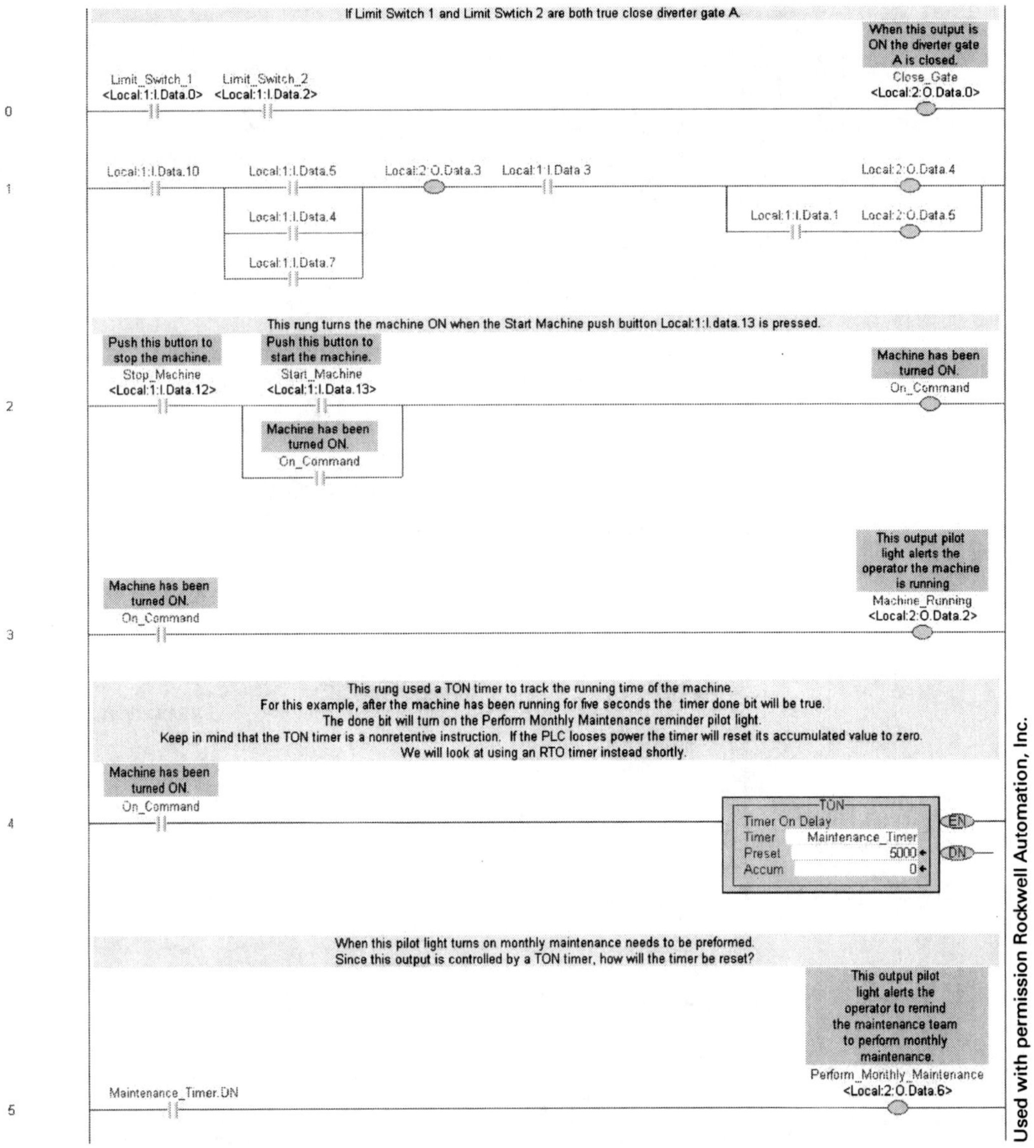

Figure 13-20 Rung comments to add.

7. _____ Next we will edit our ladder to add a reset instruction so we can easily reset our maintenance timer. Program the rung and add the necessary documentation as illustrated in Figure 13-21. When you program the input instruction on rung 6, right-click the question mark and configure the alias tag and main operand description as in Figure 13-22. This shows an easy way to assign the alias and main operand description in one operation.

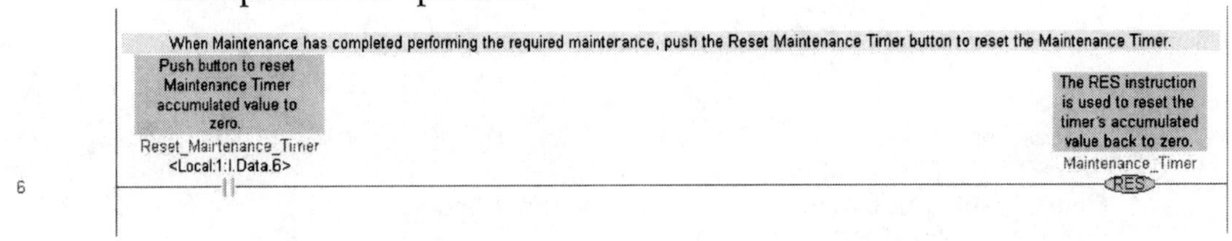

Figure 13-21 Rung added during ladder editing.
Used with permission Rockwell Automation, Inc.

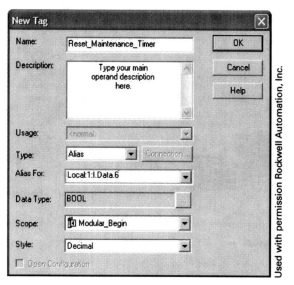

Figure 13-22 Add the main operand description as the alias tag is assigned.

8. _____ The rung to add to your ladder routine is shown in Figure 13-21.
9. _____ up your communications and download your project.
10. _____ Verify that your ladder rungs operate correctly.
11. _____ Save your project.

LAB EXERCISE 2: Programming a TOF Timer

This lab exercise creates a new project, creates rungs including a TOF timer, and adds documentation. The exercise's objectives are as follows:

- Create a new RSLogix 5000 project.
- Perform an I/O configuration.
- Create tags.
- Create alias tags.
- Create ladder rungs using RSLogix 5000 software.
- Program and understand the TOF timer.
- Verify ladder rungs and check for programming errors.
- Add main operand descriptions.
- Add rung comment descriptions.
- Set up the communications path.
- Download the project to your ControlLogix or CompactLogix controller.
- Monitor operation.

THE LAB

It is your turn to create and configure a new RSLogix 5000 project and create the ladder rungs and associated documentation, as shown in Figure 13-23.

1. _____ Create a new ControlLogix project.
2. _____ Select the correct version of your RSLogix 5000 software.
3. _____ Select the correct controller.

344 ADDING LADDER RUNG DOCUMENTATION

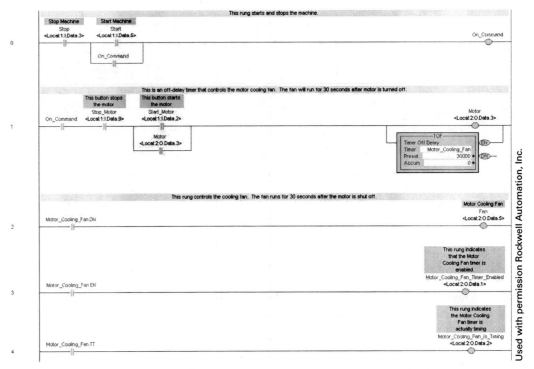

Figure 13-23 Ladder logic to program for Lab Exercise 2, Programming a TOF Timer.

4. _____ If applicable, select the correct controller slot number.
5. _____ Perform an I/O configuration.
6. _____ Decide when you will assign your alias tags and main operand descriptions. Create them accordingly.
7. _____ Program your rungs.
8. _____ Enter remaining documentation.
9. _____ After completing the project, set up a communications path and download your project.
10. _____ After downloading, place your controller into Run mode and go online.

As you monitor your rungs and tags, answer the following questions. Make sure your two stop buttons are physically closed, or true, before you begin.

11. _____ Start the machine.
12. _____ Start the motor.
13. _____ With the motor running, what happened to the timer enable bit?
14. _____ Explain what happened to the done bit.
15. _____ Explain what happened to the timer timing bit.
16. _____ Press the Stop Motor button to stop the motor.
17. _____ Explain what happens to the motor cooling fan.

LAB EXERCISE 3: Interpretation Ladder Rungs and the Monitor Tags Window

1. _____ Figure 13-24 contains rungs from our current application. Answer the questions as you interpret the ladder rungs.
2. _____ Explain what the timer is currently doing.
3. _____ Figure 13-25 shows the Monitor Tags view for our project. The value column has no information displayed for our tags. As you interpret the ladder rungs in Figure 13-24, fill in what the current values should be in Figure 13-25.

ADDING LADDER RUNG DOCUMENTATION 345

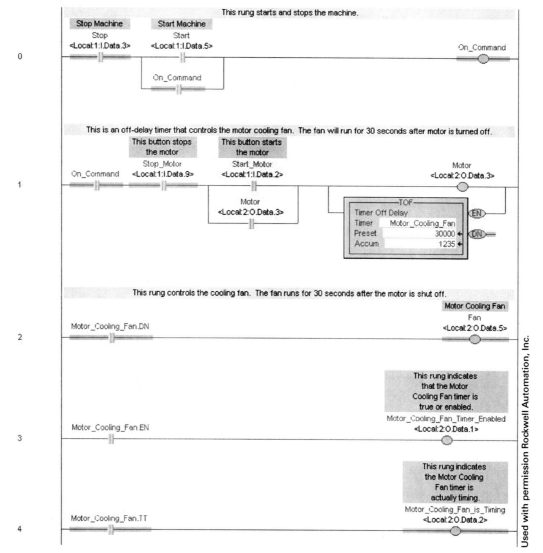

Figure 13-24 Ladder rungs for Lab Exercise 3 questions.

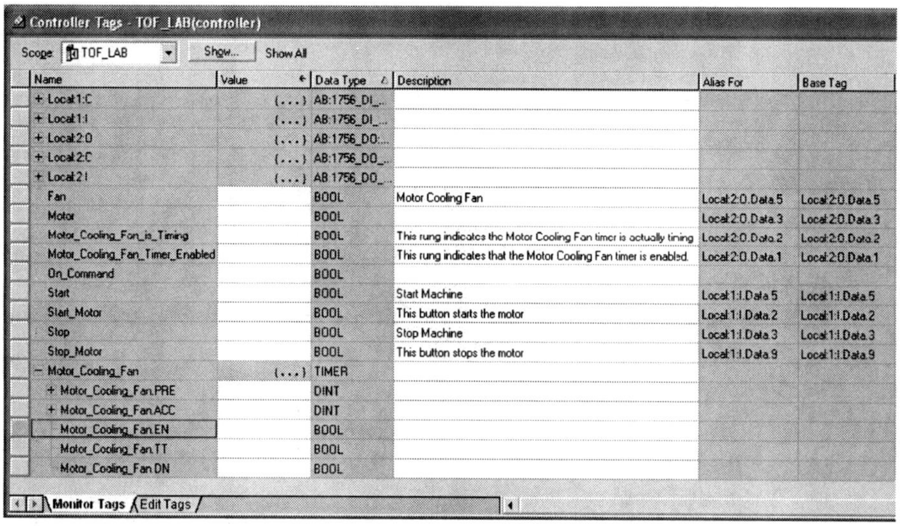

Figure 13-25 Fill in value column with the expected information from ladder rungs in Figure 13-24.

346 ADDING LADDER RUNG DOCUMENTATION

LAB EXERCISE 4: Programming an RTO Timer

For this lab, we add a retentive timer on delay and documentation to the rungs created for the last exercise. When completed, we will download the project and monitor the timer as it runs. In this lab, we perform the following:

A. Refer to Figure 13-26 and add an RTO timer to rung 1.

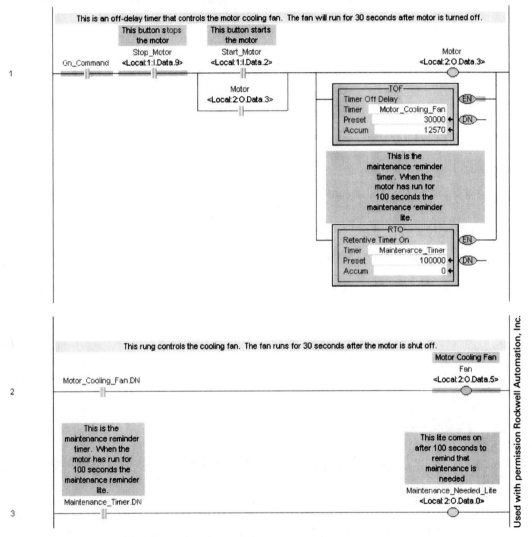

Figure 13-26 Modified ladder rungs for exercise.

B. Delete rung 4.
C. Edit rung 3 to examine the done bit for our new timer, and turn on the maintenance needed pilot light.
D. Add required documentation.

1. _____ We modify the rungs and add documentation from Figure 13-23's ladder rungs, as listed and illustrated in Figure 13-26. Refer to the figure as you follow the steps given here.
2. _____ Right-click the rung number 4.

3. _____ Select delete from the right-click menu.
4. _____ Right-click the bottom-left corner of the output branch on rung 1.
5. _____ Select Add Branch Level from the right-click menu, as illustrated in Figure 13-27.

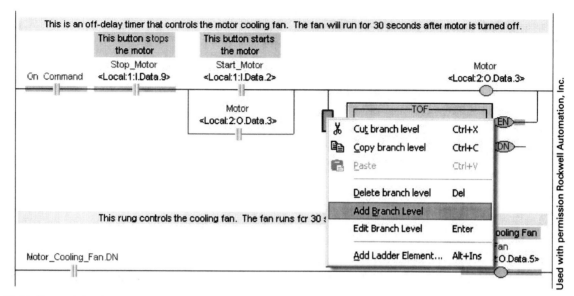

Figure 13-27 Extend the branch.

6. _____ Rung 1 should display a new branch level as shown in Figure 13-28. Note the small e's outside the left power rail. They signify that the rung has been edited, but not verified for errors.

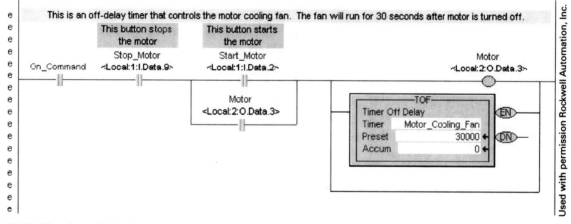

Figure 13-28 New branch level.

7. _____ Program an RTO timer on the branch, as illustrated in Figure 13-26.
8. _____ Add a main operand description, as shown in the figure.
9. _____ Modify rung 3, as illustrated in Figure 13-26.
10. _____ Create a new rung 4.
11. _____ Add an XIC instruction with the tag Reset_Maintenance_Timer. Select an unused push button for the base tag for this input.

12. _____ Program a reset instruction so that when the reset maintenance timer push button is pressed, the maintenance timer's accumulated value resets to 0.
13. _____ Add a main operand description on your own.
14. _____ Add your own rung comments to rungs 3 and 4.
15. _____ Verify your edited project.
16. _____ Verify the path is correct.
17. _____ Download and run the project.

As your project runs, monitor rungs and data and answer the following questions.

18. _____ Start the machine.
19. _____ Start the motor.
20. _____ After 5 seconds or so, stop the motor. Explain what happened to the accumulated value of the RTO timer and why.
21. _____ If this had been a TON, explain what would have happened to the accumulated value.
22. _____ If your PLC were to lose power, what would determine whether the RTO would retain its status?
23. _____ Restart the motor and after a few seconds stop the motor. Verify that the RTO still retains its accumulated value.
24. _____ Start the motor again and let the timer time out.
25. _____ If you push the Stop Motor button, what happens to the timer accumulated value?
26. _____ How do you signal the timer that maintenance has been performed and the machine is ready to go again?
27. _____ Do you remember how to turn your documentation on and off? If so, where do you go in the RSLogix 5000 software to do this?
28. _____ Go to the appropriate screen and practice turning your documentation on and off.
29. _____ Practice changing the colors of your documentation background and text color. Where do you find these options?
30. _____ Ask your instructor whether your project needs to be checked.
31. _____ Save your project.
32. _____ Ask your instructor whether you need to print out your project and hand it in for credit.

CHALLENGE

Assume management has decided that if maintenance is not completed before 120 seconds has transpired, the machine needs to be shut down, and a machine shutdown pilot light should flash on and off every 1 second until maintenance is completed and a push button is pressed. Create your own rungs to accomplish this request. Remember to add alias tags, main operand descriptions, and rung comments to explain exactly how the rungs operate.

SUMMARY

This lesson introduced us to adding text documentation to our ladder rungs. The available documentation includes alias tags, main operand descriptions, and rung comments. We looked at a couple of different ways to add our documentation. We created a new project and added TOF and RTO timers. We edited our ladder logic or added and modified our previously created ladder rungs. At this point, you should have a good idea how all timers work and how to add text documentation to ladder rungs. Rungs and timer data were monitored from the Monitor Tags window so we could see the instructions and their associated status bits as the rungs executed.

REVIEW QUESTIONS

Note: For ease of handing in assignments, students are to answer using their own paper.

1. Each rung comment can display as many as _____ lines of text that explain the function of the ladder rung.
2. The _____ is a retentive timer.
3. Documentation can be turned on or off under _____.
4. A _____ is text that can be added by the programmer to describe the function of the main tag or main operand of the instruction.
5. The TOF timer is a _____ timer.
6. The main operand is the _____ tag for any instruction that has source and destination parameters.
7. What is the RES instruction used for? _____
8. A single main operand description can have as many as _____ characters and _____ lines.
9. The _____ description is displayed directly above the base tag and its associated alias.
10. A project with errors cannot be _____.
11. After the project is downloaded and running, timers and counter data can be viewed in the _____ view.
12. An _____ is another name for a tag.
13. The alias tag and its base tag can be displayed or only the _____ tag.
14. The TOF is a timer off delay. Explain how the timer works. _____
15. Explain the difference between the TON and RTO timer. _____
16. The _____ instruction is used to reset the instructions accumulated value to 0.
17. Before a project can be downloaded, it must be checked for errors or _____.
18. Alias tags are assigned in the _____ view or _____ window.
19. When verifying your project, errors are displayed in the _____ window.
20. Before downloading, uploading, or going online, the current _____ must be verified.

CHAPTER

14

ControlLogix Counter Instructions

OBJECTIVES

After completing this lesson, you should be able to

- Identify RSLogix 5000 counter instructions.
- Program the count up (CTU) instruction and understand the associated status bits.
- Program the count down (CTD) counter and understand the associated status bits.
- Create an array of counters.
- Monitor counters, counter data, and status bits online.
- Modify and correlate counter project ladder rungs to operator interface screen objects.

INTRODUCTION

The instructions for ControlLogix counters are very similar to those for Rockwell Automation or Allen-Bradley SLC 500 and PLC 5 counters. The lab exercises in this chapter step you through programming various counters that count up (CTU) and those that count down (CTD). After programming the counters, we will download and run the project, monitoring the counters and their status bits.

INSTRUCTOR DEMONSTRATION

Follow along as your instructor demonstrates an RSLogix 5000 project containing the counter instructions. Your instructor will also demonstrate variable addressing when using arrays.

UNDERSTANDING THE COUNTER INSTRUCTIONS

If you wish to count from 0 to a predetermined value, use a count up counter. To count down from a predetermined value to 0, use a count down counter. Most counters count on the false-to-true transition of the rung's input logic. The count up counter increments each time the input logic changes from false to true, whereas the count down counter decrements each time the rung transitions from false to true. As an example, when a count up counter counts from 0 up to a 100, 100 is referred to as the "preset value." The "accumulated value" is the current, or accumulated, count. Because counters are retentive, a reset (RES) instruction is used to reset the accumulated value back to 0. Figure 14-1 lists and describes the RSLogix 5000 counter instructions.

A ControlLogix counter consists of a three-DINT structure. DINT0 of the structure contains the counter's status bits. DINT1 contains the preset value, and DINT2 contains the

CONTROLLOGIX COUNTER INSTRUCTIONS

COUNTER INSTRUCTIONS			
Instruction	Mnemonic	Use This Instruction to	Functional Description
Count up	CTU	Counts from 0 up to a desired value	CTU is used for counting the number of parts produced during a specific work shift or batch and also for counting the number of rejects from a batch.
Count down	CTD	Counts down from a desired value to 0	An operator interface device shows the operator the number of parts remaining to be made from a batch of 100.
Reset	RES	Resets a counter's accumulated value to 0	Reset is used to reset the counter's accumulated value to 0 so that another counting sequence can begin.

Figure 14-1 RSLogix 5000 counter instructions.
© Cengage Learning 2014

accumulated value. Status bits, as well as the preset value and accumulated value, can be addressed individually. Figure 14-2 illustrates the format of the counter structure.

DINT 0	Status Bits	
DINT1	Counter Preset Value (.PRE)	0 to 2,147,483,647
DINT2	Accumulated Value (.ACC)	0 to –2,147,483,648

Figure 14-2 Counter structure data format.
© Cengage Learning 2014

A counter's counting range is the numerical range within which a counter can count. ControlLogix counters can count within the range of –2,147,483,648 to +2,147,483,647. This is the range of a 32-bit signed integer. (Refer to Chapter 3 for a review of signed integers.) If an up counter counts above +2,147,483,647, an overflow is detected. Conversely, if a down counter counts below –2,147,483,648, an underflow is detected. As an example, suppose an up counter overflows, the overflow (OV) status bit is set, and the counter begins counting up from –2,147,483,648 back to 0. When a down counter counts below –2,147,483,648, the underflow (UN) status bit is set, and the counter continues counting down from +2,147,483,647 back to 0. To avoid overflowing or underflowing a counter, a RES instruction is used to reset the counter's accumulated value back to 0 when the maximum count is reached.

Counters are retentive, assuming that the controller's battery or energy storage module is in good condition. A counter then retains its accumulated value and the current state of the status bits through a power loss.

One important consideration when working with counters is the speed or frequency with which an input device is sending the false-to-true signals to the controller. The speed of the controller scan must be taken into consideration. As an example, if parts are passing an inductive proximity sensor on a conveyor faster than the controller scan can see them, some parts will not be counted. A high-speed counter module could be used when counts come in faster than the controller can see them.

THE COUNT UP (CTU) INSTRUCTION

Use the count up instruction if you want to keep an accumulated tally of the number of times some predetermined event transpires. The count up counter counts from 0 up to a predetermined value, called the preset value. As an example, if you wanted to count from 0 to 100, you would be counting up and would use a CTU counter. The predetermined value of 100 is the preset value. The accumulated value is the current, or accumulated, count. If our counter counted

45 pieces that passed a photo switch on a conveyor, the accumulated count, or value, would be 45. When all 100 pieces had passed, the counter accumulated value and preset value would be equal. The counter would then be considered done and the done status bit (DN) would be set. At this point, the counter would signal other logic within the PLC program, typically using the done bit, that the batch of 100 was completed and some action should be taken. As an example, the PLC might move the box containing the 100 parts on to the next station for carton sealing. To start counting the next batch, a RES instruction would be used to reset the counter's accumulated value back to 0. The counter instruction is comprised of the following components. Refer to Figure 14-3 to identify the components of the instruction.

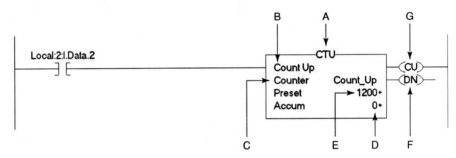

Figure 14-3 RSLogix 5000 counter parts.
Used with permission Rockwell Automation, Inc.

A. Counter mnemonic. The mnemonic is a three-letter abbreviation used to identify the counter instruction.
B. Counter instruction identification.
C. The counter parameter identifies the tag where the counter and its associated status information are stored. The tag name must be unique for each counter. The tag name can be whatever the programmer desires.
D. The accumulated value, a DINT, is our current position in the counting cycle.
E. The preset is a DINT identifying how many increments the counter is to count before the counter is considered done.
F. DN (done) is a status bit indicator displaying when the counter is done counting. The DN bit is one of the counter's status bits.
G. The CU (enable) is a status indicator that displays whether the counter instruction is true, or enabled. The CU is one of the counter's status bits.

COUNT UP STATUS BITS

In order for the PAC and human operator to monitor the current counter status, special bits, called status bits, are used. Count up counters have three status bits; they are listed in Figure 14-4.

COUNT UP STATUS BITS			
Status Bit	Identified as	Example Tag	Functional Description
Enable	CU	Count_Up.CU	Bit is true when counter instruction is true, or enabled.
Done	DN	Count_Up.DN	The done bit is true when the accumulated value is equal to or greater than the preset value.
Overflow	OV	Count_Up.OV	The counter overflow bit is true when the counter overflows above +2,147,483,647.

Figure 14-4 RSLogix 5000 count up counter status bits.
© Cengage Learning 2014

Figure 14-5 illustrates a CTU with the tag "Product_Produced." The three counter status bits are illustrated on the rungs below the counter. The programmer can decide which counter status bits are required for the application. Rungs with each of the status bits are shown in the figure, but are not required.

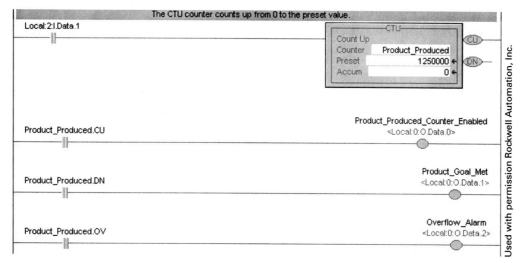

Figure 14-5 Illustration of a sample count up (CTU) counter and its associated status bits.

Ladder Explanation

Each time input Local:2:I.Data.1 transitions from false to true, the Product_Produced counter increments its accumulated value by one decimal value. When rung 1 is true, the counter is considered enabled. The Product_Produced.CU, enable bit is true when the counter is true, or enabled. At such time as the accumulated value is equal to the preset value, the counter is done, and the Product_Produced.DN, done bit will be true. Counters will continue to count past the preset value. If the accumulated value is greater than or equal to the preset value, the done bit will be set. Figure 14-6 illustrates the tags associated with the counter in Figure 14-5.

SAMPLE COUNTER TAGS			
Structure Member	**Sample Tag**	**Data Type**	**Explanation**
Preset Value	Product_Produced.PRE	DINT	Preset is the target count, say 12 pieces in a case
Accumulated Value	Product_Produced.ACC	DINT	Tracks how many counts have registered or accumulated
Count Up Enable	Product_Produced.CU	BOOL	True only when the counter is true, or enabled
Done Bit	Product_Produced.DN	BOOL	True when the counter accumulated value is equal to or greater than the preset value
Overflow Bit	Product_Produced.OV	BOOL	True when the up counter overflows

Figure 14-6 Examples of count up (CTU) counter tags.
© Cengage Learning 2014

Counters are retentive instructions. If the controller loses power for any reason and the controller battery or energy storage module is in good condition, the counter will remember, or retain, its accumulated value when power is restored. Because counters are retentive, the same reset instruction used to reset the timers is also used to reset the accumulated value of the counter back to 0. Figure 14-7 illustrates a rung containing the RES instruction to reset this counter's accumulated value back to 0. Any time the RES instruction becomes true, the counter resets. The example shows one input on the reset rung. The programmer can program whatever logic is needed as inputs on a reset rung.

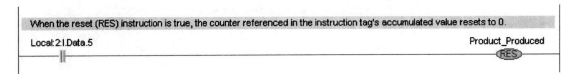

Figure 14-7 Reset (RES) instruction used to reset the Product_Produced counter accumulated value to 0.
Used with permission Rockwell Automation, Inc.

Typically a counter counts to a predetermined value and is reset. However, an up counter continues to count until the accumulated value reaches +2,147,483,647. On the next count, the accumulated value will wrap to −2,147,483,648 and start counting toward 0. The counter overflow bit (Product_Produced.OV) is true when the counter overflows above +2,147,483,647. The overflow bit is only informational and does not fault the controller. Refer to Figure 14-5 to view the overflow status bit rung.

Monitoring the CTU Counter

Tags and their values can be monitored by going to the Monitor Tags tab for the tag collection where the tag is stored. Refer to Figure 14-8 as you right-click the tag and select Monitor "Product_Produced."

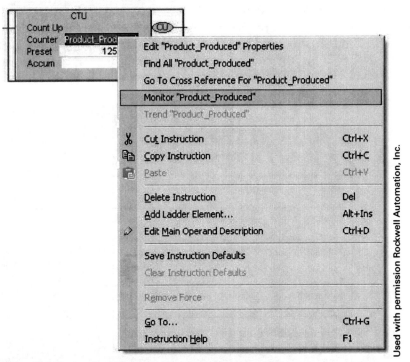

Figure 14-8 Right-click and select Monitor "Product_Produced."

Figure 14-9 is a Monitor Tags view showing the CTU associated tags and status bits.

CONTROLLOGIX COUNTER INSTRUCTIONS 355

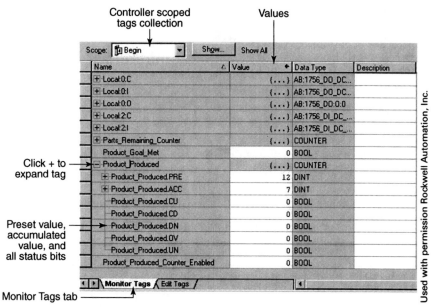

Figure 14-9 Monitoring the CTU counter.

THE COUNT DOWN (CTD) INSTRUCTION

The down counter works a little differently. Each time a count down counter sees a false-to-true rung transition, the counter is considered enabled, and the count down (CD) enable status bit is set. The accumulated value is decremented by 1. Because the accumulated value is decremented each time the input logic changes the rung from false to true, the accumulated value is typically the starting point of the count. In this example, assume we are counting a batch of 100 parts. Each time a part is made, the remaining total is displayed on an operator interface display device so that the operator can see how many more parts must be manufactured. The accumulated value is programmed with the value of 100, whereas the preset value starts at 0. As each part is made, the accumulated value is decremented by 1. When all 100 parts have been made, the accumulated value and the preset value are 0.

Use this instruction to count down over the range of +2,147,483,647 to 0. Each time the instruction sees a false-to-true transition, the accumulated value is decremented by 1 count. Assume that you want to display the remaining number of parts to be built for a specific order of 100 parts. The remaining parts to be built are displayed on an operator display device so that workers can see how many parts are needed to complete the lot. For this example, the accumulated value is set at 100 and the preset value is 0. Each time a part is completed and passes the sensor, the accumulated value is decremented by 1 decimal value. The CTD counter instruction is comprised of the following components. Refer to Figure 14-10 and identify the components of the instruction.

A. Counter mnemonic. The mnemonic is a three-letter abbreviation used to identify the counter instruction.
B. Counter instruction identification.

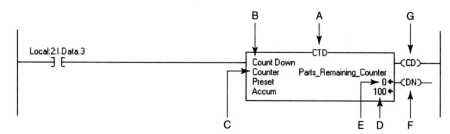

Figure 14-10 RSLogix 5000 CTD counter components.
Used with permission Rockwell Automation, Inc.

C. The counter parameter identifies the tag where the counter and its associated status information are stored.
D. The accumulated value, a DINT, is our current position in the counting cycle. This value, typically between 0 and 2,147,483,647, identifies how many increments the counter is to count before the counter is considered done.
E. When using a CTD counter, the preset is usually set to 0. Each time the counter instruction becomes true, the accumulated value is decremented toward 0 by 1.
F. DN (done) is a status bit indicator displaying when the counter is done counting. The DN bit is one of the counter's status bits. Remember the done bit is set, or a 1, whenever the accumulated value is greater than or equal to the preset value. Keep this in mind when selecting the XIC or XIO instruction associated with a down counter's done bit.
G. The CD (count down enable) is a status indicator that displays whether the counter instruction is true, or enabled. The CD is one of the counter's status bits.

Down counters also have three status bits; they are listed in Figure 14-11.

		DOWN COUNTER STATUS BITS	
Status Bit	Identified as	Example Tag	Functional Description
Enable	CD	Parts_Remaining_Counter.CD	Bit is true when counter instruction is true, or enabled.
Done	DN	Parts_Remaining_Counter.DN	The done bit is true when the accumulated value is equal to or greater than the preset value.
Underflow	UN	Parts_Remaining_Counter.UN	The counter underflow bit is true when the underflows below −2,147,483,648.

Figure 14-11 RSLogix 5000 count down counter status bits.
© Cengage Learning 2014

Figure 14-12 illustrates a CTD with the tag of Parts_Remaining_Counter. The three counter status bits are illustrated on the rungs below the counter. The programmer can decide which counter status bits are required for the application. Rungs with each of the status bits are shown in the figure, but are not required.

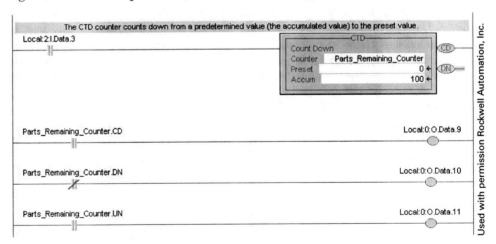

Figure 14-12 RSLogix 5000 CTD, or count down, counter.

Ladder Explanation

The accumulated value for count down counter Parts_Remaining_Counter is set at 100. The preset value is 0. Each time input Local:2:I.Data.3 transitions from false to true, the accumulated value will decrement 1 decimal value. The done bit is set, or true (1), during the entire count from 100 to 0 when programmed as an examine-if-closed instruction. If it is desired for the done bit instruction to be true when the accumulated value is less than the preset value, program an examine-if-open (as in Figure 14-12); the instruction will be true when the counter's accumulated value transitions from 0 to −1. If the preset is programmed to a value of 1, the done bit transitions when the preset goes from a value of 1 to 0. Figure 14-13 illustrates the tags associated with the counter in Figure 14-12.

SAMPLE COUNTER TAGS			
Structure Member	Sample Tag	Data Type	Explanation
Preset Value	Parts_Remaining_Counter.PRE	DINT	Preset is the target count, say 12 pieces in a case
Accumulated Value	Parts_Remaining_Counter.ACC	DINT	Tracks how many counts have registered or accumulated
Count Down Enable	Parts_Remaining_Counter.CD	BOOL	True only when the counter is true, or enabled
Done Bit	Parts_Remaining_Counter.DN	BOOL	True when the counter accumulated value is equal to or greater than the preset value
Underflow Bit	Parts_Remaining_Counter.UN	BOOL	True when the down counter underflows

Figure 14-13 Examples of count up (CTD) counter tags.
© Cengage Learning 2014

Resetting the CTD

As you remember from the CTU counter, the RES instruction resets the counter's accumulated value to 0. When using the CTD counter, we typically do not wish to put a 0 into the accumulated value when we reset the counter. Instead, we need to return the value we will start decrementing from. As a result, the RES instruction is not of much value in this application. We need to return the number 100 back to the accumulated value, as our application specifies that we count down from 100 to 0. If we were to use the RES instruction after each complete cycle, the RES instruction would reset the accumulated value to 0. The Move instruction illustrated in Figure 14-14 can be used to replace the accumulated value with 100. The Move instruction has two parameters, the source and the destination. The source indicates where the information is coming from. The source could be a tag or a constant. In this case, the number 100 was entered, a constant. The destination states where the information is to go. The Move instruction's destination for the integer 100 is Parts_Remaining_Counter.ACC. When Local:2:I.Data.4 becomes true, the Parts_Reaming_Counter.ACC tag will contain the number 100 and our counter is ready to count down from 100 to 0.

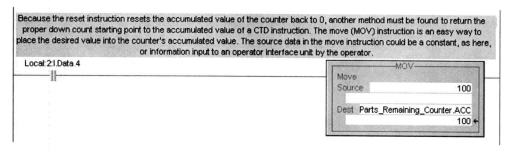

Figure 14-14 Move instruction to reset counter's accumulated value.
Used with permission Rockwell Automation, Inc.

The Move instruction does not really move a value from one location to another, thus the instruction's name is a bit deceptive. The Move instruction copies the single value from the source and pastes that value to the destination. The original data remain in the source. The Move instruction is found under the Move/Logical tab on the Language Element toolbar.

Monitoring the CTD Counter

To monitor the tags and their values, go to Monitor Tags. Remember, you can right-click the tag and select Monitor Parts_Remaining_Counter. Figure 14-15 is a Monitor Tags view showing the CTD-associated tags and status bits.

358 CONTROLLOGIX COUNTER INSTRUCTIONS

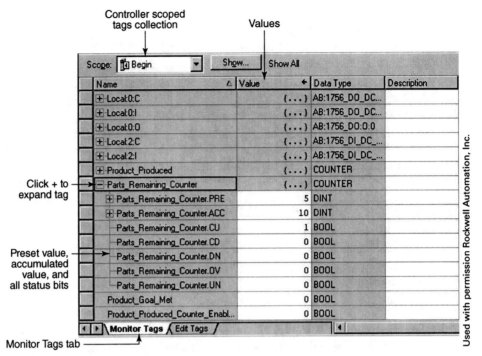

Figure 14-15 Monitoring the CTD counter.

Now that we have introduced the two counters, let's do some programming and monitoring of each counter.

LAB EXERCISE 1: Programming Counters

For this exercise, you program the counter rungs just introduced. You get a chance to create a few tags and alias tags. When you have completed setup communications, download the project, put the controller in Run, test our rungs, and monitor the tags. Figure 14-16 contains the rungs we are programming for this exercise. Adding the documentation provides practice entering rung comments but also reinforces the operation of the counters and their associated status bits.

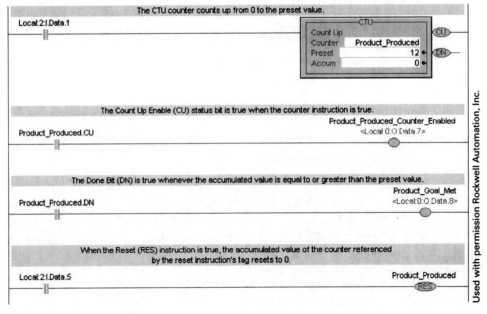

Figure 14-16 Counter rungs for exercise.

1. _____ Open RSLogix 5000 software.
2. _____ Open Begin Project.
3. _____ Verify your lab hardware's I/O configuration matches the project.
4. _____ Create a new rung.
5. _____ Program the required input instruction and tag.
6. _____ From the Timer/Counter tab on the Language Element toolbar, program the CTU instruction on the rung.
7. _____ An easy way to create a new counter tag is to right-click the counter tag parameter inside the instruction and select New Tag from the drop-down box. See Figure 14-17.

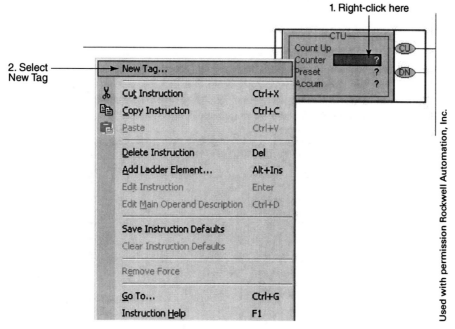

Figure 14-17 Creating a new counter tag.

8. _____ Create a new tag, as illustrated in Figure 14-18.

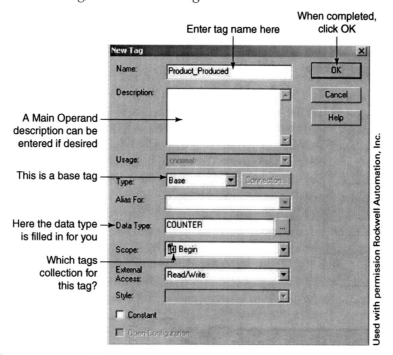

Figure 14-18 Create new tag dialog box.

360 CONTROLLOGIX COUNTER INSTRUCTIONS

9. _____ Enter the tag name.
10. _____ This is a base tag, so no modifications are required to the data type.
11. _____ Note the software already knows the data type.
12. _____ The scope of the tag is currently controller scoped.
13. _____ Enter your own Main Operand description.
14. _____ Click OK when this is completed. If using RSLogix 5000 version 20, click on create.
    Note: When a tag is created that is a structure, all of the associated members and status bits are automatically created.
15. _____ To view the tags just created, right-click the Product_Produced tag inside the counter instruction.
16. _____ From the right-click menu, click Monitor Product_Produced.
17. _____ The tags collection should open and display the Monitor Tags view, as in Figure 14-19.
18. _____ Click + in front of the Product_Produced tag to expand the counter structure. The tag should be selected from the list. Note that in the figure there is a box drawn around the tag to identify it. All of the counter's tags and status bits should display in a view similar to that shown in Figure 14-19.

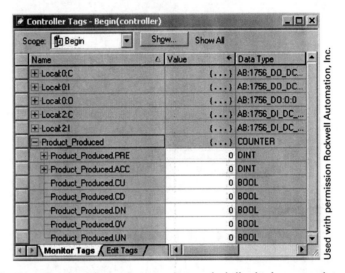

Figure 14-19 The Product_Produced tag structure expanded displaying associated tags.

19. _____ Because this is an up counter, which of the status bits pertain to this type counter?
20. _____ When you have completed this, close the Monitor Tags window.
21. _____ Click the ? in the counter instruction block to open the preset parameter.
22. _____ Enter the preset value and press the Enter key.
23. _____ The accumulated value is typically left at 0. Press Enter to accept the 0 and complete this parameter. The counter instruction programming is complete.
24. _____ Refer to Figure 14-16 as you create a new rung.
25. _____ Program the XIC instruction.
26. _____ To add the Product_Produced.CU tag, double-click the ? above the instruction.
27. _____ Click the down arrow, #1 in Figure 14-20.
28. _____ Scroll to find the Product_Produced tag, #2 in the figure. Pressing the first letter of the tag, "P" in this case, takes you to tags beginning with "P."
29. _____ Click + to expand the structure. See #3 in the figure.
30. _____ Refer to Figure 14-21 and select the tag from the list by double-clicking the Product_Produced.CU tag.

CONTROLLOGIX COUNTER INSTRUCTIONS 361

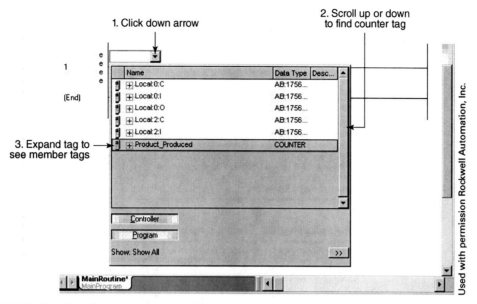

Figure 14-20 Scroll to the tag and expand the tag to see the associated member tags.

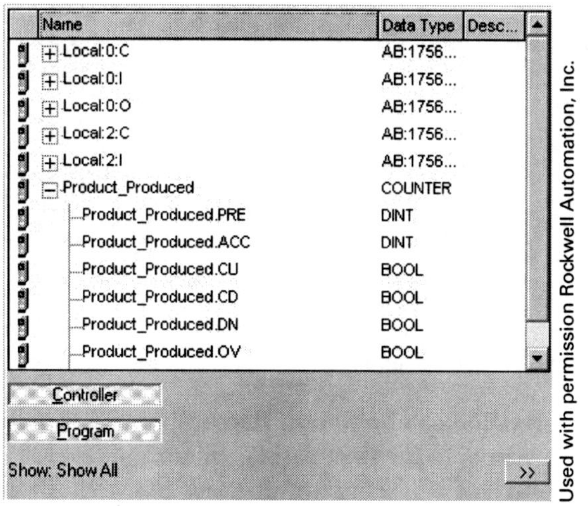

Figure 14-21 Double-click Product_Produced.CU to select the tag.

31. _____ This should take you back to the ladder rung. Press Enter to complete the programming of this instruction.
32. _____ Complete programming for the remaining rungs. Don't forget to add the documentation.
33. _____ With the CTU rungs programmed, refer to Figure 14-22 and program the rungs for the CTD counter.
34. _____ Add documentation as you go along.
35. _____ Start programming the Move instruction rung by adding the XIC instruction and associated tag.
36. _____ To program the Move instruction in order to reset our counter, refer to Figure 14-23 as you select the instruction from the Move/Logical tab of the Language Element toolbar.

   The Move instruction is used to replace the accumulated value into the Parts_Remaining_Counter.ACC tag. The source parameter can be either a tag name or a constant. The source is the value, or the tag where the value to be placed in the

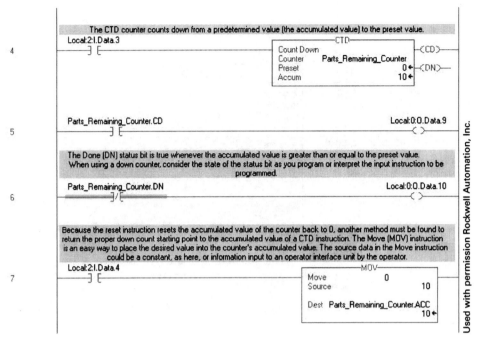

Figure 14-22 Counter ladder rungs for lab exercise.

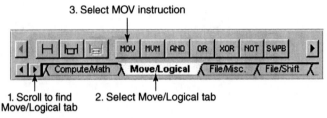

Figure 14-23 Selecting the MOV instruction.
Used with permission Rockwell Automation, Inc.

destination can be found. Remember, the source value is not literally moved from the source to the destination. In our application, we use the Move instruction with a source of 10, a constant, to place the value 10 into the Parts_Remaining_Counter.ACC tag. To program the MOV instruction:

37. _____ Double-click the single ? for the source parameter.
38. _____ Enter the value 10. Refer to Figure 14-24.

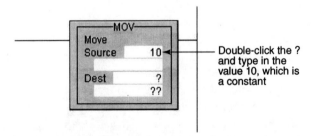

Figure 14-24 Programming the Move instruction source parameter.
Used with permission Rockwell Automation, Inc.

39. _____ Press Enter on your computer to accept the value.
40. _____ Double-click the destination parameter single ?.
41. _____ Click the drop-down arrow on the right.
42. _____ Select the Parts_Remaining_Counter.ACC tag in the same manner as you selected the status bits tags. Make sure you select the accumulated value tag. Refer to Figure 14-25.

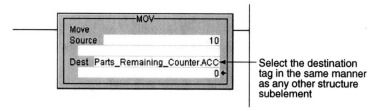

Figure 14-25 Programming the Move instruction's destination parameter.
Used with permission Rockwell Automation, Inc.

43. _____ Press Enter to complete the instruction.
44. _____ Add the documentation, as illustrated in Figure 14-22.
45. _____ Move instruction parameter information displayed is outlined in Figure 14-26.

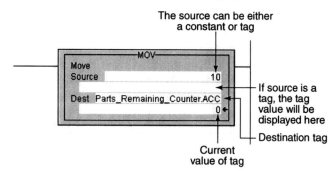

Figure 14-26 Programming the Move instruction's destination parameter.
Used with permission Rockwell Automation, Inc.

46. _____ Verify the project.
47. _____ Save the project as "Counter Exercise 1."
48. _____ Connect communications cable between the personal computer and ControlLogix.
49. _____ Check to see that the desired RSLinx driver is correctly configured.
50. _____ Reconfigure driver if necessary.
51. _____ Verify that the communication path is correct; if not, create the proper path.
52. _____ Download the project.
53. _____ Put the controller in Run mode.
54. _____ To test our counters, close the appropriate switch to make Local:2:I.Data.1 input become true.
55. _____ Explain what happens to the CU status bit indicator on the CTU instruction on rung 0.
56. _____ Make the input go false.
57. _____ Go to the Monitor Tags window and monitor the Product_Produced tag.
58. _____ Expand the structure to display the status bits.
59. _____ Make the input go true again.
60. _____ As you monitor the Product_Produced.CU bit, what do you see?
61. _____ Make the input transition from false to true a number of times. Explain what happens to the Product_Produced.ACC tag.
62. _____ Make the input transition from false to true until the accumulated value is equal to 12. Explain what happens and why.
63. _____ If you continue to increment the counter, what happens to the status bits?
64. _____ What is the maximum value the up counter will count to?
65. _____ When the counter reaches its maximum count, explain what happens to the counter and its status bits.
66. _____ When the counter overflows, will this fault the controller?
67. _____ What will happen if Local:2:I.Data.5 goes true?
68. _____ Make the rung go true. As you monitor the counter and the status bits, does it work as you expected?

Monitoring the Count Down Counter

69. _____ Close the appropriate switch to make Local:2:I.Data.3 input become true.
70. _____ Explain what happens to the CD status bit indicator on the instruction?
71. _____ Make the input go false.
72. _____ Go to the Monitor Tags window and monitor the Parts_Remaining_Counter tag.
73. _____ Expand the structure to display the status bits.
74. _____ Make the input go true again.
75. _____ As you monitor the Parts_Remaining_Counter.CD bit, what do you see?
76. _____ Make the input transition from false to true a number of times.
77. _____ Make the input transition from false to true until the accumulated value is equal to 0. Explain what happens and why.
78. _____ If you decrement the counter to –1, what happens to the done status bit?
79. _____ Explain what just happened to the rung containing Local:0:O.Data.10. Monitor the tags in the Monitor Tags window as you monitor the rungs to verify how this works.
80. _____ How many counts did we actually have before the done bit went true?
81. _____ What can we do to have the done bit change state when the counter actually reaches an accumulated value of 0?
82. _____ What is the maximum value the down counter will count down to?
83. _____ Explain what happens to the counter and its status bits if we decrement the counter once more after the counter reaches its maximum down count.
84. _____ If we program a RES instruction to reset this counter, explain what will happen when the reset goes true.
85. _____ What is the function of the Move instruction?
86. _____ What can the Move instruction source parameter be programmed as?
87. _____ By its name, the Move instruction takes a value defined as the source and physically moves that value to the destination. After executing the Move instruction, will the source value then be 0, as the starting value was moved to the destination? Explain your answer.
88. _____ To prove the answer to the question above, go offline.
89. _____ Edit the Move instruction so that the source is now a tag rather than a constant. Create a new tag and give it whatever name you choose.
90. _____ In order to count the full range of the counter, what data type must the new move source tag be?
91. _____ Go to the Monitor Tags window and enter the number 10 in the value column for the newly created tag.
92. _____ Verify the project and download.
93. _____ Put the controller back into Run mode.
94. _____ Decrement the counter to 0.
95. _____ Return to the Monitor Tags window.
96. _____ Monitor the new Move Source tag and the counter Accumulated Value tag.
97. _____ Make the Move instruction true.
98. _____ When the Move instruction became true, explain what happened to the Move Source tag and the counter's Accumulated Value tag.
99. _____ Save your project. This completes this portion of this exercise.
100. _____ Ask your instructor whether he or she needs to check off your completion of the exercise.
101. _____ Go offline.

PROGRAMMING COUNTERS THAT WORK TOGETHER

Counters can be connected, or cascaded, together to increase counting range. Figure 14-27 illustrates an example of how one counter counts machine steps while the second counter tracks the number of machine cycles.

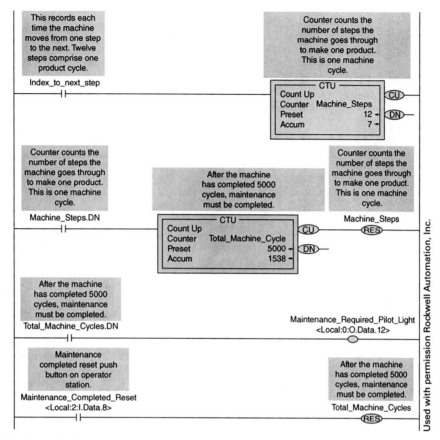

Figure 14-27 Cascaded counters.

COMBINING UP AND DOWN COUNTERS TO CREATE A BIDIRECTIONAL COUNTER

Figure 14-28 illustrates a separate count up instruction and a separate count down instruction working together to give us a counter that counts in both directions. The key is that both counters have the same tag. Input Local:2:I.Data.0 is the photo switch that senses each part as it is produced. This input increments the CTU with the tag Total_Good_Parts. Input Local:2:I.Data.11, the bad parts sensor, also decrements the CTD with the tag Total_Good_Parts. Because each of these counters has the same tag, the accumulator is shared between the two counter instructions. Being shared, the accumulator reflects the count seen by either counter instruction. When the order is completed, the RES instruction can be used to zero out the counters.

LAB EXERCISE 2: Programming an Up–Down Counter and Creating Arrays to Store Production Data

INTRODUCTION

This lab exercise provides practice programming counters and incorporates arrays for recording data. Management has requested that the PAC record the number of cases filled as well as the number of rejected cases for each work shift. Currently the plant is running three shifts every day. The PAC is to store the number of cases filled for each shift in an array. The total number of cases rejected for each shift is stored in a separate array. Data is extracted when needed by management; however, the PAC must store the data for up to a 90-day period.

Start with the "Total Cases" counter to populate a single-dimensional array to record the total number of cases manufactured during each work shift for 100 days. We use 100 days to provide a buffer in case the data are not removed promptly after the 90-day period. Remember that if an array is overflowed, the controller will fault.

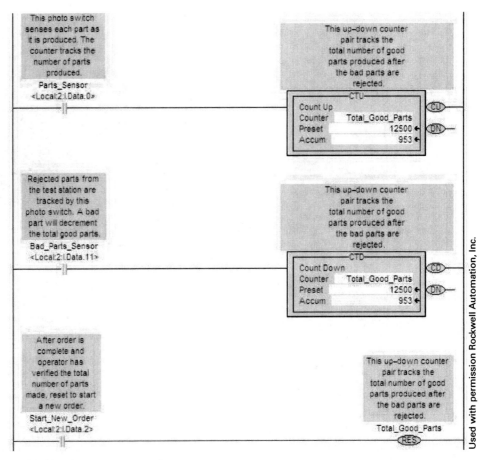

Figure 14-28 Up–down counter pair.

THE LAB

From the specifications given here, draw out the ladder logic on paper.

Add your new logic to a new project. Don't forget to document using alias tags, main operand descriptions, and rung comments. Application specifications are listed here:

- A photo switch signals the Total Cases counter each time a case of parts passes.
- A photo switch decrements the total "Total Cases" counter as cases are rejected. A weigh scale is used to weigh each case to determine whether the proper number of parts is present. The weigh scale sends a discrete signal to the PAC as the result of weighing each case. Use this input to reject cases that are underweight at the diverter.
- To store 100 days of three shifts per day data, create a single-dimensional array that contains 300 elements.
- A third counter tracks the number of parts diverted.
- Program an input from the weigh station to trigger the diverter and decrement the total cases count. Use a push button on your PAC to simulate the weigh station trigger. Because the conveyor does not move too fast, and to keep the lab simple, energize the diverter to reject bad cases 3 seconds after any improper weight case leaves the weigh scale.
- At the conclusion of each shift, the operator logs off the operator interface device. As the operator logs off, an input tag to the PAC named "save shift totals" becomes true. To keep the lab simple, use a physical input switch to simulate the trigger from the operator interface. Use this tag to trigger a move (MOV) instruction to save each of the shift totals and the number of rejected cases to their respective arrays.

- What is the Move instruction's source tag for the total cases produced data?
- List here the Move instruction's destination tag for the total cases data to provide the desired variable addressing?

- What is the Move instruction's source tag be for the rejected cases data?

- What is the Move instruction's destination tag for the rejected cases data to provide the desired variable addressing?

- Create a one-dimensional array with variable addressing for each data set to store the current total cases into the next unused array element.
- A shift counter is needed to index the data to be stored into the array to the next position. Use the Save Shift Totals input from the operator interface to increment the shift counter to the next count. The counter does not count each of the three shifts for each day and then reset. Instead, use the shift counter to count from 0 to 299, the size of the array. As an example, day 1 is shifts 0, 1, and 2; day 2 is shifts 3, 4, and 5; and so on.
- Figure 14-29 illustrates the Total Cases tag properties.

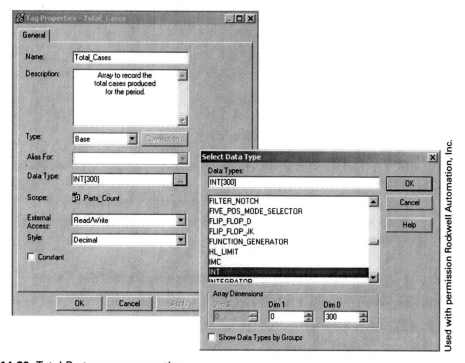

Figure 14-29 Total Parts array properties.

- Figure 14-30 is an example of what the Total Cases array might look like.
- Create a one-dimensional array to store the current number of cases rejected value. Refer to Figure 14-31 as an example of what the number of cases rejected array might look like.
- After 90 days, the array must be reset and returned to its starting position. Do not worry about creating this logic for this lesson. We will create the required logic in a future lesson.

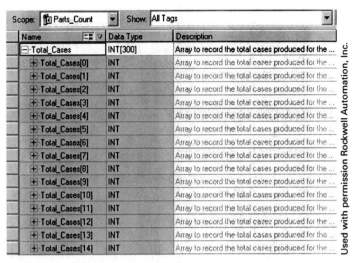

Figure 14-30 Example of Total Cases array.

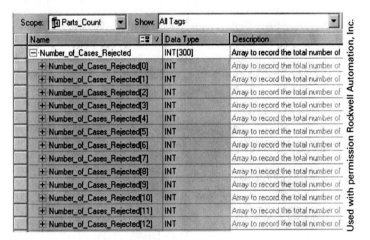

Figure 14-31 Example of rejected parts array.

- When you have completed creating your modified project, download and verify its correct operation.
- Ask your instructor whether project operation check off is required.
- Save your project as "Case counter array," as we will be adding additional logic to it in future chapters.

LAB EXERCISE 3: Maintenance Reminder Timer

This lab exercise modifies Lab Exercise 4: Programming an RTO Timer, from Chapter 13.

1. _____ Open the project you saved as Lab Exercise 4: Programming an RTO Timer, from Chapter 13.
2. _____ Modify the lab exercise to include the following specifications:
    Let's assume the Maintenance timer for regular scheduled maintenance was set up to have maintenance performed after every 720 hours (for the last lab, we used 100 seconds as a more realistic value for testing). This time, 720 hours, is about one month's running time. Management wants to have major maintenance performed after the maintenance timer has cycled six times, or six months. Modify the project to include the additional ladder rungs to signal when six-month maintenance is due.

3. _____ Don't forget to add needed reset instructions and documentation.
4. _____ Program a six-month major maintenance needed flashing reminder pilot light.
5. _____ Add ladder logic to reset the month counter after maintenance has been completed.
6. _____ Program logic so that if the major maintenance is not completed within 24 hours (for lab testing program use a more realistic value), the machine will be shut down and cannot be restarted until maintenance has been completed.
7. _____ When you have completed modifying this exercise, download the project and verify operation.
8. _____ Ask your instructor whether proper operation validation of your project is required.
9. _____ Save your project.

LAB EXERCISE 4A: Up–Down Counters Tracking Cars in a Parking Garage

INTRODUCTION

For this exercise, you have been hired to develop a PAC ladder project to control the operation of a parking garage. The parking lot attendant closes up and turns everything off when the ramp closes each night at midnight. The day shift operator powers up everything when opening each morning. Each time a car enters, the driver takes a ticket and proceeds to find an empty parking spot. You need to develop a system to know when there are parking spaces available and when there are none. If there are spots available, the ticket machine issues a ticket and allows a car into the lot. Assume the parking lot holds 20 cars. If the lot is full, no ticket is issued, the arm across the entry is not raised, and the LOT FULL sign is illuminated.

An up–down counter pair in conjunction with a couple sensors can be used to count each time a car enters and exits the lot. Using counters in conjunction with the sensors allows tracking of how full the lot is.

The following devices connect to our PAC:

- Ticket-dispensing machine signal to dispense ticket if there is space in the lot
- Because management wants to know how many cars are parked in the lot each day, a counter used in conjunction with the ticket-dispensing machine, to record the number of cars that have entered
- An arm operation allowing or restricting entry to the lot. Assume power is needed to lower the arm and hold it down. When power is removed, a counter weight on the arm helps the arm raise and stay up.
- LOT FULL sign designed to turn on or off
- Sensor looking for the presence of a car at the ticket dispenser
- Sensor watching for cars to exit the lot

Fill in the table in Figure 14-32, indicating whether a device is an input or output. Assign base tags and aliases you intend to use as you develop your project.

THE LAB

1. _____ Open the RSLogix 5000 software.
2. _____ Create a new project.
3. _____ Configure the PAC's I/O.
4. _____ Create needed tags and alias tags.
5. _____ Create ladder logic.
6. _____ Document each ladder rung.

370 CONTROLLOGIX COUNTER INSTRUCTIONS

PARKING GARAGE I/O ASSIGNMENT SHEET			
Device	Input or Output?	Base Tag	Alias
Ticket-dispensing machine			
LOT FULL sign			
Sensor looking for presence of a car			
Arm allowing or restricting entry into the garage			
Sensor watching for cars to exit the lot			

Figure 14-32 I/O assignments for parking garage project.
© Cengage Learning 2014

7. _____ When you have completed this, save your project with a name of your choosing. We will be adding more to this project in future lessons.
8. _____ Configure communications and download the project.
9. _____ Verify operation of the parking garage project.
10. _____ Check with your instructor to see whether operation of your project needs to be checked off for the completion of this exercise.

LAB EXERCISE 4B: Modifying the Parking Lot Application

Management has requested modification of the application so that the total number of cars parked in the lot for up to 31 days is recorded. Saving the total cars parked could be completed at the end of each day or at the beginning of a new day. There is a status bit or flag that can be used to easily trigger saving the total cars parked, at startup each day. We will research and program this bit in the lab exercise. At the start of each day when the PAC goes into Run mode, the following is to happen:

- The total cars parked counter accumulated value will be transferred to the next position in the array.
- The total cars parked counter will be zeroed out.

THE LAB

1. _____ Create a single-dimensional array that will contain the total cars parked in the lot for each day. You might wish to refer to Figure 14-33 as you create the array.
2. _____ We need to find a way to trigger the data transfer of total cars parked for yesterday to the next array element, as well as to clear out the total cars parked counter automatically at the beginning of each day. ControlLogix has a status bit that is true only on the first scan whenever the PAC goes into Run mode from any other operating mode. This bit can be used to trigger the necessary actions when the PAC is powered up and begins its first scan in Run mode. You need to do some research to find what this bit is. *Hint:*
   - Go to RSLogix 5000 help.
   - Select Contents.
   - Select Index.
   - Enter Status.
   - Double-click Status Flags.
   - Depending on your version of software, the Access Run Time Controller Configuration and Status window may display, or you may have to double-click on it to open it.

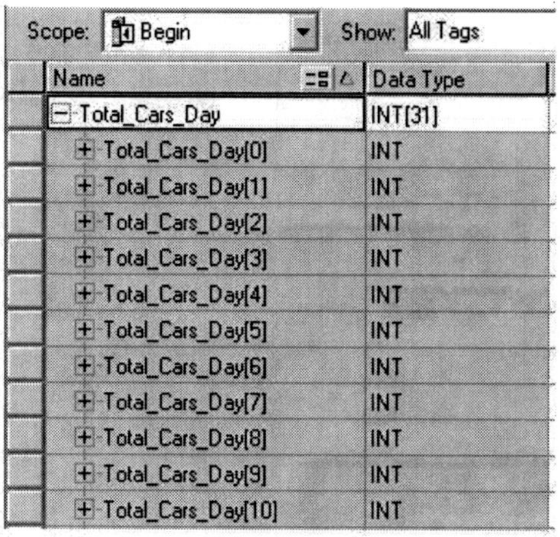

Scope: Begin	Show: All Tags	
Name		Data Type
⊟-Total_Cars_Day		INT[31]
⊞-Total_Cars_Day[0]		INT
⊞-Total_Cars_Day[1]		INT
⊞-Total_Cars_Day[2]		INT
⊞-Total_Cars_Day[3]		INT
⊞-Total_Cars_Day[4]		INT
⊞-Total_Cars_Day[5]		INT
⊞-Total_Cars_Day[6]		INT
⊞-Total_Cars_Day[7]		INT
⊞-Total_Cars_Day[8]		INT
⊞-Total_Cars_Day[9]		INT
⊞-Total_Cars_Day[10]		INT

Used with permission Rockwell Automation, Inc.

Figure 14-33 Example of total cars array.

3. _____ What status bit, also called a status flag, did you discover for use in this situation?
4. _____ Explain what this bit does.
5. _____ Create the rungs necessary to satisfy the specifications.
6. _____ The example array in Figure 14-33 was designed to have 31 elements. On second thought, you decide to expand the array to 32 elements in order to add 1 extra element to help guard against overflow if the data were not extracted until the first day of the new month. Go back and change the size of the array. At this point, do not worry about guarding against overflowing the array or clearing out the array. We will add the required logic to monitor the current position and clearing out the array in a future lesson.
7. _____ Document each ladder rung.
8. _____ When it is completed, save your project with a name of your choosing. We will be adding more to this project in future lessons.
9. _____ Configure communications and download the project.
10. _____ Verify operation of the parking garage project.
11. _____ Check with your instructor to see whether operation of your project needs to be checked off for the completion of this exercise.

LAB EXERCISE 5: Counter Modification of the Traffic Light Application

As you remember from programming our traffic light sequence from Lab Exercise 4 of Chapter 12, the north and south lights are always green until a vehicle is seen by the sensors in the pavement. Currently, if pedestrians want to cross, either they have to cross on a red light or wait until a vehicle is seen by the sensor. To make the intersection safer for pedestrians crossing with the green light, Walk–Don't Walk signs have been ordered. Each light has a push button for a pedestrian wishing to cross. The push button is evaluated by the PAC the same as if a vehicle were sensed by the pavement sensor. The Walk–Don't Walk signs and push buttons are only installed for pedestrians wishing to cross on the north side going east to west or west to east along with those on the south side wishing to go east to west or west to east. Because the north and south lanes are usually green, no additional hardware will be installed. Refer to Figure 14-34. You have been assigned to modify the current project to add the needed logic to control the new Walk–Don't Walk signs.

PROJECT MODIFICATIONS

- Pedestrians wishing to cross from east to west, or west to east on either the north or south side will be provided a push button to press to change the light to green and the Don't Walk signs to Walk so they can safely cross.

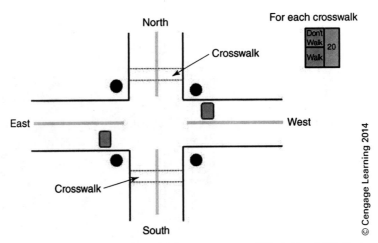

Figure 14-34 Intersection with crosswalk, vehicle sensors, and Walk–Don't Walk signs.

- Each Walk–Don't Walk sign is comprised of three pieces:
  - Red Don't Walk sign light. Light is either on or off.
  - Green Walk sign light. The light is either on or off.
  - Numeric display showing time before the traffic light turns red. This display times down from 20 seconds to 0, alerting the pedestrian of the remaining time before the light turns red.
- The city traffic engineer wants to keep a running total of how many times a pedestrian pushed one of the buttons to cross the intersection in either east to west or west to east. To keep the programming basic, create logic to keep a running total of each of the four possible crossing directions. Program a single reset input to clear the totals. Each time a technician performs preventative monthly maintenance on the lights, he or she will record the values and clear the totals. We will expand the programming on this feature in a future lesson. The four crossing directions counts that are needed are listed here.
  - West to east—north side
  - East to west—north side
  - West to east—south side
  - East to west—south side

WALK–DON'T WALK SEQUENCE

- The red Don't Walk sign is normally illuminated.
- When a pedestrian presses any of the four push buttons to cross the intersection, or when a vehicle is detected at either pavement sensor, the following transpires in addition to the normal sequence already programmed.
- When the traffic light turns green:
  - The green Walk sign illuminates for 5 seconds.
  - The Don't Walk sign then blinks on and off every 1 second for 20 seconds.
  - The time left until red display decrements each second, starting at 20 seconds.
  - When time expires, the traffic light transitions from green to yellow and then to red.
  - The numeric display only displays information when timing is down from 20 seconds to 0. This display is normally off.

  Note: To keep the lab basic, store the time down value in a tag named "OUT_to_walk_dont_walk_time_left_display." Do not worry about the mechanics of transferring the value to the display.

The Walk–Don't Walk display hardware is illustrated in Figure 14-35.

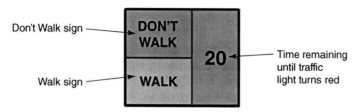

Figure 14-35 Walk–Don't Walk sign with display.
© Cengage Learning 2014

THE LAB

1. _____ Open the RSLogix 5000 software.
2. _____ Open Chapter 12's Lab Exercise 4: Modification of Traffic Light Application.
3. _____ Continue to add to your tag assignment sheets.
4. _____ It is usually best to draw your rungs out on paper before you start programming.
5. _____ Create needed tags.
6. _____ Create ladder logic.
7. _____ Document each ladder rung.
8. _____ When it is completed, save your project. We will be adding to the project in future lessons.
9. _____ Configure communications and download the project.
10. _____ Verify operation of the project.
11. _____ Check with your instructor to see whether a check off is needed for the completion of this exercise.
12. _____ Check with your instructor to see whether a printout of your project needs to be generated and handed in for credit.

LAB EXERCISE 6: Programming Using Operator Interface Tags

INTRODUCTION

This exercise provides additional timer and counter programming experience; however, we are going to introduce a new challenge: Our ControlLogix interfaces to an operator interface terminal. If your lab has operator interface terminals, your instructor may have you create and configure the terminal and interface it to the ControlLogix. If your lab does not have the terminals or the instructor does not wish to use them in this lab, the PAC programming can still be completed. Or your instructor may already have the operator interface screens created and connected to your PAC, so all you have to complete is the PAC programming. Check with your instructor to see how he or she wishes to proceed.

Operator interface terminals are very popular as a way to replace mechanical switches, push buttons, and indicator lights. The terminal is used to input information from the human operator into the PAC and system as well as to provide current information from the system through the PAC to the operator interface for viewing by the operator. There are many different names for these terminals. A couple common terms you may have heard are EOI (electronic operator interface), HMI (human–machine interface), or simply operator interface. Terminal screens are created using the specific software packages associated with your specific terminal. When you have finished creating the screens, the completed screen project will be downloaded to the terminal, much like a PAC project is downloaded to a PAC. As an example, if using the standard Rockwell Automation PanelView used for these labs examples, the software required is Rockwell's PanelBuilder 32 Software. If using a newer Rockwell Automation PanelView Plus terminal, RSView ME Studio software is required. Screen object creation is very similar among the different terminal brands. Even though operator terminals are available for most current industrial networks, newer systems use Ethernet/IP as the network of choice for connecting the operator interface and PAC.

This lab provides the opportunity to develop an operator interface application and interface it to the ControlLogix. Because there are many different manufacturers of and types of operator interface, we do not include the actual programming of the terminal in this lab exercise, as we do not know what brand and type you are using. For our examples, we use a Rockwell Automation standard PanelView terminal. See your instructor for information associated with your specific operator interface terminal and how to proceed with this lab.

- If you do not have an operator interface terminal, this programming lab can still be completed. You are only modifying the RSLogix 5000 project.
- You may modify the operator interface screen objects as part of this lab. See your instructor for additional information for creating the terminal's screens.
- The instructor may have the operator interface screens already created, and you only need to make the PAC programming modifications and then test them on the operator interface. If you are using operator interface with screens already created, make sure you have a reference list of the screen object tags. If the operator interface has its own tag database, the tags you create for your ControlLogix project must point to the internal operator interface tags. Check with your instructor for additional information.

If you create the operator interface screens, be sure to save your work for both the operator interface screens and PAC project modifications, as we will modify the terminal's screens and PAC ladder logic in future labs.

VARIABLE FREQUENCY DRIVE INTERFACE

This lab also provides the opportunity to interface a variable frequency drive to the ControlLogix PAC. Because there are many different manufacturers of and types of drives, we do not include the setup and programming of the drive in this lab exercise, as we do not know what brand and type you will be using. For our examples, we provide the general information. Even though in newer applications, operator interface and drives typically interface to their PAC via Ethernet/IP, there are many options here also. The operator interface and drive could be on different networks and use the PAC as a bridging device. See your instructor for information associated with your specific drive and how to precede with this lab.

- If you do not have a drive, this programming lab can still be completed. You will only be modifying the RSLogix 5000 project.
- Your instructor may have already programmed and connected the drive to your PAC. You may need to configure the drive and set up network communications. See your instructor for additional information for configuring and interfacing the drive.

APPLICATION OVERVIEW

- An operator interface device like a Rockwell Automation PanelView is used to provide the operator an easy way to enter and view process data.
- The operator interface has its own separate software used to create the screens.
- A variable frequency drive is used to provide different mixing speeds.
- If you are programming the operator interface terminal, see your instructor for documentation associated with programming and interfacing the PAC. This lab introduces the operator interface terminal screen objects and how they operate for the

standard Rockwell Automation PanelView. Even though operator interface screen objects are very similar among different manufacturers, always refer to your specific hardware and software when working with these products. Our task here is to create ladder rungs and understand the correlation among the RSLogix 5000 Controller Scoped tag collection, ladder logic, and PanelView screen objects.

PROJECT SPECIFICATIONS

This exercise assumes that you have already created the operator interface screens or they have been developed by others. Our job is to understand and program the PAC to interface to the operator interface, using the tags provided by the operator interface programmer. The program for this exercise is a simulated mixing application where an operator manually weighs and adds ingredients to a mixer as directed by the current recipe. After ingredients are added, the operator selects one of three mixing times and speeds and the number of batches to make and then presses start to begin the first batch. When each batch is completed, the mixing tank is drained and manually cleaned. After the tank is cleaned, the operator can start the next batch.

1. Operator adds the required gallons of solution to the tank, as specified by the recipe.
2. There are three mixing times, depending on the product being produced. Times used for mixing are 1 minute, 2½ minutes, or 5 minutes. The operator terminal has three touch-screen objects identified as Time 1, Time 2, and Time 3 for selecting the mixing time. The mixing times are associated with the screen objects, thus the operator needs only to select the correct button. As the operator sets up for the next batch, he or she selects the proper time by pressing the desired button. When the operator presses one of the time selection buttons, the screen object button has been programmed in the terminal software so that the button blinks to identify that it has been selected.
3. The ladder logic uses only one timer instruction and a Move instruction to provide the desired time value for the timer. Use the tags: "Time_1," "Time_2," and "Time_3" as the inputs controlling each Move instruction. Figure 14-36 is an example of how the rungs might be programmed. Keep in mind when using an operator interface device, real-world input tags are not used to address screen objects.

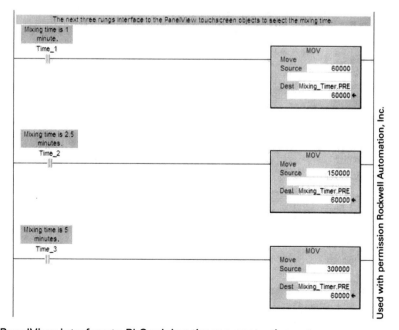

Figure 14-36 PanelView interface to PLC mixing timer preset value entry.

If you do not have an operator interface terminal, assign time tags as aliases to hardwired push buttons, similar to the rungs in Figure 14-37.

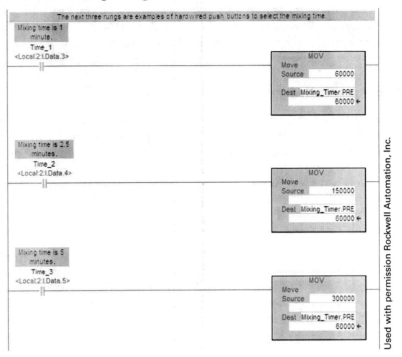

Figure 14-37 Hardwired push-button interface to PAC mixing timer preset value entry.

4. There are three mixing speeds, depending on the product being produced. Speeds used for mixing are slow (20 Hz), medium (40 Hz), or high (60 Hz). If the drive is configured for a standard induction motor, 60 Hz equals 1,750 rpm. The operator terminal has three touchscreen objects for selecting the mixing speed. As the operator sets up for the next batch, he or she selects the proper mixing speed by pressing the desired button. When the operator presses one of the speed buttons, the input signal interfaces with PAC logic, as illustrated in Figure 14-38. When selected, the screen object button has been programmed

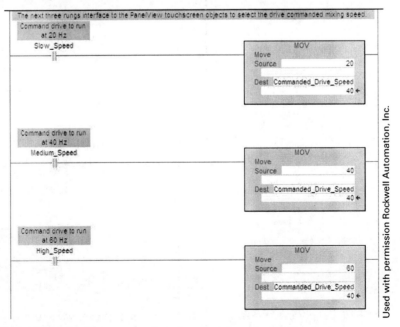

Figure 14-38 Sample rungs for operator interface touchscreen logic.

so that the button will blink to identify that it has been selected. Figure 14-38 is an example of how the rungs might be programmed. Keep in mind that when using an operator interface terminal, real-world input tags are not used to identify screen objects.

If you do not have an operator interface terminal, assign tags to hardwired push buttons, similar to the rungs illustrated in Figure 14-39.

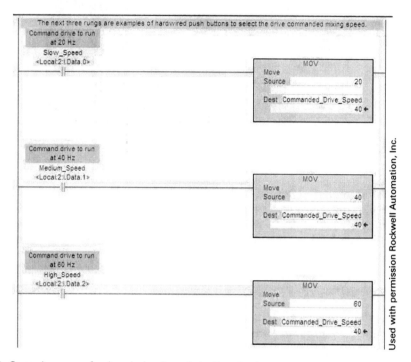

Figure 14-39 Sample rungs for hardwired push-button logic.

5. At the beginning of a batch, the operator enters the total number of batches to be made. Use a down counter, which gets its input from the operator interface device, to track the remaining number of batches. The tag to populate the counter with the value is Batches_to_Make. If you do not have an operator interface, you will have to manually enter values in the Batches to Make tag.
6. Start and stop push buttons to begin or stop the process are also on the operator terminal. Each time a new batch has successfully started the CTD decrements by one count. The remaining count is sent via Ethernet to the operator interface device so that the operator can monitor the number of batches left to be produced for this order. The tag Batches_Remaining is used for this information.
7. Drain the tank for 2 minutes or until the tank empty sensor becomes true.
8. When the current batch is completed, the Batch Complete terminal indicator intensifies.
9. The operator cleans the tank and starts the next batch.

Figure 14-40 is the Mixing Setup screen of the Rockwell Automation PanelView operator terminal screen we are using for this exercise.

Operator interface terminals can be ordered as function key only, touchscreen only, or a combination of both. The figure illustrates a combination of touchscreen and keypad terminal. Function keys are positioned along the bottom left of the unit. Push buttons, indicators, data entry, or display items on the screens are referred to as screen objects. Screen objects can be assigned a function key rather than be used as touchscreen objects. Function keys might be used when operators wear gloves, making touching screen objects difficult. For the exercise, we use the terminal as a touchscreen terminal. This means the operator touches the screen object to select it.

378 CONTROLLOGIX COUNTER INSTRUCTIONS

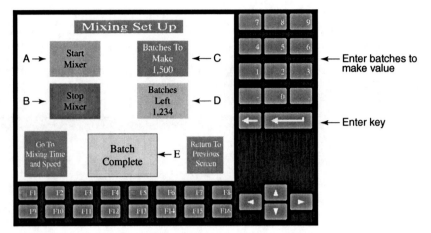

Figure 14-40 Mixing Setup screen for lab exercise.
© Cengage Learning 2014

Notice the numerical keypad in the upper right. This keypad is used after the operator selects an object allowing entering of numerical data, such as the number of batches to be made. Select the data entry screen object by touching it, then key in the numerical information using the keypad. After the data has been entered, the operator presses the Enter key to enter the data.

MIXING SETUP SCREEN OBJECTS

The first screen is the Mixing Setup screen. Refer back to Figure 14-40. For our mixing application, the following screen objects are specified (if you do not have an operator interface, use push buttons and indicator lights on your demo):

A. Touch Start screen object to start the mixer.
B. Touch Stop screen object to stop the mixer.
C. Touch the screen object and then enter the number of batches to be made on the keypad. Press Enter to accept and complete the data entry.
D. Screen object D displays the number of batches left to produce.
E. Indicator object intensifies when each batch is complete.

The two screen objects in the bottom left and right of the screen are navigation keys used to move from one screen to another. Figure 14-41 lists technical information regarding the touchscreen objects.

The numeric keypad on the upper-right side of the PanelView is used to enter numeric data. To enter the number of batches to make, the operator touches the Batches to Make screen object to select it and then keys in the number of batches required, using the numeric keypad. When the value has been entered, press Enter. The value is sent to the PLC tag.

Because there is not enough space on the first screen to include all specified objects, a second screen entitled Mixing Time and Speed has been created on the terminal. This screen is illustrated in Figure 14-42.

Screen 2 contains mixing time and mixing speed screen objects. The operator selects the desired speed and time settings by pressing the proper screen object. (If you do not have an operator interface, a couple three-position selector switches could be used.) The selected screen object has been programmed to blink when selected. As you create your ladder rungs, consider how you will handle one of these objects being pressed after the mixing process begins. The screen object in the lower-right-hand corner is a return to the previous (Mixing Setup) screen. In future lessons, we will modify our operator interface screens to allow the operator to enter a specific time and speed value. Figure 14-43 lists technical information regarding the Mixing Time and Speed touchscreen objects.

CONTROLLOGIX COUNTER INSTRUCTIONS 379

	OPERATOR INTERFACE MIXING SETUP SCREEN OBJECT IDENTIFICATION				
Figure Letter	Description	Object Type	PLC Input or Output	Data Type	Tag
A	Start mixer screen object	Normally open momentary push button	In	BOOL	Start_Mixer
B	Stop mixer screen object	Normally closed momentary push button	In	BOOL	Stop_Mixer
C	Number of batches to make. Touch the screen object and enter number of batches to make. This information is sent to the PLC and moved into the counter preset value. This is a numeric data entry object.	Numeric data entry	In	INT	Batches_to_Make
D	Displays number of batches made. The down counter sends the remaining batches to make to this numeric data display object. The value displays as shown.	Numeric data display	Out	INT	Batches_Remaining
E	Batch complete indicator light object. This screen object illuminates when the batch is complete.	Indicator	Out	BOOL	Batch_Complete

Figure 14-41 Operator interface Mixing Setup screen object technical data.
© Cengage Learning 2014

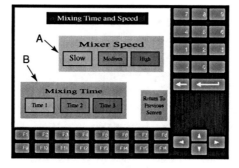

Figure 14-42 Mixing Time and Speed setup screen.
Used with permission Rockwell Automation, Inc.

Operator Interface Tag Database

Many operator interface terminals programming software contains a tag database that is filled in with each screen object's tag name, data type, PLC tag correlation, and PLC node address each object associates with. Figure 14-44 illustrates a portion of the tag database listing our tag names and data types. This database example is from the Panel Builder 32 software. The programmer has to create the same tags in his or her PLC tags collection. The tag names must be exactly the same in the PLC as in the operator interface in order for the data to transfer.

380 CONTROLLOGIX COUNTER INSTRUCTIONS

OPERATOR INTERFACE MIXING TIME AND SPEED SCREEN OBJECT IDENTIFICATION						
Figure Letter	Description	Object Type	PLC Input or Output	Time or Speed	Data Type	Tag
A	Slow Speed	Normally open push button	Input	20 Hz	BOOL	Slow_Speed
A	Medium Speed	Normally open push button	Input	40 Hz	BOOL	Medium_Speed
A	High Speed	Normally open push button	Input	60 Hz	BOOL	High_Speed
B	Time 1	Normally open push button	Input	1 min	BOOL	Time_1
B	Time 2	Normally open push button	Input	2.5 min	BOOL	Time_2
B	Time 3	Normally open push button	Input	5 Min	BOOL	Time_3

Figure 14-43 Operator interface Mixing Time and Speed screen object technical data.
© Cengage Learning 2014

	Tag Name	Data Type
1	Batch_Complete	Bit
2	Batches_Remaining	Signed Integer / Int
3	Batches_To_Make	Signed Integer / Int
4	High_Speed	Bit
5	Medium_Speed	Bit
6	Slow_Speed	Bit
7	Start_Mixer	Bit
8	Stop_Mixer	Bit
9	Time_1	Bit
10	Time_2	Bit
11	Time_3	Bit

Figure 14-44 Partial view of a PanelView tag database.
Used with permission Rockwell Automation, Inc.

DIRECT TAGS WHEN USING CONTROLLOGIX AND PANELVIEW PLUS

One advantage when using newer technology such as ControlLogix and the Rockwell Automation's PanelView Plus is the ability to directly associate tags between the PanelView Plus and the RSLogix 5000 tag collection. RSView ME software is the development software for the PanelView Plus operator interface. Screen objects in the RSView ME software can be directly associated via Ethernet and RSLinx Enterprise software directly with tags in the RSLogix 5000 tag collection. This direct association is referred to as direct tags. Figure 14-45 shows an overview of the operator interface, ControlLogix, and the controller's tag database. In the example, the operator interface is interfaced with the ControlLogix via Ethernet/IP using a 1756-ENBT module, then across the chassis backplane and into the ControlLogix controller scoped tags collection.

Using direct tags and ControlLogix eliminates the need for creating a separate database in the operator interface software and in the PAC. Figure 14-46 illustrates the Time 1 screen object pointing to the Time_1 tag in RSLogix 5000 controller scoped tags collection. As the operator interface programmer creates each screen object, he or she can use Ethernet communications directly to the ControlLogix controller and associate the tag directly to the RSLogix 5000 tag collection. The logic programmer uses the same tag when creating ladder logic.

Figure 14-47 shows the correlation of the number of batches to make screen object and the controller scoped tags collection.

CONTROLLOGIX COUNTER INSTRUCTIONS 381

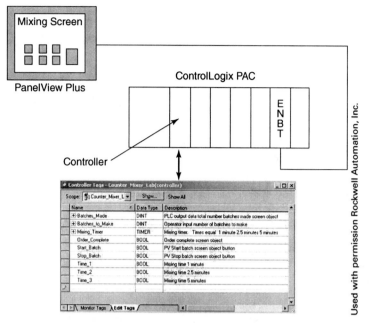

Figure 14-45 The PanelView Plus accesses the tags directly in the controller's tag collection via Ethernet.

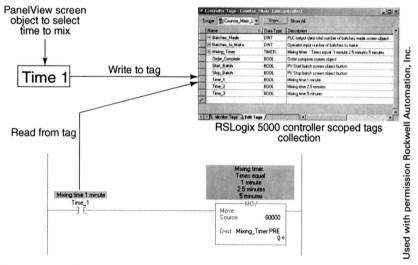

Figure 14-46 Operator interface direct tag association with RSLogix 5000 controller scoped tags collection.

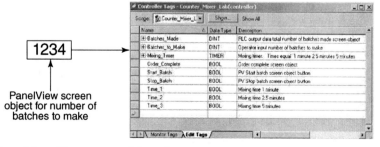

Figure 14-47 PanelView Plus screen object direct correlation to RSLogix 5000 controller tags.
Used with permission Rockwell Automation, Inc.

THE LAB

Now it is your opportunity to turn the project specifications into ladder logic.

1. _____ Create a new project or reuse the Begin project.
2. _____ Enter the specified tags in the controller scoped tags collection. To keep this lab simple, we assume we are not using direct tags. The PanelView programmer has already entered the tags in the PanelView software as he or she created the application for the operator terminal. Remember, the tag name must be an exact match to correspond with the screen objects.
3. _____ Only the application specifications have been provided to you from the operator interface developer. It is your task to complete the ControlLogix project to satisfy the supplied specifications. There are many answers to this lab. It is your decision as how to proceed from here.
4. _____ Don't forget to add complete documentation.
5. _____ Remember to save your work.

SUMMARY

This chapter covered a lot of ground. We learned about the two counter instructions and the reset instruction. We edited and modified earlier lab exercises by adding additional timers and counters. A couple of ways in which counters can be paired to track each other's operating cycles or to lengthen counting ranges were incorporated into lab exercises. Although we illustrated only a couple of possibilities, there are many other applications in which timers and counters can be made to work together. Two labs incorporated variable addressed arrays to record system counter data. Future labs will investigate guarding against array overflow, resetting the array, and clearing data after they have been sent to management's database.

We introduced operator interface and variable frequency drives to the programming environment, as well as introducing programming considerations. If operator interface terminals were available, you may have had the opportunity to create the terminal's screens and interface it to your PAC and verify proper operation of the ladder logic to the terminal's screen objects. A variable frequency drive was an optional add-on to the lab exercise as a way to learn about operator interface to PAC ladder logic to drive integration. The next chapter introduces comparison instructions.

Summary of Counter Features

- A counter instruction is an output instruction.
- Counters count the false-to-true transitions of the rung's input logic.
- As an output, the counter instruction counts each time the input logic changes from a false to a true rung.
- Input logic can be a signal coming from an external device, such as a limit switch or sensor, or from internal logic.
- Each time the counter instruction registers a false-to-true rung transition, a count up counter's accumulated value is incremented by 1.
- Each time a count down counter registers a false-to-true rung transition, its accumulated value is decremented by 1.
- One important consideration when working with counters is the speed or frequency at which an input device is sending the false-to-true signals to the controller. The speed of the controller scan must be taken into consideration. As an example, if parts are passing an inductive proximity sensor on a conveyor faster than the controller scan can see them, some will not be counted.
- Counting range is the numerical range within which a counter can count. The RSLogix 5000 counters can count within the range of −2,147,483,648 to +2,147,483,647. This is the range of a 32-bit signed integer.

CONTROLLOGIX COUNTER INSTRUCTIONS 383

REVIEW QUESTIONS

Note: For ease of handing in assignments, students are to answer using their own paper.

1. A counter instruction is a(n) _____ instruction.
2. As an output, the counter instruction counts each time the _____ changes from false to true.
3. A counter's input signal can come from an external device, such as a limit switch or sensor, or from _____.
4. Each time the counter instruction sees a _____, a count up counter's accumulated value is incremented by 1.
5. One important consideration when working with counters is the _____, or _____, in which an input device is sending the false-to-true signals to the controller.
6. The ControlLogix counters can count within the range of negative _____ to positive _____.
7. A ControlLogix counter's counting range is the range of a 32-bit _____.
8. If a ControlLogix counter counts above _____, an overflow is detected and the overflow bit is set.
9. If a ControlLogix down counter counts below _____, an underflow is detected.
10. To avoid overflowing or underflowing counter, the _____ instruction resets the counter accumulated value back to 0.
11. ControlLogix counters are retentive. Assuming that the controller _____ is in good condition, a counter retains its _____, _____, and _____ bits, through a power loss.
12. The two types of counters are _____ and _____.
13. Fill in the missing information in the table in Figure 14-48.

COUNTER INSTRUCTIONS		
Instruction	Mnemonic	Functional Description
Count up		
	CTD	
Reset		

Figure 14-48 Counter instructions.
© Cengage Learning 2014

14. Each time a count down counter sees a _____, its accumulated value is decremented by 1.
15. When the counter instruction is true, the instruction is _____.
16. When the _____ and _____ are equal, the counter is considered done.
17. Fill in the missing information in the table in Figure 14-49.

COUNTER STATUS BITS				
Status Bit	Identified as	Instruction Used	Example Tag	Functional Description
CU			Parts_Counter.CU	
	Count down enable			True when the instruction is enabled, or true
	Done			
OV				
			Parts_Counter.UN	

Figure 14-49 RS Logix 5000 Counter Status bits.
© Cengage Learning 2014

18. Identify parts A–K shown in Figure 14-50.

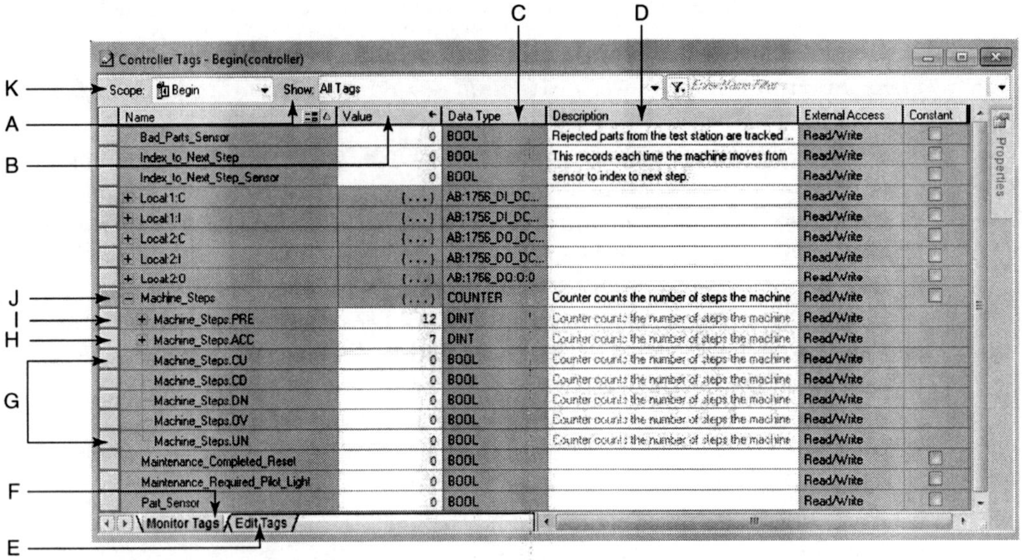

Figure 14-50 Monitoring tags.

19. Identify parts A–J shown in Figure 14-51.

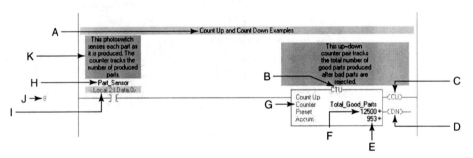

Figure 14-51 Interpreting counter ladder rungs.
Used with permission Rockwell Automation, Inc.

20. If a counter is overflowed or underflowed, will the controller fault?

CHAPTER 15

Comparison Instructions

OBJECTIVES

After completing this lesson, you should be able to

- Describe the function of the common comparison instructions.
- Identify RSLogix comparison instructions.
- Explain two ways to program the Limit Test instruction.
- Create and interpret ladder logic containing comparison instructions.
- Incorporate comparison instructions into the traffic light project.

INTRODUCTION

Comparison instructions are input instructions that test two values to determine whether the instruction is true or false. Comparison instructions include equal, not equal, less than, less than or equal, greater than, greater than or equal, masked comparison for equality, and the limit test, all to test one value against another. Most comparison instructions have two parameters, source A and source B. Typically, source A must be a tag, whereas source B can be a tag or a constant. As an example, using an equal instruction, source A could be a counter's accumulated value and source B could be a constant such as 10. When the counter accumulated value is equal to 10, the instruction is true. On the other hand, if source B is a tag, then when the value stored in the tag is equal to the counter's accumulated value, the instruction is true. Comparison instructions can be used to pick multiple presets off instructions like timers and counters. The limit test instruction can be used to limit the range of data an operator can enter for time, temperature, pressure, or speed. Figure 15-1 lists the comparison instruction we work with in this chapter.

CONTROLLOGIX COMPARISON INSTRUCTIONS		
Instruction Name	**Mnemonic**	**Description**
Equal	EQU	Source A is equal to Source B
Not Equal	NEQ	Source A is not equal to Source B
Less Than	LES	Source A is less than Source B
Less Than or Equal	LEQ	Source A is less than or equal to Source B
Greater Than	GRT	Source A is greater than Source B
Greater Than or Equal	GEQ	Source A is greater than or equal to Source B
Compare	CMP	Use an expression to test for equality
Limit Test	LIM	Test two limits to determine if they are within a predetermined range

Figure 15-1 Comparison instructions for this chapter.
© Cengage Learning 2014

THE EQUAL (EQU) INSTRUCTION

The equal instruction is an input instruction used to test when two values are equal. Programming the equal instruction consists of two steps. The equal instruction is placed on the ladder rung as an input instruction. Then the two parameters, source A data and source B data, are entered.

When source A and source B are equal, the instruction is true; otherwise, the instruction is false.

Source A: This is the tag of the data to test for equality.
Source B: Source B can be either a constant or a tag.

If you use a constant as source B, the constant will be tested for equality with the data residing in the tag specified in source A. If you use a tag as source B, the data contained in that tag will be tested for equality with the data residing in the tag specified in source A.

Sample Ladder Rung

Figure 15-2 illustrates an equal instruction controlling an output instruction.

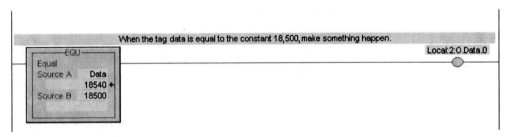

Figure 15-2 An equal instruction controlling output.
Used with permission Rockwell Automation, Inc.

Ladder Explanation

The ladder rung in Figure 15-2 illustrates an equal instruction where source A is the tag named Data and source B is a constant, the number 18,500. Directly under the tag named Data is the value 18,540. This is the value stored in the tag. The value stored in source B is a constant. Because the value in source A is not equal to the value in source B, the instruction is false. As it is, the output instruction is also false.

THE NOT EQUAL (NEQ) INSTRUCTION

The not equal (NEQ) instruction is an input instruction used to test two values for inequality. Use this instruction to determine whether two specified sources of data are not equal. The not equal instruction is true when the data stored in the tag specified as source A is not equal to either the data stored in the tag specified as source B or a constant entered in source B.

Source A: This is the tag of the data to test for inequality.
Source B: Source B can be either a constant or a tag.

If you use a constant as source B, the constant will be tested for equality with the data residing in the tag specified in source A. If you use a tag as source B, the data contained in that tag will be tested for equality with the data residing in the tag specified in source A.

Ladder Explanation

Figure 15-3 illustrates a not equal instruction where source A is the tag Counter_A.ACC, which contains the number 560. Source B has a constant, the number 1,234. Looking directly under the tag Counter_A.ACC, you find the value of the DINT stored in the counter's accumulated value. Because the value in source A is not equal to the value in source B, the instruction is true. When source A is anything other than the constant 1,234, as in source B, the instruction is true. In this example, as a result of the instruction being true, the output instruction is also true. The rung is shown in the off line state.

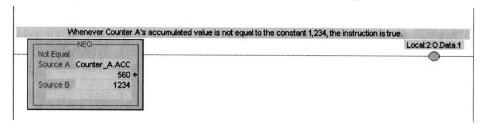

Figure 15-3 Not equal input instruction.
Used with permission Rockwell Automation, Inc.

THE LESS THAN (LES) INSTRUCTION

The less than (LES) instruction is an input instruction used to test whether one value (one source of data) is less than another. The less than instruction is true when the data stored in the tag specified as source A are less than either the data stored in the tag specified as source B or a constant entered in source B.

Source A: This is the tag of the data to test to see whether the value is less than that in Source B.
Source B: Source B can be either a constant or a tag.

Figure 15-4 illustrates a less than instruction ladder rung.

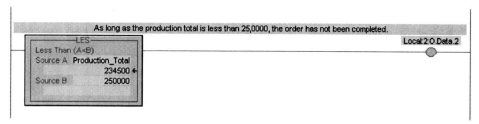

Figure 15-4 The less than instruction tests to see whether the value in source A is less than the value in source B.
Used with permission Rockwell Automation, Inc.

The less than instruction in Figure 15-4 tests to see whether the value in source A, Production_Total is less than the value represented in source B. The tag Production_Total contains the value 234,500. Source B contains the constant 250,000. Because A < B, the less than instruction is true. Because the input instruction is true, the output instruction will also be true. Figure 15-5 provides an overview as to how the instruction works. The rung is shown in the off line state.

If	Then the Instruction Will Be	Example
A is equal to B (A = B)	False	If A = 250,000 and B = 250,000
A is greater than B (A > B)	False	If A is 250,000 or greater, this rung will be false.
A is less than B (A < B)	True	If A is 249,999 or less, this rung will be true.

Figure 15-5 Table showing A and B relationship for A less than B logic.
© Cengage Learning 2014

THE LESS THAN OR EQUAL (LEQ) INSTRUCTION

Use this instruction to determine whether one source of data is less than or equal to another. This instruction is true when the data stored in the tag specified as source A are less than or equal to either the data stored in the tag specified as source B or a constant entered when programming this instruction.

Source A: This is the tag of the data to test to see whether it is less than or equal to source B.
Source B: Source B can be either a constant or a tag.

The ladder rung in Figure 15-6 illustrates a less than or equal input instruction. This rung has the current value of 1,234 stored in source A, Mixing_Timer.ACC. Because the tag Mixing_Timer.ACC is less than the tag Number entered into source B, which is 2,750, the instruction will be true (refer to Figure 15-6). The rung is shown in the off line state.

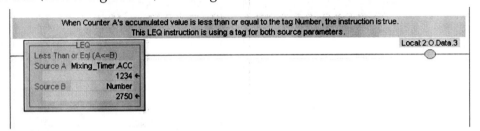

Figure 15-6 Less than or equal instruction tests to see whether Source A is less than or equal to Source B.
Used with permission Rockwell Automation, Inc.

Figure 15-7 provides an overview as to how the instruction works.

If	Then the Instruction Will Be	Example
A is equal to B (A = B)	True	If A = 2,750 and B = 2,750
A is greater than B (A > B)	False	If A is 2,751 or greater, this rung will be false.
A is less than B (A < B)	True	If A is 2,749 or less, this rung will be true.

Figure 15-7 The less than or equal instruction tests to see if A is less than or equal to B.
© Cengage Learning 2014

THE GREATER THAN (GRT) INSTRUCTION

The greater than (GRT) instruction is an input instruction that tests to see whether one source of data is greater than another source of data. The greater than instruction is true when the data stored in the tag specified as source A is greater than either the data stored in the tag specified as source B or a constant entered as source B when programming this instruction.

Source A: This is the tag of the data to test to see whether it is greater than source B.
Source B: Source B can be either a constant or a tag.

Figure 15-8 illustrates a greater than instruction.

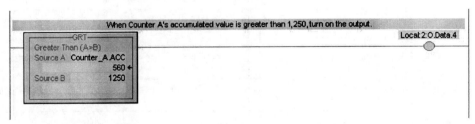

Figure 15-8 The greater than instruction tests to see if A > B.
Used with permission Rockwell Automation, Inc.

Source A, Counter_A.ACC (currently containing 560), is not greater than the constant 1,250 (entered as source B). Because source A is not greater than source B, the instruction is false and the OTE instruction will be false.

Figure 15-9 illustrates how the greater than instruction works.

If	Then the Instruction Will Be	Example
A is equal to B (A = B)	False	If A = 1,250 and B = 1,250
A is greater than B (A > B)	True	If A is 1,251 or greater, this rung will be true.
A is less than B (A < B)	False	If A is 1,250 or less, this rung will be false.

Figure 15-9 The greater than instruction tests to see if A > B.
© Cengage Learning 2014

THE GREATER THAN OR EQUAL (GEQ) INSTRUCTION

The greater than or equal (GEQ) instruction determines whether one source of data is greater than or equal to another.

The instruction is true when the data stored in the tag specified as source A is greater than or equal to either the data stored in the tag specified as source B or the constant entered. Figure 15-10 illustrates a greater than or equal instruction on a rung. Rung is in offline state.

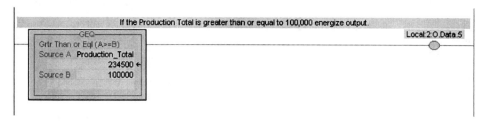

Figure 15-10 The greater than or equal instruction tests to see if A is greater than or equal to B.
Used with permission Rockwell Automation, Inc.

The greater than or equal instruction has the value 234,500 stored in source A. Source B has 100,000. Because 234,500 is greater than 100,000, the greater than or equal instruction will be true. Figure 15-11 illustrates how the instruction works.

If	Then the Instruction Will Be	Example
A is equal to B (A = B)	True	If A = 100,000 and B = 100,000
A is greater than B (A > B)	True	If A is 100,000 or greater, this rung will be true.
A is less than B (A < B)	False	If A is 99,999 or less, this rung will be false.

Figure 15-11 Table showing A and B relationship for A greater than or equal to B.
© Cengage Learning 2014

THE COMPARE (CMP) INSTRUCTION

The compare (CMP) instruction can be used to enter an expression like "If Tag_1 plus Tag_2 is greater than Tag_3, the instruction will be true." Use this instruction if you want to program an expression using comparison operations. The instruction is true when the expression is satisfied. Basic math as well as logical operations like AND, OR, and NOT can be included in the expression. Some data conversion is also possible. Two ASCII strings can be compared with simple compare operations such as equal, not equal, greater than, and so on. Strings are equal if their characters match. ASCII characters cannot be entered into a non-ASCII comparison expression.

Figure 15-12 lists selected compare instruction operators. Refer to the General Instruction Set Reference Manual or RSLogix 5000 software Instruction Help screens for additional information.

SAMPLE OF COMPARE INSTRUCTION OPERATORS					
Operator	Symbol	Data Types	Operator	Symbol	Data Types
Add	+	DINT, REAL	And	And	DINT
Subtract	-	DINT, REAL	OR	OR	DINT
Multiply	*	DINT, REAL	Absolute value	ABS	DINT, REAL
Divide	/	DINT, REAL	Cosine	COS	Real
Equal	=	DINT, REAL	Tangent	Tan	Real
Less Than	<	DINT, REAL	SQR	Square Root	DINT, REAL
Less than or equal	<=	DINT, REAL	Integer to BCD	TOD	DINT
Not equal	<>	DINT, REAL	Log base 10	LOG	DINT
Parenthesis can be used to group characters and specify order of operation.					

Figure 15-12 Selected comparison instruction operators and data types.
© Cengage Learning 2014

Figure 15-13 illustrates a compare instruction where the instruction will be true when Counter_A.ACC is greater than Counter_B.ACC plus the value 25.

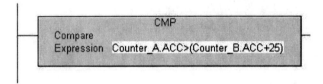

Figure 15-13 Compare instruction.
Used with permission Rockwell Automation, Inc.

Even though the compare instruction can be used to perform a simple comparison, as illustrated in Figure 15-14, this is not an efficient use of the compare instruction. The operation as illustrated in the figure will be executed without any problems; however, this comparison could be performed faster and would consume less memory by using a standard compare instruction such as the greater than instruction. Faster instruction execution means faster scan time.

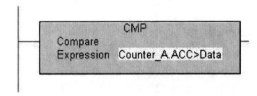

Figure 15-14 Compare instruction performing a simple comparison operation.
Used with permission Rockwell Automation, Inc.

Arithmetic Status Bits, or Flags

Compare instruction execution affects the math status bits, also referred to as math flags, if the expression contains operations that normally update these flags, such as add, subtract, multiply, and divide. Refer to the instruction set help for information regarding the math flags.

The Limit Test (LIM) Instruction

In many applications, an operator enters process information such as time, temperature, pressure, or motor speed, using an operator interface terminal. A few examples of operator-entered data are listed here:

- 0 to 60 minutes mixing time
- 0 to 100 psi
- 0 to 1,750 rpm
- 0 to 2,500 gallons per minute

The limit test (LIM) instruction is one way to set the limits of the data the operator is permitted to enter into the system. To follow our examples:

- Mixing time exceeding 60 minutes will be considered invalid data.
- Pressure exceeding 100 psi will not be allowed.
- Motor speed outside the range of 0 to 1,750 rpm will not be permitted.
- Temperature less than 325°F (163°C) or exceeding 425°F (218°C) will not be accepted by the PAC.

The limit test instruction can be used to test whether values are within or outside of the specified range. Programming the limit test instruction consists of entering three parameters: Low Limit, Test, and High Limit. Figure 15-15 illustrates a limit test instruction.

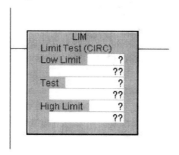

Figure 15-15 Limit test instruction (LIM).
Used with permission Rockwell Automation, Inc.

Limit Test Instruction Test Parameter

The test parameter is typically a tag, either a SINT, INT, DINT, or a REAL number, which provides the input information from the system that the high and low limits will be evaluated against.

Limit Test High and Low Limit Parameters

The high and low limit parameters are the values or limits the test parameter is evaluated against. The parameters could be a constant or a tag value.

Figure 15-16 illustrates a possible application for a limit test instruction where an operator enters a motor speed command for a variable frequency drive into the system, using an operator interface terminal. The motor speed reference is sent to the controller via a network connection. The controller uses the limit test instruction to verify that the operator input motor speed is within acceptable range. If the test input value, the tag Drive_Speed_Input, from the operator interface is within the acceptable range (0–1,750), the limit test instruction will be true. Being true, the new motor speed command can be programmed to be sent on to the variable frequency drive.

Analysis of the Limit Test Instruction from Figure 15-16

In this example, the LIM instruction has a low limit of 0, whereas the high limit is 1,750. The valid speed command includes the value 0 up to and including 1,750. If the new speed command value is between 0 and 1,750, inclusive, the instruction will be true and the new speed command can be programmed to be passed on to the variable frequency drive. If the speed reference input is outside of the 0 to 1,750 rpm range, the instruction will be false, and no speed reference will be passed on to the drive. As a result of the instruction being false, the programmer can determine how the PAC will react to this situation. As one example, the drive can be programmed to run at the last commanded speed. Analyzing Figure 15-16, the limit test instruction has the low limit parameter as a constant, 0, whereas the high limit is the constant 1,750. The test parameter is the tag Drive_Speed_Input, which is the data input from the operator. The value 1,250 directly below the test parameter is the current value of the tag. The current value is 1,250, which is within the limits of 0 to 1,750, so the data are valid and the instruction will be true.

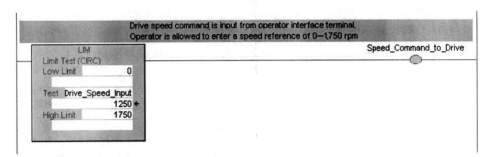

Figure 15-16 Limit Test instruction used to limit motor speed input by the operator. The rung is shown in the off line state.
Used with permission Rockwell Automation, Inc.

Figure 15-17 illustrates a limit test instruction where all parameters are tags. Note the left-pointing arrow below each parameter. The current value of the tag is displayed to the left of that arrow. Currently the tag Low_Pressure has the value of 0, whereas the High_Pressure tag is 250. If the test tag Pressure_Value is between 0 and 250, inclusive, the instruction will be true. The current value of the Pressure_Value tag is 123, so the LIM instruction is true. If the pressure were outside of these limits, the instruction would be false.

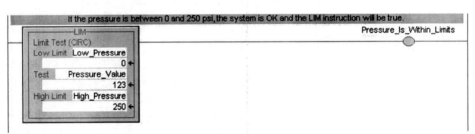

Figure 15-17 LIM instruction using tags for all parameters. The rung is shown in the off line state.
Used with permission Rockwell Automation, Inc.

By programming the LIM instruction differently, we can look at a temperature, speed, or pressure value, as an example, and determine whether they are outside a desired range. Figure 15-18 shows the same LIM instruction as in Figure 15-17; however, the low and high limit tags have been reversed. Now the low limit is 251, whereas the high limit is −1. In this case, if the pressure is anything outside the range of 0 to 250, the instruction will be true. A pressure value within the 0 to 250 range will make the instruction false. Currently the Pressure_Value tag is 123. Because the value 123 is not outside of the range 0 to 250, the instruction will be false. If an operator entered a pressure of 275 psi, which would be outside of the allowable range, the instruction would be true.

Figure 15-19 illustrates combining the two limit test instructions monitoring the operator-provided data. If the pressure value is valid, the data will be sent to the system. If the operator enters data outside of the acceptable values, the tag Bad_Pressure_Value_Reenter will be true and could be programmed to send a bit to the operator interface and display a message suggesting that the operator enter a valid pressure value.

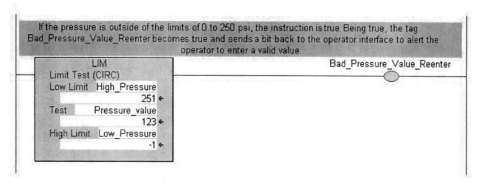

Figure 15-18 LIM instruction testing to see if a value is outside of the limits.
Used with permission Rockwell Automation, Inc.

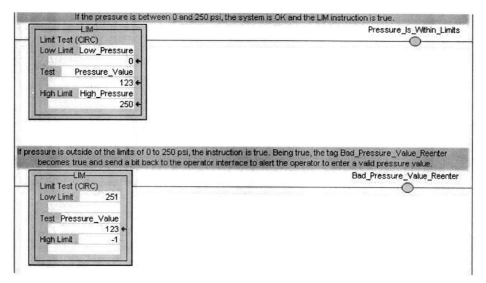

Figure 15-19 Combining the limit test instructions to test operator entered pressure data.
Used with permission Rockwell Automation, Inc.

SUMMARY

Comparison instructions are input instructions that test two values to determine whether the instruction will be true or false. Comparison instructions include equal, not equal, less than, less than or equal, greater than, greater than or equal, masked comparison for equality, and the limit test, all to test one value against another. The Masked equal, or MEQ, instruction is much like the equal instruction, with a filter called a mask. The mask is a hexadecimal value used to determine the specific bits in the source that are to be tested for equality. We will work with this instruction a bit later when we learn about hexadecimal masking. Most comparison instructions have two parameters, source A and source B. Typically, source A must be a tag, whereas source B can be a tag or a constant.

The limit test instruction is used to test for values whether they are within or outside of the specified range. Programming the limit test instruction consists of entering three parameters: low limit, test, and high limit. The test parameter provides the information the high and low limits will be evaluated against. The limit test instruction can be programmed to determine whether information entered by an operator is either within the specified limits or outside the limits. Information found to be outside acceptable limits can make the limit test instruction true and trigger a bit to an operator interface terminal to alert the operator to re-enter valid data.

REVIEW QUESTIONS

Note: For ease of handing in assignments, students are to answer using their own paper.

1. The _____ instruction is an input instruction used to test when two values are equal.
2. When executing the equal instruction, when source A and source B are equal, the instruction is _____; otherwise, the instruction is false.
3. When programming the equal instruction, if you use a constant as source B, the constant will be tested for equality with the data residing in the tag specified in source _____.
4. The _____ instruction is used to test two values for inequality.
5. The _____ instruction is true when the data stored in the tag specified in source A is not equal to either the data stored in the tag specified as source B or a constant entered when entering this instruction.
6. The _____ instruction is used when you want to test whether one value is less than another value.
7. What instruction would be true when Source A is less than source B?
   a. Equal
   b. Not equal
   c. Greater than
   d. Less than
   e. All of the above
8. If you use a tag as source B, this instruction will cause the data contained in the source B tag to be tested to see whether they are greater than the data residing in the tag specified as source A. What instruction is used to accomplish this?
9. The _____ instruction is true when the data stored in the tag specified as source A is less than or equal to either the data stored in the tag specified as source B or a constant entered.
10. If you use a constant as source B, this instruction will cause the tag in source A to be tested to see whether they are equal to or greater than the data in source B. What instruction did you program?
11. The greater than instruction, or _____, is used to determine whether one source of data is greater than another.
12. The LES instruction is an input instruction that tests to see whether one source of data is _____ than another source of data.
13. The greater than or equal instruction's programming abbreviation is _____.
14. The _____ is used to determine whether motor speed data to be passed on to a variable frequency drive is within a valid range.
15. Suppose the greater than or equal instruction has the value 19 stored in source A and a value of 23 stored in source B. Is this instruction true or false?
16. How would you configure an instruction to test to determine whether an operator has input valid temperature data within the range of 325°F to 375°F (163°C to 191°C)? Draw the ladder rung and fill in the required tags and parameters. You may need to create tags for the rung and its output.
17. Configure a rung to test to determine whether an operator has input invalid temperature from question 16. Draw the ladder rung and fill in the required tags and parameters. You may need to create tags for the rung and its output.
18. To evaluate an expression using a single instruction, the _____ instruction could be used.
19. Create the expression to be programmed in a compare instruction to determine whether the daily production goal has been exceeded. Assume the plant is working three shifts each day.
20. To test in order to see whether only bits 12, 13, 14, and 15 in a source tag are equal to _____ same bits in another tag, _____ instruction could be used.

LAB EXERCISE 1: Evaluating Comparison Instructions

OBJECTIVES

- Understand the operation of comparison instructions.
- Determine whether comparison instructions are true or false.
- Evaluate ladder rungs with comparison instructions.

THE LAB

Interpret each of the following instructions. Is the instruction currently true or false? Explain your answer.

1. Is the instruction shown in Figure 15-20 true or false? Explain your answer.

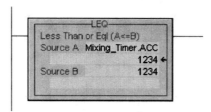

Figure 15-20 Instruction to evaluate for question 1.
Used with permission Rockwell Automation, Inc.

2. Evaluate the instruction shown in Figure 15-21, and explain its current state.

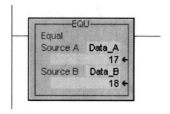

Figure 15-21 Instruction to evaluate for question 2.
Used with permission Rockwell Automation, Inc.

3. Is the instruction shown in Figure 15-22 true or false? Explain your answer.

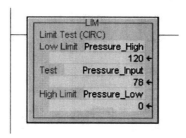

Figure 15-22 Instruction to evaluate for question 3.
Used with permission Rockwell Automation, Inc.

4. Is the instruction shown in Figure 15-23 true or false? Explain your answer.

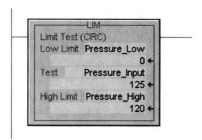

Figure 15-23 Instruction to evaluate for question 4.
Used with permission Rockwell Automation, Inc.

5. Evaluate the instruction shown in Figure 15-24, and explain its current state.

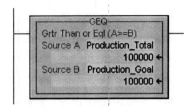

Figure 15-24 Instruction to evaluate for question 5.
Used with permission Rockwell Automation, Inc.

6. Evaluate the instruction shown in Figure 15-25, and explain when this instruction will be true. Assume Production Goal is 25,000.

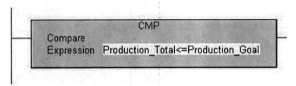

Figure 15-25 Instruction to evaluate for question 6.
Used with permission Rockwell Automation, Inc.

7. Is the instruction shown in Figure 15-26 currently true or false?

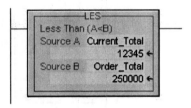

Figure 15-26 Instruction to evaluate for question 7.
Used with permission Rockwell Automation, Inc.

8. When the instruction shown in Figure 15-27 is executed, will it be true or false?

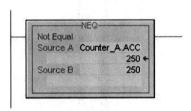

Figure 15-27 Instruction to evaluate for question 8.
Used with permission Rockwell Automation, Inc.

9. Evaluate the instruction shown in Figure 15-28, and explain what needs to happen to make the instruction true.

```
                           CMP
    Compare
    Expression  (Shift_1_Production+Shift_2_Production+Shift_3_Production)>Production_Goal
```

Figure 15-28 Instruction to evaluate for question 9.
Used with permission Rockwell Automation, Inc.

10. Referring to the compare instruction for the last question and using the information from Figure 15-29, in what state will the compare instruction in Figure 15-28 be after it is executed?

Tag	Value
+ Counter_B	{...}
+ Data	76540
+ Drive_Speed_Input	12750
+ High_Pressure	750
+ Local:0:C	{...}
+ Local:0:I	{...}
+ Local:0:O	{...}
+ Local:2:C	{...}
+ Local:2:I	{...}
+ Local:4:C	{...}
+ Local:4:I	{...}
+ Local:4:O	{...}
+ Low_Pressure	0
+ Mixing_Timer	{...}
+ Number	987
+ Pressure_value	146
+ Production_Goal	32550
+ Production_Total	765
+ Shift_1_Actual	874
+ Shift_1_Goal	1000
+ Shift_1_Production	9256
+ Shift_2_Actual	764
+ Shift_2_Goal	1000
+ Shift_2_Production	8674
+ Shift_3_Actual	995
+ Shift_3_Goal	1000
+ Shift_3_Production	1032
+ Timer_1	{...}
+ Value	980

Figure 15-29 Tags collection for reference for question 10 CMP instruction.

11. Is the instruction given in Figure 15-30 true or false? Explain your answer.

Figure 15-30 Instruction to evaluate for question 11.
Used with permission Rockwell Automation, Inc.

12. When the instruction shown in Figure 15-31 is executed, will it be true or false?

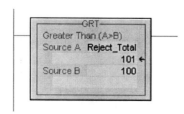

Figure 15-31 Instruction to evaluate for question12.
Used with permission Rockwell Automation, Inc.

13. When the instruction in Figure 15-32 tests the temperature input, what will the instruction output be?

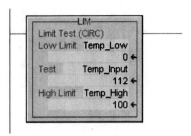

Figure 15-32 Instruction to evaluate for question 13.
Used with permission Rockwell Automation, Inc.

14. Evaluate the rung in Figure 15-33, and answer the following questions.

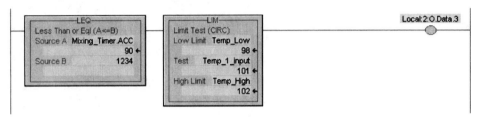

Figure 15-33 Rungs to evaluate for question 14.
Used with permission Rockwell Automation, Inc.

   a. If these rungs were displayed on a computer screen, would this rung be executed by the PLC?
   b. Referring to the answer from question 14a, how did you determine your answer?
   c. If the rung were being executed, explain why the LEQ instruction would be either true or false.
   d. Explain the function of the limit test instruction.
   e. If the rung were being executed, would the limit test currently be true or false?
   f. True or false: The LIM instruction would go false if the Temp_High value reached 102°F (39°C), as the temperature is now out of limits.
   g. If the rung were currently being executed, in what state would the output be?

15. Evaluate the rungs in Figure 15-34, and answer the following questions.

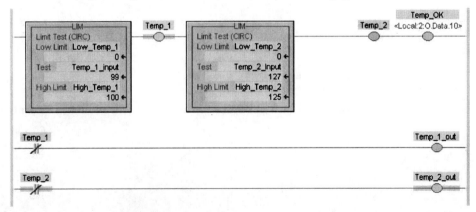

Figure 15-34 Evaluation rungs for question 15.
Used with permission Rockwell Automation, Inc.

   a. Is the first rung a verifiable rung, using RSLogix 5000 software?
   b. What makes this rung different from a traditional PLC ladder rung?

c. If you have worked with earlier Rockwell Automation PLCs such as the SLC 500 or PLC 5, would this logic be verifiable and downloadable?
d. In what operating mode was the PLC with these rungs in their current state? How do you know?
e. Explain the current status of these rungs.

16. _____ Refer to Figure 15-35 and Figure 15-36 to determine the current state of the CMP instruction. What value would Production Goal have to be to make the instruction true?

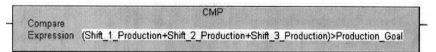

Figure 15-35 CMP instruction for evaluation for question 16.
Used with permission Rockwell Automation, Inc.

Tag	Value	Type
+ Counter_A	{...}	COUNTER
+ Counter_B	{...}	COUNTER
+ Data	18540	DINT
+ Drive_Speed_Input	12500	DINT
+ High_Pressure	250	DINT
+ Local:0:C	{...}	AB:1756_DO_DC...
+ Local:0:I	{...}	AB:1756_DO_DC...
+ Local:0:O	{...}	AB:1756_DO:0:0
+ Local:2:C	{...}	AB:1756_DI_DC_...
+ Local:2:I	{...}	AB:1756_DI_DC_...
+ Local:4:C	{...}	AB:1756_DO_DC...
+ Local:4:I	{...}	AB:1756_DO_DC...
+ Local:4:O	{...}	AB:1756_DO:0:0
+ Low_Pressure	0	DINT
+ Mixing_Timer	{...}	TIMER
+ Number	75	DINT
+ Pressure_value	123	DINT
+ Production_Goal	30000	DINT
+ Production_Total	25000	DINT
+ Shift_1_Production	9256	DINT
+ Shift_2_Production	8674	DINT
+ Shift_3_Production	10324	DINT
+ Timer_1	{...}	TIMER
+ Value	3450	DINT
+ weight_1	125	DINT
+ weight_2	99	DINT

Figure 15-36 Tags collection for evaluation for question 16.
Used with permission Rockwell Automation, Inc.

17. _____ Evaluate the instruction in Figure 15-37. Explain how the instruction will evaluate the tag Data. *Hint:* RSLogix 5000 Instruction Help can be useful here.

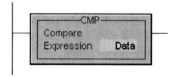

Figure 15-37 CMP instruction for question 17.
Used with permission Rockwell Automation, Inc.

18. _____ The next questions pertain to the CMP instruction in Figure 15-38. How will the instruction act on the expression?

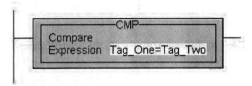

Figure 15-38 CMP instruction for question 18.
Used with permission Rockwell Automation, Inc.

19. _____ How is this instruction used in this manner, and what is its impact on scan time and program efficiency?
20. _____ The next two questions pertain to the following situation: Assume you had an analog input module channel wired to a 4 to 20 ma signal. A rung must be programmed to turn on an alarm bit if the input signal falls below 4 ma or goes above 20 ma. What instruction would you use?
21. _____ How would you program the instruction's parameters?

LAB EXERCISE 2: Programming Comparison Instructions

OBJECTIVES

Upon completion of this laboratory exercise, you should be able to

- Program a free running timer and understand its operation.
- Program and interpret the operation of comparison instructions.
- Describe the usage and programming of the limit test instruction.

INTRODUCTION

Comparison instructions can be used to pick multiple presets or actions to be triggered from variable data such as a timer or counter accumulated value. This exercise introduces incorporating comparison instructions to trigger events at counter accumulated values other than when the accumulated and preset values are equal.

THE LAB

For this exercise, we program a free-running timer to send a 1-second pulse into a counter. The counter provides the changing numerical value that is used to evaluate the comparison instructions. We view the comparison instructions as their source A value represented by the counter accumulated value changes as the instructions become true or false.

1. _____ Open the Begin project.
2. _____ Create the ladder logic, as illustrated in Figure 15-39.
3. _____ As the rungs are programmed, create the needed tags. Enter the values to the tags as you create them, with the exception of the CMP instructions. We manually add values as we test the operation of the instructions.
4. _____ Add rung comments describing how the rung and instruction works.
5. _____ When you have finished programming, download and save the project as "Compare."

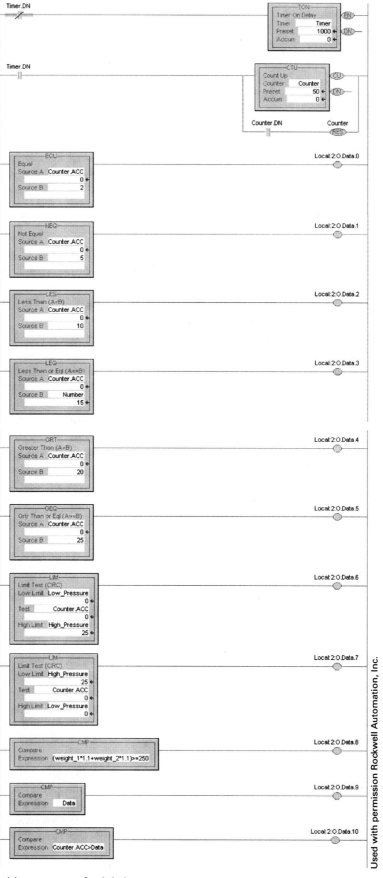

Figure 15-39 Ladder program for lab 2.

6. _____ Go into run mode.
7. _____ As the rungs execute, observe the counter's accumulated values change and note the true or false state of each of the instructions.
8. _____ Answer the following questions.
   a. _____ When is the equal instruction true?
   b. _____ When is the not equal instruction true?
   c. _____ When is the less than instruction true?
   d. _____ When is the less than or equal instruction true?
   e. _____ When is the greater than instruction true?
   f. _____ When is the greater than or equal instruction true?
   g. _____ Explain the operation of the first limit test instruction.
   h. _____ Explain the operation of the second limit test instruction.

The following four questions relate to the CMP instruction controlling Data.8.

9. _____ Write out the expression.
10. _____ Referring back to the tags collection in Figure 15-36, if the instruction were executed using the data found in the figure, would the instruction be true or false?
11. _____ As your project runs, go to the tags collection and manually enter values to the tags to make the first compare instruction true.
12. _____ Manually enter a value to any tag, and watch the first compare instruction go false.
13. _____ Explain how the CMP instruction on the next rung with the tag Data will operate. Experiment by manually entering values into the tag to make the instruction go true and then false.
14. _____ Explain how the last rung and its compare instruction works.
15. _____ Go offline, as we are going to modify the ladder logic.
16. _____ Create a new tag Number_A with a value of 2.
17. _____ Create a new tag Number_B. Give this tag the value of 3.
18. _____ Create another new tag Number_C with a value of 17.
19. _____ Program a CMP instruction to multiply Number_A times Number_C, then Subtract Number_B. If the answer is greater than the tag counter.ACC, turn on output Data.11.

LAB EXERCISE 3: Incorporating Comparison Instruction into the Traffic Light Project

OBJECTIVE

In Chapter 14, we added the Walk and Don't Walk signs to our traffic light project. We also included counters to track the number of times a pedestrian pressed the button to cross the intersection. In this lesson, we use counters and comparison instructions to record how many times the north–south lights and the east–west lights cycle on and off. The Public Works Department has been receiving complaints that bulbs are burning out and not being replaced in a reasonable time. Your job is to add the necessary logic to track the number of cycles for each of the two directions.

Our traffic engineer has been conducting some tests and has determined that if the green, yellow, and red bulbs are replaced after 5,000 cycles, the problem with burnt-out bulbs disappears. Modify the program so as to turn on and flash a yellow light inside the enclosure for the maintenance technician to monitor when 4,750 cycles have been met or exceeded for each direction. Program a red light to flash for each direction when 5,000 cycles have been recorded. The goal is to have the maintenance technician monitor the notification indicators and change the lightbulbs before the maximum cycle life of 5,000 has been reached. In order to test your program, you will probably wish to adjust the values to something more realistic for the lab.

For pedestrian safety, we need to find a way to notify central control if any of the Walk or Don't Walk bulbs burn out prematurely. A radio modem will be installed to send this notification. Engineering has suggested you order a 1756-OB16D module to set up a test. Explain what specific feature prompted the selection of this particular module.

Draw one rung so that when the bulb prematurely burns out for Walk light 2, the Walk_Light_2_Out_Notification tag will become true to indicate the problem. Assume the output module is in slot 6.

INCORPORATING A 1756-OB16D INTO THE TRAFFIC LIGHT PROJECT

For this lab, we modify our traffic light project and program the required rungs to monitor each Walk and Don't Walk lightbulb. If you have a 1756-OB16D module either available or currently in your class demo, complete the following steps. If you do not have the diagnostic module or are using a CompactLogix controller, skip to the next section. CompactLogix does not have diagnostic I/O modules.

1. _____ To this point in our traffic light scenario, we assumed that we were not aware of the diagnostic features available for the 1756-OB16D module, so the modules were not incorporated into the system design. As a result of the system redesign, a diagnostic module was ordered for testing.
2. _____ Remove the standard module and insert the new diagnostic module.
3. _____ If you do not have a 1756-OB16D module currently installed in your classroom demo, after powering down place the module in an empty slot.
4. _____ After inserting the module, wire the module's removable terminal block to the output indicators.
5. _____ Refer to the Rockwell Automation PLC wiring diagram, publication number CIG-WD001B-EN-P. This publication can be viewed or downloaded from the Rockwell Automation Web site. Go to the Publications Library and search for the publication number. The number listed above was current at the time of publication. If the number comes up as invalid, try searching for "CIG-WD" or try searching by title or key words. Figure 15-40 illustrates the cover of the publication.
6. _____ Add the new 1756-OB16D diagnostic module to the I/O configuration.
7. _____ Examine the new diagnostic tags created for the diagnostic module.
8. _____ Add the new diagnostic rungs as per specification.
9. _____ Where are the module's diagnostic functions configured?
10. _____ When you have finished, save your project.
11. _____ Download the project, and put the PAC into Run mode.
12. _____ If possible, either remove one of the lightbulbs representing a Walk or Don't Walk light from your demo to simulate a bulb burning out. If you are unable to remove a bulb for testing, ask your instructor whether you should power down the PAC and how to safely disconnect one of the output wires going to the bulb. Explain what you observe and how your ControlLogix reacted to the situation.
13. _____ Monitor the appropriate tag as well as the notification light on your demo to verify that the rung operates correctly.
14. _____ Explain how we could use this same diagnostic feature to trap a loose connection for this or some other application.
15. _____ Explain how to clear the feature referred to in step 16.

Figure 15-40 I/O Modules Wiring Diagrams reference manual.

16. _____ After testing to see that the module operates as specified, reinstall the bulb or safely rewire the output wiring to simulate the bulb replacement.
17. _____ Unlatch the no load bit and verify proper operation.
18. _____ This completes this lab exercise.
19. _____ Before you go offline with your project, ask your instructor whether he or she wishes to check off your completion of this lab.

LAB EXERCISE 4: Without a Physical 1756-OB16D Module

If you have a modular ControlLogix without the required 1756-OB16D diagnostic module, the lab can be completed by offline editing of your project and creating the logic. You will not be able to download and test your logic, but basic programming can be understood by completing the following lab.

1. _____ Open up the traffic light project you have been working on.
2. _____ Save the project under a different name, as we are going to modify the renamed project by adding a 1756-OB16D module, and create the required notification ladder rungs.
3. _____ We will be adding additional logic to the traffic light project; by saving the diagnostic modifications as a separate project, you can experience creating rungs, using the diagnostic functions, and still maintain your original working project for future modifications
4. _____ Explain what would happen if you downloaded this completed project into a controller residing in a chassis that did not have the required diagnostic output module.
5. _____ Add the new diagnostic module to the I/O configuration.
6. _____ Examine the new diagnostic tags created for the diagnostic module.
7. _____ Add the new diagnostic rungs as per specification.
8. _____ Ask your instructor whether your completion of the lab needs to be recorded.
9. _____ Save your project if you desire.

LAB EXERCISE 5: Operator Interface Project Modification from Chapter 14

OBJECTIVES

Upon completion of this laboratory exercise, you should be able to

- Modify an existing project to include comparison instructions.
- Program and interpret the operation of comparison instructions.
- Describe the usage and programming of the limit test instruction.

INTRODUCTION

In this lab exercise, we modify the mixing application logic associated with operator interface screen objects programmed in the ControlLogix Counters Instructions in Lab Exercise 6: Programming Using Operator Interface Tags, from Chapter 14. This lab modifies the project to incorporate comparison instructions.

- If you have an operator interface terminal available, check with your instructor to see whether to incorporate the operator interface into this lab.
- If you do not have an operator interface terminal, this programming lab can still be completed. You will complete the PAC programming portion and understand how the PAC logic associates with the operator interface objects.
- You may modify the operator interface screen objects as part of this lab. The instructor may have the operator interface screens already modified so that you only need to make the PAC programming modifications and then test them on the operator interface. Check with your instructor for additional information.
- If you have a variable frequency drive integrated into your system, you will be able to test your modifications.

The motor speed screen objects will be modified to provide the operator more flexibility. Rather than having three preset speeds—slow, medium, and high—the objects have been replaced with a single screen object for numeric entry. Figure 15-41 shows the new speed selection screen objects. Not only will this modification save valuable screen space, but it will also allow the operator to enter a numeric value for the drive speed rather than the three current separate buttons and their associated fixed speeds. For this application, the operator will be able to touch the numeric data entry screen object and enter a value ranging from 0 to 1,750 rpm on the terminal's keypad. Once the new motor speed has been entered, the Enter key is pressed to send the new speed to the ControlLogix. See Figure 15-42 for the modified mixing speed screen object.

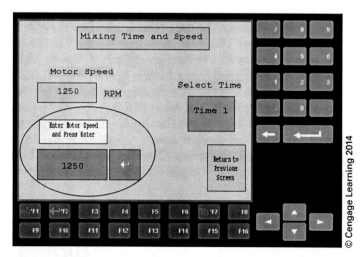

Figure 15-41 New mixing speed screen objects.

406 COMPARISON INSTRUCTIONS

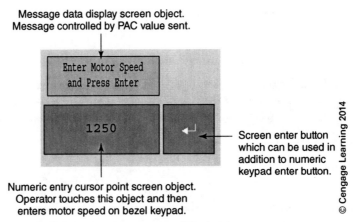

Figure 15-42 Features of mixing speed screen modified objects.

One issue you have to solve when modifying the ladder rungs is how the PAC will determine and react if the operator attempts to enter a faster motor speed than is acceptable. The acceptable values are from 0 up to 1,750 rpm. The operator will not be allowed to overspeed the mixer. If the operator enters a value that is invalid, the ControlLogix will send a tag back to the operator interface and display a "Speed Out of Range Reenter" message, as illustrated in Figure 15-43. Invalid mixing speeds will be ignored by the PAC and the last speed will be maintained.

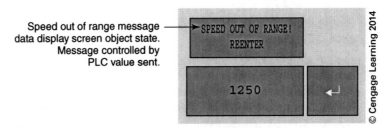

Figure 15-43 Speed out of range message screen object will be displayed upon invalid mixing speed entry. Note original valid speed value was not modified.

Each speed entry will be evaluated by the PAC logic to determine whether the value is valid. The PAC tag controlling the message display screen object contains a 0 under normal conditions. This tag is evaluated by the operator interface in order to display the first state of the message display (see Figure 15-44). If the operator enters an invalid mixing speed, the PAC will trigger logic to send a value of 1 to the Speed_Out_of_Range_Message tag. When the operator interface message display object sees a 1 in the Speed_Out_of_Range_Message tag, the message display shows the message associated with message 1. Keep in mind that the message display is triggered by a number, not a bit. The value 0 triggers the display of message text associated with 0, whereas a 1 in the same tag triggers the display of the message text associated with message 1 (see Figure 15-44). If we had another message, that text would be triggered by the number 2. As an example, the message display for an Allen-Bradley Standard PanelView can display up to 255 different messages, each message triggered by its specific integer value. We will create this logic as part of this lab.

Figure 15-44 Message display numeric triggers.
© Cengage Learning 2014

MOTOR SPEED DISPLAY SCREEN OBJECT MODIFICATION

A new screen object will be programmed on the operator interface screen to display the current motor speed, as illustrated in Figure 15-45. The PAC will receive the speed feedback from a variable frequency drive. We will not program logic for this new feature in this lesson. We reserve an introductory discussion of working with this information until later in this lab, and the programming of this information for a future lab exercise.

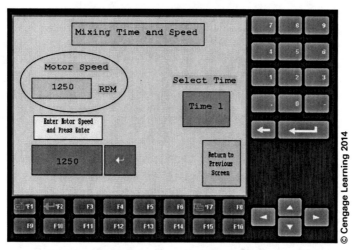

Figure 15-45 Motor speed feedback displayed on operator interface screen.

For your reference, the mixer motor speed in rpm will be displayed on a Numeric Data Display screen object as a whole number between 0 and 1,750. Figure 15-45 shows the data display screen object along with the associated screen text. There are three pieces of these new screen features on our operator interface screen. Besides the numerical data display object, there are two pieces of screen text. Screen text is simply text placed on the screen to help the operator understand how the screen object operates. See Figure 15-46 for features of the mixing motor speed display objects.

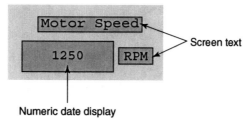

Figure 15-46 Motor speed display screen objects features.
© Cengage Learning 2014

The value to populate the numeric data display typically comes in from the drive on an Ethernet/IP network, to be stored in a PAC tag. The value coming from the drive is referred to as the drive feedback speed and in many cases is represented as a number from 0 to 32,767 in binary. The value 0 to 32,767 typically represents a motor speed of 0 to 1,750 rpm. The 0 to 32,767 value must be scaled in the PAC logic and sent on to the operator interface device via an Ethernet/IP network. Because we will not explore the scaling of this information until a later chapter, we will not incorporate the motor speed data transfer into this lab exercise.

SELECT MIXING TIME MODIFICATIONS

To save precious screen space on our operator interface terminal, we will improve the select time buttons from three buttons to a single button. The single Time button will be a multistate push button on our operator interface. Refer to Figure 15-47 for the new Time push button.

408 COMPARISON INSTRUCTIONS

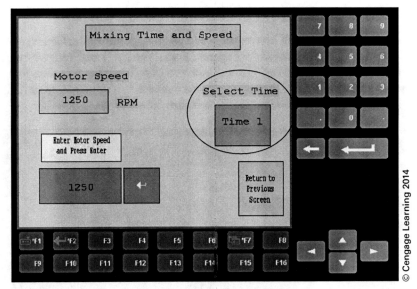

Figure 15-47 Modified Time entry operator interface screen object.

Each time the operator pushes the button, it goes to another state. In our application, the first state is Time 1. Push the button and the button state goes to state 2, or Time 2, and a third push of the button goes to state 3, or Time 3. The next button push will wrap around to state 1 again. Each push-button state sends a value into the PLC. Figure 15-48 illustrates each time state and its associated values sent to the PLC. Rather than having single XIC instruction associated with an input tag for each push button, as in the last lesson, now each press of the button sends a value to the PLC. We still use the three rungs and their associated MOV instructions to change the mixing time, but we modify the rungs and use comparison instructions to test which button state is active and which MOV instruction to execute.

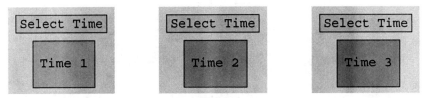

Figure 15-48 Time selection states and the associated values sent to the PAC program.
© Cengage Learning 2014

THE LAB

For this exercise, we modify Lab Exercise 6: Programming Using Operator Interface Tags, created in the ControlLogix Counters Instructions in Chapter 14.

1. _____ Open Lab Exercise 6: Programming Using Operator Interface Tags, the project you completed in Chapter 14.
2. _____ Modify the ladder logic to satisfy the new specifications.
3. _____ A motor speed of 0 to 1,750 rpm entered by the operator will be considered valid data. Even though a speed command outside of the specified range could be entered, the PAC will reject the invalid entry. What instruction should you program to ensure that the motor will never be commanded to overspeed?
    a. List the instruction's parameters. Include the parameter data to be programmed as you create the rung.

4. _____ Program the rung so that when the instruction from step 3 is true as the result of valid data entry, the tag named "Speed Command to Drive" will become true.
5. _____ If the operator enters a motor speed that is out of range, what instruction should you use to trap this action?
6. _____ Program the logic so when the instruction from step 5 is true, the Speed_Out_of_Range_Message tag goes true and triggers the speed out of range screen object on the operator interface device. If the rung is false, a 0 will be sent to the operator interface device to display the message "Enter Motor Speed and press Enter."
7. _____ Because the three Time buttons have been replaced with a single screen object button that will send the values listed below to the PAC, rather than a bit from each XIC instruction, how should you modify the rung with the MOV instructions?

Select Time 1, and a 0 is sent to the PLC.
Select Time 2, and a 1 is sent to the PLC.
Select Time 3, and a 2 is sent to the PLC.

8. _____ We will add logic for the motor speed display screen object in a future lesson.
9. _____ After making the modifications to your project, download the project and verify proper operation.
10. _____ If you have an operator interface device, monitor the tags coming into the ControlLogix from the operator interface.
11. _____ Try entering different motor speeds to verify proper integration between the operator interface and the ControlLogix.
12. _____ Test the Select Time multistate push buttons for proper integration to the ControlLogix tags and ladder logic.
13. _____ Ask your instructor whether your project completion needs to be checked off.
14. _____ Save your project.
15. _____ This completes this lab exercise.

LAB SUMMARY

In this lab, we learned more about how operator interface devices interface with PAC logic. Even if you did not have the opportunity to actually connect an operator interface device or drive to your ControlLogix, you should have a better understanding of the PAC programming associated with the peripheral devices.

CHAPTER 16

Data-Handling Instructions

OBJECTIVES

After completing this lesson, you should be able to

- Zero tags data using the Clear instruction.
- Clear array tags containing old data, using the File Fill instruction.
- Explain the difference between a Move instruction and a Copy instruction.
- Determine the data resulting after execution of a Masked Move instruction.
- Extract mixed data coming in from a network.
- Extract data using the Bit Field Distribute instruction.
- Program a Copy instruction to copy recipe data from one array to another.
- Program a Swap instruction to correct extracted data value errors.

INTRODUCTION

One of the main advantages of programmable controllers over relay control systems is the ability to easily program the PAC to meet the current control needs. As microprocessor power was incorporated into the PAC, adding instructions that allowed data-handling capabilities opened a whole new horizon of functionality. Major advances in PAC flexibility and control were now possible compared to earlier PAC relay replacers. Today's PACs have vast instruction sets, including some that allow the PAC to store huge amounts of information pertaining to the manufacturing process. For example, stored data could include numerous recipe sets, one set for each of the different products produced; logs of production data from many products or batches; and operator production data. Data-handling instructions allow stored data to be moved or copied from controller memory for use in ladder logic instructions for the process currently being controlled. Network input or output data can be systematically organized into a block for efficient transfer across the network. To be used in the PAC project, data that were blocked together may have to be extracted and values corrected before data can be used.

This chapter introduces the following RSLogix 5000 instructions:

- Clear
- Move
- Masked Move
- Bit Field Distribute
- Copy
- File Fill
- Swap

THE CLEAR INSTRUCTION

The Clear (CLR) instruction is used to clear the value of the destination tag to 0. The Clear instruction in Figure 16-1 currently contains the destination tag named Destination Data, which contains a value that will be zeroed when the instruction goes true. The destination data type can be any of the following: SINT, INT, DINT, and REAL. The 2# on the far left of the tag value signifies that the tag value or style is displayed in binary. The style can be modified in the tags collection to display the tag value in the desired radix.

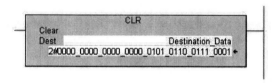

Figure 16-1 The Clear instruction.
Used with permission Rockwell Automation, Inc.

THE MOVE INSTRUCTION

The Move instruction is an output instruction that moves a copy of a value or a single tag to a specified destination. When the instruction is true, a copy of the information specified as the source is moved to the destination. The Move instruction only moves a copy of the source data. The term *move* is a bit deceiving, as it implies that data is physically moved from one location to another, which would mean that the original data would not remain in the original location. However, this is not the case with the Move instruction. This instruction moves a copy of the source data to the destination. It is important to understand that only a *copy* of the source data is moved to the destination and that the original data still remain in the source. The source and destination data type can be any of the following: SINT, INT, DINT, and REAL. The instruction parameters are listed here:

- **Source:** The source contains a single number, a constant, or the tag where the information that is to be moved to the destination is found.
- **Destination:** The destination is the tag where the Move instruction sends a copy of the information specified in the source.

Figure 16-2 illustrates a ladder rung; when true, the one-shot instruction triggers the Move instruction to execute one time. The Move instruction moves a copy of the data contained in the source, Tag A, to the destination, Tag B. Currently, Tag A contains the value 1,234 and the destination currently contains a 0. After executing the instruction, a copy of the Tag A data resides in Tag B.

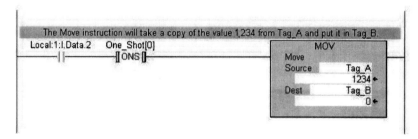

Figure 16-2 Ladder rung containing a Move instruction.
Used with permission Rockwell Automation, Inc.

Associated Status Bits

Upon completion of the execution of Math, Move, or Logical instructions, status bits are available to track the operation of the instruction by providing a status bit to the controller if there is an overflow condition, whether the result is negative, positive, or a 0, and whether instruction execution generated a carry or borrow. The Move instruction has the same status bits as math instructions. Move instruction status bits include an overflow bit, a 0 bit, and a sign bit. The user must program logic to interrogate the desired status bits immediately after the instruction executes, as these status bits are updated immediately after each Math, Move, or Logical instruction is executed. The four common status bits, sometimes called arithmetic flags, are defined as follows.

Overflow Bit (S:V)

The tag of the overflow bit is S:V and is set by the controller whenever the result of executing a Math, Move, or Logical instruction produces a destination result too large to fit in the destination tag. If the destination contains a value that does not overflow, the overflow bit is reset. As you know, a 16-bit integer can store a value up to 32,767. If the result of a move operation where the destination is an INT data type and the destination is greater than 32,767, an overflow has occurred, and the S:V bit will be set. This bit is information only and will not cause the controller to fault. Figure 16-3 shows a Move instruction that currently has a value of 50,000 in Tag A. The next rung is programmed to trap an overflow as the result of executing this instruction. Assuming the destination, Tag B, was an INT, an overflow would be generated when the move is executed.

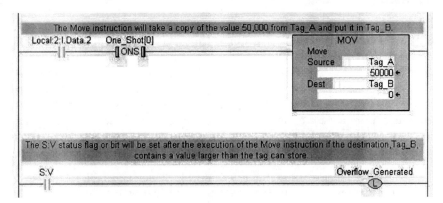

Figure 16-3 Logic to trap an overflow.
Used with permission Rockwell Automation, Inc.

Because the overflow status bit is only true for the instant the rung is executed, when an overflow occurs, a latch instruction is used to trap, or latch, that bit for our reference. Because there is only one set of math flags in the controller, the status of the bits is updated immediately after instruction execution.

0 Bit (S:Z)

The controller sets the 0 bit when the result of a Math, Move, or Logical instruction produces a 0 value in the destination. If the destination contains a value other than 0, the bit will be reset. Figure 16-4 shows the 0 status bit S:Z controlling the Destination_Is_Zero latch.

Negative (S:N)

If the result of a Math, Logical, or Move instruction produces a negative value in the destination, the sign bit (S:N) will be set. The sign bit will be reset if the result in the destination is a positive value. Figure 16-5 shows logic to trap a negative destination value as the result of instruction execution.

Carry (S:C)

The carry status bit (S:C) is set as the result of arithmetic operations where a carry or borrow was generated. An example is when a math operation on the lower word of a DINT is performed in which a carry is generated out of the most significant bit of the lower word and into the upper

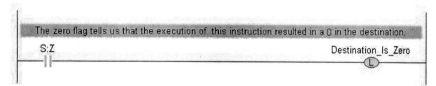

Figure 16-4 Logic to trap a 0 as the result of instruction execution.
Used with permission Rockwell Automation, Inc.

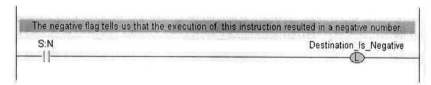

Figure 16-5 Logic to trap a negative destination value as the result of instruction execution.
Used with permission Rockwell Automation, Inc.

word. A borrow is the opposite, where a borrow is generated out of the upper word as the result of a lower word-based math operation. The carry status bit is typically not associated with Move-type instructions.

Example Using the Move Instruction

Each position of a three-position selector switch is used to select the current length specification of a product to be cut for a manufacturing process. The length parameter is stored as a constant in the source parameter of each of three Move instructions. Switch position 1 equals 6 inches (15¼ cm), position 2 equals 12 inches (30½ cm), and position 3 equals 24 inches (61 cm). Contacts for each selector switch position are separate inputs to a PLC input module. One ladder rung is programmed for each switch position. When a particular rung is true, the appropriate Move instruction on that rung sends a copy of the length information to the length tag for use in the program. Figure 16-6 shows the selector switch in position for the length of 12 inches (30½ cm). Note that currently all the Move instruction destination length tags contain 12.

THE MASKED MOVE (MVM) INSTRUCTION

Use the Masked Move instruction to select specific (typically, not all) bits from an INT or DINT source, and extract only those bits by moving them through a filter, called a mask, to the destination tag. The mask is the filtering device that determines, bit by bit, whether a copy of the bit will be allowed to move from the source to the destination. When using a DINT source and destination, the mask is comprised of 32 bits. An easy way to understand the mechanics of masking is to think of the mask as a series of doors, one for each bit. If the door is open, a copy of the source bit can move through the door and replace the current destination bit. If the door is closed, a copy of the bit is not allowed to move from the source to the destination. We say that the bit has been masked. When a bit is masked, the destination bit is undisturbed. Masks are typically represented in base 16, or hexadecimal, as hexadecimal is a shorthand method of entering mask values when programming.

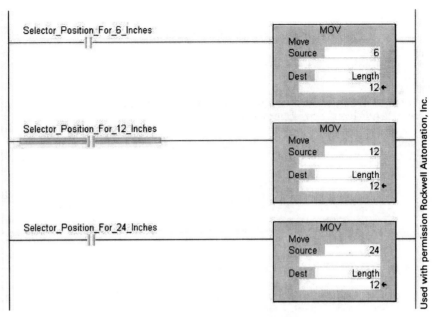

Figure 16-6 Move instruction providing variable length information.

Example Using the Masked Move Instruction

DeviceNet network has four nodes wishing to send the following data to the controller:

Node A = 123
Node B = 59
Node C = 248
Node D = 48

DeviceNet is a byte-based network, so the minimum data transfer from any node to the controller is at least a byte. A single byte can represent the decimal values 0 to 255. If numeric data to be transferred across the network is in the range of a byte, then in order to conserve memory, the data need not be stored in a larger data type. When configuring the network, the programmer has options as to the data size and where the data will be stored in the controller scoped tags collection. The determination as to where and how the network I/O data are stored in the controller is referred to as data mapping. Because ControlLogix is a 32-bit PAC, data are mapped into 32-bit tags that directly correspond to controller scoped I/O tags for the slot in which the DNB module resides. Data mapping is part of the network configuration that is set up in the network configuration software called RSNetWorx for DeviceNet.

One option is to store each piece of data in a separate tag, as illustrated in Figure 16-7. The advantage of storing each tag as a separate DINT is ease of access. When programming, select the I/O tag as normal for use in the logic. However, because controller memory is finite, in this example, 24 bits of each DINT are wasted.

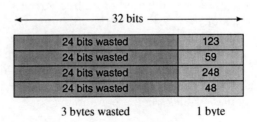

Figure 16-7 Each node's data is stored in a separate tag.
© Cengage Learning 2014

DeviceNet I/O Structure

A DeviceNet network can support up to 64 nodes. All network node data are allocated in an array of 124 input DINTS and 123 output DINTS that are assigned to the network I/O structure in the controller tags collection when the 1756-DNB module is added to the RSLogix 5000 project I/O configuration. Input base tags are grouped together into an array of 124 elements, and the output base tags are grouped together in an array of 123 elements. Figure 16-8 shows the basic tag structure for a DNB residing in slot 6 of the chassis. The array element numbers are in the square brackets.

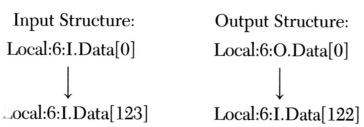

Figure 16-8 DeviceNet input and output structure.
© Cengage Learning 2014

Figure 16-9 shows Figure 16-7 with base tags assigned to each node's input data. Here four input tags, or DINTS, of the input structure have been consumed to store only four bytes of data.

Tag	Wasted	Value
Local:6:I.Data[2]	24 bits wasted	123
Local:6:I.Data[3]	24 bits wasted	59
Local:6:I.Data[4]	24 bits wasted	248
Local:6:I.Data[5]	24 bits wasted	48

Figure 16-9 Base tags associated with node input data.
© Cengage Learning 2014

Possibly, a better input data mapping option, to conserve controller memory and get more efficient data transfer by having less network traffic, is to consolidate the data of the four tags into a single DINT, as illustrated in Figure 16-10.

| Local:6:I.Data[2] | 48 | 248 | 59 | 123 |

Figure 16-10 Four nodes of data consolidated into a single DINT.
© Cengage Learning 2014

As you remember from earlier studies, data are stored in binary and as a 32-bit signed integer inside the controller. The controller does not know that there are four pieces of information represented in this tag and therefore interprets these 32 bits as the decimal value of 821,574,523. To make the data useable, we need to separate, or extract, the data, as the value of 821,574,523 is meaningless because there are four separate pieces of information contained in the tag. This data can be individually extracted using a Masked Move instruction. Figure 16-11 illustrates extracting Data A from the DINT and putting the data, 123 into its own tag called "DeviceNet Node 12 Data." From here, each additional piece of data can be extracted using a separate Masked Move instruction.

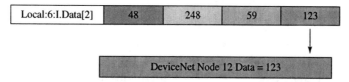

Figure 16-11 Data A from DeviceNet node 12 extracted to their own tag.
© Cengage Learning 2014

The Masked Move instruction used for the data extraction is displayed in Figure 16-12. Note the source tag is Local:6:I.Data[2], whereas the destination tag is DeviceNet Node 12 Data. The mask is the filtering device and in this case allows only the lower byte of the lower word to transfer from the source to the destination. We explore the mechanics of the mask next. Keep in mind as we investigate masking principles that inside the controller all tag data such as the instruction source and destination tags are stored as binary. Even though we view numbers in a decimal format, the PAC works only with bits.

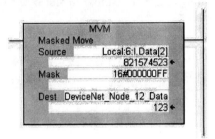

Figure 16-12 Masked Move instruction for Data A extraction.
Used with permission Rockwell Automation, Inc.

INTRODUCTION TO HEXADECIMAL MASKING

As mentioned earlier, an easy way to understand how a mask works is to think of the mask as a series of doors, one door for each bit in the source. Referring to Figure 16-13, the source bit is a 1 and the door is open, so one could pretend the bit has little legs and a copy of that bit could run through the door to the destination. The door represents the mask. When a copy of the source bit passes from the source to the destination, the source bit replaces whatever bit was in the destination. The original source bit is never affected.

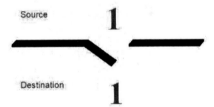

Figure 16-13 If the door is open, a copy of the source bit can pass through the door and replace the starting destination bit.
© Cengage Learning 2014

In Figure 16-14, the source bit is a 1 and the door is closed, so a copy of the source bit cannot get through the door and replace the destination bit. Again, the door represents the mask. Here we are going to say the bit has been masked out. When a bit is not allowed to pass from the source to the destination, the destination bit is not affected.

Figure 16-14 If the door is closed, a copy of the source bit cannot pass through the mask (door) and replace the original destination bit.
© Cengage Learning 2014

Mask Determination

To keep the example simple, rather than work with the normal 16 or 32 bits, let's work with just 4 bits and demonstrate how to determine the mask. Figure 16-15 shows 4 source bits, with all of the doors representing the mask as currently open. Assume the destination tag started out as

DATA-HANDLING INSTRUCTIONS 417

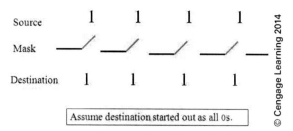

Figure 16-15 An open door representing the mask means a copy of the bit can pass.

all 0s. When the Masked Move instruction is executed, copies of the 4 bits pass through their respective doors and replace the original destination 0 bit.

To determine the mask for Figure 16-15, refer to the table in Figure 16-16 and assign a Yes representing each open door where the bit can pass. If the bit will not be allowed to pass, the door is closed. Assign a No in that bit position in the mask. For this example, all doors are open, so there will not be any No's assigned. Figure 16-16 is the same as Figure 16-15 except that the doors have been removed and replaced with the word Yes.

Source	1	1	1	1
Mask	Yes	Yes	Yes	Yes
Destination	1	1	1	1

1. To determine the mask, ask yourself, one bit at a time, is the door open or closed?
2. If the door is open, put a Yes in that mask position.
3. If the door is closed, put a No in that mask position.

Figure 16-16 Source bits are allowed to pass to destination.

When working with computers, we are working with binary, which only has two states, on or off, yes or no, represented by a 1 for on, or Yes, and a 0 for an off state, or No. As a result, a Yes will become a 1, whereas a No will become a 0. In Figure 16-17, we have replaced each Yes with a 1. Our mask is now 1111. Referring to the table in Figure 16-18, convert 1111 in binary to F in hexadecimal.

The table in Figure 16-18 can be used as an easy cross-reference for converting among decimal, binary, and hexadecimal when determining masks.

Source	1	1	1	1
Mask	1	1	1	1
Destination	1	1	1	1

1. Change each Yes to a 1.
2. Change each No to a 0.
3. The current mask will be 1111.
4. Refer to table and convert 111 binary to hexadecimal.
5. 1111 in binary equals F in hexadecimal.

Figure 16-17 Change each Yes to a 1.

Figure 16-19 shows 4 source bits, with only two of the doors representing the mask as currently open. Assume the destination tag started out as all 0s. When the Masked Move instruction is executed, only a copy of bits 0 and 3 will pass through their respective doors and replace the original destination 0 bit.

418 DATA-HANDLING INSTRUCTIONS

DECIMAL, BINARY, HEXADECIMAL CONVERSION TABLE		
Decimal	Binary	Hexadecimal
0	0000	0
1	0001	1
2	0010	2
3	0011	3
4	0100	4
5	0101	5
6	0110	6
7	0111	7
8	1000	8
9	1001	9
10	1010	A
11	1011	B
12	1100	C
13	1101	D
14	1110	E
15	1111	F

Figure 16-18 Cross-reference for converting among decimal, binary, and hexadecimal when determining masks.
© Cengage Learning 2014

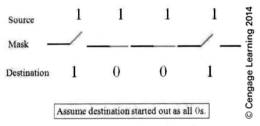

Figure 16-19 An open door representing the mask means a copy of the bit can pass, and a closed door means a copy of the bit cannot pass.

To determine the mask for Figure 16-19, refer to the table in Figure 16-20 and assign a Yes representing each open door. If a door is closed, assign a No in that bit position in the mask. For this example, two doors are open and two doors are closed. Figure 16-20 is the same as Figure 16-19 except that the doors have been removed and replaced with the word Yes or No.

Source	1	1	1	1
Mask	Yes	No	No	Yes
Destination	1	0	0	1

1. To determine the mask ask yourself one bit at a time is the door open or closed?
2. If the door is open, put a Yes in that bit position.
3. If the door is closed, Put a No in that bit position.

Figure 16-20 A Yes means a copy of the bit can pass, whereas a No means a copy of the bit cannot pass through to the destination.

A Yes becomes a 1, whereas a No becomes a 0. In Figure 16-21, we have replaced each Yes with a 1 and each No with a 0. Our mask is now 1001. Referring to the table in Figure 16-18, convert 1001 in binary to 9 in hexadecimal.

Source	1	1	1	1
Mask	1	0	0	1
Destination	1	0	0	1

1. Change each Yes to a 1.
2. Change each No to a 0.
3. Current mask will be 1001.
4. Refer to table to convert 1001 binary to hexadecimal.
5. 1001 in binary equals 9 in hexadecimal.

Figure 16-21 Change each Yes to a 1 and each No to a 0; then convert the binary value to hexadecimal.
Used with permission Rockwell Automation, Inc.

Masked Move Example 1

As an example, let's examine the Masked Move instruction in Figure 16-22. Assume the destination started as all 0s. When true, the Masked Move instruction moves a copy of the data contained in the source tag Data_To_Be_Masked and filter it through the mask 16#0000FFFF, and replace any destination bits that were allowed to pass into the Destination_Data tag. The figure shows the instruction after it has been executed.

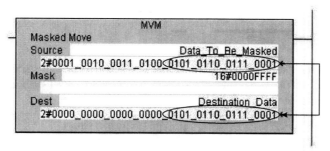

Figure 16-22 Masked Move instruction to evaluate for example 1.
Used with permission Rockwell Automation, Inc.

Analysis of Figure 16-22 Instruction Execution

Below we analyze a text representation of Figure 16-22. Again assume we are starting with the instruction's destination tag as all 0s. Note that where there are 1s in the mask, a copy of the source bit was allowed to move through the mask to the destination bit, and where there are 0s in the mask, the source bit was held back, or masked out. Remember to evaluate each source bit one at a time in order to evaluate how that specific source bit's mask bit will affect its behavior.

Source:	0001 0010 0011 0100 0101 0110 0111 0001
Mask: 0000 FFFF	0000 0000 0000 0000 1111 1111 1111 1111
Destination:	0000 0000 0000 0000 0101 0110 0111 0001

Because the mask contained 0000 0000 0000 0000 1111 1111 1111 1111, which equals (0000 FFFF), only the lower word of the 32-bit tag was allowed to pass through the mask from the source to the destination.

Masked Move Example 2

Let's assume the controller provides 16 bits of status information, but only bits 12, 13, 14, and 15 contain the information that we are currently interested in. Even though bits 0 through 11 contain information, this information is not pertinent at this time. Refer to Figure 16-23 to see how we can use a Masked Move instruction to extract the four bits of information we desire.

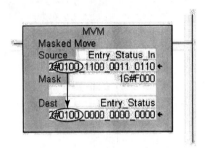

Figure 16-23 Extracting bits 12, 13, 14, and 15 using a Masked Move instruction.
Used with permission Rockwell Automation, Inc.

To understand how the instruction works, let's look at the source bit pattern, the mask bits, and determine what the destination tag will contain after instruction execution. Assume the destination tag began as all 0s.

Source:	0100 1100 0011 0110
Mask: F000	1111 0000 0000 0000
Destination:	0100 0000 0000 0000

Because the mask contained 1111 0000 0000 0000 (F000), only a copy of the upper nibble of the word was allowed to pass through the mask from the source to the destination.

Masked Move Example 3

As a brainteaser, we have a source tag and mask as listed following. Assume the destination tag is starting with all 0s. What bit pattern will the destination tag contain after executing this Masked Move instruction?

Source = 0001 0010 0011 0100 0101 0110 0111 0001

Mask = 2BA7 A1CF

Evaluate the Masked Move instruction. Remember to evaluate each source bit one at a time and to evaluate each mask bit as either an open or closed door.

Source:	0001 0010 0011 0100 0101 0110 0111 0001
Mask: 2BA7 A1CF	0010 1011 1010 0111 1010 0001 1100 1111
Destination:	????? ????? ???? ???? ???? ???? ???? ????

Instruction after execution:

Source:	0001 0010 0011 0100 0101 0110 0111 0001
Mask: 2BA7 A1CF	0010 1011 1010 0111 1010 0001 1100 1111
Destination:	0000 0010 0010 0100 0000 0000 0100 0001

Did you come up with the correct answer? Don't let evaluating the Masked Move instruction throw you. Simply evaluate one bit at a time against the mask bit to determine what the destination tag will contain after instruction execution.

Masked Move Example 4

The Rockwell Automation E3 Electronic Motor Overload is a very popular device found on many DeviceNet networks because of the wealth of information associated with motor operation. Two pieces of information, or parameters, available from the electronic overload are the time before the overload relay will trip (Time 2 Trip) and the overload relay's thermal capacity. Each parameter provides its current status as a value, which is stored as a word. Because each parameter consists of 16 bits, to conserve controller memory, these two parameters' 16-bit values could be combined into one DINT. The upside to doing this is that we have been efficient regarding memory allocation; however, two pieces of information have been combined in one 32-bit tag. Refer to Figure 16-24 and note that if each parameter contained the decimal value of 1, the tag Local:1.I.Data[3] value would equal 65,537. The value 65,537 contains unusable information as Time 2 Trip currently equals 1, and the Thermal Capacity value also contains the decimal value of 1. This is another example for the need to extract the two pieces of data and get them into a usable format.

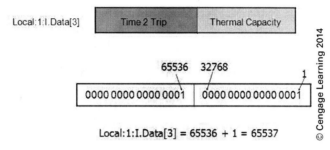

Figure 16-24 Combined E3 electronic overload parameter data is unusable in current form.

From the last section, you can see that we could use a Masked Move instruction to extract the Thermal Capacity. The destination of the Masked Move instruction would only contain the Thermal Capacity value. Figure 16-25 illustrates the Masked Move extraction operation.

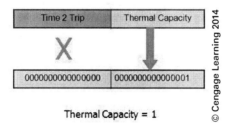

Figure 16-25 The Masked Move instruction could be used to extract the Thermal Capacity parameter and store the extracted data in its own tag.

Figure 16-26 shows the Masked Move instruction programmed to do the extraction of the Thermal Capacity parameter. The destination tag equals a 1, as it should.

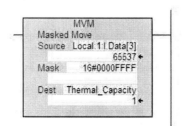

Figure 16-26 The Masked Move instruction for extracting Thermal Capacity information.
Used with permission Rockwell Automation, Inc.

422 DATA-HANDLING INSTRUCTIONS

The Time 2 Trip parameter extraction is going to be more challenging because the information is in the upper word of the DINT. Refer to Figure 16-27 to see that extracting the Time 2 Trip information using a Masked Move instruction results in a tag value of 65,536 rather than the actual data, which is a decimal value 1.

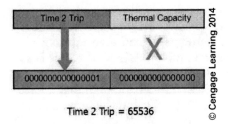

Figure 16-27 Extracting the Time 2 Trip using a Masked Move instruction results in a tag value of 65,536.

Figure 16-28 is the Masked Move instruction to extract the Time 2 Trip, as illustrated in Figure 16-27.

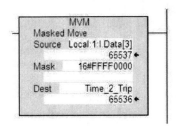

Figure 16-28 Masked Move instruction for extracting Time 2 Trip data.
Used with permission Rockwell Automation, Inc.

Additional programming would be required to shift the bits so the first bit of the tag is in the proper bit position to represent the correct value. Figure 16-29 shows the bit redistributing to represent the correct information. The Bit Field Distribute (BTD) instruction can be used to extract and shift the data in one operation. We will look at the BTD instruction in the next section.

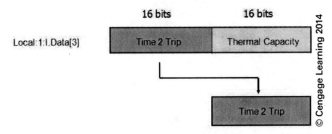

Figure 16-29 Upper word bits would need to be shifted, or redistributed into the proper position, so as to provide the correct data.

Masked Move Status Bits

The Masked Move instruction has math status bits similar to other Move and Logical instructions. Masked Move instruction status bits include a carry bit, an overflow bit, a 0 bit, and a sign bit. Here is a listing of the Masked Move instruction's status bits:

- The carry bit is not affected by the Masked Move instruction and is always reset.
- The overflow bit is not affected by the Masked Move instruction and is always reset.
- The 0 bit is set if the result is 0; otherwise, it is reset.
- The sign bit is set if the result is negative; otherwise, it is reset.

Mask Rules

1. The mask is typically either binary or a hexadecimal value.
2. Data are passed through the mask bit by bit. The mask bit in the same position as the source bit determines whether a copy of the data is to pass. To pass data through the mask, set the appropriate bit (setting a bit means making it equal 1). To mask data from passing from the source to the destination, reset the appropriate bit (resetting a bit means making it equal 0).
3. Destination bits that correspond to 0s in the mask are not changed.
4. Mask bits can be either a constant or the tag where that mask is found.

THE BIT FIELD DISTRIBUTE (BTD) INSTRUCTION

Use the Bit Field Distribute instruction when you need to not only extract data but also redistribute the consecutive bits into their proper position so the data is represented correctly. In the last section, we had a situation where we used a Masked Move instruction to extract the Time 2 Trip information but still needed additional programming to redistribute the resulting bits so as to correct the value. The Bit Field Distribute instruction can extract the desired data and shift, or redistribute, the bits to the correct location, as illustrated in Figure 16-30.

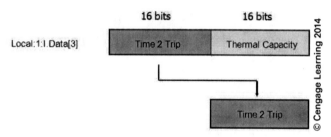

Figure 16-30 The Bit Field Distribute instruction can extract the Time 2 Trip data and shift the bits into the proper position.

Figure 16-31 shows a Bit Field Distribute instruction to accomplish the extraction, as illustrated in Figure 16-30. Next, we look into configuring the instruction.

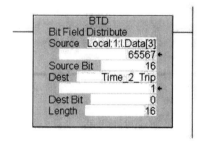

Figure 16-31 The Bit Field Distribute instruction extracted the Time 2 Trip data and redistributed the bits into the proper position.
Used with permission Rockwell Automation, Inc.

Programming the Bit Field Distribute Instruction

The instruction has the following parameters:

- **Source:** The tag containing the data to be operated on. For our example, select the tag Local:I:1[3] by clicking the drop-down area at the right of the data entry field, and select the tag from the drop down as normal.

- **Source Bit:** We are going to select a consecutive group of bits to extract. What is the bit number for the first bit to be extracted? Remember to start counting at the right and start with bit number 0. In our example, we are going to extract the upper word, or 16 bits. Bits 0 through 15 are included in the lower word of the DINT. Bits 16 through 31 comprise the upper word. In this example, we want to start with bit 16. See Figure 16-32. In Figure 16-31, the number 16 was entered into the source bit data entry field.
- **Destination:** Where is the instruction to store the result? The extracted and redistributed data for our example will be stored in the Time 2 Trip tag.
- **Destination Bit:** Where is the source bit to be redistributed to? For our example, starting bit 16 needs to be moved into bit position 0. See Figure 16-32.
- **Length:** How many bits is the instruction to work with? For our example, we are taking the upper word, or 16 bits.

Figure 16-32 shows a graphical representation of using the BTD instruction for our example.

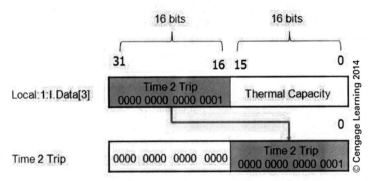

Figure 16-32 Graphical representation of execution of Bit Field Distribute instruction.

THE COPY INSTRUCTION

Use the Copy instruction to copy a consecutive array of tags from the specified source to the specified destination. The difference between a Move instruction and a Copy instruction is that although the Move instruction moves a copy of one tag to a new location, the Copy instruction copies an array of tags to the specified destination. If you had a recipe containing 20 values representing ingredients to make a certain product, one Copy instruction could copy all 20 values at the same time, whereas 20 Move instructions would be required. There are three parameters when programming a Copy instruction.

1. **Source:** The source is the first tag or element in the array to be copied.
2. **Destination:** Destination is the starting tag or element, where the number of elements specified in the length parameter will be copied.
3. **Length:** The length parameter is the number of array elements that are to be copied.

There are no associated status bits associated with the Copy instruction.

Applying the Copy Instruction

When working with multiple recipes to make different products such as different kinds of cookies, drink mixes, or different colored paints, there is typically a common area of controller memory in which the recipe for the product to be made will be stored after being copied from the collection of stored recipes. There could be over a hundred different colored paint recipes. Copy instructions could be used to copy the blue paint recipe to the common area of memory so that the PAC will make blue paint for the next batch, as demonstrated in Figure 16-33.

DATA-HANDLING INSTRUCTIONS   425

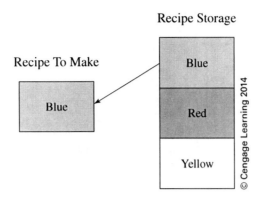

Figure 16-33 Copy paint color recipe for next batch.

Copy Instruction Example 1

Figure 16-34 has one Copy instruction to copy Recipe A to a destination array named Recipe To Make, and a second Copy instruction to copy a Recipe B to the same destination array. Note the length of the array to be copied is six elements, or values. As you remember from our study of arrays, the [0] at the end of a tag identifies that we are working with an array and current element 0 of that array. For this example, we are going to copy Recipe A, starting with element 0, for a length of six elements: Recipe A element 0 through Recipe A element 5. The one-shot instruction is included, as the recipe only needs to be copied once. Without the one shot, the recipe will copy every scan for which the rung is true.

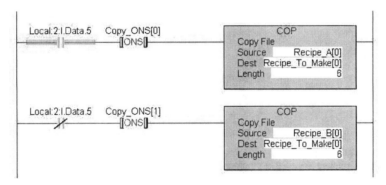

Figure 16-34 Rung to copy Recipe A.
Used with permission Rockwell Automation, Inc.

The six elements to be copied will be copied to the tag Recipe To Make, starting with element 0, also for a length of six elements. Figure 16-35 shows three arrays of tags: Recipe To Make, Recipe A, and Recipe B. Note Recipe A contains the values 0 to 5. Currently, the first rung of Figure 16-34 is true, and the values 0 through 5 will be copied only once to replace any old data in the Recipe To Make array, starting at element 0. Refer to the figure to see that the Recipe To Make array contains the same numbers as Recipe A.

Copy Instruction Example 2

For this example, we are going to copy Recipe B, starting with element 0 for a length of six elements: Recipe B element 0 through Recipe B element 5. The one-shot instruction is included, as the recipe only needs to be copied once. Figure 16-36 shows the second rung true, so the Copy instruction for Recipe B will execute.

The six elements to be copied from Recipe B will be copied to the tag Recipe To Make element 0 (Recipe_To_Make[0]) also for a length of six elements. Figure 16-37 shows three arrays of tags: Recipe to Make, Recipe A, and Recipe B. Note that Recipe B contains the values

426 DATA-HANDLING INSTRUCTIONS

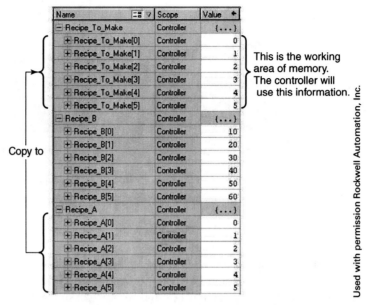

Figure 16-35 Arrays containing recipe data being copied to a working area of memory.

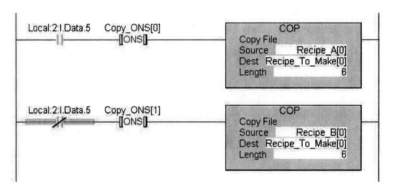

Figure 16-36 Logic to copy Recipe B.
Used with permission Rockwell Automation, Inc.

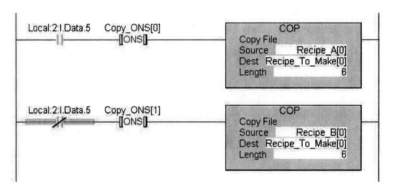

Figure 16-37 Recipe B copied to the Destination array.

10 to 60. When the second rung is true, as illustrated in Figure 16-36, the Copy instruction will be executed, and the values 0 through 60 will be copied to and replace any old data in the Recipe To Make array, starting at element 0.

One last note regarding reading and writing to array elements: Attempting to read from or write to an array element that does not exist causes the controller to fault. This is referred to as an array subscript error and results in a major fault on the controller. In our example, we have array elements 0 to 5. Many people run into trouble here because they assume we have array elements 1 to 6. Because element 6 does not exist, attempting to read from that element or write to that element results in an array subscript error. An array subscript error is a major recoverable fault.

THE FILL FILE (FLL) INSTRUCTION

When working with the Copy instruction, we populated arrays with recipe data. Another example of using an array to store data in consecutive locations might be the collection of daily production data for each process, operator, or work shift. Figure 16-38 could be an example of a portion of an array used to store production data. The array tag names can be aliased if so desired so that Monday First Shift could be an alias for Production Data[0], Monday Second Shift could be an alias for Production Data[1], and so on.

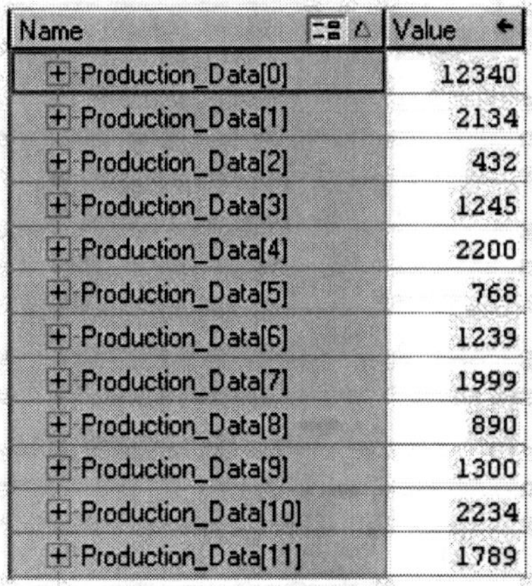

Name	Value
+ Production_Data[0]	12340
+ Production_Data[1]	2134
+ Production_Data[2]	432
+ Production_Data[3]	1245
+ Production_Data[4]	2200
+ Production_Data[5]	768
+ Production_Data[6]	1239
+ Production_Data[7]	1999
+ Production_Data[8]	890
+ Production_Data[9]	1300
+ Production_Data[10]	2234
+ Production_Data[11]	1789

Used with permission Rockwell Automation, Inc.

Figure 16-38 Example of using an array to store production data.

Production data are typically extracted from PAC memory on a regular basis via Ethernet and stored in a database on a server for management use. After the production data have been extracted from the PAC, each element of the array is typically zeroed out so a new group of data can begin to be collected. An easy way to clear out an array is to use the Fill File instruction, as illustrated in Figure 16-39. The source parameter of the instruction is either a constant or the tag where the value, typically a 0, is found to populate each element of the array. The destination parameter is the beginning element of the array that will be operated on by the Fill File instruction when the instruction goes true. In most cases, one would wish to execute the Fill File instruction only once when the instruction goes true, so a one-shot instruction would be part of the input logic. The length parameter defines how many elements in the array will be operated on. To avoid errors, the Source and Destination data types should always be the same data type. If the Source and Destination are dissimilar data types, the instruction may convert the source to match the Destination data type as part of instruction execution, which could lead to data errors. Refer to instruction help for additional information. The Source and Destination can operate on SINT, INT, DINT, and REAL data types or a structure. The length is a DINT. There are no arithmetic status flags associated with this instruction.

428 DATA-HANDLING INSTRUCTIONS

Figure 16-39 Fill File instruction to zero out 24 elements in the Production Data array.
Used with permission Rockwell Automation, Inc.

Figure 16-40 shows the same portion of the array as is shown in Figure 16-38. Let's assume the array contained 24 elements to be zeroed after the production data has been extracted. The Fill File instruction in Figure 16-39 could be used to zero out all elements in the array. Figure 16-40 could represent a portion of the array zeroed out as the result of executing the Fill File instruction.

Name	Value
⊞ Production_Data[0]	0
⊞ Production_Data[1]	0
⊞ Production_Data[2]	0
⊞ Production_Data[3]	0
⊞ Production_Data[4]	0
⊞ Production_Data[5]	0
⊞ Production_Data[6]	0
⊞ Production_Data[7]	0
⊞ Production_Data[8]	0
⊞ Production_Data[9]	0
⊞ Production_Data[10]	0
⊞ Production_Data[11]	0

Used with permission Rockwell Automation, Inc.

Figure 16-40 A portion of the Production Data array showing the elements being zeroed out.

THE SWAP BYTE INSTRUCTION

While working with data extraction, we looked at an example where we needed to extract the Time 2 Trip data from a DINT, using a Masked Move instruction, and then the need to shift the extracted bits into the proper bit value position in the resulting DINT so as to represent the correct value. As an easy way to accomplish the same operation with one instruction, we used the Bit Field Distribute instruction. Refer back to Figure 16-31. Keep in mind that when using the Masked Move instruction, you could extract whichever bits were needed, not necessarily consecutive bits. When extracting using the Bit Field Distribute instruction, a consecutive group of bits has to be operated on. In more advanced applications, there may be a time where the Bit Field Distribute instruction might not work, but there is still the need to shift bits into position after the extraction.

RSLogix 5000 software has a Swap Byte instruction (SWPB) that may be used in some data extraction situations. Figure 16-41 shows the Swap Byte instruction.

Programming the Swap BYTE Instruction

The Swap Byte instruction has three parameters:

1. **Source:** The tag of the information to be operated on. The instruction operates on INT, DINT, and REAL data types.
2. **Destination:** The tag for storing the result of the Swap operation.

DATA-HANDLING INSTRUCTIONS **429**

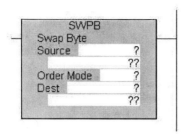

Figure 16-41 Swap Byte instruction.
Used with permission Rockwell Automation, Inc.

3. **Order Mode:** The Swap Byte instruction has three operating modes, referred to as the Order Mode. The three options include Word, High/Low, and Reverse. When programming, select the desired order mode from the drop down in the right area of the instructions data entry area. The modes operate as follows:

- **Word Mode:** Word mode swaps the upper and lower word of a DINT, as illustrated in Figure 16-42. The top DINT of the graphic shows the starting point, whereas the bottom DINT shows the result after executing the instruction.

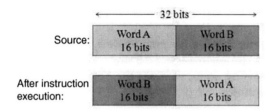

Figure 16-42 Word mode.
© Cengage Learning 2014

Figure 16-43 illustrates a Swap instruction using Word mode. For this example, the source tag upper word contains all 1s, whereas the lower word contains all 0s. Refer to Figure 16-43 to see the result. After executing the instruction, the source upper word contents will be swapped to the destination bottom word.

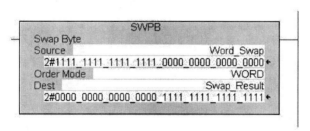

Figure 16-43 Result of Swap instruction swapping the upper and lower word between the source and destination.
Used with permission Rockwell Automation, Inc.

- **High/Low Mode:** High/Low mode swaps the upper and lower bytes in each word, as illustrated in Figure 16-44. The top DINT of the graphic shows the starting point, whereas the bottom DINT shows the result after executing the instruction.

Figure 16-45 illustrates a Swap instruction using High/Low mode. For this example, the source tag upper word and upper byte contains 1111 1111, and the upper word lower byte contains 0000 0000. The lower word upper byte contains 1010 1010, and the lower byte contains 0000 1111. After executing the instruction, the upper and lower bytes of each word are swapped.

430 DATA-HANDLING INSTRUCTIONS

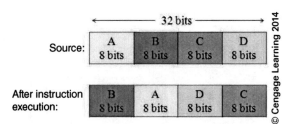

Figure 16-44 High/Low mode.

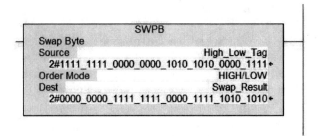

Figure 16-45 Swap instruction executing a High/Low swap.
Used with permission Rockwell Automation, Inc.

- **Reverse Mode:** Reverse mode reverses the order of the bytes contained in the DINT. Refer to Figure 16-46 to see where the top DINT of the graphic shows the starting point, whereas the bottom DINT shows the result after executing the instruction.

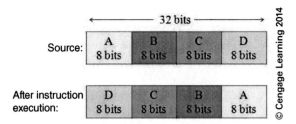

Figure 16-46 Reverse mode.

Figure 16-47 shows a Swap instruction after execution of a reverse swap.

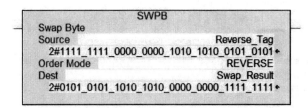

Figure 16-47 Swap instruction executing a reverse swap.
Used with permission Rockwell Automation, Inc.

Applying the Swap Byte Instruction, Using Word Mode

Next, we look at a possible application for the Swap Byte instruction in conjunction with a variable frequency drive on a DeviceNet network. By default, the drive provides two words of data to the controller across the network. The first word is drive status and the second word is the speed feedback. On many drives, the speed feedback is a binary 0000 0000 0000 0000 representing 0 Hz, or rpm up to 0111 1111 1111 1111 (32,767 decimal) representing 60 Hz, or 1,750 rpm. If we combine the two pieces of data in one DINT, to conserve controller memory we will need to extract each of the words in order to get each piece of information into separate tags. Because the drive speed feedback will be in the upper word, we could use a Masked Move for extraction of the data and the Swap Byte instruction, in Word mode, to swap the words so the value coming from the drive is correct. Figure 16-48 illustrates the Masked Move instruction for extracting the drive's speed feedback word and the Swap Byte instruction to swap the words so the data have the proper value. Granted, a Bit Field Distribute instruction would be a better choice in this situation. However, we will use this as an easy-to-understand example using the Swap instruction.

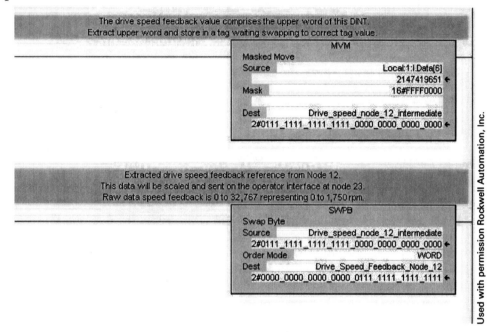

Figure 16-48 Word swap to swap drive speed reference from DeviceNet network.

Masked Move and Swap Byte Application Example

Let's expand on an earlier discussion regarding the Masked Move instruction and extracting the Time 2 Trip parameter associated with an E3 overload relay on a DeviceNet network. We learned that we could use a Masked Move instruction to extract the data contained in the Time 2 Trip parameter. However, because the data were in the upper word of a DINT, the data value was inaccurate information. The original Figure 16-30 is shown as Figure 16-49 and reviews the situation illustrating that if we did use the Masked Move instruction, additional programming would be required to shift the bits so the first bit of the tag would be in the proper bit position to represent the correct value.

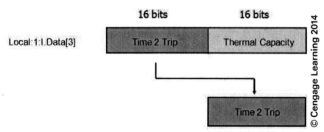

Figure 16-49 Original Figure 16-30 where the upper word bits need to be shifted, or redistributed into the proper position so as to provide the correct data.

432 DATA-HANDLING INSTRUCTIONS

The Masked Move instruction in Figure 16-50 could be used to extract the Time 2 Trip data to an intermediate storage tag named Intermediate Time 2 Trip. Using the Swap Byte instruction in Word mode, where the upper and lower words are swapped, the correct value is now in the Time 2 Trip tag.

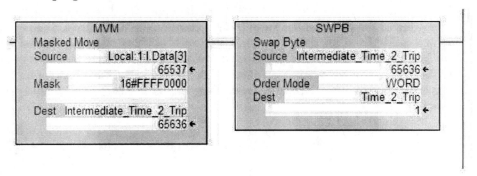

Figure 16-50 Using the Swap Byte instruction to swap the upper and lower words in the Intermediate Time 2 Trip tag to correct the incorrect value problem.
Used with permission Rockwell Automation, Inc.

LAB EXERCISE 1: DeviceNet Data Extraction Programming

INTRODUCTION

To achieve efficient memory management, multiple pieces of data from an E3 electronic overload relay data will share the same double word as mapped in RSNetWorx for DeviceNet. As a result, additional programming will be required to extract the data and place each piece of data in a separate tag. We will program the MVM, BTD, and SWPB instructions to extract and correct the data. To keep the lab simple, we will not configure a DeviceNet network. We will add the 1756-DNB module to our I/O configuration so that the network tags will be created in our project. Because we do not have an actual network providing the data, we will manually enter the data into the tags, simulating the data coming from the network.

THE LAB

1. _____ The E3 electronic overload provides the information from the DeviceNet network, as illustrated in Figure 16-51.

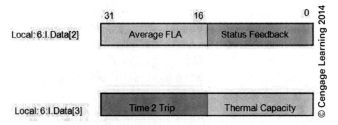

Figure 16-51 ControlLogix tags referenced for lab assuming 1756-DNB in chassis slot 6.

2. _____ Assume the following data are being stored in the following tags:
   Average FLA = 1 amp
   Status Feedback = 0010 binary (status data are typically binary)
   Time 2 Trip = 3 seconds
   Thermal Capacity = 4

3. _____ If you do not have a 1756-DNB module, leave chassis slot 6 empty. We are going to simulate a 1756-DNB in that slot for this lab. If you do have a 1756-DNB module to insert into your chassis, check the module label and record the firmware number.
4. _____ Open a new RSLogix 5000 project.
5. _____ Add the 1756-DNB into your I/O configuration for slot 6. If you are unable to configure the 1756-DNB module into slot 6, place it into an available chassis slot. Adjust the DeviceNet I/O address in the programming portion of the lab to match the actual slot number where the 1765-DNB is configured.
6. _____ Add the 1756-DNB module into the I/O configuration.
7. _____ As you begin the I/O configuration, the RSLogix 5000 software will ask for the module firmware level. Enter the major revision (firmware value) you recorded in step 3. If you do not have a module to use for the lab exercise, accept the default major revision (firmware level) provided by the software.
8. _____ The 1756-DNB New Module window, similar to that shown in Figure 16-52, should display.

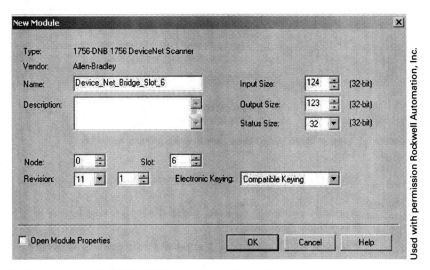

Figure 16-52 Configuring the 1756-DNB module for the lab.

9. _____ Assign the module whatever name you wish.
10. _____ Assign the 1756-DBB module a node address of 0.
11. _____ Fill in the slot number where the module resides or where you would put the module if you do not use a DeviceNet bridge module. The slot number is part of the input and output tag structure.
12. _____ Note, but do not change, the input size and output size values. This information specifies the number of input and output DINTs that are assigned to this module and the network. The input structure is an array of 124 DINTs, and the output structure is an array of 123 DINTs. The value contained in the square brackets of an input or output tag associates with the specific tag the programmer associated with each network device as he or she configured the DeviceNet network. The tag Local:6:I.Data[2] seen in Figure 16-51 identifies the module in slot 6 and the array element 2 as where the input data from the network are stored for our Average FLA and Status Feedback information for this exercise. Refer back to Figure 16-51 to make the association. The association of data coming into the network from each field device and where that specific data are stored in the ControlLogix tags are determined by the programmer when he or she configures the network. This association is referred to as mapping the data and is set up in the RSNetWorx for DeviceNet software.

434 DATA-HANDLING INSTRUCTIONS

13. _____ For this lab, we are not going to configure the rest of the DNB module, so uncheck the Open Module Properties box in the lower-left corner of the New Module window.
14. _____ Click OK to exit the configuration.
15. _____ Go to the controller scoped tags in your project and locate the input tags created.
16. _____ The lab specifications have us manually entering a 1 into the Average FLA to simulate data coming in from the network. Add the FLA simulated value in binary in the area identified in Figure 16-53. Note that the style has been changed to binary. Keep in mind that in a real-world situation, the information we are manually entering would be provided by the network.

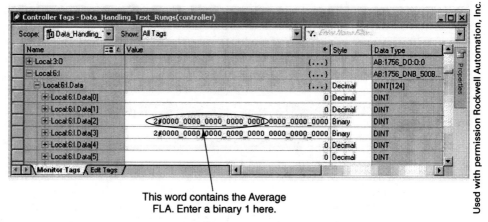

Figure 16-53 Upper input word representing FLA from network.

17. _____ Manually add the value of a binary 2 to the Status Feedback portion of the proper tag.
18. _____ Manually enter the value of 3 into the upper word representing Time 2 Trip, which is currently 3 seconds.
19. _____ Manually add the value of 4 into the lower word representing Thermal Capacity = 4.

PROGRAMMING LADDER LOGIC

Sample rungs are shown in Figure 16-54 for your reference.

20. _____ Program a rung of logic to use a MVM instruction to extract the Status Feedback value. Store the extracted value in a tag named Status Feedback.
21. _____ Use an input push button of your choice and a one shot to trigger the MVM instruction.
22. _____ Program a rung of ladder logic with a one shot and a BTD instruction to extract the Average FLA value, and store it in the lower word in a tag named Average FLA.
23. _____ Use an input push button of your choice and a one shot to trigger this BTD instruction.
24. _____ Program a rung of ladder logic with a one-shot instruction and a BTD instruction to extract the Thermal capacity value, and store it in a tag named Thermal Capacity.
25. _____ Use an input push button of your choice and a one shot to trigger this BTD instruction.
26. _____ Program a rung of logic with a one-shot instruction and a MVM instruction to extract the Time 2 Trip data.
27. _____ Use the SWPB instruction to shift the data to the proper position so that the value will be the actual current value of the tag, which is 4. Store the extracted value in a tag named Time 2 Trip.

28. _____ Use an input push button of your choice and a one shot to trigger this logic.
29. _____ Your basic rung structure should look similar to the rungs illustrated in Figure 16-54.

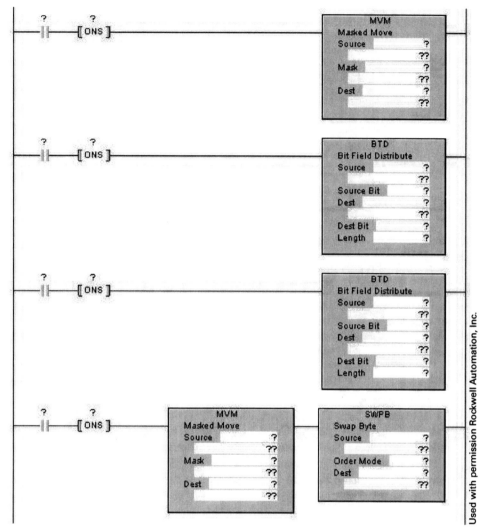

Figure 16-54 Sample rungs for lab exercise.

30. _____ As you program the required rungs, don't forget to document your rungs.
31. _____ When you have completed the project, download it and verify operation.
32. _____ When you have finished extracting and shifting the data, your tags should show the values illustrated in Figure 16-55. The tags style is decimal except for the Status Feedback, which is typically binary information.

Name	Scope	Value
+ Average_FLA	Controller	1
+ Status_Feedback	Controller	2#0000_0000_0000_0000_0000_0000_0000_0010
+ Thermal_Capacity	Controller	3
+ Time_2_Trip	Controller	4

Figure 16-55 Controller tags showing extracted data.
Used with permission Rockwell Automation, Inc.

LAB EXERCISE 2: Programming the Copy Instruction

One popular use of the Copy instruction is for setting up recipe data for different products just before they are manufactured. Let's assume we manufacture four different types of soft drinks: fruit punch, tropical punch, citrus punch, and orange. Each product has a different recipe, and each recipe contains data regarding the amount of each ingredient that is needed to produce the product. Figure 16-56 lists the ingredients, in gallons, in each drink recipe.

Ingredients	Fruit Punch	Tropical Punch	Citrus Punch	Orange
Water	100	120	130	150
Sweetener	25	25	25	20
Grape Flavor	8	9	0	0
Orange Flavor	0	25	75	90
Pineapple Flavor	10	25	30	0
Apple Flavor	2	8	0	0
Pear Flavor	1	0	1	0
Strawberry Flavor	8	0	0	0
Passion Fruit Flavor	0	10	0	0

Figure 16-56 Recipes for our drink manufacturing process.
© Cengage Learning 2014

Each list of ingredients for each drink product is its recipe. Because each amount is a numeric value, we store these values as DINTS in an array. Referring to the table in Figure 16-56, create a one-dimensional array to store each recipe. Don't forget to create the destination array. Name the destination whatever you wish.

The operator selects which recipe will be manufactured by touching an operator interface touchscreen object. We can simulate the touchscreen objects using push-button inputs on our lab trainer.

Because we have four recipes, we need to program four Copy instructions. Use ONS instructions to ensure that the Copy instruction is executed only once as the result of the input switch going from false to true. If making fruit punch for 12 hours, the recipe must be copied only one time, not at every scan. When you have completed creating your logic, download and test the project. Make sure the Copy instructions copy the correct data.

SUMMARY

This chapter introduced some of the common data-handling instructions. The instructions covered in this chapter included the following:

- The Clear (CLR) instruction to clear out a single tag
- The Move (MOV) instruction to copy a single tag
- A Masked Move (MVM) instruction to move selected bits through a hexadecimal mask to the destination
- An instruction for extracting a consecutive group of bits and redistributing them, the Bit Field Distribute instruction (BTD)
- The Copy (COP) instruction for copying a consecutive array of elements
- The File Fill (FLL) instruction to clear out a consecutive array of elements
- The Swap BYTE (SWPB) instruction to allow for rearranging words or bytes within a tag

Today's PACs have extensive instruction sets that include instructions that allow the PAC to store vast amounts of information pertaining to the manufacturing process.

Data-handling instructions allow stored data to be moved or copied from processor memory into the currently executing PAC ladder program. In one of our lab exercises, we stored data in a single-dimensional array, which could simulate numerous recipes for different products to be produced. The flavored drink lab exercise illustrates recipe data storage and using the Copy instruction to copy the current recipe to a common area of PAC memory for use.

Data extraction instructions such as the Masked Move and Bit Field Distribute are used to extract network input or output data when those data have to be consolidated into a single DINT for efficient memory. A Swap instruction may be necessary to swap words or bytes to correct for data consolidation value errors due to the physical position of the information within a DINT.

The Clear instruction is used to clear, or zero out, a single tag, whereas the Fill File instruction is typically used to zero out a consecutive array of values.

We learned that all information used with computers like PACs always begins with bit, word, DINT, or element 0. For example, an array that consists of 100 elements includes addressable elements 0 through 99. Reading or writing to a nonexistent array element results in a major controller fault.

REVIEW QUESTIONS

Note: For ease of handing in assignments, students are to answer using their own paper.

1. The Move instruction does which of the following?
   a. Moves a copy of data from source tag A to source tag B
   b. Moves a copy of data from the source tag to the designated destination
   c. Moves a copy of data from source B to source A
   d. Moves the data in source B into source A
   e. Any of the above depending on how the instruction is programmed
2. The _____ instruction zeros out the tag specified in the destination parameter of the instruction.
3. To extract consecutive bits in a tag and redistribute them in the destination tag, use a single instruction program, the _____ instruction.
4. When the Move instruction is true, a _____ of information specified as the source is moved to the tag specified as the Destination.
5. What will the destination contain after the execution of the following instruction? The destination starts out with all 0s, as shown in Figure 16-57.

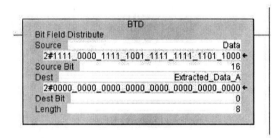

Figure 16-57 BTD instruction for question 5.
Used with permission Rockwell Automation, Inc.

6. The _____ instruction can be used to clear out a consecutive array of values.
7. The _____ instruction moves or copies one tag from the source to the destination.
8. Explain what is contained in the Source tag after executing a Move instruction.
9. The _____ instruction is an output instruction that moves a copy of one tag to a specified destination through a mask that filters out bits that are not to be transferred from the source to the destination.

10. What will the Destination tag contain after execution of the instruction in Figure 16-58? The instruction is shown before it is executed.

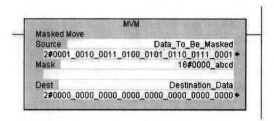

Figure 16-58 MVM instruction interpretation for question 10.
Used with permission Rockwell Automation, Inc.

11. Use the _____ instruction to copy the contents of a consecutive array of elements to another location.
12. Complete the rules of masking:
    a. The mask is typically either a(n) _____ or _____ value.
    b. Data are passed through the mask _____. The mask bit in the same position as the source bit determines whether the data are to pass. To _____ data through the mask involves setting the appropriate mask bit. Setting a bit means making the bit equal to a(n) _____. To _____ data from passing from the source to the destination, you reset the appropriate bit. To _____ a bit means to make the bit a 0.
    c. Destination bits that correspond to _____ in the mask are not changed. They remain as they were.
    d. Mask bits can be either a(n) _____ or the _____ where that mask is found.
13. A mask bit of _____ allows a copy of the bit to pass from the source to the destination.
14. What will the destination tag contain after execution of the instruction in Figure 16-59? The instruction is shown before it is executed.

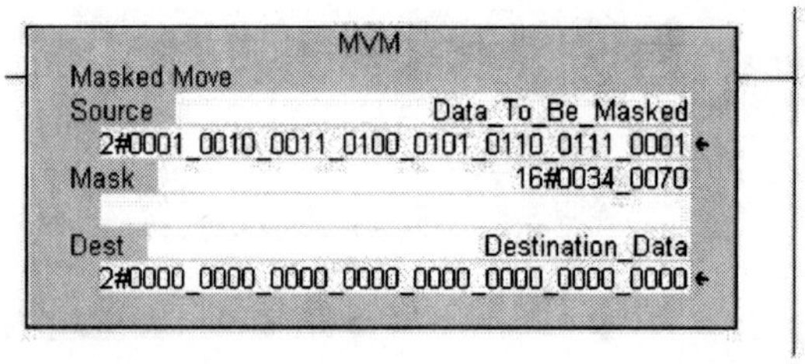

Figure 16-59 MVM instruction interpretation for question 14.
Used with permission Rockwell Automation, Inc.

15. The mask 000F allows which bits to pass?
    a. Upper nibble
    b. Lower nibble
    c. Upper byte
    d. Lower byte
    e. Bits 1 through 8
    f. Bits 8 through 15

16. List and define the three order modes for the SWPB instruction.
17. What will the Destination tag contain after execution of the instruction in Figure 16-60? The instruction is shown before it is executed.

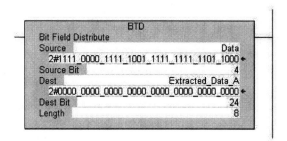

Figure 16-60 MVM instruction interpretation for question 17.
Used with permission Rockwell Automation, Inc.

18. What will the Destination tags contain after execution of the instructions in Figure 16-61? The instructions are shown before execution.

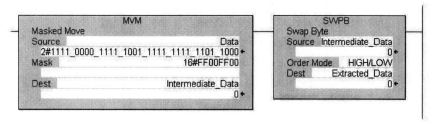

Figure 16-61 MVM and SWPB instruction interpretation for question 18.
Used with permission Rockwell Automation, Inc.

19. The mask 0FFF allows which bits to pass?
    a. Upper nibble
    b. Lower nibble
    c. Upper 8-bit word
    d. Bits 4 through 11
    e. Bits 0 through 11
    f. Bits 1 through 11
20. The Copy destination parameter is the starting element where the elements specified in the _____ parameter will be copied.

CHAPTER

17

Introduction to the Get System Values (GSV) and Set System Values (SSV) Instructions

OBJECTIVES

After completing this lesson, you should be able to

- Identify the GSV and SSV instructions.
- Explain usage of the GSV and SSV instructions.
- Interpret GSV and SSV instructions.
- Program ladder logic incorporating the GSV and SSV instructions.
- Create a fault routine and use GSV and SSV instructions in a fault routine.
- Introduce the user-defined data type.

INTRODUCTION

Programmable logic controllers PLCs like the Rockwell Automation PLC 5 and SLC 500 have a separate data file named the Status File (S2) for the storage of controller status information. For a programmer, the status file may be used as the project is developed for configuring certain advanced programming instructions. After the project is complete, unless modification of those instructions is necessary, access to the status file for these programming features is rarely done. If the machine is running without faulting, many times status file information is not required as a part of everyday operations. As a result, the controller scan time used to update rarely used information wastes valuable scan time, thus decreasing PLC throughput.

To help increase efficiency, ControlLogix does not have a status file to update. If a user requires information that traditionally would have been stored in a status file, a Get System Value, or GSV, instruction is programmed to get the desired information from the controller. A user who wishes to send the controller information programs a Set System Value, or SSV, instruction. Both of these instructions are outputs and execute each scan that the rung remains true. Consideration as to how often the instruction is executed is important because execution of these instructions each and every scan is probably an inefficient use of scan time. Programming input logic that includes a push button, a one-shot instruction, or a timer to trigger instruction execution only when the information is needed is a more efficient use of scan time.

This chapter introduces the GSV and SSV instructions. Lab exercises interpreting the GSV and SSV instructions provide understanding as to how the instructions work. Programming exercises provide an opportunity to create ladder rungs using these instructions. To complete the exercises, you need to access GSV and SSV information either from the RSLogix 5000 software instruction set help screens or a hard copy or an electronic copy of the instruction set reference manual, the last of which is downloadable from the Rockwell Automation Web site (ab.com).

Select Literature Library under the Quick Links heading. Search for 1756-RM, which lists all RSLogix 5000 reference manuals. Select the RSLogix 5000 Controllers General Instructions Reference Manual, and download.

THE GSV AND SSV INSTRUCTIONS

The GSV instruction is a ladder logic output instruction that, when it becomes true, gets the requested information from the controller and places it in the specified destination. Figure 17-1 shows the RSLogix 5000 GSV instruction.

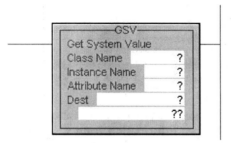

Figure 17-1 RSLogix 5000 GSV Instruction.
Used with permission Rockwell Automation, Inc.

The SSV instruction is also a ladder logic output instruction; when it goes true, it sends information from the source tag or tag structure to the controller in order to update the desired system parameter(s). Figure 17-2 shows the RSLogix 5000 SSV instruction.

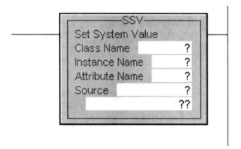

Figure 17-2 RSLogix 5000 SSV instruction.
Used with permission Rockwell Automation, Inc.

When you work with these instructions, the instructions documentation must be reviewed to ensure understanding of operation and proper programming. The documentation specifies whether the instruction source or destination parameter is a tag, an array, or a user-defined data type. When using either instruction, the programmer must make certain the source or destination data type matches the size and format required for proper execution of the instruction. For example, when using the GSV instruction, if instruction documentation specifies that an array of seven DINTs will be provided, the destination parameter must be programmed as specified, even though only a portion of the provided information is desired. The user can select the desired tag or tags from the destination after it is populated. For instance, let's say the programmer wishes to incorporate the GSV instruction's local date/time information in the project. The date and time information is referred to in the documentation as the attribute. The attribute provides an array of seven DINTs ranging from the year to microseconds. Let's assume the programmer did not need the microsecond value. Figure 17-3 shows this scenario where the local date time attribute is getting seven DINTs from the controller. However, the destination, which is an array, does not include an element to store the microsecond's value. Because the size of the source and destination do not match, the instruction will not execute and a minor fault will be logged.

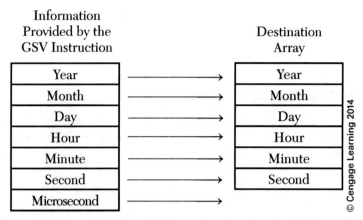

Figure 17-3 GSV instruction providing more information than the destination tag will accept.

The SSV instruction works in a similar manner. Rather than getting information from the controller, the SSV sends information from a tag or tag structure to the controller. Like the GSV, the SSV instruction source tag must match the size and format of the data the controller expects.

GSV and SSV Objects

Because ControlLogix does not have a status file like the PLC 5 or SLC 500 PLCs, controller system data is stored in objects. GSV and SSV instructions get data and set data that is stored in these objects. When programming either instruction, the object name is selected from the drop down associated with the instruction's Class Name, as illustrated in Figure 17-4.

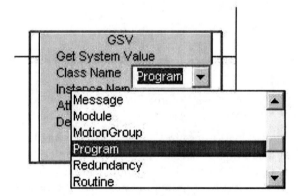

Figure 17-4 Selecting GSV Program Class Name, or object.
Used with permission Rockwell Automation, Inc.

If you refer to GSV and SSV Instruction Set Manual pages, there is a page similar to a table of contents entitled GSV/SSV Objects near the front of the chapter listing the available objects. RSLogix 5000 instruction set help also displays a list of the GSV/SSV objects that you can click to obtain additional information. A few of the currently available GSV and SSV objects (Class Name selections) for RSLogix 5000 software are listed in Figure 17-5. Refer to the instruction set documentation for the complete list.

The available objects are related to software revision and hardware firmware level. As with other software and hardware features, the GSV and SSV available objects are likely to expand as the RSLogix 5000 software versions progress.

INTRODUCTION TO THE GET SYSTEM VALUES (GSV) AND SET SYSTEM VALUES (SSV) INSTRUCTIONS

GSV AND SSV OBJECTS	
Object Name	**Use this Object to**
Axis	Obtain servo module axis status information
Controller	Get status information about the controller's execution
ControllerDevice	Identify controller physical hardware
DF1	Monitor controller's serial port driver status
FaultLog	Obtain current controller fault status information
Message	View status information associated with a specific message instruction
Module	Obtain status information from an I/O module or communication module in the chassis
Program	View or modify the current status about a specific program
Routine	View or modify the current status of a specific routine
Serial Port	Obtain serial port configuration and status information
Task	View or modify the current status about a specific task
Wall Clock Time	Time stamping

Figure 17-5 Description of selected GSV and SSV Class Name selections.
© Cengage Learning 2014

Instance Name

For example, Figure 17-6 is a GSV instruction showing the drop down Instance Name options available for the program class. In this example, we use the GSV instruction to get some information regarding the main program. The number of Instance Name selections depends upon the specific class name selected when programming the instruction.

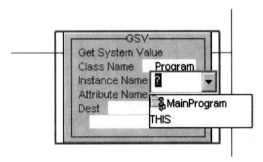

Figure 17-6 Program class instance selections.
Used with permission Rockwell Automation, Inc.

Examples for the Instance parameter could include the following:

- Task name
- Program name
- Routine name
- I/O module name
- The word *This* could be entered into this parameter to get information pertaining to the current task, program, or routine.
- This parameter could be left blank.

Keep in mind that the options for the Instance parameter are specific to each class name. The list above shows some of the possibilities. Refer to instruction documentation for the available options for the specific class you are working with.

Attribute Name

The attribute is the specific information you are looking to send to or receive from the controller. Figure 17-7 illustrates the last scan time attribute being selected from the drop down while programming the GSV instruction. Here we want to get the last scan time for the main program.

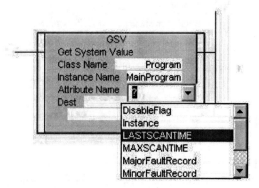

Figure 17-7 Selecting LASTSCANTIME attribute.
Used with permission Rockwell Automation, Inc.

Finding GSV Instruction Information from RSLogix 5000 Help

Let's do a little research on the GSV instruction and the LASTSCANTIME attribute, using the RSLogix 5000 help screens. An electronic or hard copy of the GSV and SSV from the RSLogix 5000 Instruction Set Reference manual, either supplied by your instructor or downloaded from the Rockwell Automation publications library from the Internet, could also be used.

To find the instruction information from RSLogix 5000 help, follow the steps outlined here:

1. _____ In the RSLogix 5000 software, click Help on the far right of the RSLogix 5000 menu bar.
2. _____ From the drop down, select Instruction Help.
3. _____ One easy way to find help on the GSV instruction is to click Alphabetical Listing underlined text near the top of the right pane, as shown in Figure 17-8. The figure is from RSLogix 5000 software version 19. If you have a different software version, the display may look a bit different.

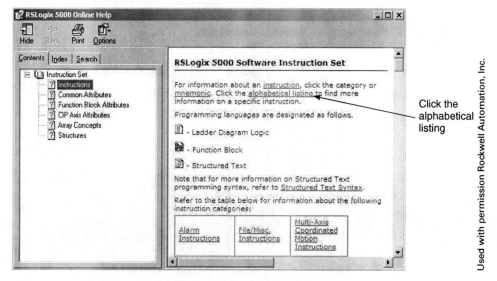

Figure 17-8 RSLogix 5000 version 19 Software Instruction Set help.

4. _____ The next window has buttons listed as A–Z near the top of the screen. Click G to select instructions starting with the letter G.
5. _____ The next view should display a list of the instruction three-character abbreviations, called mnemonics. Click GSV.
6. _____ You should see the GSV and SSV instruction help information, as displayed in Figure 17-9.

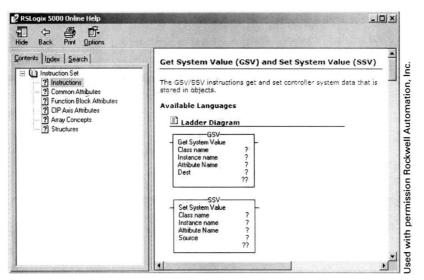

Figure 17-9 GSV and SSV help.

7. _____ Scroll to the very bottom of the window to find the related topics section.
8. _____ Click the GSV/SSV objects' underlined text.
9. _____ Figure 17-10 shows how the GSV/SSV Objects window should display.

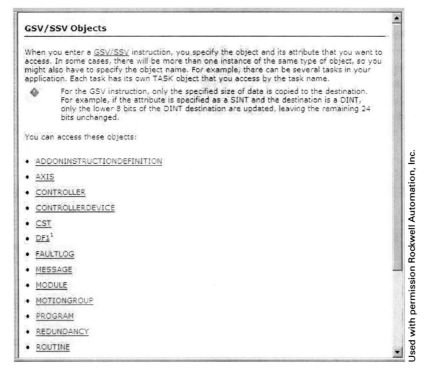

Figure 17-10 GSV and SSV Objects listing.

10. _____ The available objects, or class name selections available for your particular software revision, should be listed.
11. _____ Select the class name you need help for. To complete the table shown in Figure 17-11, click the word Program.
12. _____ The program object and its associated attributes should be listed for your review.
13. _____ The information needed to complete the table in Figure 17-11 should be displayed. Note the following:

   - The attribute name
   - Data type for the source or destination
   - Identification as to whether the instructions are used in a standard task (standard ControlLogix controller) or in a safety task when using ControlLogix as a safety PAC, also known as GuardLogix.

14. _____ Fill in the information in Figure 17-11. Filling in the needed information in the table provides you hands-on experience in digging out information required to program the instruction either from the help screens or from the Instruction Set Reference manual.

Attribute	Data Type	Instruction in a Standard Task	Instruction within a Safety Task	Description
LastScanTime				

Figure 17-11 Fill in the information for the last scan time attribute definition from instruction set help or the Instruction Set Reference manual.
© Cengage Learning 2014

Source or Destination

The next parameter to program for a GSV instruction is the destination for the data the controller provides or the source of the data the SSV instruction sends to the controller. The source or destination data types are either a simple tag, array, or user-defined data type, which is defined by instruction documentation. When an SSV instruction is programmed, the source of the information to be sent to the controller is created by the programmer to match the size and layout as described in the instruction documentation. When a GSV instruction is programmed, the destination of the information provided is created to match the size and layout as described in the instruction documentation. Figure 17-12 is the completed GSV instruction we have been using as our example. The GSV instruction provides the last scan time for the main program this instruction is programmed in. The destination of the information is the tag Main_Program_Scan_Time, which is a DINT.

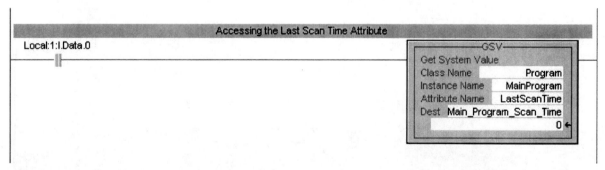

Figure 17-12 A GSV instruction programmed to get the last scan time of the main program.
Used with permission Rockwell Automation, Inc.

INTRODUCTION TO THE GET SYSTEM VALUES (GSV) AND SET SYSTEM VALUES (SSV) INSTRUCTIONS

INTERPRETING GSV AND SSV INSTRUCTIONS

The following exercise provides practice interpreting GSV and SSV instructions from the figures in the exercise. We will not download the project file at this time. Refer to your GSV and SSV documentation or the help screens to answer the following questions.

Figure 17-13 shows the project Controller Organizer for your reference.

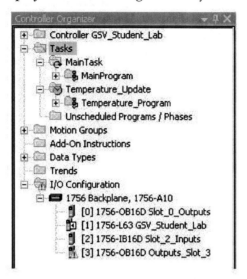

Figure 17-13 Controller Organizer for the project in this lab exercise.
Used with permission Rockwell Automation, Inc.

1. _____ Refer to Figure 17-14 as you answer the following questions.
   a. _____ The first rung in the figure contains a TON instruction. Explain why we would have this instruction here and why its done bit is used to trigger the two GSV instructions.
   b. _____ The top GSV instruction's class is Program. The instance is MainProgram. Explain what the instance specifies.
   c. _____ What information does the attribute tell us?
   d. _____ What data type does the destination tag need to be?
   e. _____ What time value does the data represented in the destination tag have?
   f. _____ What is the difference between the information requested by the top and bottom GSV instructions?

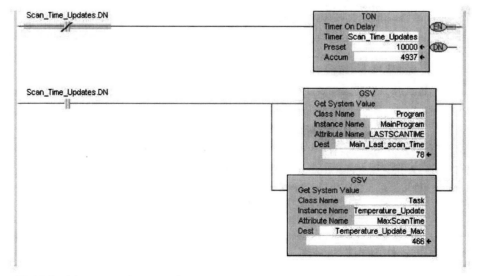

Figure 17-14 Ladder rungs for question 1.
Used with permission Rockwell Automation, Inc.

2. _____ The following questions refer to the ladder rungs in Figure 17-15.
   a. _____ Explain what information this GSV instruction will provide.
   b. _____ Using your instruction documentation, fill in the table in Figure 17-16 with the information provided by the GSV instruction for the LED_Status attribute.
   c. _____ Explain what the EQU instruction tests for on the second rung.
   d. _____ What does the EQU instruction test for on the third rung?
   e. _____ The fourth rung's EQU instruction tests for what condition?
   f. _____ Looking at the ladder logic, what is the current status of our PAC and the I/O modules?

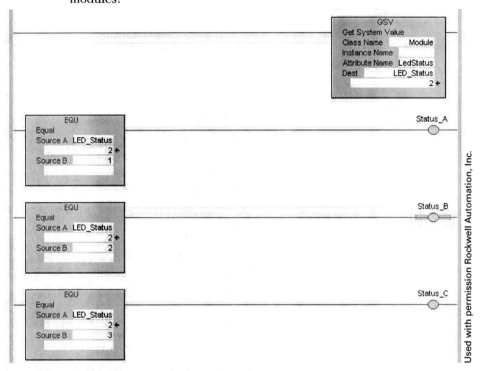

Figure 17-15 GSV instruction accessing the LED_Status attribute.

	LED STATUS ATTRIBUTE DESCRIPTION	
Value	I/O Status Indicator State	Description

Figure 17-16 Answer to question 2b.
© Cengage Learning 2014

3. The following questions refer to the ladder rungs in Figure 17-17.
   a. _____ The Instance Name is Temperature_Update. Explain what this refers to.
   b. _____ Refer to your documentation and describe what the function of this SSV instruction is.
   c. _____ What is the data type of the source?
   d. _____ Define the meaning of the source value sent by the SSV instruction to the controller.

INTRODUCTION TO THE GET SYSTEM VALUES (GSV) AND SET SYSTEM VALUES (SSV) INSTRUCTIONS

Figure 17-17 Interpreting the Output attribute of the Disable Update.
Used with permission Rockwell Automation, Inc.

    e. _____ Explain the function of the OTE instruction with the tag Output_Processing.0.

    f. _____ Explain the purpose of the second rung in Figure 17-17.

LAB EXERCISE 1: Programming a GSV Instruction

This lab steps you through the programming of a GSV instruction to get the last scan time of the Main Program, as illustrated in Figure 17-12.

LAB OVERVIEW

The RSLogix 5000 instruction set help provides general information; or refer to the GSV/SSV chapter for information from the RSLogix 5000 General Instructions Reference Manual for more detailed information on programming this instruction.

Steps for programming a GSV instruction to get the Main Program's last scan time are listed below.

1. _____ Open the Begin project saved at the end of the I/O configuration labs.
2. _____ Add the Temperature_Update periodic task, as in Figure 17-13.
3. _____ Refer to Figure 17-14 and program the timer logic so we can sample the scan time information every 10 seconds.
4. _____ Select the GSV instruction from the I/O tab on the Language Element toolbar.
5. _____ To program the Class Name, double-click the question mark to the right of the Class Name text.
6. _____ Click the down arrow.
7. _____ Because we are working with the Program Object, double-click Program from the list. See Figure 17-18.

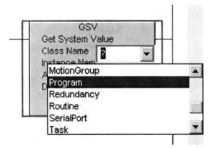

Figure 17-18 Select the Program object from the drop down.
Used with permission Rockwell Automation, Inc.

8. _____ To select the Instance Name, double-click the question mark to the right of the text.
9. _____ Click the down arrow.
10. _____ Because we are looking for the last scan time of the MainProgram, double-click the MainProgram to select it from the list, as shown in Figure 17-19.

450 INTRODUCTION TO THE GET SYSTEM VALUES (GSV) AND SET SYSTEM VALUES (SSV) INSTRUCTIONS

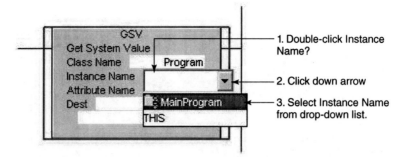

Figure 17-19 Select MainProgram as the instance parameter.
Used with permission Rockwell Automation, Inc.

Note another selection is the word THIS. If the attribute being programmed for the GSV or SSV instruction pertains to the specific task, program, or routine the instruction is programmed in, the THIS selection can also be used. Here, we are looking for the last scan time of this program (the MainProgram), so either selection for this parameter would be acceptable.

11. _____ Refer to Figure 17-20 as you double-click the Attribute Name parameter question mark.

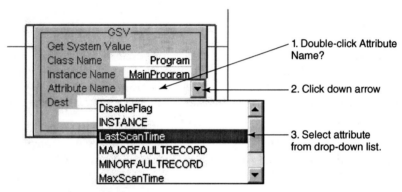

Figure 17-20 Selecting the LastScanTime attribute.
Used with permission Rockwell Automation, Inc.

12. _____ Double-click LastScanTime from the list.
13. _____ The last parameter is the destination (Dest) for the information GSV instruction is going to provide. Referring back to the instruction documentation, what data type does the destination need to be?
14. _____ To create a new tag for our scan time destination, either go to the controller scoped tags collection or right-click the Destination parameter question mark.
15. _____ If you right-click the Destination parameter question mark, select New Tag. You should see the selection at the top of the list.
16. _____ Create a tag named Main_Last_scan_Time. Make sure the tag data type is correct. If desired, a main operand description can be entered from that window.
17. _____ When completed, your GSV instruction should look like Figure 17-21.

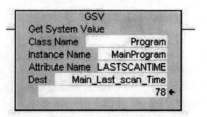

Figure 17-21 Completed GSV instruction providing the last scan time information.
Used with permission Rockwell Automation, Inc.

INTRODUCTION TO THE GET SYSTEM VALUES (GSV) AND SET SYSTEM VALUES (SSV) INSTRUCTIONS

When programming the Instance Name, the word THIS could have been selected, as illustrated in Figure 17-22. Both rungs in this particular case would accomplish the same thing.

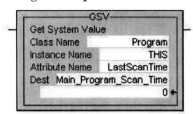

Figure 17-22 GSV instruction programmed to get the last scan time of THIS Program.
Used with permission Rockwell Automation, Inc.

18. _____ Refer to Figure 17-14 and add the logic to monitor the maximum scan time for the Temperature Update task.
19. _____ Download your project and go to Run mode.
20. _____ What is the last scan time for the MainProgram? The scan time value will vary.
21. _____ In what time format is the Last Scan Time displayed?
22. _____ Record the maximum scan time for the Temperature_Update task.
23. _____ Where could you go in the RSLogix 5000 software to view this scan time information in the project for the main program?
23. _____ Monitor the Last Scan Time for your project in RSLogix 5000 software. Does it match with the information provided by the GSV instruction?

LAB EXERCISE 2: Programming a SSV Instruction

For this lab exercise, we modify the properties of the periodic task for the project we worked with in the last exercise. We will program a SSV instruction to modify the properties of the periodic task to allow us to manually enable or disable the Disable Automatic Output Processing To Reduce Task Overhead feature. For some applications, it is not necessary to update the outputs at the end of task execution, so to improve scan time this feature can be disabled. By default, the Disable Automatic Output Processing To Reduce Task Overhead check box is unchecked, which means outputs will be updated at the end of the execution of this task and at the RPI. Figure 17-23 displays the Periodic Task Properties.

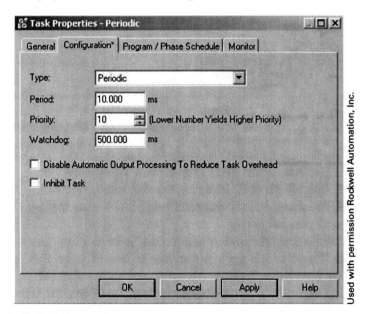

Figure 17-23 Periodic Task Properties.

To modify the task's properties, use an SSV instruction to direct the controller to modify the task object. Turn to the Task Object page in your GSV/SSV documentation to answer the following questions.

1. _____ What class name will be programmed for this instruction?
2. _____ What attribute would be used to Disable Automatic Output Processing to Reduce Task Overhead feature?

Fill in the table in Figure 17-24 with the proper information.

Attribute	Data Type	Description

Figure 17-24 Attribute definition.
© Cengage Learning 2014

We want the ability to command the ControlLogix to either enable or disable automatic processing at the end of the task, so we use the SSV instruction. As we are sending the controller information, we create a source tag, which, according to the documentation, is a DINT. Parameters to program this specific SSV instruction are listed here.

- The Class or Object Name is TASK.
- The Instance Name, or the specific task we wish to modify, is named Temperature Update.
- The attribute to be modified is Disable Update Outputs.
- Because we are sending the controller information, we need to create a source of the information, the source tag. For this exercise, create a tag with the name Enable_Output_Processing, whose data type is DINT.

Figure 17-25 illustrates the completed SSV instruction to enable or disable output processing.

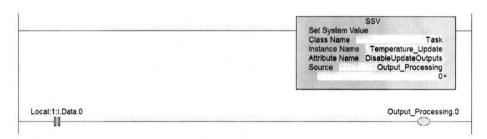

Figure 17-25 SSV instruction to enable or disable automatic output processing at task completion.
Used with permission Rockwell Automation, Inc.

1. _____ Refer to Figure 17-25 as you create a new rung.
2. _____ Program the SSV instruction as illustrated.
3. _____ Program a rung to control the status of the source tag.
4. _____ Verify your project.
5. _____ Download the project and put the controller into Run mode.
6. _____ Where in the project can you go to verify the SSV instruction did what it was commanded to do?
7. _____ Toggle the input switch and make the source tag change from false to true, and verify the proper operation of the SSV instruction.

8. _____ How do you know that the SSV instruction is operating properly?
9. _____ Does the source tag have to be true or false to check the box to Disable Automatic Output Processing To Reduce Task Overhead?

LAB EXERCISE 3: Using the GSV Instruction to Monitor the Battery Light

In Chapter 1, we discussed the purpose of the battery on the controller and talked about battery life. If you have 1756-L61, L62, or L63 controllers Series A or older, upon loss of power the battery is used 100 percent to back up the controller's volatile memory. If the battery were dead and power were lost, the project would be lost in a few seconds.

When the battery needs to be changed, the battery light on the front of the controller illuminates. Even though the L61, L62, and L63 series B and L64 and L65 controllers use the battery differently, you would still want to be alerted if the battery light were to illuminate. To make certain we would not run into a problem regarding the battery and loss of our project upon power failure, a maintenance individual might check each controller's battery light every week or month, but that is a lot of unnecessary work.

In this lab, we program a GSV instruction to check the status of the battery light. When the battery light illuminates on the controller, the GSV instruction and its associated logic could ring a bell, turn on a pilot light on an operator counsel, or flash a screen object on an operator interface device. The ladder logic we are going to program is illustrated in Figure 17-26.

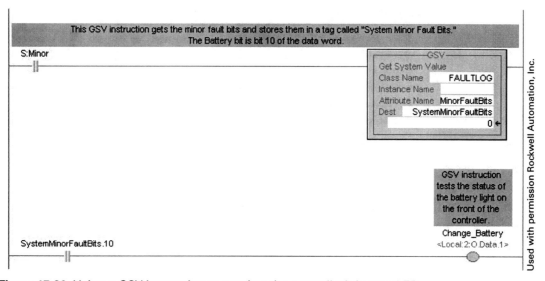

Figure 17-26 Using a GSV instruction to monitor the controller's battery LED.

Programming the GSV Instruction to Monitor the Battery Light

When the battery LED illuminates on the controller, this is considered a minor fault. A ControlLogix minor fault only sets a bit, and does not cause a hard fault on the controller. Fault information can be viewed from the Controller Properties windows. Figure 17-27 shows the Controller Properties window and the Minor Faults tab. Note the battery minor fault has a check in the box, and the description under the Recent Faults heading describes the current problem. The information in this window is read only. In order to use the minor fault information in our ladder for notification, we use a GSV instruction and have the controller provide this information.

Before we begin programming, let's do some preliminary research on the GSV instruction. Referring to the GSV instruction documentation, you will find the minor fault bit information under the FaultLog object.

454 INTRODUCTION TO THE GET SYSTEM VALUES (GSV) AND SET SYSTEM VALUES (SSV) INSTRUCTIONS

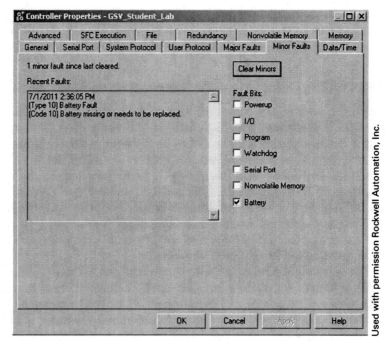

Figure 17-27 Minor Faults tab from Controller Properties.

1. _____ Referring to the GSV instruction programmed in Figure 17-26, note the input instruction's tag is S:Minor. S:Minor is an RSLogix 5000 status bit that is true only when a minor fault is encountered. Because there is no need to have the GSV instruction execute when there is no minor fault, we use this status bit to trigger the GSV instruction only when there is a minor fault.
2. _____ As you program the input, enter S:Minor as the tag above the XIC instruction.
3. _____ Open the GSV/SSV instruction documentation to the section on the FaultLog object.
4. _____ Insert a new rung below the SSV instruction programmed for the last exercise.
5. _____ Program a GSV instruction.
6. _____ Select the class name as FaultLog.
7. _____ In this case, the Instance Name is left blank.
8. _____ What attribute will you select?
9. _____ Locate the attribute in the instruction documentation. Note the following information associated with the attribute:

    - Data type is _____.
    - The instructions are associated with _____.
    - Bits identifying the current minor fault are listed as _____.
        - Bit 4 _____
        - Bit 6 _____
        - Bit 9 _____
        - Bit 10 _____

10. _____ The Destination tag can be whatever name you wish to assign. For our example, the tag name is SystemMinorFaultBits, which is a DINT.
11. _____ The next rung XIC instruction tag references back to the destination tag, SystemMinorFaultBits and specifically bit 10, which is the battery bit. The XIC tag is SystemMinorFaultBits.10.

INTRODUCTION TO THE GET SYSTEM VALUES (GSV) AND SET SYSTEM VALUES (SSV) INSTRUCTIONS

12. _____ Program the OTE instruction with a base tag and alias tag of your choosing to turn on a light on your ControlLogix trainer.

This OTE instruction can control anything the programmer wishes to use to alert either an operator or maintenance individual that the battery needs to be changed. Because operator interface devices are very popular today, the OTE typically is a BOOL reference used to send a bit to a screen object on the operator interface device. The screen object could flash to announce which PAC needs its battery changed.

13. _____ When completed, verify your project.
14. _____ Download the project and put the controller into Run mode.
15. _____ If you have a modular ControlLogix, unplug the battery on the controller. The CompactLogix battery is behind the controller end plate. To unplug the battery, the CompactLogix must be powered down, the end plate removed, and the battery disconnected.
16. _____ The BATT LED should illuminate.
17. _____ Did the output programmed to alert you of the battery light energizing turn on?
18. _____ Go offline.
19. _____ Save the project.

LAB EXERCISE 4: Programming the GSV instruction to Monitor the Controller I/O LED

When configuring a module's I/O configuration, the user has the option to cause a hard fault on the controller if communications between the I/O module and the controller are lost while in Run mode. If the box is left unchecked and communications are lost, the controller will not fault, but the I/O LED will flash green to alert the user of an I/O problem. Because the PAC is typically in an enclosure out of sight of the operator or maintenance personnel, a GSV instruction can be used to get the status of the I/O LED. Using this information, logic can be programmed to display the information on an operator interface screen so the operator can be alerted to any module problems if the Major Fault on Controller If Connection Fails While In Run Mode is unchecked in the module properties. For this lab, we will program a GSV instruction to get the status of the I/O LED. We then test the information received to determine the status of the I/O LED and can then determine what to do about the situation. Referring to the instruction set reference materials, find the required information to answer the following questions.

1. _____ The I/O LED attribute will be found under what object?
2. _____ What is the attribute name?
3. _____ List the destination tag data type.
4. _____ Define the function of the attribute.
5. _____ Refer to the documentation for the I/O LED value definitions.
6. _____ After reviewing the LED status attribute, what will be programmed in the Instance Name parameter?
7. _____ Explain your answer to the last question.
8. _____ Before we program the logic, list the parameter information you will program into the GSV instruction.

   - Class Name:
   - Instance Name:
   - Attribute Name:
   - Destination: Module Not Running; this will be what data type?

456 INTRODUCTION TO THE GET SYSTEM VALUES (GSV) AND SET SYSTEM VALUES (SSV) INSTRUCTIONS

9. _____ To keep the lab simple, we only want to know whether the I/O LED is any state other than solid green. If the LED is anything other than solid green, we want to turn on one of the lights on your trainer. We have to program a rung of logic referencing the GSV destination tag testing to see whether the tag contains a number other than 3. What instruction will you use to determine whether the destination tag is any number other than 3?
10. _____ Draw the logic containing the GSV instruction and the logic to test the destination tag value. Add the output instruction to signify an I/O problem by energizing a light on your ControlLogix. Add rung comments to explain how the rungs work. Also include an alias tag and main operand description for the output instruction.
11. _____ Program the GSV rungs below the rungs in the last exercise.
12. _____ Make sure you save your project.
13. _____ Download the project and put the controller into Run mode.
14. _____ In our laboratory environment, if you are using a modular ControlLogix how could we simulate an I/O module failure? If you are using a CompactLogix, testing this feature easily will not be possible. Check with your instructor as to how to continue.
15. _____ Verify the GSV instruction for the I/O light operating correctly.
16. _____ Check with your instructor to see whether your exercise needs to be checked off as completed.
17. _____ Save the project.
18. _____ Go offline.

LAB EXERCISE 5: Programming an instruction to Modify a Periodic Tasks Interrupt Rate

For this programming exercise, we need to modify the interrupt rate for our Temperature Update periodic task. By default, the rate is 10 milliseconds. Under condition A, the rate must be 250 milliseconds, whereas the rate for condition B must be 500 milliseconds.

1. _____ Before we program the logic, determine and list the following information.
2. _____ Explain why you will select a GSV or SSV instruction.
3. _____ Class Name is _____.
4. _____ The Instance Name is programmed as _____.
5. _____ Attribute Name is _____.
6. _____ How is the rate value represented?
7. _____ Program the source Tag as New_Periodic_Rate, which is a data type.
8. _____ Create the required logic. You might consider the rungs in Figure 17-28 as a possible starting point.

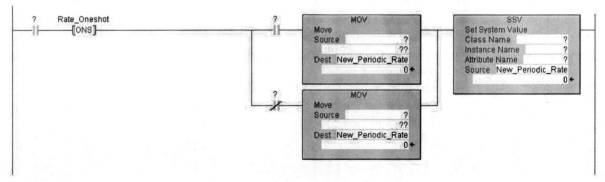

Figure 17-28 Possible starting rung for variable task rate logic.
Used with permission Rockwell Automation, Inc.

9. _____ When completed, download and run the project.
10. _____ Open the task property page and verify the interrupt rate (period) changes when the logic is executed.
11. _____ Save your project, as we will be adding additional logic to it.
12. _____ Ask your instructor whether your work needs to be checked off for credit.

LAB EXERCISE 6: Programming SSV Instruction to Change a Modular ControlLogix Controller Mode with the Loss of a Critical I/O Module

For this exercise, we assume the modular ControlLogix has an output module in slot 2 that is critical for production of a very specialized product. If for any reason we lose communications (the connection) to the module while making this particular product, we wish to generate a major fault on the controller. If we are making any other product, we do not want to fault the controller due to a connection loss, but only notify of the module failure using the controller's I/O LED. For this lab, create logic to command the controller to check the box Major Fault On Controller If Connection Fails While In Run Mode for the slot 2 output module when an input of your choosing goes true, and uncheck the box when that same input goes false. Figure 17-29 illustrates the box being checked.

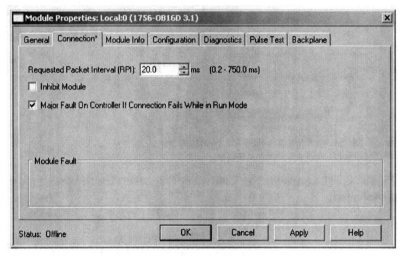

Figure 17-29 Slot 2 Module Properties Connection tab.
Used with permission Rockwell Automation, Inc.

Program a SSV instruction to send the controller the command to check or uncheck the box as needed.

1. _____ Continue adding logic to the project from the last exercise.
2. _____ Before we program the logic, do a little research on the SSV instruction. List here the parameter information you will program into the SSV instruction.

   - Class Name
   - Instance Name:
   - Attribute Name:
   - Source Tag: _____, which will be an _____ data type.

3. _____ Create the required logic.
4. _____ Download, run, and test your logic.
5. _____ Save your project, as we will be adding additional logic to it.
6. _____ Ask your instructor whether completion of this lab is to be checked off.

458 INTRODUCTION TO THE GET SYSTEM VALUES (GSV) AND SET SYSTEM VALUES (SSV) INSTRUCTIONS

CHALLENGE LAB EXERCISE 7: The GSV Entry Status Attribute

The Entry Status attribute uses a GSV instruction to monitor the current operating status of any desired module configured in the project's I/O configuration. The module for which we want status information is identified by programming the module's name in the Instance Name parameter of the GSV instruction. When configuring each module in the I/O configuration, we assigned each module a unique name. Here we will use that name to specify which module the GSV instruction will retrieve status information from. When we programmed the GSV instruction earlier and used the LED Status attribute, the Instance Name parameter was left blank because the I/O LED was providing status information by way of the GSV instruction for all of the modules currently configured. In this situation, the GSV instruction is used to provide information from a specific module. This exercise tests your ability to understand the GSV instruction and correlate instruction set documentation to actual ladder rungs. For this lab exercise, you need the GSV/SSV instruction set documentation. Evaluate the logic in Figure 17-30 and answer the questions.

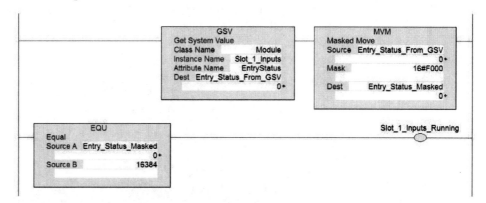

Figure 17-30 Entry Status rungs for evaluation.
Used with permission Rockwell Automation, Inc.

Figure 17-31 illustrates the Controller Organizer for the project associated with the rungs being evaluated.

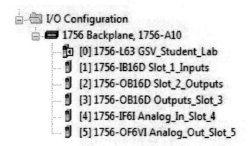

Figure 17-31 Controller Organizer for evaluation.
Used with permission Rockwell Automation, Inc.

Rung Evaluation Questions

1. _____ Using your documentation, which GSV object are we accessing with the GSV instruction?
2. _____ What is the attribute programmed in the GSV instruction?
3. _____ The attribute uses the _____ data type.
4. _____ Text under the description states the following: "Specifies the current state of the specified map entry."

INTRODUCTION TO THE GET SYSTEM VALUES (GSV) AND SET SYSTEM VALUES (SSV) INSTRUCTIONS

This sentence states that the Entry Status attribute will provide status information for the module whose name is programmed in the Instance Name parameter of the GSV instruction. For this example, the module residing in slot 1 is named Slot 1 Inputs. As we program the instruction, we will select the module's name from the drop down to the right of the Instance Name parameter.

5. _____ The description text continues, saying, "The lower 12 bits should be masked when performing a comparison operation. Only bits 12–15 are valid."

The Entry Status attribute's text states this attribute is an INT data type, or 16 bits. For this application, only bits 12–15 provide information we are interested in. Because the lower 12 bits of the 16-bit word do not pertain to this application, we have to find a way to separate the upper 4 bits from the lower 12 bits. The Masked Move (MVM) instruction can be used to extract bits 12–15 from the Masked Move instruction's source parameter, which is the GSV destination tag. The instruction's mask filters out, or masks, bits 0–11. As a result of the masking, only the desired bits, 12–15 will be allowed to transfer from the source to the destination. Figure 17-32 illustrates the operation of the Masked Move instruction. For the sake of illustration, assume the Masked Move instruction source tag currently contains all 1s and the destination tag started with all 0s. After executing the Masked Move instruction, only bits 12–15 were transferred from the Masked Move instruction's source to the destination.

```
                        15 12                 0 ← Bits
          Source:       1111 1111 1111 1111
                        ↓↓↓↓
          Destination:  1111 0000 0000 0000
```

Figure 17-32 Masked Move instruction operation for our GSV data extraction.
© Cengage Learning 2014

The Masked Move instruction uses a hexadecimal mask as the filtering device. Figure 17-33 shows the mask as 16#F000. Remembering your hexadecimal to binary conversion, F equals 1111. A 1 in a mask position allows the source bit to pass through to the destination, whereas a 0 in the mask bit position prevents the bit from passing. Bits that do not pass from the source to the destination are said to be masked out. Figure 17-33 illustrates the mechanics of the Masked Move instruction.

```
          Source:       1111 1111 1111 1111
                        ↓↓↓↓
          Mask:         1111 0000 0000 0000
                        ↓↓↓↓
          Destination:  1111 0000 0000 0000
```

Figure 17-33 Source bits with a 1 in their associated mask bit position will be allowed to pass from the source to the destination.
© Cengage Learning 2014

6. _____ Referring to your GSV instruction documentation for the Entry Status attribute, the description information contains two columns. The Value column lists the values of the four bits extracted from the GSV destination tag using the Masked Move instruction. The values 0000 through the 7000's radix are identified by the leading 16#. What does the 16# in front of the value signify?

7. _____ The meaning column from the attribute documentation correlates the numeric code with text defining the current state of the module accessed by the GSV instruction. For instance, if the GSV instruction reported the module was faulted, a 1,000 in hexadecimal would be reported.

8. _____ The equal (EQU) instruction in Figure 17-30 is used to determine the specific value being provided to the controller from the module via the GSV instruction and the extraction of bits 12–15 using the Masked Move. When programming a comparison

instruction, such as the equal to use the extracted hexadecimal data provided from the Masked Move instruction, the hexadecimal data is converted to a decimal value. The equal instruction's Source B is also programmed as a decimal value.

Let's refer to Figure 17-30. Here we want to specifically test to see whether the module is running using the GSV instruction and Entry Status attribute. According to the Entry Status attribute, a running module will report 16# 4000 to the controller. Converting 16# 4000 to a decimal value equals 16,384 decimal. After executing the Masked Move instruction, you should expect to see a 16,384 in the Masked Move instruction's destination tag Entry Status Masked, as in Figure 17-30. To test for a running module, the value 16,384 would be the value to test for using the equal instruction. With a running module, the Entry Status Masked tag in both the Masked Move destination and the equal instruction's Source A would contain 16,384. If source A, which would be 16,384 is equal to Source B, the constant 16,384, the equal instruction will be true, thus making the rung true. If the OTE instruction tag were called Module Slot_1_Inputs_Running, this output would be true, alerting us to the module's current status.

9. _____ The table in Figure 17-34 lists the hexadecimal values and their decimal equivalent that could be reported by way of the GSV instruction and Entry Status Attribute.

ENTRY STATUS ATTRIBUTE CODE CONVERSION		
Hexadecimal Value	**Decimal Value**	**Meaning**
16#0000	0	Standby
16#1000	4,096	Faulted
16#2000	8,192	Validating
16#3000	12,288	Connecting
16#4000	16,384	Running
16#5000	20,480	Shutting down
16#6000	24,576	Inhibited
16#7000	28,672	Waiting

Figure 17-34 Entry Status codes converted to decimal.
© Cengage Learning 2014

10. _____ To keep our exercise simple, we will use the equal instruction to only determine whether the Slot_1_Inputs module is running. Referring to the converted values from Figure 17-34, note that the value 16,384 means the module is reporting that it is running. If the module is running, the equal instruction will be true. With the instruction is true, the output bit could be sent to an operator interface device to display the current status as running. Using multiple GSV instructions, data from other modules could also sent to an operator interface screen to provide troubleshooting information for engineering or maintenance personnel. More complex programming could provide the current state of each module in your system to a multistate indicator on the operator interface. A multistate indicator could be programmed on an operator interface screen to display the specific text from the module for whatever state is reported from standby to waiting.

PROGRAMMING THE ENTRY STATUS LAB

1. _____ Program the ladder rungs from Figure 17-30 to monitor the status of your slot 1 input module.
2. _____ Download the project and verify the GSV instruction operates as expected. If you are using a modular ControlLogix, remove the input module in slot 1 from the chassis; this will simulate a module failure. You will remember from Chapter 1 that with

INTRODUCTION TO THE GET SYSTEM VALUES (GSV) AND SET SYSTEM VALUES (SSV) INSTRUCTIONS

modular ControlLogix it is OK to remove a running I/O module from the chassis. If you are using a CompactLogix, you will not be able to remove the module.

3. _____ Modify your project and create similar logic to use a GSV instruction to determine whether the output module in chassis slot 2 is running.

4. _____ Program logic for both the input module and the output module to test to see whether either module is faulted. Use the GSV instruction and Fault Code attribute to get the fault code for the faulted module.

CHALLENGE LAB EXERCISE 9: Programming a User Fault Routine

OBJECTIVES

- Create a user fault routine.
- Program the GSV instruction.
- Program an SSV instruction.
- Introduce a User-Defined Data Type.

INTRODUCTION

Faults fall into two categories: major and minor. A major fault faults the controller. In fault mode, the controller stops and real-world outputs are put in their fault mode or program state mode, depending on the type of fault, as defined in the module's I/O configuration. A minor fault only sets a bit, and the controller continues to run. An example of a minor fault is the battery light alerting that the battery needs to be changed. As you remember from the GSV/SSV section, the minor fault bits can be accessed using a GSV instruction. Armed with that information, the programmer can program the appropriate logic and take the desired action. Major faults can be further separated into major recoverable and major nonrecoverable. If a major nonrecoverable fault occurs, the controller will fault. If a major recoverable fault does occur, the programmer can access fault information using a GSV instruction and attempt to recover from the fault in a fault routine.

We introduced the fault routine in Chapter 4. This lab exercise teaches you how to create and program a user fault routine. The term that is used as we attempt to identify and recover from the fault is to "trap a fault" in the fault routine. Trapping a fault simply means to identify and attempt to recover from the fault. We have to program a GSV and an SSV instruction to interact with the controller as we identify the problem and attempt to recover from it.

A reference publication you might find useful as you complete this lab is the *Logix 5000 Controllers Major, Minor, and I/O Faults* publication shown in Figure 17-35. This publication is a good reference that can be downloaded from the Web. Some of the topics covered in this publication include information on minor faults, major faults, and fault codes; how to program a fault routine and the Controller Fault handler; and how to write Power-Up Handler logic.

We are going to program two timers and logic simulating an operator entering a negative value into one of the timers by moving a negative value into one timer when a button is pressed. If a timer instruction executes with a negative value in either the preset or accumulated value, the controller will log a major recoverable fault and shut down if the problem is not cleared. Being an instruction execution error, which the faults programming manual defines as a type 4 recoverable fault, the negative value in a timer can be trapped in a local fault routine and recovered from. If the local fault routine does not exist or cannot recover from the fault, the Controller Fault Handler is then scanned to see whether logic has been programmed there, and an attempt is made to recover from the fault. Leaving the Controller Fault Handler with the problem solved, the

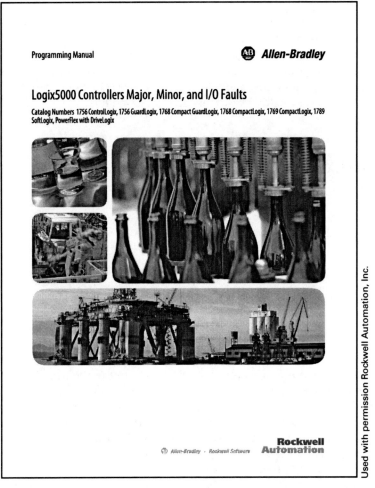

Figure 17-35 Rockwell Automation *Logix 5000 Controllers Major, Minor, and I/O Faults* programming manual.

controller continues to run. Exiting the Controller Fault Handler with the fault still intact causes the controller to fault. Figure 17-36 illustrates a portion of a controller organizer showing the Controller Fault Handler and a local program fault routine named Timer Fault.

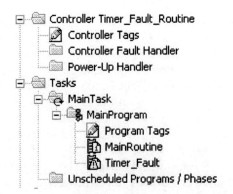

Figure 17-36 Controller organizer showing the local fault routine and the Controller Fault Handler.
Used with permission Rockwell Automation, Inc.

INTRODUCTION TO THE GET SYSTEM VALUES (GSV) AND SET SYSTEM VALUES (SSV) INSTRUCTIONS

DESIGNING A FAULT ROUTINE

The following points pertain to creating our fault routine:

- When a type 4 fault is declared, the controller automatically executes the local fault routine, if it exists. No Jump to Subroutine instruction is required to access the fault routine.
- GSV instruction is programmed to get fault information from the controller.
- Program logic to identify the specific problem. When a fault is declared, ControlLogix provides two tags of information for fault identification: the fault type and fault code.
- If the specific fault code and fault type have been identified in the fault routine, create logic to record the problem has been identified.
- Identify the specific timer causing the problem.
- Fix the problem.
- Clear fault type and code tags.
- Program a SSV instruction to communicate to the controller the problem has been solved.

THE LAB PART 1: Faulting the Controller

First, we are going to program the logic needed for our negative value in a timer scenario. After creating our logic, we will download and test our project to verify the logic works correctly and also faults the controller.

1. _____ Refer to the faults programming manual, major fault code section, and find the fault type and fault code the controller will declare as a result of a negative value in a timer instruction. Note there are many pages of fault types and codes.
2. _____ Add the logic in Figure 17-37 in the Main Task and in the Main Routine in the project you have been working with for this chapter's exercises. If you have already used the input tags earlier, you may wish to assign unique tags to the input push buttons.
3. _____ Make sure the inputs for your newly programmed rungs are off, or false.
4. _____ Save and then download the project. Go into Run mode.

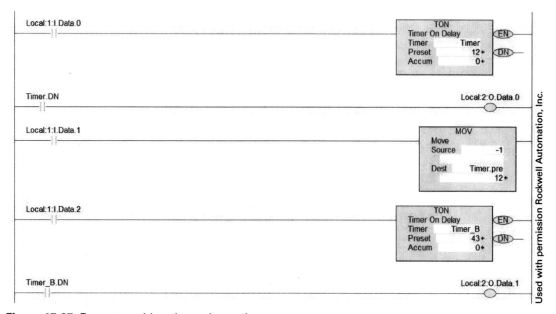

Figure 17-37 Rungs to add to the main routine.

5. _____ Before we test this logic, let's synchronize our personal computer clock with our controller, so when we view the major faults information regarding the fault, the date and time will be accurately represented. To set date and time, go to controller Properties.
6. _____ Refer to Figure17-38 as you click the Date/Time tab.

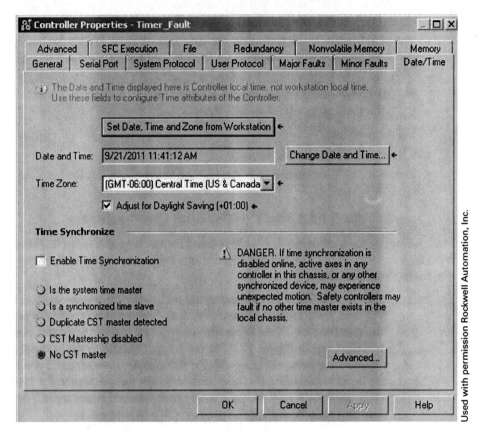

Figure 17-38 Setting the controller date and time.

7. _____ Click Set Date, Time and Zone from Workstation. The Date and Time and Time Zone should update.
8. _____ Execute both timers to verify that they operate properly.
9. _____ Make the rungs go false so the timers reset themselves.
10. _____ Make the rung go true, and move the negative value into the timer. After the value has updated, make the rung go false.
11. _____ How did the controller react to the negative value in the timer?
12. _____ Enable the timer with the negative value.
13. _____ The controller should be faulted, as illustrated in Figure 17-39.

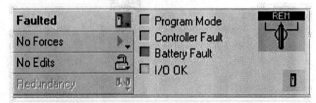

Figure 17-39 Faulted controller as the result of the negative value in the timer.
Used with permission Rockwell Automation, Inc.

INTRODUCTION TO THE GET SYSTEM VALUES (GSV) AND SET SYSTEM VALUES (SSV) INSTRUCTIONS

14. _____ On the Online toolbar, note the word Faulted. To the right of the word Faulted, click the red icon with the down arrow to display the drop-down box, as illustrated in Figure 17-40.

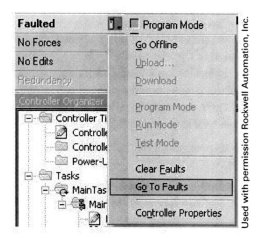

Figure 17-40 Drop down showing Clear Faults and Go To Faults selections active.

15. _____ Click Go To Faults.
16. _____ Fault information should be displayed, similar to that shown Figure 17-41.

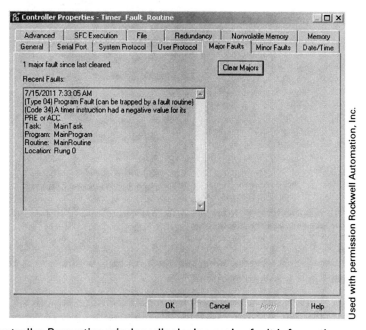

Figure 17-41 Controller Properties window displaying major fault information.

17. _____ Record the following information.
    - Fault Type _____
    - Fault Code: _____
    - Task fault declared: _____
    - Program where fault occurred: _____
    - Routine where fault occurred: _____

18. _____ What does the fault information tell us about the recoverability regarding this particular fault?
19. _____ Clear the fault by clicking the Clear Majors button.
20. _____ What operating mode did the controller go into?
21. _____ Remove the negative value from the timer instruction's preset by typing in the value 5000. Note that there is a blue arrow to the right of the preset value. What is the significance of the blue arrow?
22. _____ Put the controller back into Run mode.
23. _____ Test that the timer operates normally.
24. _____ Turn off the input switches so all of the rung inputs go false.
25. _____ Make sure you save your project and go offline.

Our next task is to program a fault routine and the necessary logic to recover from our recoverable timer fault.

26. _____ Create a fault routine in your main program, as illustrated in Figure 17-42. Refer to Chapter 1 of the *Logix5000 Controllers Major, Minor, and I/O Faults* publication, and review sections on these topics:

- Major Fault State
- Placement of Fault Routines
- Choose where to Place the Fault Routine
- Creating a Fault Routine for a Program

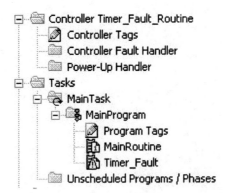

Figure 17-42 Controller organizer showing a user-created fault routine named Timer Fault.
Used with permission Rockwell Automation, Inc.

Before we begin programming our fault routine, we have to define what the logic needs to do to identify and solve our problem.

27. _____ To get fault information from the controller, we use a GSV instruction. A GSV instruction is programmed in the fault routine to get the major fault information from the controller. Refer to the GSV/SSV instruction set reference to determine what class name will be selected as we program the GSV instruction.
28. _____ What will the Instance Name be?
29. _____ List the attribute name.
30. _____ Locate the data type information. List the information provided by the GSV instruction and the data type in the table in Figure 17-43.

The Major Fault Record attribute is providing multiple pieces of information in a block, referred to as a structure. Considering this is a block of data, explain why we could or could not create an array to store the data the GSV is providing us.

INTRODUCTION TO THE GET SYSTEM VALUES (GSV) AND SET SYSTEM VALUES (SSV) INSTRUCTIONS   467

MAJOR FAULT RECORD DATA	
Tag	Data Type

Figure 17-43 Major fault record attribute data type.
© Cengage Learning 2014

Introduction to User-Defined Data Types

When we have a number of tags we wish to group together as a block of data, as long as they are all the same data type an array can be created. If we wish to store dissimilar data types together as a block or structure, a User-Defined Data Type, also referred to as a UDT, will be created. Remember, when working with GSV/SSV instructions, the size and layout of the data must match what the GSV is providing or what the controller is expecting from the SSV instruction. What happens if the size and layout do not match? Hint: Refer to GSV / SSV documentation.

31. _____ To create our User-Defined Data Type, go to the Controller Organizer and expand the data types folder, as illustrated in Figure 17-44.

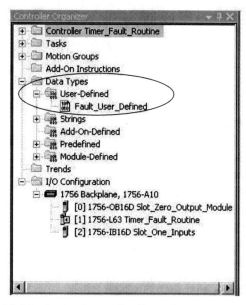

Figure 17-44 User-Defined folder in the Controller Organizer.
Used with permission Rockwell Automation, Inc.

32. _____ Right-click User Defined and select New Data Type. The window displayed should look as in Figure 17-45, without the information added.

33. _____ Enter the name "Fault User Defined," as shown in Figure 17-45. Not only is this the name of the UDT, but this is also the data type we will select as we create the GSV instruction destination tag. Refer back to Figure 17-44 and note the Fault User Defined text under the user-defined folder. When you have completed creating this user-defined data type named "Fault User Defined," the Data Types folder in your project should look the same as that in the figure. "Fault User Defined" is the name we choose for the lab, but you could name it whatever you want.

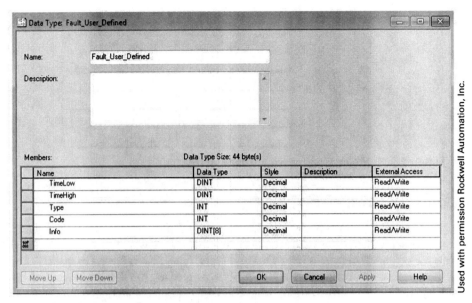

Figure 17-45 Fault User Defined data type for use in this lab exercise.

34. _____ Add a description if you wish.
35. _____ Type in the member names and their data types exactly as they are in the figure and instruction documentation. Refer to the documentation, as the size and layout of the data entry must match for the instruction to work.
36. _____ The member description text is optional, even though it is good practice to enter for future reference by you or others.
37. _____ Again, note the name of the data type is "Fault User Defined". You will be selecting this as the data type as we go to our controller scoped tags collections and create our destination tag structure for the GSV instruction we are about to program.
38. _____ Click OK when completed.
39. _____ Examine the Controller Organizer Data Types folder and verify that yours looks the same as that shown in Figure 17-44.
40. _____ With our structure to accept the data supplied from the GSV instruction created, we will go to the controller scoped tags collection and create the destination tag.
41. _____ Open the controller scoped tags collection.
42. _____ Go to the Edit Tags tab and create a new tag called "Faults." This will be the GSV destination tag name for the structure the GSV instruction will populate.
43. _____ Because we will be populating the UDT just created, what should the data type be?
44. _____ Select the data type as normal. It should be in the drop-down list.
45. _____ When completed, you should see a + to the left of the tag name, signifying this is a structure. Click + to expand the tag.
46. _____ Figure 17-46 shows the controller scoped tags collection with the GSV destination tag, named Faults. Note that the data type for the Faults tag is Faults_User_Defined, the name of the UDT. All of the members from the UDT were automatically transferred when the tag was created.
47. _____ Make a special note of the resulting tag names, because you will be programming some of them shortly as we program our fault routine logic. Note that each member from the UDT was merged with the tag name faults. The resulting member names are as follows:
    - Faults.TimeLow
    - Faults.Time High
    - Faults.Type (This is the reported fault type. We expect a type 4 fault.)

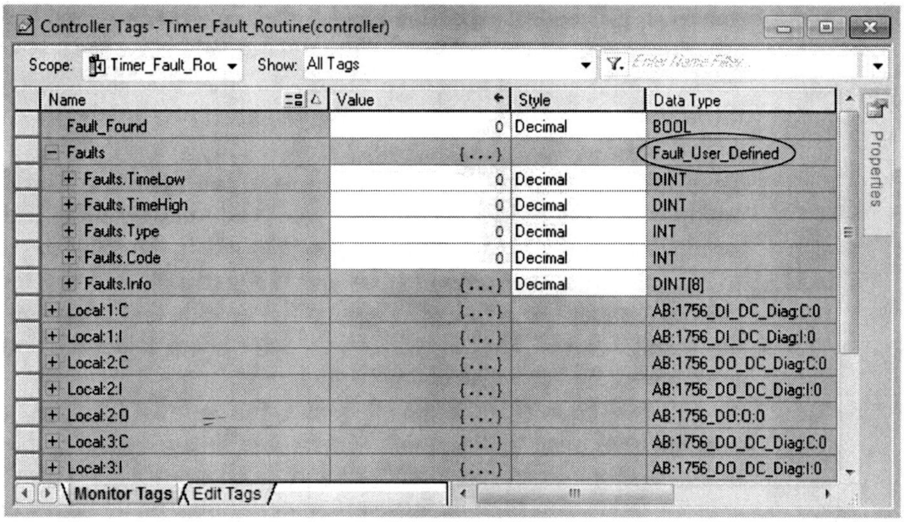

INTRODUCTION TO THE GET SYSTEM VALUES (GSV) AND SET SYSTEM VALUES (SSV) INSTRUCTIONS 469

Figure 17-46 The controller scoped tags collection with the GSV destination tag, named Faults. Note that the data type for the faults tag is Faults_User_Defined, the name of the UDT.

- Faults.Code (From our instruction documentation, we expect a code 34.)
- Faults.Info (In some cases, there may be additional information provided. For this attribute, there is no additional information provided.)

48. _____ Now that we have the preliminary work completed, we can create the logic in our fault routine.

PROGRAMMING THE FAULT ROUTINE

49. _____ Open the fault routine.
50. _____ Referring to Figure 17-47, program the GSV instruction in your fault routine. Make sure the GSV destination tag is the first member of the UDT, Faults.TimeLow.

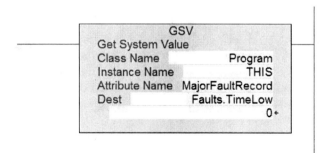

Figure 17-47 GSV instruction.
Used with permission Rockwell Automation, Inc.

51. _____ The GSV instruction will populate the UDT with the information needed to program our fault routine.
52. _____ The next rung to program has to determine what specific fault we are attempting to trap, record it if we did find the problem, and then determine which timer has the negative value.
53. _____ First, let's determine what specific fault we are trying to trap. To identify the current fault information, there are two things we have to determine: Is the current fault

due to an instruction execution problem, classified as a type 4 fault? And what specific fault code are we looking for? A negative value in a timer is a fault code 34.

54. _____ We have to program a rung to test to see whether the fault type tag contains a 4 and if the faults code tag contains a 34. We will need two instructions. What instructions will we use?

Keep in mind that in most applications, a fault routine would be used to trap multiple fault conditions. Then there would be multiple rungs similar to this testing for different fault codes. Create the rungs to satisfy the following specification.

55. _____ If the fault type instruction is true and the instruction to test for the fault code is true, a bit should be set identifying that the specific fault has been found. Name this tag "Fault Found." This bit will be used later in our logic to identify that a fault has been found. Add this instruction to your rung.

56. _____ Now that we know that the specific fault is the code 34, identified as a negative value in a timer, we need to determine which timer is causing the problem. Typically, only timers that operators can modify should be changeable and possibly end up with a negative value. For this lab, let's assume either time could be the problem. Continue programming this same rung with the instructions to test both timers to determine which timer contains the negative value.

57. _____ Knowing which timer has the negative value, to fix the problem we need to replace the negative preset value with either a 0 or another positive number. Program the necessary instruction for each timer to complete programming this ladder rung.

58. _____ Next, we will start programming the third rung.

59. _____ With the problem solved, the fault type and fault code tags must be zeroed out. We only want to zero out these tags if the fault has been found and corrected. Remember there could be numerous rungs testing to see what the specific problem is. If any rung—or rungs, in the case of multiple faults—takes care of the problem, we zero out the fault type and fault code tags. If we were unable to determine and fix the problem, we wish to leave the fault type and fault code tags with their values intact. Program the third rung to satisfy this specification.

60. _____ If the fault has not been resolved, we wish to exit the fault routine with the fault.type and fault.code values still in those tags. Assuming the fault was not resolved in the fault routine, and we exited the fault routine with these tag values intact, what does the controller do next?

61. _____ The current status is either the faults.type tag has been cleared and the faults.code tag has been cleared, or if the fault has not been resolved, those tag values will still be intact. In either situation, we need to send the major fault record back to the controller. The controller will examine those two tags and decide what the next step is. The SSV instruction is used to send the major fault record back to the controller. How do you think that SSV instruction would be programmed if we were sending the same major fault record back to the controller that we get from the controller with the GSV instruction?

62. _____ After programming the SSV instruction, the fault routine programming is complete.

63. _____ Check your project for errors. Correct any errors you find.

64. _____ Now we can download and test to see whether all works properly.

TESTING THE FAULT ROUTINE

1. _____ Download the project and go into Run mode.
2. _____ Run the timer and verify proper operation.
3. _____ Make sure all input buttons are false.

4. _____ Close the switch momentarily and make the Move instruction become true, and move the negative value into the timer.
5. _____ Verify that the rung moving the negative value to the timer is false.
6. _____ Verify the negative value in the timer instruction. Because the timer is not being executed, the controller should not have faulted.
7. _____ Make the timer go true.
8. _____ Explain what happened to your controller.
9. _____ Where will you look to observe the current value of the controller?
10. _____ What happened to the negative value in the timer?
11. _____ Save your project and go offline.
12. _____ This completes this lab exercise.

SUMMARY

To help increase efficiency, ControlLogix does not have a status file to update each scan, as many traditional PLCs do. If a user requires information that would have traditionally been stored in a status file, then a Get System Values, or GSV, instruction, is programmed to get the desired information from the controller. A user who wishes to send the controller information programs a Set System Values, or SSV, instruction. These instructions are outputs and execute each scan that the rung remains true. As we completed the exercises, we referred to GSV and SSV documentation either from the RSLogix 5000 software instruction set help screens, a hard copy, or an electronic copy from the instruction set reference manual, to gain better understanding of the instructions and how to program them.

We also referenced the *Logix5000 Controllers Major, Minor, and I/O Faults* publication to see where to gather information on writing and placement of fault routines as well as error codes and their definitions. In the real world, you will find yourself researching topics such as these to understand how the logic works so you can create or interpret the logic required to fulfill the specifications provided.

When working with the GSV and SSV instructions, the source and destination parameters of the instruction must be carefully programmed to match the size and layout of the data either coming from the controller via the GSV instruction or being provided to the controller with the SSV instruction. Data formatting information is found in the instruction documentation. The data types for the source and destination parameters for these instructions fall into one of three categories: a simple tag, a simple array, or a user-defined data type.

Our first GSV examples, such as getting the last scan time, had a single tag, a DINT, for the destination. In our fault routine application, multiple tags were supplied via the GSV instruction. Because the data types for the Major Fault Record attribute provided by the GSV instruction were dissimilar, we were unable to group the data together in a standard array, as an array must be 100 percent the same data type. To group dissimilar data types together, as represented by the Major Fault Record attribute, a user-defined data type was created.

Faults fall into two categories major and minor. A minor fault only sets a bit, and the controller continues to run; however, a major fault will fault the controller, shutting it down. In fault mode, the controller stops and real-world outputs are put in their major or minor fault mode state, as defined in the module's I/O configuration. Major faults can be further separated into major recoverable and major nonrecoverable. If a major nonrecoverable fault occurs, the controller will fault. If a major recoverable fault occurs, the programmer can attempt to recover from the fault in a fault routine.

The fault routine we created was as to recover from a negative value in a timer, which is defined as an instruction execution problem, or type 4 fault. An instruction execution problem could be the result of bad programming or invalid data entry. If the fault is not cleared, the controller will shut down. With a type 4 major recoverable fault declared, the controller automatically enters the local fault routine, if it exists. If after executing the logic in a fault routine, the controller sees the problem has been corrected, the controller will continue to run as if the problem never existed. If it is desired not to fix the problem, or logic has not been written to trap the specific problem, the controller will exit the fault routine with the fault still active. If the controller exits the local fault routine with the fault still present, the Controller Fault Handler will be executed next to see whether the problem can be recovered from there. Remember, the Controller Fault Handler contains no logic until someone creates it. When the controller enters the Controller Fault Handler it executes the logic in an attempt to recover from the problem. Upon exiting the Controller Fault Handler with the problem resolved, the controller continues to run. If the problem persists, the controller faults.

REVIEW QUESTIONS

Note: For ease of handing in assignments, students are to answer using their own paper.

1. The _____ instruction is a ladder logic output instruction that, when becoming true, gets the requested information from the controller and places it in the specified destination.
2. Because ControlLogix does not have a status file like the PLC 5 or SLC 500 PLCs, controller system data will be stored in _____.
3. The _____ instruction is an output instruction when true; information is sent from the instruction's source tag to the controller to update the desired system parameter(s).
4. When programming either the GSV or SSV instruction, the object name is selected from a drop down as the _____.
5. List the different categories that Source and destination tags used for GSV and SSV instructions generally fall into.
6. When programming a GSV instruction, the destination of the information provided is created to match the _____ and _____, as described in the instruction documentation.
7. The _____ is the name of the information sent to or received from the controller.
8. If the controller exits the local fault with the fault still present, the _____ will be executed next to see whether the problem can be recovered from there.
9. Faults fall into two categories: _____ and _____.
10. If using an array as a source or destination, we need to remember all array elements must be the same _____.
11. Explain what the instance specifies if a GSV instruction's class is Program and the instance was This.
12. The _____ instruction with _____ class and _____ attribute can be programmed to change priority of the task.
13. The temperature of one solution used in our process is critical in order to make a quality product. How could you trigger an alarm if the temperature update periodic task experienced an overlap? Draw out the logic and answer the following questions.
    a. Would you use a GSV or SSV instruction to accomplish this?
    b. Class Name: _____
    c. Instance Name: _____
    d. Attribute Name: _____
    e. Will the tag be the source or destination?
    f. What data type will the tag be?
    g. If the tag for the information provided by the controller were called Temp_Overlap_Trap, list the complete tag below that contains the information that an overlap has occurred.

INTRODUCTION TO THE GET SYSTEM VALUES (GSV) AND SET SYSTEM VALUES (SSV) INSTRUCTIONS

    h. The instruction documentation states that the status bit must be manually reset once the controller sets the bit. Explain how you accomplish this reset.

    i. When the An Overlap Occurred In This Task bit has been cleared, will this alone alert the controller that the bit has been cleared, or do we have to do more?

    j. Draw out the logic to reflect your answer to question I.

14. A _____ data type can be used to group dissimilar data together as represented by the Major Fault Record attribute.

15. When the controller encounters a major recoverable fault, a _____ can be programmed to attempt to recover from the fault rather than shutting the controller down.

16. In fault mode, the controller stops and real-world outputs are put in their major recoverable or nonrecoverable fault mode state, as defined in the module's _____.

17. When using either the SSV or GSV instruction, the programmer must make certain the source or destination data type matches the _____ and _____ required for proper execution of the instruction.

18. For example, when using the GSV instruction, if instruction documentation specifies that an array of seven DINTs will be provided, the destination parameter must be programmed as _____, even though only a portion of the provided information is desired.

19. The source or destination data types will be a(n) _____, a(n) _____ or a(n) _____ _____ _____.

20. To determine the source or destination data type required when programming one of these instructions, refer to _____.

21. If the GSV or SSV rung were left _____, the instructions would be updated each scan.

22. Rather than spend scan time updating a GSV instruction each scan, have a _____ instruction sample the GSV instruction data every few seconds.

23. Tags entered into the user-defined tag structure are referred to as _____.

24. A minor fault sets a bit, and the controller continues to run. How can we program logic and get information associated with the fault?

    - What instruction would you program?
    - What will you program for the instance parameter?
    - What should the class parameter be?
    - What attribute will you program?

25. How can we trigger a GSV instruction to only execute and alert us when the controller battery light illuminates?

CHAPTER

18

Introduction to the RSLogix 5000 Function Block Programming Language

OBJECTIVES

After completing this lesson, you should be able to

- Correlate relay ladder instructions to their function block equivalent.
- Identify basic function blocks and understand their operation.
- Identify the basic function block organizational elements.
- Interpret basic logical function block logic.
- Interpret basic analog function blocks.
- Download and monitor function blocks.

INTRODUCTION

Of the current Rockwell Automation programmable logic controllers (PLCs), which include the PLC 5, SLC 500 family, and ControlLogix, only ControlLogix supports the function block programming language. The standard package of RSLogix 5000 software contains only the ladder logic programming language. Other languages such as function block can be purchased as an add-on. Higher end RSLogix 5000 software packages include the Function block programming language as well as other features.

Function block diagram programming language is commonly referred to as FBD or simply function block. Function block programming is one of the PLC programming languages standardized as the result of the IEC 1131-3 international standard for programmable controller programming languages. Function block is graphical language where program instructions appear as blocks, called function blocks, which are wired together and resemble a circuit diagram. Function block diagram programming is commonly used in applications involving a high degree of data flow between control components, such as proportional integral derivative (PID), variable frequency drive systems, and process control.

This chapter introduces some of the basic function block instructions. After a brief introduction, we will open a project and download and interpret basic function block routines. Later labs provide the opportunity to modify function block diagrams and to download, run, and monitor the projects.

Figure 18-1 illustrates a main program with a main ladder logic routine and a function block subroutine. Note the function block icon to the left of the routine named Process. Remember that any routine must be completely in the same programming language. In this example, Process is a subroutine and can be accessed from the main ladder routine using the Jump to Subroutine instruction.

Figure 18-2 shows a main program with the main routine as function block. The main routine can be any of the four RSLogix 5000 programming languages.

INTRODUCTION TO THE RSLOGIX 5000 FUNCTION BLOCK PROGRAMMING LANGUAGE 475

Figure 18-1 RSLogix 5000 program showing a ladder logic main routine and a function block subroutine.
Used with permission Rockwell Automation, Inc.

Figure 18-2 RSLogix 5000 program showing a function block diagram main routine.
Used with permission Rockwell Automation, Inc.

- To create a new routine, right-click the program and select New Routine to open a window similar to that shown in Figure 18-3.
- Enter the routine name.
- Enter a description, if desired.

Select the routine type.

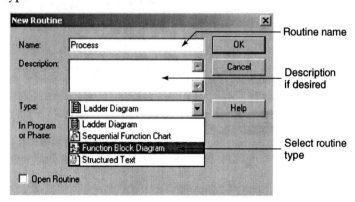

Figure 18-3 Creating a new function block diagram routine.
Used with permission Rockwell Automation, Inc.

- Refer to Figure 18-4 and click the Assignment drop-down arrow to assign the new routine. If the routine is to be a subroutine, select <none>.

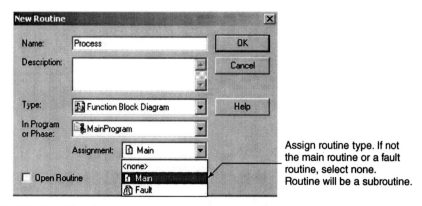

Figure 18-4 Select routine type.
Used with permission Rockwell Automation, Inc.

- Click OK when completed.
- A main function block routine should have been created, similar to the one shown in Figure 18-2.

FUNCTION BLOCK COMPONENTS

The function block workspace is referred to as a sheet, and the number of sheets the project can have is unlimited. When creating a function block routine, the size paper you can print is specified as the sheet size. This is set up by right-clicking a blank spot of a sheet and selecting properties. Select the Sheet Layout tab, as illustrated in Figure 18-5, then select the sheet size and orientation. Click Ok when completed. Sheet sizes are selectable for standard metric or English paper sizes.

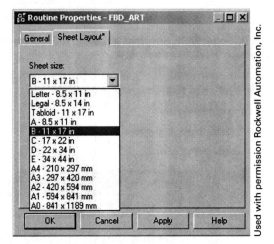

Figure 18-5 Selecting sheet size and orientation from the sheet properties.

Figure 18-6 is that of a function block sheet identifying the basic components. The figure shows a sheet being monitoring with the controller in Run mode.

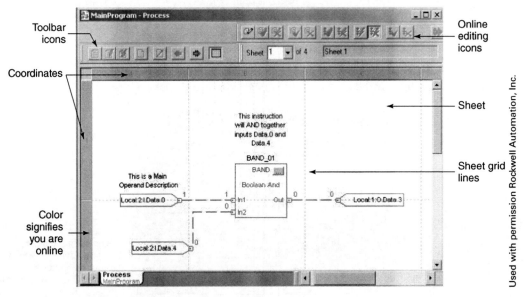

Figure 18-6 Function block basic sheet components.

INTRODUCTION TO THE RSLOGIX 5000 FUNCTION BLOCK PROGRAMMING LANGUAGE 477

Note the following identified sheet components:

- Controller Run mode online monitoring is signified by the color in the coordinate's area at the top and left side of the sheet.
- Letters A, B, and C across the top of the sheet and the numbers 1 and 2 on the left side of the sheet are the sheet coordinates. Coordinates are used for identifying the location of components on a sheet, as well as the position of function block objects when navigating from one sheet to another.
- Sheet grid lines break the sheet up into sections used in conjunction with the coordinates.
- See Figure 18-7 for identification of toolbar icons for online editing.

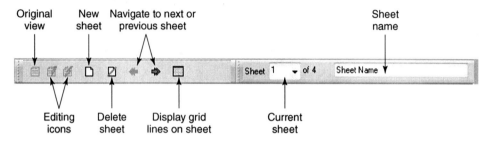

Figure 18-7 Toolbar icons.
Used with permission Rockwell Automation, Inc.

CONFIGURING WORKSTATION OPTIONS

Configuring the way your function block sheet will display is set up in Workstation Options in much the same way as we did for our ladder logic view.

1. _____ Select Tools from your Windows Menu Bar and then select Options from the drop down, as illustrated in Figure 18-8.

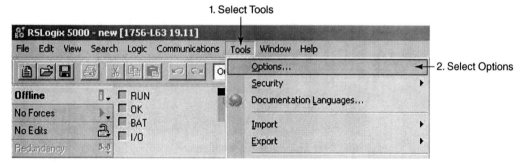

Figure 18-8 Select Options from the Tools drop-down menu.
Used with permission Rockwell Automation, Inc.

2. _____ The Workstation Options window should display, similar to the one in Figure 18-9.
3. _____ Select the Function Block Diagram Editor (FBD Editor).
4. _____ The Function block preferences window should open. Here you can turn on or off features such as the grid on your sheet, Show Tag Descriptions, and display Show Execution Order as on or off. Refer to Figure 18-9 to review the available options.
5. _____ To modify font and color preferences, select FBD Editor Font/Color. See Figure 18-10.
6. _____ Select the item for which you wish to change font and color.
7. _____ Select Foreground Color or Background Color and change to suit.

478 INTRODUCTION TO THE RSLOGIX 5000 FUNCTION BLOCK PROGRAMMING LANGUAGE

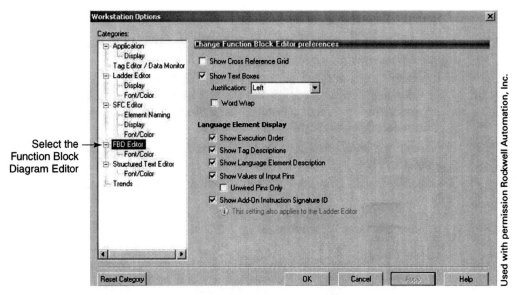

Figure 18-9 To edit function block display options, select FBD Editor.

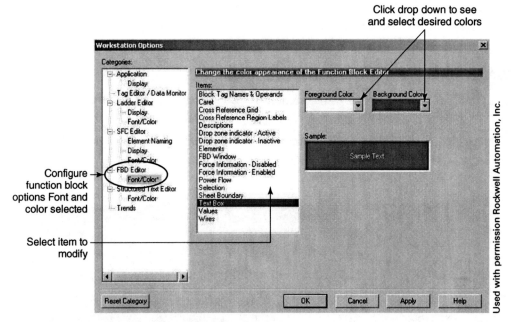

Figure 18-10 Configuring font and color preferences.

Figure 18-11 shows the input components to a function block instruction. Note the shape of the input reference. In the figure, a standard base tag is displayed in each input reference. These tags are the same controller scoped or programmed scoped tags that were used with ladder programming. The lines, referred to as wires, connect the input reference to the function block instruction. The small square box containing a dot to the right of the input reference is the connection point. The connection point is called a pin. The dot in the pin signifies that this reference is BOOL. The other end of the wire connects to a similar pin on the function block instruction. The current tag value is shown to the upper-right area of the input reference. Input pins are on the left side of a function block, whereas output pins are on the right side. The dotted line (wire) also signifies bit data. Figure 18-12 shows an instruction with a BOOL output connected to an output reference symbol with the tag Local:0:O.Data.4.

INTRODUCTION TO THE RSLOGIX 5000 FUNCTION BLOCK PROGRAMMING LANGUAGE 479

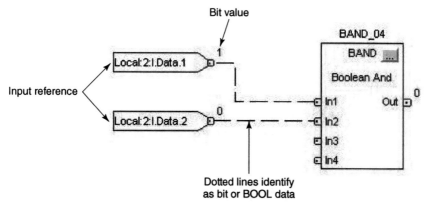

Figure 18-11 Function block input components.
Used with permission Rockwell Automation, Inc.

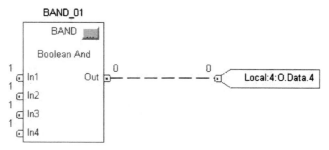

Figure 18-12 BOOL output wire and output reference.
Used with permission Rockwell Automation, Inc.

Figure 18-13 is that of a counter function block. Note that the input tag Enable_Counter_Bit value is currently 0. This bit input is BOOL, thus the dotted line from the input reference to the counter instruction. As you remember, a counter preset is a value. In the figure, the input tag Counter_Preset is currently 237,465, a DINT, which is being input from the input reference Counter_Preset tag. Because this tag is a value, there is no dot in the pin, and the connecting wire is a solid line. The absence of the dot in a pin, in conjunction with the solid wire, identifies the tag as a value. The current value output from the Instruction Accumulated Value (ACC pin) is 0. Note the 0 value to the right of the ACC output pin and the solid wire to the output reference containing the tag Counter_Accumulated_Value. The counter's Done bit pin is on the lower-right side of the function block. The Done bit is being fed back into the Reset pin. When the counter is done, the Done bit goes true and resets the counter.

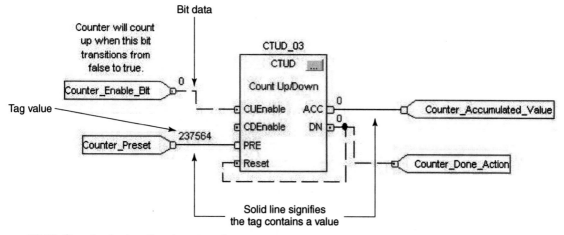

Figure 18-13 Counter instruction input and output components.
Used with permission Rockwell Automation, Inc.

Linking Function Block Logic between Sheets

Input and output wire connectors are used as the link between sheets. The top left function block in Figure 18-14 is sheet 2, whereas the bottom-right function block is sheet 3. The top-left function block in Figure 18-14 shows an output wire connector with the reference tag And_Output. The coordinates directly below the connector are 3-A1. These coordinates identify the opposite end of the link as sheet 3 coordinates A1. The bottom-right function block in the figure is sheet 3. The input connector contains the same Add_Output tag with the coordinate's 2-C1. Go to sheet 2 coordinates C1 to reference the other end of the reference.

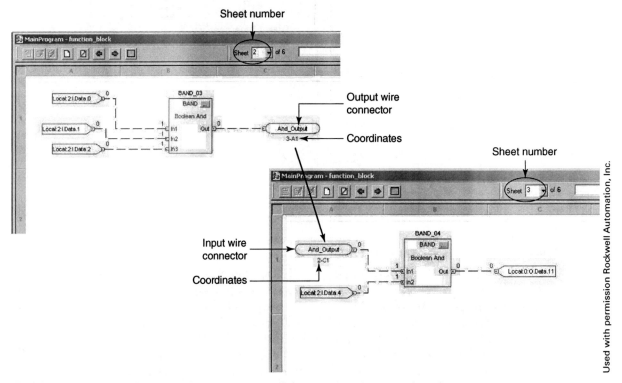

Figure 18-14 Input and output wire connector to link components to another sheet.

Language Element Toolbar

Figure 18-15 shows the Language Element toolbar for the function block language. Function block instructions are found under their appropriate tabs just like ladder programming.

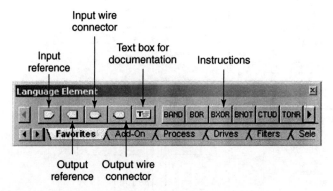

Figure 18-15 Function Block Language Element toolbar showing the Favorites tab.
Used with permission Rockwell Automation, Inc.

The four programming components—input reference, output reference, input wire connector, and output wire connector—along with a few instructions are identified on the Favorites tab of the toolbar. The function block Favorites tab can be customized similarly to the ladder toolbar. Near the center of the toolbar is the text box, which provides a way to enter text documentation to the sheet. Because there are no ladder rungs in the function block, there are no rung comments. Use a text box in place of a rung comment. The text box can be floating or anchored to a specific instruction.

Assigning Tags and I/O References

Assigning a tag to an input reference is the same as associating the tag to a ladder instruction. Figure 18-16 illustrates clicking the lower input reference and then the down arrow to display the tags collection. Select the tag and bit in the same manner as you do with a ladder instruction. Tag assignment to an output reference is in the same manner.

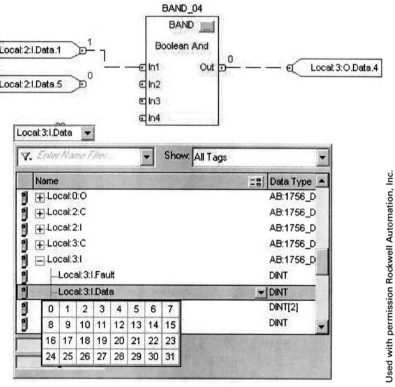

Figure 18-16 Assigning a tag to an input reference.

Connecting I/O References to a Function Block

Connect a wire from an input reference to the desired function block pin by dragging and dropping the wire, as illustrated in Figure 18-17. Remember that a pin with a dot signifies that BOOL data. Wires must be connected to like pins on each end. The software does not allow connecting dissimilar data types; that is, a BOOL reference cannot be connected to a pin expecting a value. Programming errors or incomplete instructions displays an X in the problem areas, as illustrated in Figure 18-18. All programming errors must be corrected before the project can be downloaded.

Function Block Documentation

Function block documentation features include a tag name and main operand description along with a text box. Function block documentation features are identified in Figure 18-19. Because the function block programming language does not contain ladder rungs, there are no rung comments.

482 INTRODUCTION TO THE RSLOGIX 5000 FUNCTION BLOCK PROGRAMMING LANGUAGE

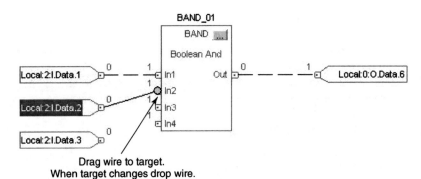

Figure 18-17 Drag wire to pin.
Used with permission Rockwell Automation, Inc.

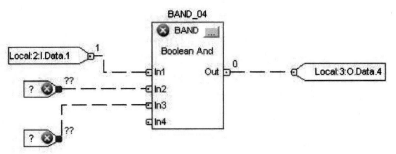

Figure 18-18 Programming errors identified with an X on problem areas.
Used with permission Rockwell Automation, Inc.

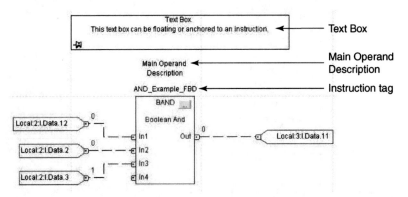

Figure 18-19 Function block documentation.
Used with permission Rockwell Automation, Inc.

A text box is provided to enter documentation associated with the function block instruction. Program a text box by dragging the text box from the Language Element toolbar and dropping the text box in the desired area. The main operand description functions similarly to its ladder logic counterpart. The function block can also have a user-specified instruction tag name associated with the instruction. Refer to Figure 18-20 to rename the instructions tag as you review the following steps.

1. Right-click the default instruction tag, as in example BAND_01. Refer to Figure 18-17.
2. Select Edit BAND_01 properties.
3. Enter the desired tag name.
4. Click OK when completed.
5. The tag name AND_Example_FBD, as illustrated in Figure 18-19, should display above the function block.
6. Refer to Figure 18-20 to enter a Main Operand Description or alias tag in the same manner as you did when working with ladder logic.

INTRODUCTION TO THE RSLOGIX 5000 FUNCTION BLOCK PROGRAMMING LANGUAGE   483

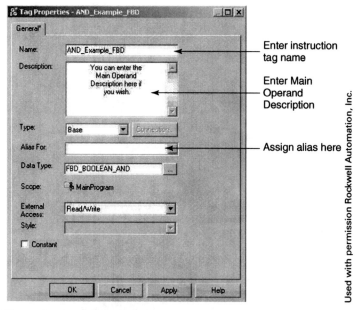

Figure 18-20 Modifying the instruction tag name.

Figure 18-21 is the program scoped Monitor Tags view. BAND_01 tag is illustrated and expanded to display the function block instruction's tags. Note that the tag's associated parameters are displayed.

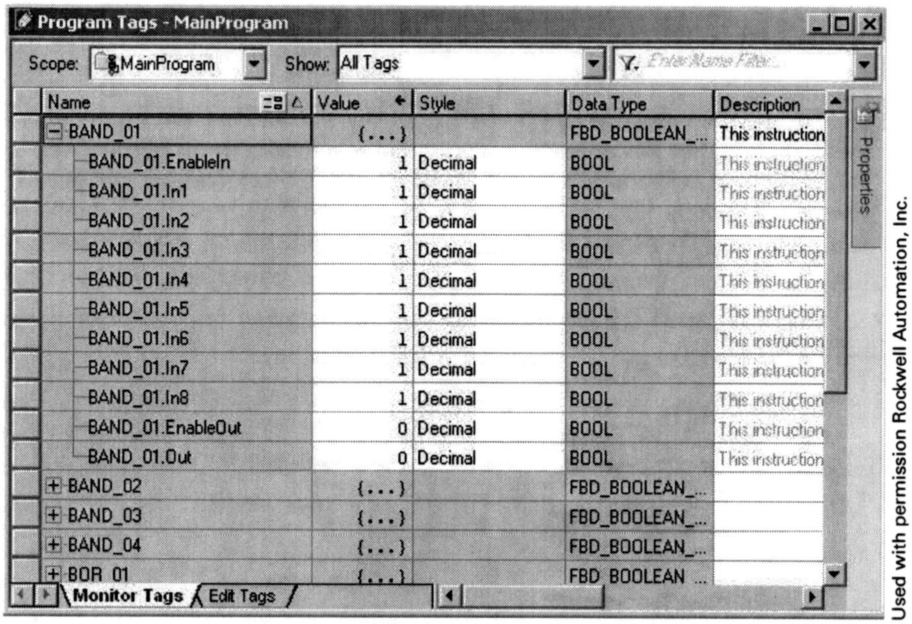

Figure 18-21 Monitor Tags view of instruction tag monitoring.

A text box can be either floating or anchored to a function block instruction. Figure 18-22 illustrates using a mouse to click the pushpin and drag a link to the instruction the text box will be anchored to. When anchored, the pushpin will appear as if it were picked up and pushed into the sheet.

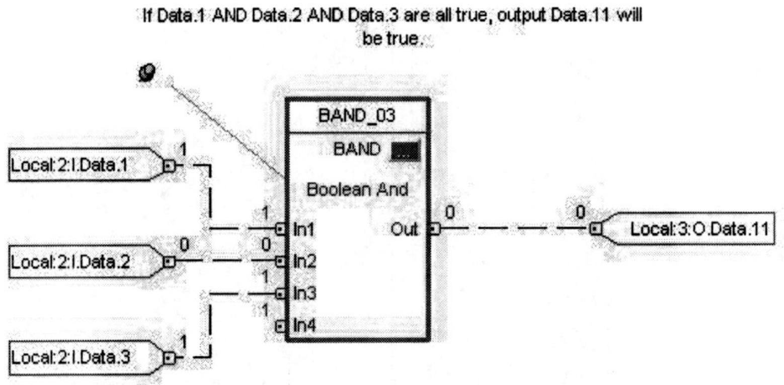

484 INTRODUCTION TO THE RSLOGIX 5000 FUNCTION BLOCK PROGRAMMING LANGUAGE

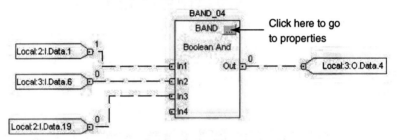

Figure 18-22 Attaching a floating text box to an instruction.
Used with permission Rockwell Automation, Inc.

Function Block Properties

The properties of a function block can be viewed, monitored, and modified by clicking the box icon on the instruction, as identified by Figure 18-23.

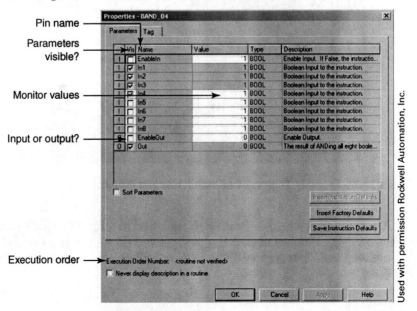

Figure 18-23 Click on the box to go to the instruction's properties.
Used with permission Rockwell Automation, Inc.

Figure 18-24 displays the properties of the instruction from Figure 18-23. The first column identifies whether the instruction parameter is an input or an output. The Vis (visible) column contains check boxes to turn on or off the function block instructions pins. Unused pins can be turned off—that is, not visible—as a way to simplify the instruction by not displaying unused features. The Name column identifies the pin's function, and the Value column displays the current value, or state, of the pin. The Value column is used to monitor the current function block status.

Figure 18-24 Viewing a function block's properties.

The execution order identifies the order in which the function block instructions will be executed. We explore execution order later in this chapter. The execution order is displayed after the routine has been verified.

LOGICAL FUNCTION BLOCK INSTRUCTIONS

This section introduces the basic logical function block instructions. The principles of combining function block inputs are basically the same as those with ladder logic. Instead of ANDing and ORing normally open and normally closed ladder logic symbols on ladder rungs, AND along with OR function blocks are connected together using wires.

The AND Function Block

Figure 18-25 illustrates three-input AND ladder logic. If Data 1 AND Data.2 AND DATA.3 are all true, the output will be true. The ladder logic in the figure does not show logical continuity, as Data.2 is false; thus the rung is false. The function block diagram represents the same logic using a Boolean AND (BAND) function block. Figure 18-26 illustrates a BAND function block. Notice the three input references to the right of the function block; they have Data.1, Data.2, and Data.3 as the tags referenced. Even though it is in a little different format, the logic in Figure 18-26 is exactly the same as that for the ladder rung in Figure 18-25. The input references, as we discussed earlier, represent the input tags where the data is coming from. In this example, we are going to AND the same three input references together. We are looking to see whether Data.1 AND Data.2 AND Data.3 are all true. That is the job of the BAND block. On the left side of the function block are the input pins. In the figure, pins 1 through 4 are visible. Even though only four pins are visible, this function block can AND up to eight inputs. In this example, only three pins are used.

Figure 18-25 Three-input AND ladder logic.
Used with permission Rockwell Automation, Inc.

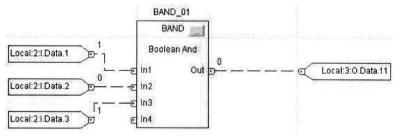

Figure 18-26 BAND function block.
Used with permission Rockwell Automation, Inc.

Notice the bit just to the right of each of the input references. The bit identifies the current logical state of its associated input reference. When the inputs represented by Data.1 AND Data.2 AND Data.3 are true, the BAND function block will be true. With these being true, the output pin will be true. The dotted line, or wire connected between the output pin and the output reference symbol identified as DATA.11, represents the output data as a bit.

Notice the bit just to the right of function block's output pin. This identifies the current logical state of the output pin for the instruction. In this example, input Data.1 and input Data.3 are true.

Input reference Data.2 is false, as there is a 0 to the right of the input reference. Because we do not have logical continuity, when executed, the BAND block will mark its output as false, a 0. As a result, the associated output reference and its tag will also be false, as illustrated in the figure.

Clicking the View Properties box in the upper-right-hand corner of the BAND function block reveals the BAND properties view, as illustrated in Figure 18-27. Note that there are eight input pins for this function block. They are named In1 through In8. Checking the boxes in the visibility (Vis) column turns on the four pins and displays them on the function block. Because In5 through In8 are not checked, these pins will not be visible. This illustrates how the programmer can select the function block options that are specifically needed for his or her application. Notice that the Out visibility check box has been checked. For this example, other parameters and their associated pins are not checked and will not be visible on the function block.

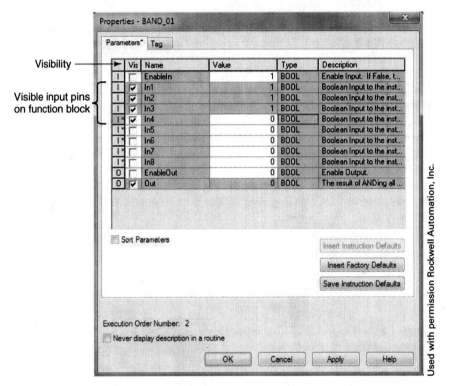

Figure 18-27 Function block BAND properties box.

OR Function Block

Figure 18-28 is that of OR ladder logic. If Data.1 OR Data.2 OR Data.3 is true, the output will be true. Currently, the rung is true because Data.2 as well as Data.3 are true. Remember, with OR logic if one or more, or all, OR inputs are true, the output is true.

Figure 18-28 Or ladder logic.
Used with permission Rockwell Automation, Inc.

The function block Boolean OR (BOR) is illustrated in Figure 18-29. As with the BAND function block, up to eight inputs may be logically ORed together using the BOR block. For this example, if Data.1 OR Data.2 OR Data.3 are true, the output of the BOR function block will be true, or a 1. As with ladder OR logic, if any combination of input references are true, the function block will be true. With the function block true, the output reference tag Data.11 will be true. This BOR function block shows the same logic illustrated in Figure 18-28.

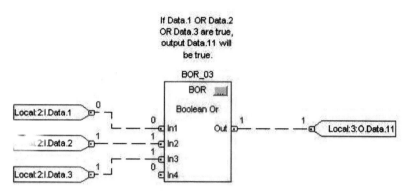

Figure 18-29 Boolean OR function block.
Used with permission Rockwell Automation, Inc.

Exclusive OR Function Block

With Exclusive OR logic if one or the other input is true, but not both, the output will be true. Figure 18-30 shows input ladder logic where if Data.1 OR Data.2 are true, the output is true. If both Data.1 and Data.2 are true, the output will be false. The figure shows Data.1 as true. Data.2 is false because the XIC instruction is not highlighted and the XIO instruction is highlighted. As a result there is a path of true instructions on the top branch of the rung, making the output true. If Data.2 were also to become true at the same time as Data.1, can you see that the output would go false?

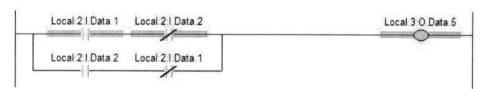

Figure 18-30 Exclusive OR ladder logic.
Used with permission Rockwell Automation, Inc.

Figure 18-31 illustrates the Function Block Boolean Exclusive OR (BXOR). If the input reference representing Data.1 OR the input reference representing Data.2 is true, but not both, the BXOR output will be true. The figure illustrates that Data.1 is true, whereas input Data.2 is also true. Note the 1s to the right of the input reference symbol as well as to the left of the BXOR input pins. This is similar logic as that shown in the ladder rung in Figure 18-30. However, in this case the output of the BXOR will be false because both inputs are true. With the output false, output reference Data.5 will also be false. The 0 to the right of the BXOR out pin and the 0 to the left of the output reference symbol signify the instruction as false.

The NOT Function Block

NOT logic is used when the state of the input needs to be inverted, or changed to the opposite state. The function block Boolean NOT (BNOT), illustrated in Figure 18-32, shows the input

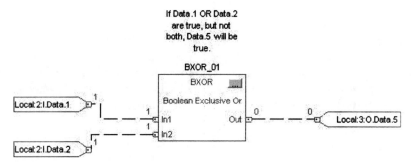

Figure 18-31 XOR function block.
Used with permission Rockwell Automation, Inc.

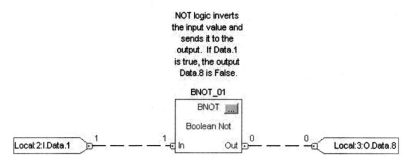

Figure 18-32 NOT function block.
Used with permission Rockwell Automation, Inc.

reference representing input tag Data.1 is true, so the output pin of the BNOT function block will be false, or a 0. The output is the opposite of the input.

SET DOMINANT AND RESET DOMINANT

This section introduces the Set Dominant (SETD) function block as well as the Reset Dominant (RESD) Block. As an introduction to these function block instructions, let's review the ladder logic latch (OTL) and unlatch (OUT) instructions. Figure 18-33 illustrates two ladder rungs, the first with the output latch and the second with the output unlatch. Normally these instructions are used as a pair. If the output tag is latched on, it will need to be turned off, or unlatched. In the figure, if the Start Fan tag is true for one scan or longer, the output latch instruction's Fan tag will go true and retain its true state even though the input logic goes false. If the Stop Fan input tag goes true for a minimum of one scan, the Fan output will unlatch, or turn off. The latch and unlatch instructions are retentive instructions. As retentive instructions, if power is lost to the PAC, the rung output conditions will be retained as its pre–power loss state when power is resumed. One issue regarding the latch and unlatch instructions is what happens when both instructions are true at the end of the logic scan. In the case of Figure 18-33, because the unlatch instruction is scanned after the latch instruction, the Fan tag unlatches, or turns off. Figure 18-34 has the unlatch instruction programmed before the latch instruction. In this case, if both inputs are true, at the end of the logic scan the latch instruction will latch, or turn the Fan tag on. This is an example where the order in which the instructions are programmed can impact how they behave. This is commonly referred to as the last rung rule. The last rung rule states that the last time an instruction is evaluated takes precedence over any previously executed occurrence of the same instruction. In this situation, the last time the tag is evaluated will take precedence over an earlier instruction referencing the same tag.

INTRODUCTION TO THE RSLOGIX 5000 FUNCTION BLOCK PROGRAMMING LANGUAGE 489

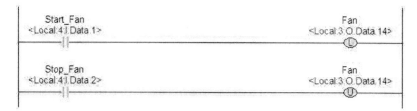

Figure 18-33 Ladder logic latch and unlatch instructions.
Used with permission Rockwell Automation, Inc.

Figure 18-34 Latch and unlatch instructions with the unlatch instruction programmed before the latch.
Used with permission Rockwell Automation, Inc.

The function block Set Dominant and Reset Dominant instructions provide the programmer better control as to how latching and unlatching operations will behave. Both instructions have set and reset input pins incorporated into the function block. The block's outputs are labeled as Out and OutNot. The OutNot logical state is the opposite of the Out pin state. Figure 18-35 illustrates using the Set Dominant instruction to latch on our ventilation fan. The Set Dominant instruction is programmed when the Set input has precedence over the Reset input. Use this instruction to ensure that the output tag has been latched, overriding the status of the reset parameter. We describe here how the Set Dominant function block operates. We assume input push buttons are normally open, momentary.

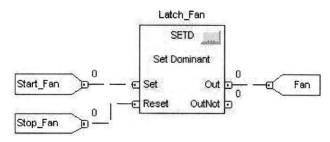

Figure 18-35 Set Dominant function block.
Used with permission Rockwell Automation, Inc.

1. Start Fan input transitions from false to true, so the Fan turns on and stays on.
2. If Fan is on and the Stop Fan input transitions from false to true, the Fan will go false, or turn off.
3. If the Stop Fan push button is true and the Start Fan button is pressed, the Fan will turn on.
4. If both input push buttons are pressed at the same time, the Fan will be true as the Set command dominates over the Reset.
5. If both input push buttons are true and Stop Fan goes false, the Fan will turn off.

Figure 18-36 shows the Properties window for the Set Dominant instruction from Figure 18-35.

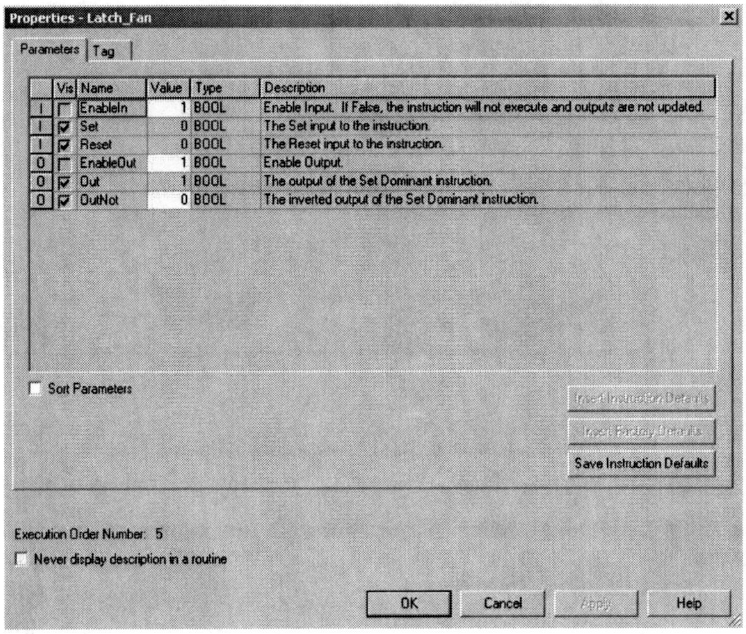

Figure 18-36 Set Dominant properties view.

The Reset Dominant instruction works in the opposite manner to the Set Dominant instruction. If both the set and reset parameters were true, the Out pin would be false, or unlatched. With the Out parameter false, the OutNot pin is true. This instruction is programmed when the Reset input takes precedence over the Set input. In this situation, we need to override the Set input to ensure the Out pin is false. In order to latch the Out pin, the Reset pin must be false and the Set input pin must transition from false to true. Figure 18-37 illustrates the Reset Dominant instruction to latch and unlatch our ventilation fan. The figure illustrates that both the Set and Reset inputs are true. The output is false because the Reset dominates, or overrides, the set command.

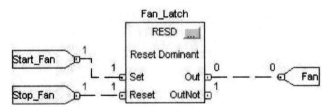

Figure 18-37 Reset Dominant instruction.
Used with permission Rockwell Automation, Inc.

Next, we describe how the Reset Dominant function block operates. We assume input push buttons are normally open, momentary.

1. Start Fan input transitions from false to true, so the Fan turns on and stays on.
2. If Fan is on and the Stop Fan input transitions from false to true, the Fan will go false, or turn off.
3. If the Stop Fan push button is true and the Start Fan button is pressed, the Fan will stay in its off state.
4. If the Fan were on and both input push buttons were pressed at the same time, the Fan would turn off, as the Reset command is dominant, or dominates over the set.
5. If both input push buttons are true and Stop Fan goes false, the Fan will turn on.

Figure 18-38 is the Set Dominant Fan Latch properties view. Note the visibility check boxes and value columns.

INTRODUCTION TO THE RSLOGIX 5000 FUNCTION BLOCK PROGRAMMING LANGUAGE 491

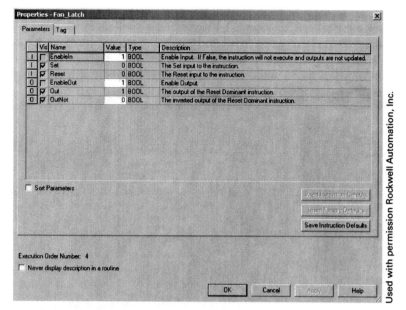

Figure 18-38 Reset Dominant properties.

TIMER FUNCTION BLOCKS

Timer function blocks are similar to those used in ladder programming. One difference is function block timers have the timer reset incorporated into the block. The blocks also have an input pin associated with the preset value, whereas the preset can originate from an input reference either as a tag or a constant. The preset can also be entered as a constant in the block's Property screen. The accumulated value is available from its output pin. Each of the three timers has the same preset value, accumulated value, and status bits as their ladder counterparts. Viewing each timer's Property screen shows additional features and parameters that ladder instructions did not contain.

Timer instructions are found under the Timer/Counter tab on the Language Element toolbar.

Timer On Delay with Reset (TONR)

The TONR is basically a ladder TON instruction with a built-in reset. The timer begins timing when the Timer Enable pin becomes true. The Done bit is true and the timer stops timing when the accumulated value and preset value are equal. Figure 18-39 is the Timer On Delay with Reset function block.

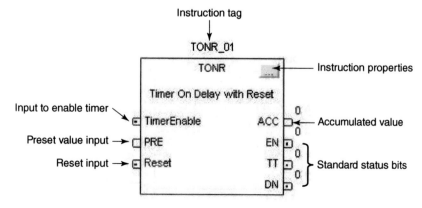

Figure 18-39 Timer On Delay with Reset instruction.
Used with permission Rockwell Automation, Inc.

Figure 18-40 shows the TONR properties view. Notice the visibility check boxes. Here you can turn on or off the function block's pins. All of the familiar ladder logic timer instruction status bits are available in function block and work in the same manner. If you do not need the timer's Enable status bit or Timer Timing status bit for your application, turn the pin off by unchecking the Visibility box. Data in the Value column can be used for monitoring.

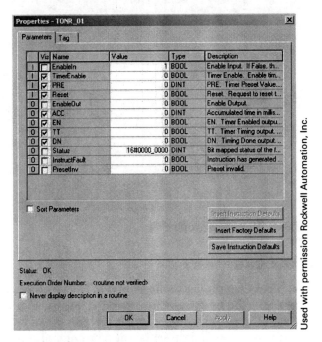

Figure 18-40 TONR properties view.

Timer Off Delay with Reset (TOFR)

The TOFR is a nonretentive off-delay timer with a built-in reset. This instruction works very similar to that of the Timer Off-Delay (TOF) ladder instruction. The timer starts its timing cycle when the Timer Enable input pin goes false. Again, ladder logic and function block timer status bits work in the same manner. Figure 18-41 illustrates the Timer Off Delay with Reset block.

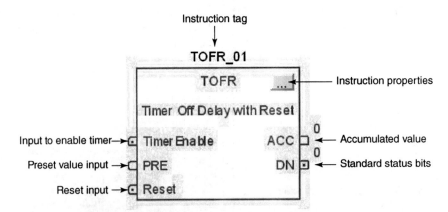

Figure 18-41 Timer Off Delay with Reset.
Used with permission Rockwell Automation, Inc.

Retentive Timer On Delay with Reset

The Retentive Timer On Delay with Reset (RTOR) is a retentive timer on delay with a built-in reset. This timer is also very similar to the ladder RTO instruction except the function block has a

INTRODUCTION TO THE RSLOGIX 5000 FUNCTION BLOCK PROGRAMMING LANGUAGE   493

built-in reset. This timer begins timing when the Timer Enable pin transitions from false to true. Figure 18-42 shows the Timer On Delay with Reset function block instruction. Being retentive, the instruction remembers, or retains its accumulated value until the Reset input pin becomes true. The Done bit is true when the accumulated value and preset value are equal. To reset the timer block, simply make the Reset pin go true.

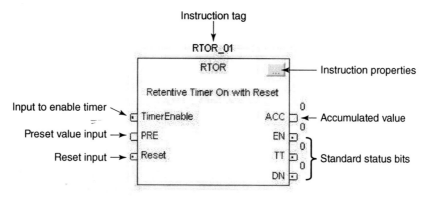

Figure 18-42 Retentive Timer On Delay with built-in reset.
Used with permission Rockwell Automation, Inc.

The Counter Function Block

There is a single counter function block that can be configured to count up, down, or in both directions at the same time. See Figure 18-43. Each time the Count Up Enable (CUEnable) pin transitions from false to true, the counter increments its accumulated value by 1. When the Count Down Enable (CDEnable) pin transitions from false to true, the counter decrements its accumulated value by 1. Figure 18-43 shows the Count Up/Down counter. Whereas ladder programming uses a separate reset instruction to reset the counter's accumulated value to 0, the function block counter has the reset incorporated into the block. Note the Reset pin; the dot signifies that a BOOL reference is expected. This counter has the same status bits as the ladder counter instructions. The instruction Properties view can be used to enter the instruction's properties and configure the instruction to suit the programmer. Figure 18-44 displays the CTUD properties screen.

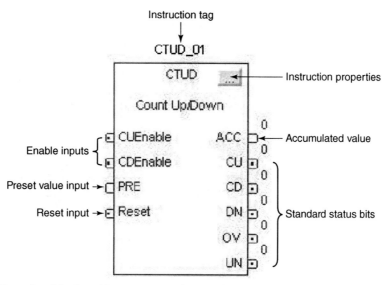

Figure 18-43 Function block up/down counter instruction.
Used with permission Rockwell Automation, Inc.

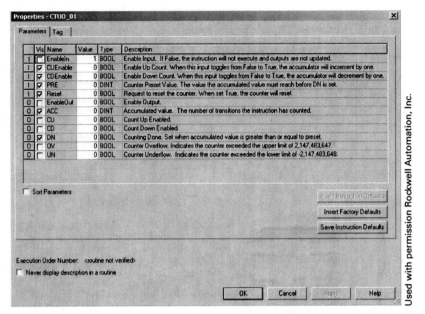

Figure 18-44 Counter properties view.

The instruction can be configured as an up counter, down counter, or an up/down counter by programming the appropriate inputs to the function block. The counter instruction is found under the Timer/Counter tab on the Language Element toolbar.

LANGUAGE ELEMENT TOOLBAR

Now we have introduced some of the basic function block instructions; function block instructions can be found on the function block Language Element toolbar in the same manner as ladder logic. The function block Favorites tab can also be customized to display the commonly used function blocks. Figure 18-45 shows an example of how the Favorites tab could be customized to display many of the instructions we have been working with. As with ladder logic, the function block Language Element toolbar Favorites tab can be customized by selecting Toolbars from the View drop down on the Windows menu bar. If the function block instructions have not been configured on the Favorites tab, the basic logical instructions can also be found on the Move/Logical tab. Timer and counter instructions are found on the Timer/Counter tab. As with ladder programming, instructions can be clicked or dragged into position on a sheet.

Now that we have introduced a few of the basic function blocks, let's look at combining function blocks to create the desired logic.

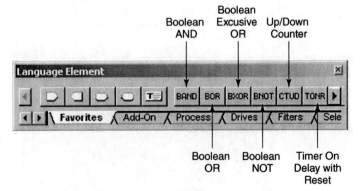

Figure 18-45 Language Element toolbar showing basic instructions.
Used with permission Rockwell Automation, Inc.

COMBINING FUNCTION BLOCKS

To solve the expression (Data.14 AND Data.4) OR (Data.12 AND Data.2) to make output Local:3.O.Data.9 true, function blocks must be combined, as illustrated in Figure 18-46. Trace through the logic. If these function blocks were being executed with their current input conditions, what would you expect the output state to be? Can you explain why? Note, for this example, that the output pins of all function blocks have been left at 0 for your evaluation.

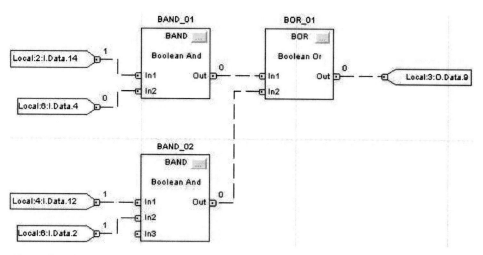

Figure 18-46 Combining AND with OR function blocks.
Used with permission Rockwell Automation, Inc.

Next, we look at a start–stop station providing inputs to start a motor. The typical start push button is normally open, momentary. A typical stop push button is usually normally closed, momentary. The ladder logic in Figure 18-47 shows a combination AND–OR start–stop seal-in ladder logic. Currently, the motor is not running. The Start button is pressed to start the motor, and the Motor Started OTE becomes true. The Motor Started XIC seals in around the Start push button.

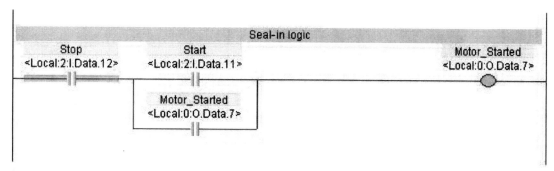

Figure 18-47 Seal-in ladder logic.
Used with permission Rockwell Automation, Inc.

496 INTRODUCTION TO THE RSLOGIX 5000 FUNCTION BLOCK PROGRAMMING LANGUAGE

Figure 18-48 shows comparable function block logic to the seal-in ladder logic in Figure 18-47, with the motor stopped.

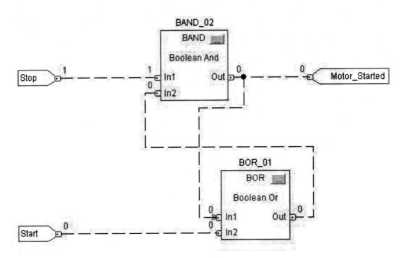

Figure 18-48 Function block seal-in logic. Motor currently not running.
Used with permission Rockwell Automation, Inc.

Figure 18-49 illustrates the function block states after the Start push button was pressed and the motor started.

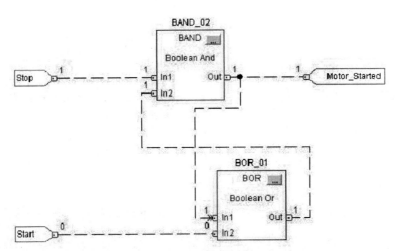

Figure 18-49 Function block seal-in logic. Motor currently is running.
Used with permission Rockwell Automation, Inc.

Another example of combining function blocks is a start–stop seal-in circuit, incorporating a counter to track the number of machine cycles. After 5,000 cycles, the maintenance reminder pilot light or operator interface screen object illuminates to alert the operator. Note the reset for the counter is on a separate rung using the RES instruction. Figure 18-50 shows basic ladder logic to satisfy the specification.

Figure 18-51 illustrates the function block equivalent to the ladder logic in Figure 18-50.

INTRODUCTION TO THE RSLOGIX 5000 FUNCTION BLOCK PROGRAMMING LANGUAGE 497

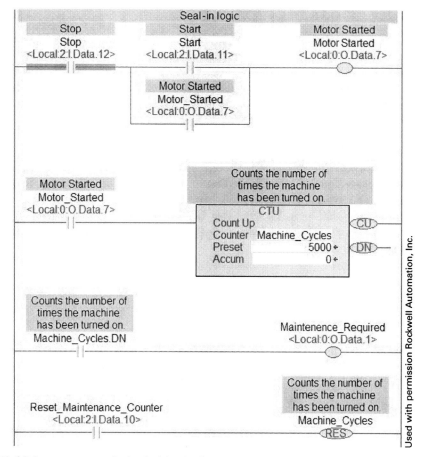

Figure 18-50 Maintenance reminder ladder logic.

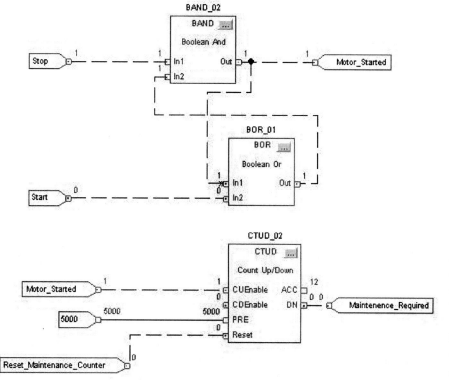

Figure 18-51 Maintenance reminder function block logic.

FUNCTION BLOCK ORDER OF EXECUTION

When the controller goes into Run mode, ladder rungs are scanned, starting with rung 0, left to right, one after the other, to the highest number. The sequence of which block is scanned first, second, third, and so on is referred to as the order of execution. Function block order of execution is somewhat similar to how ladder is scanned. Functions blocks are typically scanned left to right. Figure 18-52 shows a number of function blocks with their order of execution identified with circles. The RSLogix 5000 software determines the order of execution when the project is verified or with the verification associated with a download. On the sheet, execution order is defined by the order in which the function blocks are wired. Typically, the order flows from input to output. The numbers in the upper-left-hand corner of each function block identifies the order of execution for each block. The order of execution values can be turned on or off in the Workstation Options. Refer back to Figure 18-9 to see Workstation Options, and locate the check box to show execution order. One easy way to navigate to Workstation Options is to right-click a blank spot on your sheet and select Options. Figure 18-53 illustrates order of execution for two separately wired function blocks.

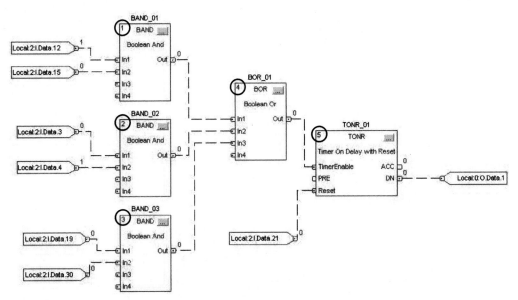

Figure 18-52 Function block order of execution.
Used with permission Rockwell Automation, Inc.

Figure 18-54 displays the same logic as is seen in Figure 18-53. Note how the execution order changes when the BXOR function block is moved slightly to the left on the sheet.

The BXOR function block is now executed first because it is the farthest block to the left. Remember, function blocks are executed left to right, similar to ladder logic.

ANALOG FUNCTION BLOCKS

After completing this section, you should understand some of the commonly used analog function blocks that allow you to scale data, configure alarms, and provide minimum and maximum value information.

The Scale (SCL) Function Block

The Scale function block can be used to scale analog values for analog I/O modules that do support scaling in the I/O configuration, or scale input or output data associated with a network. Most analog

INTRODUCTION TO THE RSLOGIX 5000 FUNCTION BLOCK PROGRAMMING LANGUAGE 499

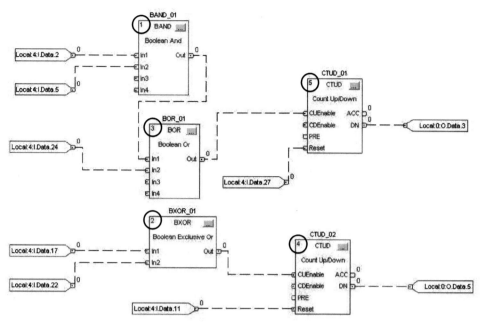

Figure 18-53 Order of execution for separate strings of blocks.
Used with permission Rockwell Automation, Inc.

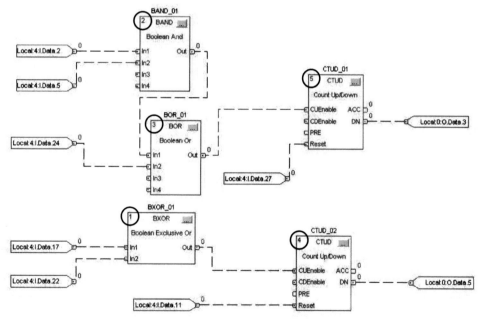

Figure 18-54 BXOR function block is now executed first.
Used with permission Rockwell Automation, Inc.

I/O cards provide scaling of analog data in the I/O configuration, as we learned in Chapter 6. From Chapter 6, we also learned that when configuring analog modules, either Integer or Float Data will be selected as the data format. When selecting Integer, scaling and process control alarming features are lost. Data coming into or going out of ControlLogix across a network may also require scaling of the raw data supplied to or from a field device. A variable frequency drive sending speed feedback information or receiving a speed command across Ethernet/IP is an example of data requiring scaling. In another example, an operator could enter a speed command of 0 to 1,750 rpm on the operator interface terminal and send the data via Ethernet/IP into a ControlLogix. The drive's speed command typically requires scaling to tell the drive how fast to go before being transferred from the PAC to the drive. In many cases, the drive expects a value between 0 and 32,767, representing the 0 to 1,750 rpm data entered by the operator. The Scale instruction takes the raw data, 0 to 1,750 rpm, and converts it (scales it) to the data expected by the drive, which is typically 0 to 32,767. Figure 18-55 illustrates using a Scale function block for the scenario mentioned above. The input data come from the Data in from PanelView tag. Currently, that value is 855.7. The scaled output data is currently 16,023.311 and is stored in the Speed_to_Drive tag. The scaled speed command is sent to the drive by way of the network. Keep in mind that additional programming not illustrated in the figure is needed to to send the scaled speed command to the drive. Notice that the text associated with the input references and function block input and output pins shows the current values. Also note the input references for the InRawMin and InRawMax, along with the Input Engineering Units Minimum (InEuMin) and Input Engineering Units Maximum (InEUMax), have been programmed as constants. Any of the input references could have been a tag where the value would have been found.

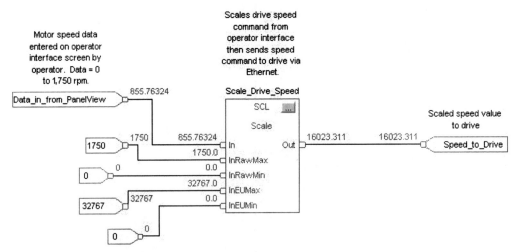

Figure 18-55 Function block scale instruction.
Used with permission Rockwell Automation, Inc.

Scale Function Block Pins:

- In = Data from operator interface terminal tag, 0 to 1,750 rpm
- InRawMax = 1,750 rpm
- InRawMin = 0 rpm
- InEUMax (engineering units) = 32,767
- InEUMin (engineering units) = 0
- Out = Tag containing the scaled data 0 to 32,767 to be sent to the drive

Figure 18-56 shows the Scale_Drive_Speed function block from the Figure 18-55 Properties window. Correlate the checked visible boxes and the pins visible in Figure 18-55. Also note the constant values entered for the input raw minimum and raw max values, as well as the input EU minimum and maximum values.

Scale Function Block Integrated Alarming

The Scale function block has two alarming parameters to monitor the function block's input value to see whether the value has fallen below the input raw minimum value or exceeded the input raw maximum value. Refer to Figure 18-56 as we introduce the Scale function block alarming features.

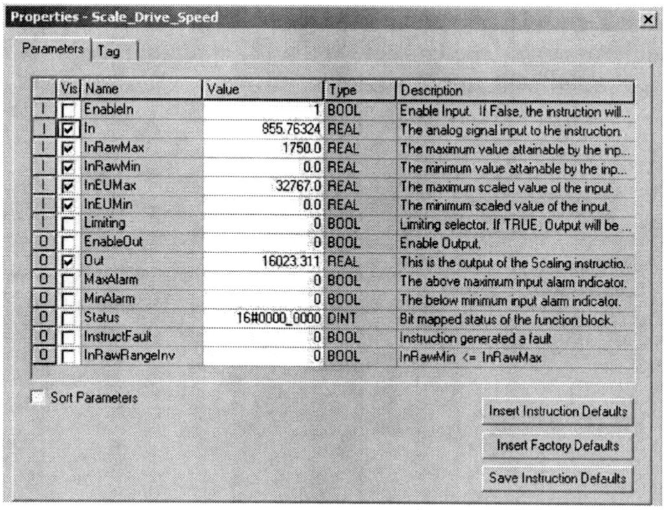

INTRODUCTION TO THE RSLOGIX 5000 FUNCTION BLOCK PROGRAMMING LANGUAGE

Figure 18-56 Scale instructions Properties window.

Maximum Alarm (MaxAlarm):

- If the block's input value exceeds the programmed value of the input raw maximum (InRawMax), the maximum alarm (MaxAlarm) out pin will go true.
- Tag data type is BOOL.

Minimum Alarm (MinAlarm):

- When the block's input value falls below the programmed value of the input raw minimum (InRawMin), the minimum alarm (MinAlarm) out pin goes true.
- Tag data type is BOOL.

Figure 18-57 shows a Scale function block used for temperature input data scaling. Note the two additional output pins not used in Figure 18-55. The MaxAlarm pin is true when the input pin value is greater than the in maximum raw value, which is programmed as 16,384. If the input temperature value in falls below 3,277, the programmed input raw minimum, the Minimum Alarm pin will be true. If you refer back to Figure 18-56, you can view the Maximum Alarm and Minimum Alarm check boxes in the Properties window.

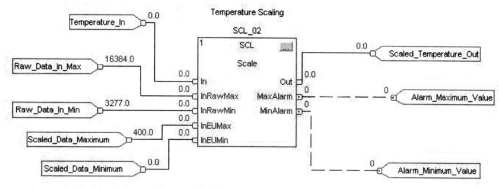

Figure 18-57 Scale block from Figure 18-55, now using the MaxAlarm and MinAlarm pins.
Used with permission Rockwell Automation, Inc.

The Alarm Function (ALM) Block

When configuring analog output modules in Chapter 6, we had the option to configure a High–High, High, Low, and Low–Low alarm for each output channel. An alarm function block, as illustrated in Figure 18-58, can be used to configure the same process alarms for analog modules that do not have alarms configurable in the I/O configuration, or when alarming is required with data not associated with an I/O module.

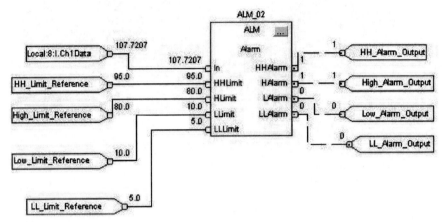

Figure 18-58 The alarm function block.
Used with permission Rockwell Automation, Inc.

The inputs on the left side of the block are the reference tags and values for the alarming. The output references on the right are the output tags that go true when the reference values are exceeded. Refer to Figure 18-58 as we introduce the alarm block operation.

- If the High–High Limit Reference tag (HH_Limit_Reference) of 95 is exceeded, the HH_Alarm_Output tag will be true.
- If the High Limit Reference tag (High_Limit_Reference) of 80 is exceeded, the High_Alarm_Output tag will be true.
- If the Low Limit Reference tag (Low_Limit_Reference) of 10 is exceeded, the Low_Alarm_Output tag will be true.
- If the Low–Low Limit Reference tag (LL_Limit_Reference) of 5 is exceeded, the LL_Alarm_Output tag will be true.

The following selected pins are displayed on the function block in Figure 18-59.

ALARM FUNCTION BLOCK PIN IDENTIFICATION FOR FIGURE 18-58		
Pin	Function	Data Type
In	Analog input signal	REAL
HHLimit	High–High Limit (HHLimit) input value to trigger the High–High Alarm (HHAlarm) output pin.	REAL
HLimit	High Limit (HLimit) input value to trigger the High Alarm (HAlarm) output pin.	REAL
LLimit	Low Limit (LLimit) input value to trigger the Low Alarm (LAlarm) output pin.	REAL
LLLimit	Low–Low limit (LLLimit) input value to trigger the Low–Low Alarm (LLAlarm) output pin.	REAL
HHAlarm	True when HHLimit value exceeds the programmed value.	BOOL
HAlarm	True when HLimit value exceeds the programmed value.	BOOL
LAlarm	True when LLimit value exceeds the programmed value.	BOOL
LLAlarm	True when LLLimit value exceeds the programmed value.	BOOL

Figure 18-59 Alarm function block pin functionality.
© Cengage Learning 2014

Figure 18-60 shows a Scale function block where the scaled output value is input into an alarm block. The Alarm block has High–High, High, Low, and Low–Low alarm features similar to the ControlLogix analog modules when Floating Point is selected as the communication format. Referring to Figure 18-60, let's assume that if the temperature tag exceeds 380°F (193.3°C), an alarm will be declared. The scale block scales the raw input data of 0 to 16,384 to 0°F to 500°F (–17.7°C to 260°C). Currently, the output from the function block is 386.05°F (196.7°C), which is over the acceptable limit. The alarm block is programmed so that if its HHLimit parameter input exceeds 380°F (193.3°C), it will make the HHAlarm pin true. Figure 18-58 shows the HHLimit pin visible on the block, with a tag providing the reference information. In this example, the programmer entered a constant directly into the parameter by clicking the box in the upper-right-hand corner of the block and typing the value into the HHLimit parameter. Because the alarm block's input data is currently over the acceptable maximum, the output tag High_High_Alarm_To Panelview is true. The High, Low, and Low–Low constants were also programmed into their associated parameters as required by the application.

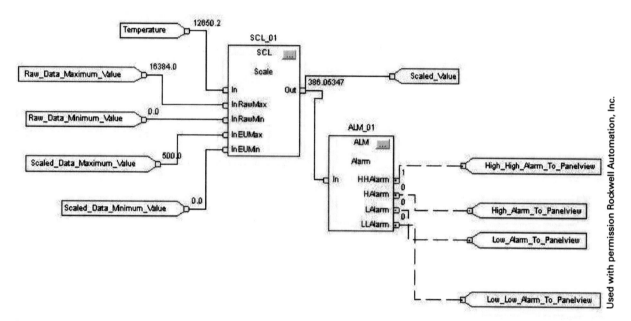

Figure 18-60 Alarm function block associated with a scale function block.

The Select Function Block

The Select function block has two inputs, In1 and In2. The state of the SelectorIn pin determines whether data input from In1 pin or In2 pin is passed to the function block's output pin.

- If SelectorIn pin is false (0), the output data will be from In1.
- If SelectorIn pin is true (1), the output data will be from In2.

Figure 18-61 shows a Select instruction where either of the two inputs, Temperature_1 or Temperature_2, can be selected using the SelectorIn pin's logical status. Referring to Figure 18-61, currently the SelectorIn pin is false and Temperature 1, currently 100°F (37.7°C), will be passed from the input tag to the block's Out pin and the Temperature_Out tag. Can you see how the Select instruction could feed into a Scale block and then on to an Alarm block?

504 INTRODUCTION TO THE RSLOGIX 5000 FUNCTION BLOCK PROGRAMMING LANGUAGE

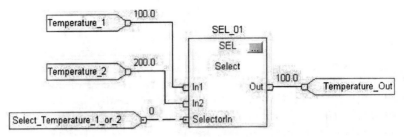

Figure 18-61 Select function block selecting either Temperature_1 or Temperature_2.
Used with permission Rockwell Automation, Inc.

The Minimum and Maximum Capture Function Blocks

Minimum and Maximum capture function blocks can be used with other blocks to record a minimum and maximum temperature of a process. Figure 18-62 shows the minimum and maximum capture function blocks in conjunction with a Scale and Alarm function block monitoring an oven temperature. The target temperature is 350°F (176.7°C). If the oven temperature exceeds

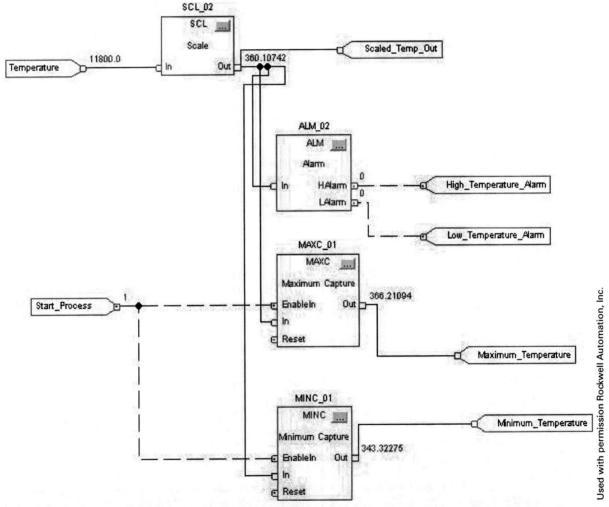

Figure 18-62 Minimum and maximum capture function blocks capturing minimum and maximum oven temperatures.

368°F (186.6°C) or falls below 340°F (171.1°C), the Alarm function block will alert us. For quality control purposes, we want to record the maximum and maximum baking temperature of each batch of cookies, which is the job of the maximum and minimum capture blocks. The maximum capture block has recorded the maximum oven temperature as 366.2°F (185.7°C), whereas the minimum capture block has recorded a minimum temperature of 343.3°F (172.9°C). Because the oven has not exceeded the maximum or fallen below the minimum acceptable oven temperature, neither of the Alarm function block outputs is true. The Reset pin on either of the capture instructions can be used to reset the displayed value to 0 or whatever value the programmer desires.

SUMMARY

This chapter was intended to only introduce the RSLogix 5000 function block programming language. Function block is a powerful language used in process control, motion control, and drive systems. Function block is a huge topic requiring a lot more exposure and experience to programming and additional features in order to effectively program function block applications. An overview of the function block language is provided here.

- Function block contains no normally open or closed instructions and no ladder rungs.
- Function block diagram language is a series of blocks connected by lines referred to as wires.
- Function block references are logically connected together using AND, OR, and NOT blocks.
- The same controller and program scoped tags are used with function block programming as with ladder programming.
- An Input Reference contains the input tag.
- The Output Reference provides the output tag data.
- Pins are the connection points on a function block.
- Pins containing a dot expect a bit or BOOL reference.
- Pins without a dot expect a value as their reference.
- An output wire connector is linked to its associated input wire connector on another sheet.

The chapter introduced logical blocks like And, Or, Not, and Exclusive Or. Because there are no ladder rungs and no normally open or closed instructions to combine to obtain the desired logic, logical function blocks are used to combine tags together in a similar fashion. The And (BAND) as well as the Or (BOR) function blocks can logically operate on up to eight inputs and provide the logical result on the output pin of the block. We looked at the Set and Reset Dominant blocks and compared them to their ladder logic relatives. We discovered function block timer and counter blocks were similar to their ladder logic equivalents. Function block timers contained an integrated reset feature within the block, whereas the single counter block could be programmed to count up, down, or in both directions at the same time. We learned we could combine function blocks to achieve the desired logical results. Analog function blocks were introduced as a way to scale analog data, select one of two input values, set up alarming, and capture the minimum or maximum process temperature. As you remember, the ControlLogix does not have analog scaling instructions in the ladder logic programming language, as scaling is typically configured as part of the I/O configuration of analog input or output modules. We learned in Chapter 6 that when the integer data type is selected as the communication format when performing an analog I/O configuration, scaling and many of the process alarm features are lost. You can jump to a function block subroutine and program the appropriate function blocks to recapture those lost features.

REVIEW QUESTIONS

Note: For ease of handing in assignments, students are to answer using their own paper.

1. Function block diagram programming is commonly used in applications involving a high degree of data flow between control components, such as _____ systems and _____.
2. The function block workspace is referred to as a _____.
3. The number of sheets when using RSLogix 5000 software is _____.
4. Input pins are on the _____ side of a function block, whereas output pins are on the _____.
5. Function block tags are the same _____ scoped or _____ scoped tags that are used with ladder programming.
6. The lines, referred to as _____, connect the input reference to the function block instruction.
7. The small square box containing a dot to the right of the input reference is the connection point for _____.
8. The dotted line (wire) signifies _____ data.
9. Function block diagram programming language is commonly referred to as _____ or simply Function Block.
10. The connection point on either an input or output reference is called a _____.
11. A pin with a(n) _____ inside signifies that this pin is expecting a BOOL reference.
12. A(n) _____ is used as a link to another sheet.
13. The absence of the dot in a pin, in conjunction with the solid wire, identifies the tag as a(n) _____.
14. Of the current Rockwell Automation programmable logic controllers, which include the PLC 5, SLC 500 family, and ControlLogix, only _____ supports function block programming.
15. The current tag value of an input reference is shown to the _____ of the input reference.
16. Function block is an optional programming language for the _____ software.
17. The standard package of RSLogix 5000 software contains only the _____ programming language.
18. The _____ provides a way to enter text documentation on a sheet.
19. When creating a function block routine, the size of paper on which you can print is specified as the _____.
20. When using the Scale function block, which pins are used to input the raw data and scaled data minimum and maximums? List and define the pin names.
21. Function block programming is one of PLC programming languages standardized as the result of the IEC 1131-3 _____ for programmable controller programming languages.
22. What function block would be used to test whether Data.21 AND Data.12 AND Data.13 AND Data.2 are all true?
23. To determine whether Data.1 AND Data.2 OR DATA.5 were true, what function blocks would you program?
24. A function block diagram is a graphical programming language where program instructions appear as blocks, called _____, which are wired together, and resemble a circuit diagram.
25. A text box can be _____ or _____ to a specific instruction.
26. Explain the function of the select function block.
27. What function block would be used if we wanted to test whether Data.21 OR Data.1 OR Data.13 were true?

28. An output wire connector displays the coordinate's 6-C4 directly below the connector. This references us to _____ coordinates _____.
29. What function block would we use to test whether Data.5 OR Data.16 were true? If both inputs were true, would the instruction be false?
30. To invert the logical state of an input tag, a _____ function block is used.
31. The Properties _____ column of the function block contains check boxes to turn on or turn off the function block instructions pins.
32. Unused pins can be turned _____ as a way to simplify the instruction by not displaying unused features.
33. The Properties window of the function block displays a _____ column, which can be used to monitor the current function block status.
34. A function block has a single counter that can be configured to count _____, _____, or _____ directions at the same time.
35. How many input pins can be displayed on an instruction like a BAND or BOR?
36. The _____ identifies the order in which the function block instructions will be executed.
37. The _____ is a nonretentive timer on delay with a built-in reset.
38. To select one of two analog input tags to be scaled and used in the current process, use the _____ function block.
39. The _____ timer begins timing when the Timer Enable pin is true.
40. The _____ instruction is programmed when the reset input would take precedence over the set input.
41. When baking bread, the oven temperature is to be 350°F (176.7°C). If the temperature were to exceed 370°F (187.8°C), you would program the _____ function block in conjunction with the scale instruction.
42. Referring to the previous question, what pin would you program with an input reference containing the tag containing 370 as the out-of-range reference?
43. Each time the _____ pin transitions from false to true, the counter increments its accumulated value by 1.
44. Program the _____ instruction when the Set input has precedence over the Reset input.
45. To trap the maximum temperature, a process has reached, the _____ block could be used.
46. When configuring analog I/O modules, if you select the _____ Communications format, analog scaling and process alarms are not available.
47. Programming errors or incomplete instructions display a(n) _____ in the problem areas.
48. Wires must be connected to like pins on each end. The software does not allow connecting _____ data types.
49. The properties of a function block can be viewed, monitored, and modified by clicking the _____ in the _____ of the function block.
50. When configuring an analog input module where scaling was not available, or data was coming in from a network where the data needed to be scaled, what would be an easy way to perform the scaling?

LAB EXERCISE A: Function Block Component Identification

1. _____ Identify the sheet components represented by letters A–G in Figure 18-63.
2. _____ Identify the components of the Language Element toolbar represented by letters A–K in Figure 18-64.
3. _____ Identify the toolbar components represented by letters A–H in Figure 18-65.

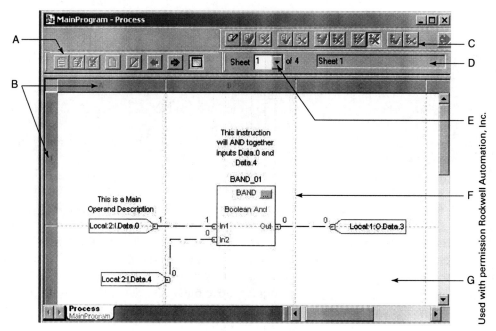

Figure 18-63 Sheet component identification.

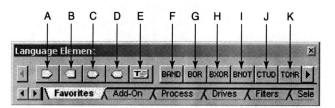

Figure 18-64 Language Element toolbar component identification.
Used with permission Rockwell Automation, Inc.

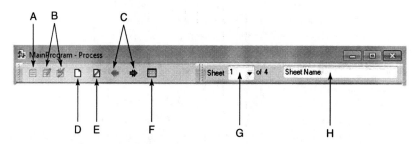

Figure 18-65 Identify toolbar components.
Used with permission Rockwell Automation, Inc.

4. _____ Identify the components of the BAND properties displayed in Figure 18-66.
5. _____ Identify the function block diagram features represented by letters A–M in Figure 18-67.
6. _____ Describe the components of the timer block in Figure 18-68. As part of your answer, include the data type of each pin.

INTRODUCTION TO THE RSLOGIX 5000 FUNCTION BLOCK PROGRAMMING LANGUAGE 509

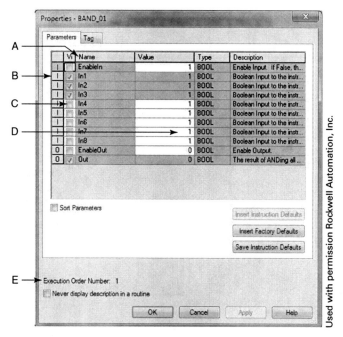

Figure 18-66 Function block properties component identification.

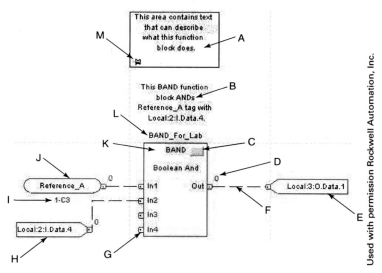

Figure 18-67 Function block component identification.

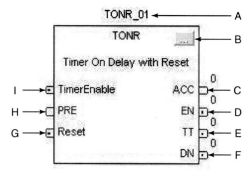

Figure 18-68 On-Delay Timer with Reset components.
Used with permission Rockwell Automation, Inc.

7. _____ Identify the counter components and list the data types for each pin (A–M) for Figure 18-69.

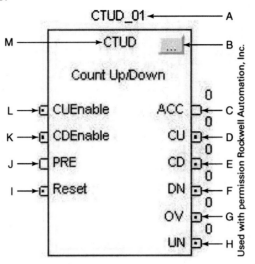

Figure 18-69 Counter Up/Down components.

LAB EXERCISE B: Function Block Diagram Logical Continuity

In this lab, we determine when function block logic is true or false. Evaluate the following function blocks to determine whether the output devices are true or false.

1. _____ Referring to Figure 18-70, explain the function of the function blocks illustrated.

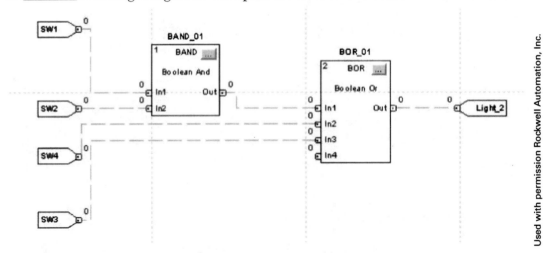

Figure 18-70 Function block diagram interpretation for question 1 and 2.

2. _____ Evaluate the logic to determine the status of Light_2. Explain your answer.
3. _____ Explain the function of the BOR function block from Figure 18-71.
4. _____ List what would have to happen to make Light_2 become true.
5. _____ How does the BNOT function block work?
6. _____ List the states of SW1, SW2, and SW9, along with SW3, to make Light_2 turn on.
7. _____ If the function block diagram in Figure 18-72 were running, what would be the current status of Light_2?
8. _____ Interpret the function block diagram in Figure 18-73 and explain how the BXOR works.

INTRODUCTION TO THE RSLOGIX 5000 FUNCTION BLOCK PROGRAMMING LANGUAGE 511

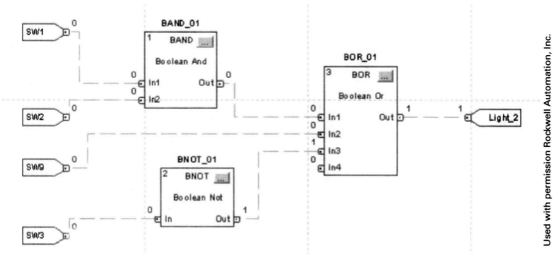

Figure 18-71 Function block diagram interpretation for questions 3 through 6.

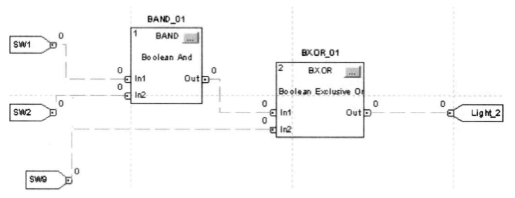

Figure 18-72 Function block diagram interpretation for question 7.
Used with permission Rockwell Automation, Inc.

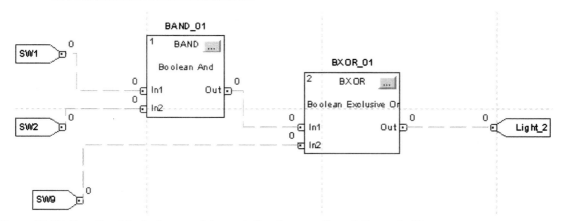

Figure 18-73 Function block diagram interpretation for questions 8 through 10.
Used with permission Rockwell Automation, Inc.

9. _____ What has to happen for Light_2 to become true?

10. _____ Is there more than one way to make Light_2 true?

11. _____ What will the status of Local:3:O.Data.9 be when this function block logic is executed in Figure 18-74? *Note*: Function block output status bits have been changed to 0 so you can determine their logical status.

512 INTRODUCTION TO THE RSLOGIX 5000 FUNCTION BLOCK PROGRAMMING LANGUAGE

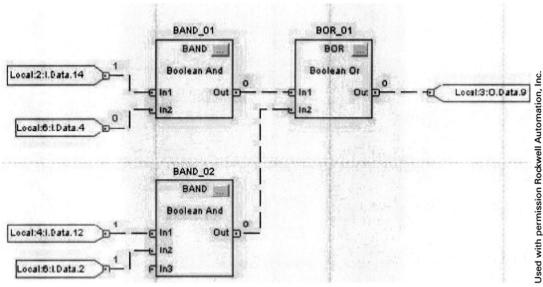

Figure 18-74 Refer to question 11. What will the status of Local:3:O.Data.9 be when executed?

12. _____ Evaluate the function blocks in Figure 18-75, and answer the following questions. Function block output status bits have been changed to 0 so you can determine their logical status.

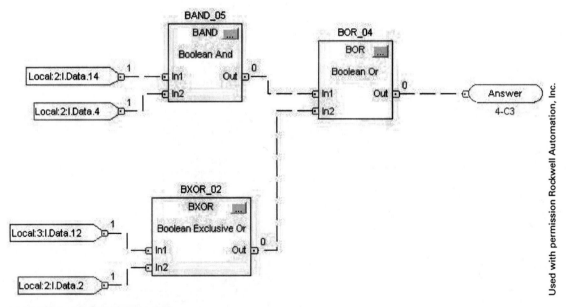

Figure 18-75 Logic for question 12.

Questions for Figure 18-75. Assume the logic is going to be executed. Answers should reflect logical states after the function blocks have been executed.

    a. What would the state of BAND_05 be?
    b. What would the state of BXOR_02 be?
    c. Explain your answer to question B.

d. What would the state of BOR_04 be?
e. What is the object containing the tag Answer?
f. Explain the function of the object referred to in question E.
g. Explain the significance of the 4-C3 text below the object containing the tag.
h. When you went to the coordinates 4-C3, you found a BNOT function block. What would the output state of that function block be if the blocks were being executed?

LAB EXERCISE C: Interpreting Function Blocks

As you interpret Figures 18-76 and 18-77, answer the following questions.

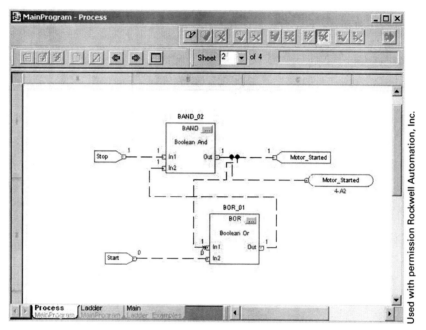

Figure 18-76 First of two sheets to interpret.

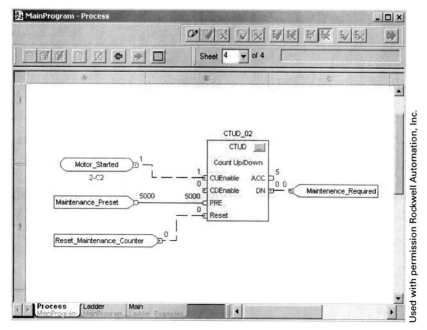

Figure 18-77 Second of two sheets to interpret.

1. _____ Viewing the sheets, how do you know whether they are online or offline?
2. _____ How many sheets are we using in this routine?
3. _____ Figure 18-77 is which sheet?
4. _____ On sheet 2, there is a reference with rounded ends near right center with the tag Motor_Started. Explain what type of reference it is.
5. _____ Referring to question 4, explain the significance of the 4-A2 displayed below the reference.

LAB EXERCISE D: Challenge Lab

In this lab, you have an opportunity to evaluate a more complex function block diagram. Currently, all inputs are false.

1. _____ As you evaluate Figure 18-78, what has to happen to make Light_4 turn on?

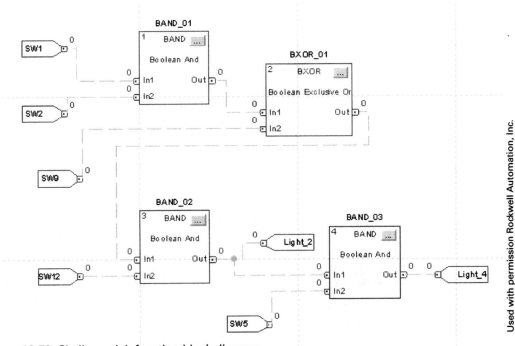

Figure 18-78 Challenge lab function block diagram.

LAB EXERCISE E: Monitoring and Interpreting Function Logical Blocks in a Logix 5000 Project

This lab exercise provides hands-on experience downloading and monitoring a project containing a function block routine. First, we correlate ladder rungs to their function block counterparts as a way to move from familiar ladder logic to function block interpretation.

1. _____ Review the ladder rungs in Figure 18-79. We correlate the rungs in the figure to function block elements in this lab exercise.
2. _____ Open the project Function Block Lab 1.
3. _____ Download the project, put the controller in Run mode, and go online.

INTRODUCTION TO THE RSLOGIX 5000 FUNCTION BLOCK PROGRAMMING LANGUAGE 515

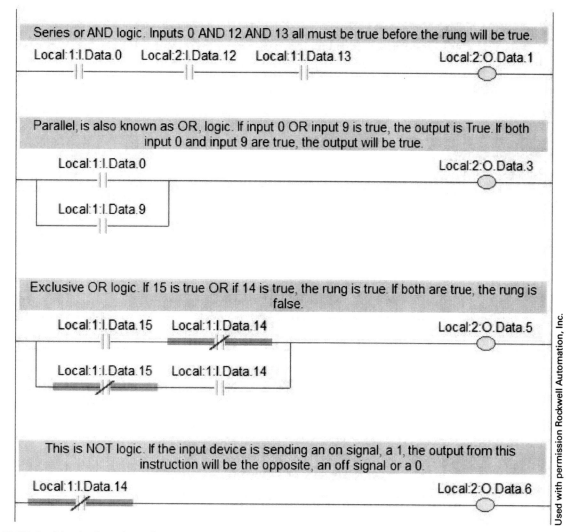

Figure 18-79 Ladder logic to be referenced for lab exercise.

4. _____ The function block diagram we work with in this exercise is sheet 1 in the function block main routine. This function block logic is illustrated in Figure 18-80.
5. _____ The first instruction on the top of the sheet is a BAND, or Boolean AND instruction.
6. _____ This instruction is the function block representation of the ladder logic on rung 0 of Figure 18-79. Test the BAND function block operation by turning the input switches on and off.
7. _____ Does the function block operate the same as the ladder logic rung?
8. _____ The next function block instruction is the BOR, or Boolean OR. This function block should work the same as the ladder logic on rung 1 in Figure 18-79.
9. _____ Test your inputs to verify that the function block works as expected.
10. _____ OR logic states that if any or all of the inputs are true, the output will be true. The principle of Exclusive OR, simply referred to as XOR, is that if either input is true, but not both, the output will be true. When using ladder logic, Exclusive Or, logic must be created through programming, as in rung 2 from Figure 18-79. If Data.14 is true or Data.15 is true, but not both, the output will be true.

516 INTRODUCTION TO THE RSLOGIX 5000 FUNCTION BLOCK PROGRAMMING LANGUAGE

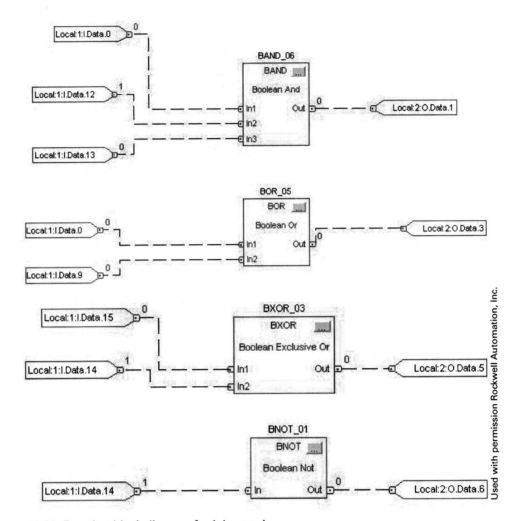

Figure 18-80 Function block diagram for lab exercise.

11. _____ The BXOR function block works in the same manner as the ladder logic. Experiment with the two input switches and verify that only one can be true at a time for the output to go true.

12. _____ The last function block instruction on this sheet is the BNOT. This is the Boolean NOT instruction. Refer to ladder logic rung 3 to see similar logic. Basically, NOT logic means that the output will be the opposite of the input. If the input was a 1, the output would be "NOTed," or a 0.

13. _____ Test the function block BNOT logic and verify that it works as expected.

14. _____ Evaluate the seal-in logic in Figure 18-81.

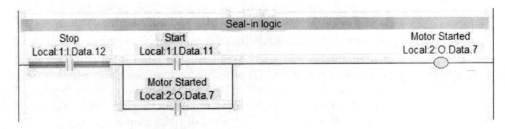

Figure 18-81 Seal-in ladder logic.
Used with permission Rockwell Automation, Inc.

15. _____ Go to sheet 2 of your function block routine. You should see BAND and BOR function blocks, as shown in Figure 18-82.

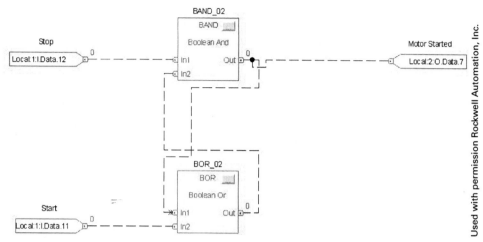

Figure 18-82 Function block seal-in logic.

16. _____ Refer back to the ladder logic in Figure 18-81 as you test your function block logic to verify that it works the same as your ladder logic seal-in rung.
17. _____ Referring to the function block diagram, explain the function of the BAND function block.
18. _____ Why does the output of the BAND feed into In1 of the BOR?
19. _____ Why does the output of the BOR feed into the In2 of the BAND?
20. _____ Explain how the BOR function block fits into the logic.

LAB EXERCISE F: Interpreting Set Dominant and Reset Dominant Function Blocks

The Set Dominant and Reset Dominant function blocks are similar to the ladder logic output latch and output unlatch instructions, but with one important difference. When selecting the function block instructions, the programmer can determine whether the set or reset input to the block will have precedence over the other. Refer to the Set Dominant and Reset Dominant portion of this chapter to review operation.

The Set Dominant function block is selected when the set input has precedence over the reset input. Use this function block to ensure that the output tag has been latched, overriding the status of the reset parameter. If both the set and reset parameters were true, the output pin would be true, or latched. Let's look at an example using the Set Dominant example in a fan application. For the example in Figure 18-83, we assume the input push buttons to control the fan are momentary, normally open.

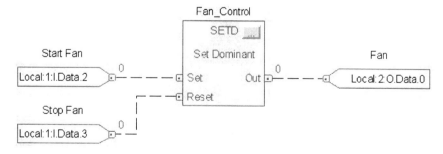

Figure 18-83 Set Dominant function block.
Used with permission Rockwell Automation, Inc.

518 INTRODUCTION TO THE RSLOGIX 5000 FUNCTION BLOCK PROGRAMMING LANGUAGE

INTERPRETING THE SET DOMINANT

Let's test the operation of the Set Dominant function block.

1. _____ Go to sheet 3 of the function block routine.
2. _____ The Set Dominant function block, as shown in Figure 18-83, should be displayed near the top of the sheet.
3. _____ Make the Start Fan tag go and stay true. Explain what you observe.
4. _____ Make the Stop Fan tag go and stay true. Explain what you observe.
5. _____ As you make Start Fan tag go false, explain what you observe.
6. _____ What do you think would happen if the Set input were to go true?
7. _____ Make the Set input go true to test your answer to question 6.
8. _____ Make both inputs go false.

Interpreting the Reset Dominant

Let's experiment with the Reset Dominant function block.

1. _____ The Reset Dominant function block logic, as shown in Figure 18-84, should be displayed just below the Set Dominant block on sheet 3.

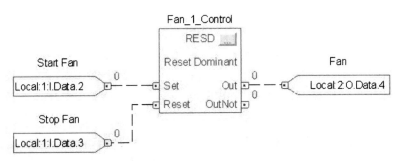

Figure 18-84 Reset Dominant function block.
Used with permission Rockwell Automation, Inc.

2. _____ Make the Start Fan tag go and stay true. Explain what you observe.
3. _____ Note how the OutNot pin works in relation to the Out pin.
4. _____ Push the button Stop Fan (Reset) and explain what you observe.
5. _____ What do you think would happen if the Set input were to go false?
6. _____ Make the Set input go false to test your answer to question 5.
7. _____ As you make the Stop Fan tag go false explain what you observe.

Set and Reset Dominant Instruction Review Questions

1. _____ Explain the operation of the Set Dominant instruction.
2. _____ How does the Set parameter work?
3. _____ Explain how the Reset parameter works.
4. _____ Explain how the OutNot pin works in conjunction with the Out pin.
5. _____ Explain the operation of the RESD instruction.
6. _____ The following questions pertain to Figure 18-85.
   a. Identify the figure.
   b. What do the I's and O's in the far left column of the figure symbolize?
   c. What do the check boxes in the Vis column signify?
   d. What is the current input status of this instruction?
   e. Describe the current output status of the instruction.

INTRODUCTION TO THE RSLOGIX 5000 FUNCTION BLOCK PROGRAMMING LANGUAGE 519

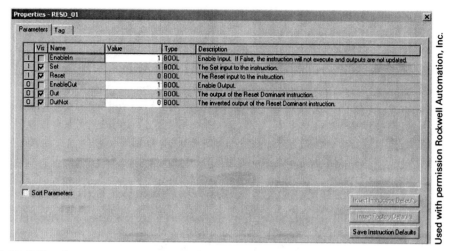

Figure 18-85 RSLogix 5000 software screen for question 6.

LAB EXERCISE G: Interpreting Timer and Counter Function Block Instructions

This exercise introduces the function block timers and counters.

1. _____ Go to sheet 4, Timers and Counters of the function block routine.
2. _____ The TONR is what type of timer?
3. _____ Explain how this timer times.
4. _____ What is the preset value for this timer?
5. _____ Go to the timer Properties screen, where you will find information such as the preset value.
6. _____ Click the button in the upper right of the TONR instruction to go to the Properties screen.
7. _____ Can you find the preset information here too?
8. _____ The Start tag starts, or enables, the timer. This is an alias tag. What does this mean?
9. _____ What is the base tag address for the Start tag?
9. _____ There are a couple ways to find this information. What are they?
11. _____ Make the Timer Enable true.
12. _____ The timer should begin timing.
13. _____ Where can you observe the accumulated values?
14. _____ How does the timer reset itself?
15. _____ Go offline with your project.
16. _____ Let's add a separate reset to this timer function block.
17. _____ Click the input reference icon.
18. _____ Position the input reference near the reset input point to the function block.
19. _____ Assign an unused input tag.
20. _____ Add the connecting wire between the input reference and the Reset input point on the TONR.
21. _____ Download your project, put the controller in Run mode, and verify that your TONR works properly.
22. _____ The next instruction is the TOFR. Explain how this instruction operates.
23. _____ What is the base tag for the Stop alias tag?
24. _____ Execute the instruction. Does it work as expected?
25. _____ Go offline and add an unused input as an input reference to reset the timer.

26. _____ Download and test your TOFR.
27. _____ The next instruction is for the RTOR timer. Explain the basic operational difference between the TONR and the RTOR.
28. _____ Download, run, and test your RTOR instruction.
29. _____ The CTUD instruction is a counter instruction that does what?
30. _____ With the program running, test this instruction to see whether it operates as you expected.
31. _____ How would you create this in regular ladder logic?
32. _____ Go offline with your controller.

LAB EXERCISE H: Analog Function Block Instructions

In this section, we introduce analog interface to function block instructions.

1. _____ Open the Analog Function Lab exercise.
2. _____ Download the project.
3. _____ Put controller in Run mode and go online.
4. _____ Sheet 1 has the scale instruction. Where would you find the raw input signal data range?
5. _____ List the raw data range.
6. _____ List the output scaled data range.
7. _____ Verify the input data and verify output data changes appropriately.
8. _____ Go to sheet 2.
9. _____ Where can you find the output scaled data range?
10. _____ Is the output value an integer or real number?
11. _____ What are the parameters of the alarm function block?
12. _____ Verify the input data. Does the alarm work as expected?
13. _____ What is the base tag for the HH_Alarm?
14. _____ H_Alarm is an alias tag for _____.
15. _____ Refer to the Help screens to determine how the minimum capture operates. Define the following parameters:
    a. EnableIn:
    b. In:
    c. Reset:
    d. Reset value:
    e. Output:
16. _____ Explain how the Batch start input fits into the operation of the instruction.
17. _____ Batch Started is an alias tag to _____.
18. _____ Why are there solid and dotted lines on the function block diagram?
19. _____ Refer to the Help screens to determine how the maximum capture operates. Define the following parameters listed:
    a. EnableIn:
    b. In:
    c. Reset:
    d. Reset value:
    e. Output:

PART

2

CONFIGURING RSLINX DRIVERS AND COMMUNICATION

INTRODUCTION

Most PAC users have difficulty with setting up communications between their personal computer and the programmable controller. The following chapters step you through configuring different communications options using Rockwell Software's RSLinx. RSLinx is used to set up communications between a computer and other hardware. Although RSLogix 5000 software is used to create, modify, and monitor the project to be downloaded into your controller, RSLinx is a second package of software used to configure the communications between the personal computer and other hardware such as a ControlLogix family member. The desired RSLinx communication driver will be part of the RSLogix 5000 project path identifying the link between the two pieces of hardware.

The following chapters step you through configuring serial, USB, and Ethernet/IP communications using Rockwell Software's RSLinx. If you have used RSLinx to configure a driver for an older Rockwell PLC, such as the SLC 500 or PLC 5, you will find the principles are very similar.

Ethernet/IP is the most popular communication method today. The steps required to establish communications with a ControlLogix family member using Ethernet/IP are included in the following overview:

1. Verify or set up personal computer IP address.
2. Select and connect proper cabling.
3. Verify or set up the Ethernet communications module IP address.
4. Configure the RSLinx driver.
5. Configure the RSLogix 5000 project path
6. Download project and put ControlLogix into Run mode

PREREQUISITES BEFORE STARTING TO CONFIGURE ETHERNET/IP COMMUNICATIONS

Before starting to configure Ethernet/IP drivers, the student should have basic knowledge of the principles of Ethernet and Ethernet/IP addressing.

COMMUNICATIONS CONFIGURATION CHAPTERS

Select the following chapters to learn how to establish communications between a personal computer and a ControlLogix family controller.

- Chapter 19 Configuring a Serial Driver Using RSLinx
- Chapter 20 Configuring a Keyspan, by Tripp Lite, High-Speed USB to Serial Adapter Model USA-19HS
- Chapter 21 Configuring an RSLinx Serial Driver Using a Rockwell Automation 9300-USBS USB to Serial Adapter
- Chapter 22 Installing and Configuring a USB driver for 1756-L7 Series Controllers
- Chapter 23 Determine and Modify a Personal Computer's I/P address
- Chapter 24 Configuring 1756 ControlLogix Modular Ethernet Hardware
- Chapter 25 Configuring an Ethernet IP Address for a CompactLogix 1769-L23E, 1769-L32E or a 1769-L35E
- Chapter 26 Configuring a 1756-ENET Ethernet Driver Using RSLinx
- Chapter 27 Configuring Ethernet/IP Drivers Using RSLinx Software
- Chapter 28 Configuring a CompactLogix 1769-L23E, 1769-L32E, or 1769-L35E Ethernet/IP Driver Using RSLinx Software
- Chapter 29 Configuring a USB Driver for a 1756 Ethernet Communications Module

CHAPTER

19

Configuring a Serial Driver Using RSLinx

Complete this lab exercise if you wish to configure a Serial RSLinx driver for either a CompactLogix controller or a modular ControlLogix using a personal computer with a standard DB9 serial port. Either a null modem serial cable or a Rockwell Automation 1756-CP3 or 1747-CP3 cable can be used to connect between the personal computer and your controller's serial port. If you wish to configure a serial driver using a personal computer with a USB port that uses a USB to serial adapter, see Chapter 21. Rockwell Automation's RSLinx software is used to configure communication drivers. RSLinx pictures in this chapter are from RSLinx versions 2.54 through 2.59. If you have a different version of RSLinx, your screens might be a bit different. The configuration steps should be the same.

INTRODUCTION TO SERIAL COMMUNICATIONS

The simplest connection between a personal computer and the PAC is a serial connection. Even though simple, connecting serially can be especially frustrating if the incorrect serial cable is used. Many ControlLogix controllers come with a RS-232 serial communications port. When establishing communication between a personal computer or industrial computer and a PAC, we need the capability to download information from the computer terminal to the PAC.

We also need to upload programs and program-related data from the PAC to the computer terminal. It might seem that any PAC with an RS-232 serial port could communicate with any other RS-232 port. This is not necessarily true. When configuring a communication link between two devices, such as a computer and a PAC, there are two important aspects of the communication link. First, there is the communication standard.

The RS-232 Communication Standard

The RS-232 communication standard defines only the physical cable connections and use for each of the nine wires inside the standard communication cable, and their associated connector pins. Remember, when referring to a communication cable, the connector pin numbers are used for wire identification. The standard does not define how many pins and wires must be used.

Minimum configuration for two-way communication only requires three wires in the 9-pin D-shell connector. In a typical RS-232 connection, the personal computer uses pin 2 for data output, and peripheral equipment, such as a modem, uses pin 2 for data input. When sending data back from the modem to the computer, pin 3 is data *output* from the modem; yet pin 3 is data *input* on the personal computer. Pin 7 is used as the ground. Figure 19-1 illustrates the minimum connections between a personal computer and peripheral equipment, like a modem. Notice that the wires go directly between the two devices, pin for pin. This is referred to as a straight-through connection.

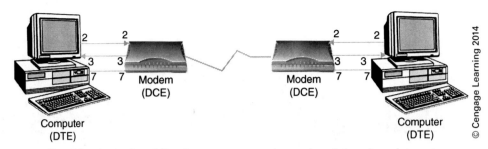

Figure 19-1 Straight-through cabling between computer and peripheral equipment.

For ease of connection, the RS-232 standard specifies that computer devices have male connectors, whereas peripheral equipment has female connectors. When communicating directly between a personal or industrial computer and a PAC, which is also a computer, there is no intermediate peripheral equipment. If the same straight-through cable were used to connect the personal computer to the PAC (refer to Figure 19-1), we would be connecting pin 2 of one computer to pin 2 of the other computer, as illustrated in Figure 19-2.

A computer's pin 2 is outgoing data. Figure 19-2 illustrates both computer devices sending output data from pin 2 and attempting to use each other's pin 2 as an input for data. Both devices are looking for input data on pin 3. In this cabling configuration, pin 3 on each computer device is connected to pin 3 on the other. This type of connection does not allow communication between the two devices. Cabling must be modified so that the output of one computer (pin 2) is connected to the input of the opposite computer (pin 3). The output from each computer (pin 2) must cross and connect to pin 3, the input of the opposite computer. Communication sent by one computer can then be received by the other computer. Data sent back by the receiving computer can be received by the originator of the transmission. Figure 19-3 illustrates the necessary minimum connection to communicate between two computers. The common name for the communication cable illustrated in Figure 19-3, where wires 2 and 3 are crossed, is a null modem cable.

The null modem cable has female 9-pin or 25-pin D-shell connectors on each end. It is called a null modem cable because it replaces two modems. Each modem is a peripheral that enables two computer devices to communicate with each other. Make certain a null modem cable is used when configuring a serial connection through RSLinx.

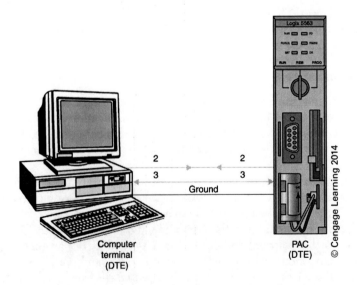

Figure 19-2 Connecting two computer devices with a straight-through cable.

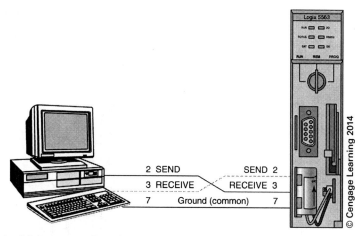

Figure 19-3 Serial wiring connections between a personal computer and serial channel of a ControlLogix controller using a null modem cable.

Communication Protocols

Even if two different PACs both support RS-232 communication, there is one other consideration when connecting two devices together: the protocol. The protocol is a set of rules that govern the way that data is formatted and timed as it is transmitted between the sending and receiving devices. Each manufacturer designs a protocol that defines data format, timing, sequence, and error checking. As a result, one controller from one PAC manufacturer will probably not be able to talk to a controller from another manufacturer, even if they both support RS-232 communication standards, as their protocols will differ. Rockwell Automation/Allen-Bradley's RS-232 communication protocol is known as RS-232 DF1. "DF1" identifies the protocol.

A serial connection to a PAC is a direct connection between the personal computer and the controller. Remember that if a modular ControlLogix is being used, then the only communication port available on most modular controllers is a serial port. The correct cable is needed in order to communicate between the personal computer and the controller. Because both are computers, we need a 1756-CP3 cable or equivalent, which is a null modem cable. Although the 1756-CP3 cable can be purchased from a Rockwell Automation distributor, a null modem cable can be purchased from many computer stores, or you can make your own cable. Do not confuse a null modem cable and a straight-through serial cable. They have different functions. For example, a straight-through serial cable is used to communicate between a personal computer and an operator interface terminal, like a Rockwell Automation PanelView. Because the PanelView is not a computer but a terminal, the null modem cable will not work. When serially connecting between two computers, a null modem cable is used. When connecting between a computer and a terminal, a straight-through serial cable is required.

Because a serial connection is a direct connection to the controller, the path is going to contain only two pieces: the driver name and the node number. Refer to Figure 19-4 for an example of a serial path to a ControlLogix controller.

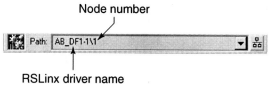

Figure 19-4 Serial communications path.
Used with permission Rockwell Automation, Inc.

LAB EXERCISE 1: Configuring a Serial Driver Using RSLinx Software

1. _____ Select a null modem serial cable or an Allen-Bradley CP3, and connect it between your personal computer's serial port and the controller's serial port. Refer to Figure 19-5 for the location of the CompactLogix serial port connection.

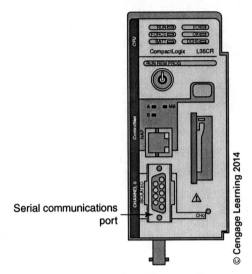

Figure 19-5 Identifying communication ports on L35CR controller.

Refer to Figure 19-6 to review a modular ControlLogix controller such as a 1756-L63 series B and serial port.

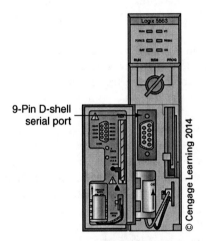

Figure 19-6 Identifying communication ports on a modular ControlLogix controller such as a 1756-L63 series B controller.

2. _____ Power up the ControlLogix.
3. _____ Power up your personal computer.
4. _____ Open the RSLinx software by clicking the icon on the computer desktop. The icon should look similar to that in Figure 19-7.

Figure 19-7 RSLinx Classic icon.
Used with permission Rockwell Automation, Inc.

CONFIGURING A SERIAL DRIVER USING RSLINX   527

As the software launches, the RSLinx splash screen, similar to the one shown in Figure 19-8 should display.

Figure 19-8 RSLinx Classic splash screen.
Used with permission Rockwell Automation, Inc.

5. _____ After RSLinx is running, click the configure driver's icon, as illustrated in Figure 19-9.

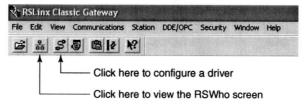

Figure 19-9 Select configure driver icon.
Used with permission Rockwell Automation, Inc.

6. _____ The Configure Drivers window should open. See Figure 19-10.
7. _____ Click the down arrow to view the list of drivers. Refer to the figure.
8. _____ Select the RS-232 DF1 devices driver.

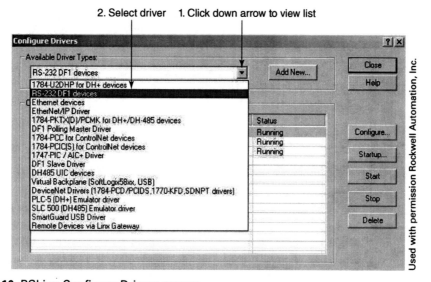

Figure 19-10 RSLinx Configure Drivers screen.

528 CONFIGURING A SERIAL DRIVER USING RSLINX

9. _____ Refer to Figure 19-11 and click Add New.
10. _____ The driver name can be changed if you desire in the Add New RSLinx Classic Driver Window. Probably the best option would be to accept the default driver name.
11. _____ Click OK to continue.

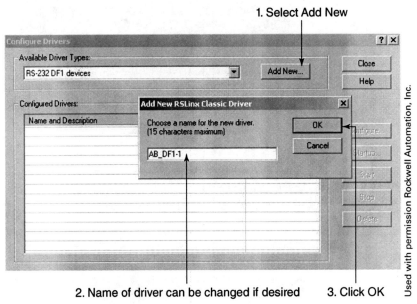

Figure 19-11 Serial driver configuration.

12. _____ The Configure RS-232 DF1 Devices window should display, as in Figure 19-12.

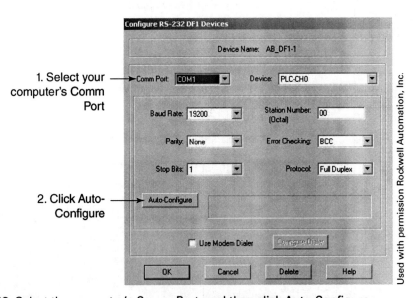

Figure 19-12 Select the computer's Comm Port, and then click Auto-Configure.

13. _____ Refer to #1 in the figure as you select the communication port the serial cable is connected to on your personal computer. If you are using a notebook personal computer, typically there is only one communication port, so COM1 would be the typical selection.
14. _____ Do not change anything else in this window.
15. _____ Click Auto-Configure, #2 in the figure.

CONFIGURING A SERIAL DRIVER USING RSLINX   529

16. _____ Figure 19-13 shows Auto Configuration Successful! in the information box to the right of the Auto Configure button. This signifies that the driver was successfully configured. Note B in the figure. The device was identified as either a ControlLogix or CompactLogix.
17. _____ Click OK, C in Figure 19-13.

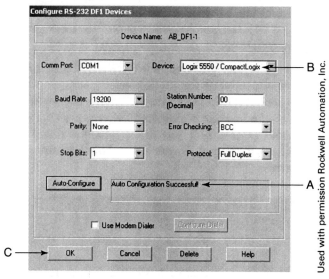

Figure 19-13 Auto Configuration Successful!

Having Communication Problems?

18. _____ Note in Figure 19-14 that the information to the right of the Auto-Configure button states: "Failed to find the baud and parity!" "Check all cables and switch settings!"

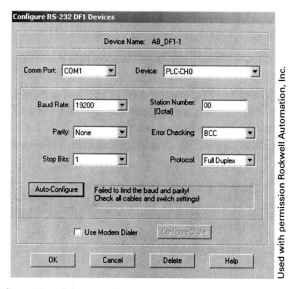

Figure 19-14 Auto configuration failed to find the controller.

This message states that the device you were attempting to communicate with did not respond. Check the following:

1. Is the PAC or device you are attempting to communicate with turned on?
2. Is Correct Communication port selected on the Configure RS-232 DF1 Devices window?

3. Are you using the correct serial cable?
4. Are you using a good serial cable?
5. Are there problems with the computer's serial port?
6. Are there problems with the PAC serial port?

Note: The error message states to check all cables and switch settings. If you are attempting to communicate between a personal computer and a ControlLogix, CompactLogix, PLC-5, or SLC 500, the switch settings part of the message does not pertain to the driver you are trying to configure.

7. _____ After fixing the problem, click OK in Figure 19-13 and the Auto Configuration Successful message should display.
8. _____ This should return you to the Configure Drivers window. See Figure 19-15.
9. _____ Under the heading Name and Description, Notice the text "RUNNING." In our exercise, there is only one driver configured and listed. In reality, there may be a number of different drivers listed. Some drivers may be running, stopped, or even in error.

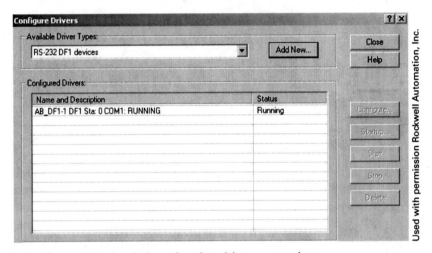

Figure 19-15 Configure Drivers window showing driver as running.

10. _____ If the text running is displayed as in the figure, close the Configure Drivers window.
11. _____ To verify that you are actually communicating with the PAC, open RSWho. Refer to Figure 19-9 to review which icon opens RSWho.
12. _____ The RSWho window should look similar to the one shown in Figure 19-16.

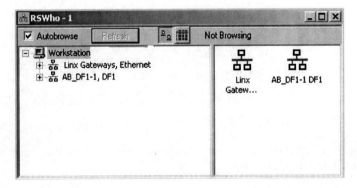

Figure 19-16 RSWho screen.
Used with permission Rockwell Automation, Inc.

13. _____ In the left pane, expand the AB_DF1-1 driver by clicking +, A in Figure 19-17.

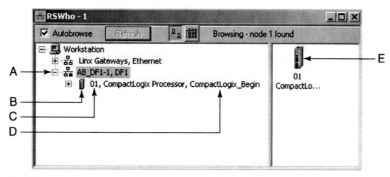

Figure 19-17 Verifying communications using RSWho.
Used with permission Rockwell Automation, Inc.

14. _____ The RSWho window should look similar to the one in the figure.
15. _____ For this example, we are communicating with a CompactLogix controller. Labels B and E in the figure display the CompactLogix controller icon. If you are configuring a modular ControlLogix controller, the icon will resemble a modular controller instead of a CompactLogix.
16. _____ Label C identifies the node address of this controller. Note, serial communications is a direct, or point-to-point, connection; so actually there are no nodes because this is not a network connection. Controllers using serial communications typically default to node 01.
17. _____ The name of the project residing in the controller is CompactLogix_Begin, as identified by D in Figure 19-17.
18. _____ The right pane of the RSWho window, label E, displays the same information as in the left pane. If your RSWho window looks similar to the one in the figure, you have successfully configured the serial driver.
19. _____ This completes the serial driver setup.

IDENTIFYING PROBLEMS

If the RSWho screen has an X through the controller, as in Figure 19-18, communications was established at one point but has currently been lost.

Check the following:

1. Is the PAC still powered?
2. Is the cable disconnected?
3. Is the cable damaged?
4. Has the PAC serial channel configuration changed?

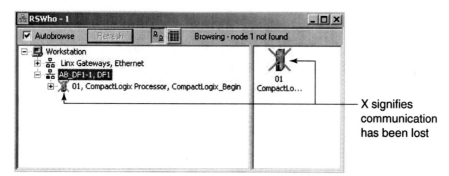

Figure 19-18 Communications, even though established at one point, is currently not available.
Used with permission Rockwell Automation, Inc.

CHAPTER

20

Configuring the Keyspan by Tripp Lite, High-Speed USB to Serial Adapter Model USA-19HS

Complete this exercise if you wish to use a personal computer USB port to establish communications with a ControlLogix family controller using the Keyspan-by-Tripp Lite High-Speed USB to Serial Adapter model USA-19HS. This lesson steps you through configuring the computer USB port and setting up an RSLinx serial driver. The RSLinx software pictures used in this chapter are from RSLinx versions 2.57 and 2.59. If you have a different version of RSLinx, the computer screens might be a bit different. However, the configuration steps are the same.

CONFIGURING THE KEYSPAN USB TO SERIAL ADAPTER

USB ports have replaced standard 9-pin D-shell serial ports on newer personal computers. To communicate with controllers with standard DB9 serial communication ports, a USB to serial adapter like the Keyspan by Tripp Lite High-Speed USB Serial Adapter, illustrated in Figure 20-1, can be used to connect between a computer USB port and the standard null modem serial cable used for PAC communications. The figure shows the USB cable and connector on the top right of the figure and the standard DB9 serial connector on the front of the adapter. The USB connector plugs into the personal computer, whereas a Rockwell Automation 1756-CP3 null modem cable plugs into the DB9 connector. The other end of the serial cable plugs into the controller's serial port as usual.

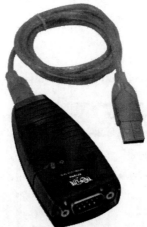

Figure 20-1 Keyspan by Tripp Lite High-Speed USB to Serial Adapter model USA-19HS. Note the 9-pin D-shell DB9 serial port.
Used with permission by Tripp Lite, Chicago, IL

Before attempting to configure a serial driver using the USB adapter, the USB adapter's software drivers must be loaded onto your computer. The CD included with the USB adapter is used to load the required drivers. The USB adapter should not be attached to the personal computer while the drivers are being loaded. Figure 20-2 shows the CD's main menu. An electronic copy of the user's manual and other important information is available by clicking the proper button in the upper left of the screen. To install the software driver, click Install Software and follow the installation steps.

Figure 20-2 Keyspan by Tripp Lite High-Speed USB Serial Adapter driver software screen.

After installation of the software, you might want to put a shortcut icon, like the one illustrated in Figure 20-3, on your computer desktop.

Figure 20-3 Keyspan USB Serial Adapter (USA19H) Assistant icon.
Used with permission by Tripp Lite, Chicago, IL

Before you can configure an RSLinx driver, you must determine the communications port assigned to the USB adapter. By double-clicking the icon, the Keyspan by Tripp Lite Serial Assistant opens, as shown in Figure 20-4. Note that the window identifies COM 3 as busy. The Keyspan by Tripp Lite software has assigned this device as COM port 3. As you configure your device, note the actual COM port assigned; it is not always COM 3, as illustrated in the figure. Remember the assigned COM port number, as it will be selected in the RSLinx driver configuration screen. If the incorrect COM port is selected, the communication driver will not configure.

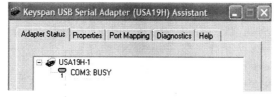

Figure 20-4 Determining the Keyspan by Tripp Lite High-Speed USB serial port assignment.
Used with permission by Tripp Lite, Chicago, IL

If you are using a different serial to USB adapter, it may not be as easy to identify the device's communications port. Another method to determine the USB serial adapter's assigned port is through Windows Device Manager. Figure 20-5 illustrates Windows XP's Device Manager, and in this case the device has been assigned COM4. Windows Vista and Windows 7 Device Manager look similar.

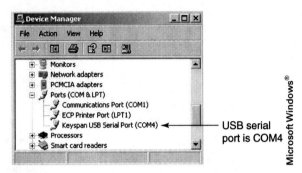

Figure 20-5 USB serial port is COM 4.

LAB EXERCISE 1: Configuring the RSLinx Serial Driver

1. _____ Select a null modem serial cable such as an Allen-Bradley 1756-CP3, and connect it between the serial DB9 connector on the USB adapter and the ControlLogix controller serial port. For example, Figure 20-6 illustrates CompactLogix communication ports.

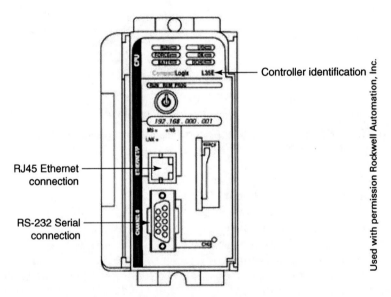

Figure 20-6 Identifying communication ports on L35E controller.

Refer to Figure 20-7 to review a modular ControlLogix controller such as a 1756-L63 series B and serial port.

2. _____ Power up the ControlLogix.
3. _____ Power up your personal computer.

CONFIGURING THE KEYSPAN BY TRIPP LITE, HIGH-SPEED USB TO SERIAL ADAPTER MODEL USA-19HS   535

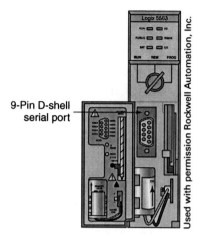

Figure 20-7 Identifying communication ports on a modular ControlLogix controller such as a 1756-L63 series B controller.

4. _____ Open RSLinx by clicking the icon on your computer desktop. The icon should look similar to the one shown in Figure 20-8 below.

Figure 20-8 RSLinx Classic icon.
Used with permission Rockwell Automation, Inc.

The RSLinx splash screen, similar to the one in Figure 20-9, should display.

Figure 20-9 RSLinx Classic splash screen.
Used with permission Rockwell Automation, Inc.

5. _____ After RSLinx launches, click the configure driver icon, as illustrated in Figure 20-10.

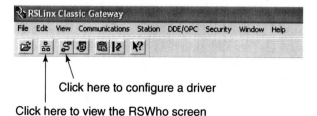

Figure 20-10 Select configure driver icon.
Used with permission Rockwell Automation, Inc.

6. _____ The Configure Drivers window should open. See Figure 20-11 below.

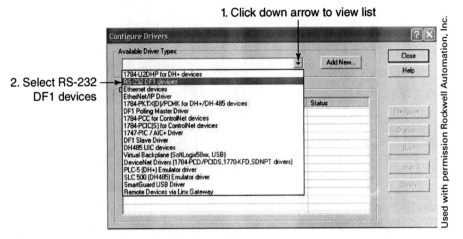

Figure 20-11 RSLinx Configure Drivers screen.

7. _____ Refer to the figure as you click the down arrow to view the list of drivers.
8. _____ Select the driver for the RS-232 DF1 devices.
9. _____ Refer to Figure 20-12, #1, and click Add New.

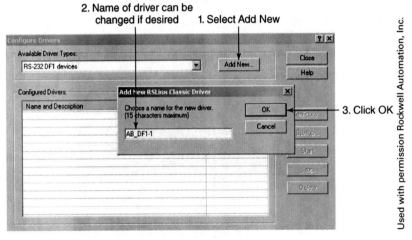

Figure 20-12 Serial driver configuration.

10. _____ Referring to the figure, the driver name can be changed if you wish, by entering a new name in #2 in the figure.
11. _____ Click OK to continue.
12. _____ The Configure RS-232 DF1 Devices window should display, as in Figure 20-13.
13. _____ Refer to #1 in Figure 20-13 as you select the USB adapter's serial port assignment.
14. _____ Verify the proper Comm Port has been selected from the USB serial port assignment. Refer to Figure 20-4 or Figure 20-5 for help on how to determine the USB serial port assignment. For our example, COM3 is selected.
15. _____ Click #2 in Figure 20-13 to Auto-Configure the driver.
16. _____ Figure 20-14 shows the text Auto Configuration Successful! in the information box to the right of the Auto-Configure button. This signifies that the driver was

CONFIGURING THE KEYSPAN BY TRIPP LITE, HIGH-SPEED USB TO SERIAL ADAPTER MODEL USA-19HS

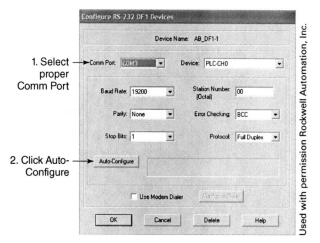

Figure 20-13 Select the computer's communications port, then Auto-configure.

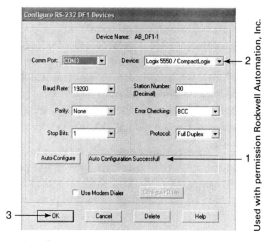

Figure 20-14 Auto Configuration Successful!

successfully configured. Note #2 in the figure. The device was identified as either a ControlLogix or CompactLogix.

17. _____ Next click OK, #3 in the figure, to exit this window.

LAB EXERCISE 2: Configuration Problems

Figure 20-15 displays information to the right of the Auto-Configure button stating "Failed to find the baud and parity! Check all cables and switch settings!" This example illustrates a common problem when configuring serial communications using a USB to serial converter with COM1 was mistakenly selected.

This message states that the device you were attempting to communicate with did not respond. Check the following:

1. _____ Is the PAC or device you are attempting to communicate with turned on?
2. _____ Is the correct Comm Port selected on the Configure RS-232 DF1 Devices window?
3. _____ Are you using the correct serial cable?
4. _____ Are you using a good serial cable?

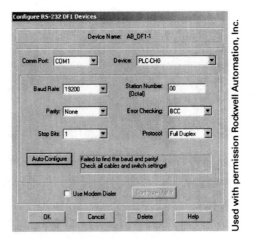

Figure 20-15 Auto configuration failed to find the controller.

5. _____ Is the USB Serial adapter working correctly? Check the user's manual and onboard LED to troubleshoot the adapter.
6. _____ If using a different USB to serial adapter, has this adapter been used successfully to configure serial PAC communications previously? Not all USB to serial adapters work reliably when attempting to establish serial communications to PACs. Some USB to serial adapters will not work at all.
7. _____ Are there problems with the computer's serial port?
8. _____ Are there problems with PAC's serial port?

Note: The error message states to check all cables and switch settings. If you are attempting to communicate between a personal computer and a ControlLogix, CompactLogix, PLC 5, SLC 500 or other Rockwell Automation PLC, the switch settings part of the message does not pertain to the driver you are currently trying to configure.

9. _____ After fixing the problem, the Auto Configuration Successful message should display; refer back to Figure 20-14.
10. _____ Click OK to return to the Configure Drivers window.
11. _____ As you refer to Figure 20-16, note the text running in two places.

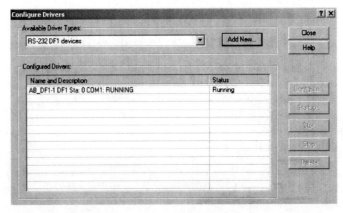

Figure 20-16 Configure Drivers window showing driver as running.
Used with permission Rockwell Automation, Inc.

12. _____ If the text "Running" is displayed, as in the figure, close the Configure Drivers window. If text other than "Running" is displayed, the driver was not successfully configured. Probably the best recovery method is to select the driver and then click Delete near the bottom right of the Configure Drivers window. Recheck everything and try to reconfigure the driver.
13. _____ Open RSWho. Refer to Figure 20-10 to review which icon opens RSWho.
14. _____ The RSWho window should look similar to Figure 20-17.

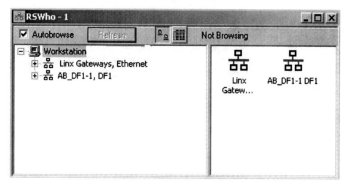

Figure 20-17 RSWho screen.
Used with permission Rockwell Automation, Inc.

15. _____ In the left pane, expand the AB_DF1-1 driver by clicking +, #1 in Figure 20-18.
16. _____ The RSWho window should look similar to the figure.
17. _____ In Figure 20-18, #2 should display either the ControlLogix or CompactLogix controller icon. The CompactLogix icon is displayed in the figure.
18. _____ In the figure, #3 identifies the node address of this controller. Serial communications is a direct, or point-to-point, connection, so actually there are no nodes because this is not a network. Controllers using serial communications typically default to node 01 for Rockwell Automation PACs.

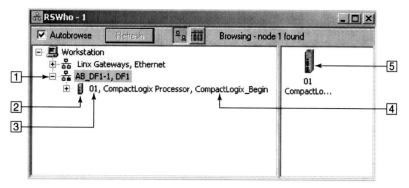

Figure 20-18 Verifying communications using RSWho.
Used with permission Rockwell Automation, Inc.

19. _____ The name of the project residing in the controller is CompactLogix_Begin, as identified by #4 in the figure.
20. _____ The right pane of the RSWho window, #5 displays the same information as in the left pane. If your RSWho window looks similar to the one in Figure 20-18, you have successfully configured the serial driver.
21. _____ This completes the serial driver setup.

IDENTIFYING PROBLEMS AFTER SUCCESSFUL CONFIGURATION

If the RSWho screen has an X through the controller, as in Figure 20-19, communications were established at one point but have currently been lost.

Check the following:

1. Is the PAC still powered?
2. Is the cable disconnected?
3. Is the cable damaged?
4. Has the PAC serial channel configuration changed?

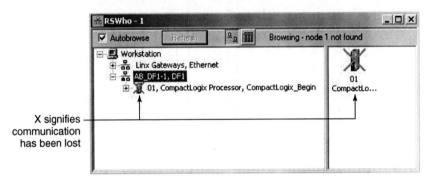

Figure 20-19 Communication, even though established at one point, is currently unavailable.
Used with permission Rockwell Automation, Inc.

CHAPTER

21

Configuring an RSLinx Serial Driver Using a Rockwell Automation 9300-USBS USB to Serial Adapter

Complete this chapter if you wish to configure an RSLinx serial driver using a Rockwell Automation 9300-USBS USB to serial adapter. The 9300-USBS is shown in Figure 21-1. This kit is used when your personal computer has a standard serial port rather than a USB port. This adapter is used to communicate between a personal computer with a standard 9-pin D-shell serial port and a PAC also with a standard serial port. The 9300-USBS kit comes with a standard USB A/B cable, the USB serial adapter, and a CD containing the USB drivers for the personal computer and a quick start card. Once the personal computer USB drivers are loaded, RSLinx software is used to configure the standard DF-1 serial driver.

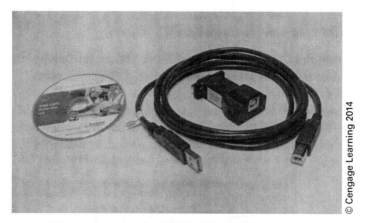

Figure 21-1 Rockwell Automation 9300-USBS USB to Serial Adapter.

Follow these procedures if you want to install the USB driver on your personal computer and use the standard RSLinx DF-1 serial driver to establish communications with a CompactLogix or 1756-L6 series or earlier modular controller, using the controller's integrated 9-pin D-shell serial port. Once configured, this adapter can also be used to communicate with the SLC 500 family processors with serial ports, as well as with PLC 5 processors. The appropriate serial cables are required for these PAC platforms.

The first section of this chapter steps you through the initial install of the USB driver on a personal computer. This has to be completed before you can go into RSLinx and configure the driver. The process only has to be completed once. The second section steps you through configuring RSLinx drivers, setting up the communication path in RSLogix 5000 project, and then downloading the project.

STEP 1: INITIAL INSTALL OF USB DRIVERS ON PERSONAL COMPUTER

If the USB drivers for the 9300-USBS are already installed on your personal computer, go on to step 2 to connect the cable and configure the RSLinx driver. This step is only required for the initial installation of the USB drivers on your computer and typically does not have to be repeated. In addition to the 9300-USBS adapter, a standard null modem serial cable, 1747-CP3 or 1756-CP3 cable, is necessary to connect between the 9300-USBS adapter and the 9-pin D-shell connector on your controller.

There are two USB drivers that must be installed on your personal computer.

Install First USB Driver

1. _____ Connect the USB A/B cable and 9300-USBS adapter to a personal computer USB port.
2. _____ The PAC does not need to be powered up at this time.
3. _____ The Found New Hardware Wizard should open, as illustrated in Figure 21-2.

Figure 21-2 The Found New Hardware Wizard.

4. _____ Insert the CD included in the 9300-USBS kit into your personal computer.
5. _____ Select Install the software automatically (Recommended).
6. _____ Click Next to continue.
7. _____ The software will begin the install. Multiple windows and messages will be displayed in the bottom of the window. Wait until the installation completes.
8. _____ When the installation is completed, a window similar to that shown in Figure 21-3 displays, informing that the wizard has installed the first USB driver.

Figure 21-3 The Found New Hardware Wizard has installed the first USB driver.

9. _____ Click Finish to continue.

Install Second USB Driver

10. _____ The Found New Hardware Wizard window to install the USB Serial Port will open, as in Figure 21-4.

Figure 21-4 The Found New Hardware Wizard to install the USB Serial Port.

11. _____ Select Install the software automatically (Recommended).
12. _____ Click Next.
13. _____ The software will begin to install. Multiple windows and messages will be displayed in the bottom of the window. Wait until the installation completes. See Figure 21-5.

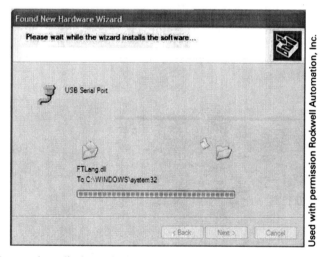

Figure 21-5 One of many installation windows displayed as software installs.

14. _____ When the software is installed, click Finish as shown in Figure 21-6.
15. _____ Click Finish to close the Wizard. Your personal computer now has the USB Serial Port driver installed.
16. _____ You will probably get the message shown in Figure 21-7, asking you to reboot your personal computer. Click Yes to continue.
17. _____ Remove the USB driver CD from your computer.
18. _____ Next, open RSLinx to view the drivers installed, and get ready to configure communications for download.

544 CONFIGURING AN RSLINX SERIAL DRIVER

Figure 21-6 The Found New Hardware Wizard has installed the USB Serial Port driver.

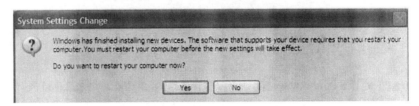

Figure 21-7 The Found New Hardware Wizard has installed the USB Serial Port driver. Click Yes to reboot your computer.
Used with permission Rockwell Automation, Inc.

RSLINX DRIVER SETUP

1. _____ Power up your ControlLogix.
2. _____ Select a null modem serial cable or an Allen-Bradley 1747-CP3 or 1756-CP3, and connect it between your personal computer's serial port and the controller's serial port. Refer to Figure 21-8 for the location of a CompactLogix serial port connection.

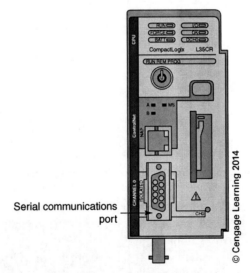

Serial communications port

Figure 21-8 Identifying communication ports on the L35E controller.

Refer to Figure 21-9 to review a modular ControlLogix controller such as a 1756-L63 series B and serial port.

CONFIGURING AN RSLINX SERIAL DRIVER 545

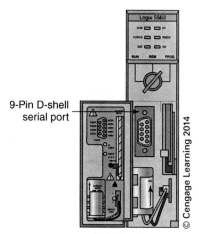

Figure 21-9 Identifying communication ports on a modular ControlLogix controller such as a 1756-L63 series B controller. (Used with permission of Rockwell Automation, Inc.)

3. _____ Before you open RSLinx and configure the driver, you need to determine what communications port the personal computer has assigned to the 9300-USBS serial adapter. Figure 21-10 illustrates Windows 7's Device Manager, and in this case the device has been assigned COM5. The most common problem people have when configuring this driver is figuring out the communications port number. The port number could be almost anything. In this example, the port was assigned as 5. You need to determine what communications port was assigned by your computer and use that communications port as you configure the driver. Windows XP and Windows Vista Device Manager look similar.

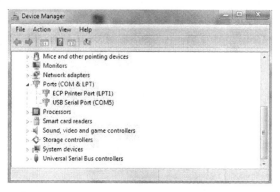

Figure 21-10 Windows 7's Device Manager window showing the serial to USB converter as COM5.
Used with permission Rockwell Automation, Inc.

4. _____ Open the RSLinx software by clicking the icon on the computer desktop. The icon should look similar to Figure 21-11.

Figure 21-11 RSLinx Classic icon.
Used with permission Rockwell Automation, Inc.

As the software launches, the RSLinx splash screen, similar to the one depicted in Figure 21-12 should display.

546 CONFIGURING AN RSLINX SERIAL DRIVER

Figure 21-12 RSLinx Classic splash screen.
Used with permission Rockwell Automation, Inc.

5. _____ After RSLinx is running, click Configure Driver icon, as illustrated in Figure 21-13.

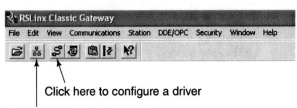

Figure 21-13 Select the Configure Driver icon.
Used with permission Rockwell Automation, Inc.

6. _____ The Configure Drivers window should open. See Figure 21-14.

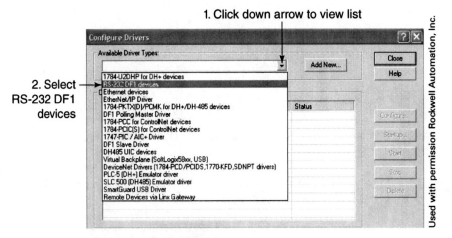

Figure 21-14 RSLinx Configure Drivers screen.

7. _____ Refer to the figure as you click the down arrow to view the list of drivers.
8. _____ Select the RS-232 DF1 devices driver.
9. _____ Click Add New.
10. _____ The driver name can be changed, if you desire, in the Add New RSLinx Classic Driver window (Figure 21-15). The best option is to accept the default driver name.
11. _____ Click OK to continue.

CONFIGURING AN RSLINX SERIAL DRIVER 547

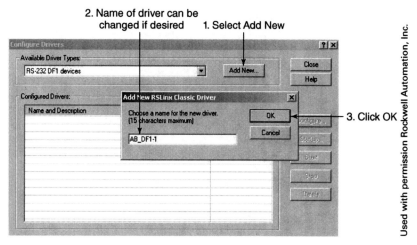

Figure 21-15 Serial driver configuration.

12. _____ Refer to #1 in Figure 21-16 as you select the communication port assigned to the USB port the serial cable is connected to on your personal computer. Remember, for this example the computer is assigned COM5, but your port number may be different. Be sure to select the correct port assignment for your computer.

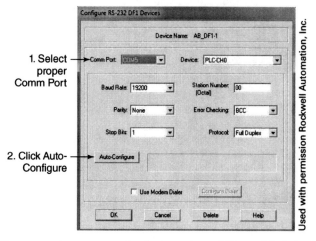

Figure 21-16 Select the computer communications port, and then click Auto-Configure.

13. _____ Do not change anything else in this window.
14. _____ In Figure 21-16 click Auto-Configure, #2.
15. _____ Figure 21-17 shows Auto Configuration Successful! in the information box to the right of the Auto-Configure button. This signifies that the driver was successfully configured. Note #1 in the figure. Callout #2 identifies the controller as either a ControlLogix or CompactLogix.
16. _____ Click OK, #3 in the figure to continue.
17. _____ If you are having configuration problems, continue to the next step. If your communication configuration was successful, as illustrated in Figure 21-17, continue to step 18.

Having Communication Problems?

18. _____ Note in Figure 21-18 that the information to the right of the Auto-Configure button states: "Failed to find the baud and parity! Check all cables and switch settings!"

548 CONFIGURING AN RSLINX SERIAL DRIVER

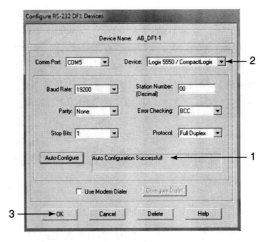

Figure 21-17 Auto Configuration Successful!
Used with permission Rockwell Automation, Inc.

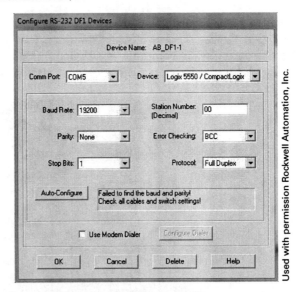

Figure 21-18 Auto configuration failed to find the controller.

This message states that the device you were attempting to communicate with did not respond. Check the following:

a. Is the PAC or device you are attempting to communicate with turned on?
b. Is the correct communication port selected on the Configure RS-232 Devices window? When using a serial to USB adapter, this is the most common problem.
c. Are you using the correct serial cable?
d. Are you using a good serial cable?
e. If you are using a different serial to USB adapter, is it compatible with the PLC you are attempting to configure?
f. Are there problems with the computer's serial port?
g. Are there problems with the PACs serial port?

Note: The error message states to check all cables and switch settings. If you are attempting to communicate between a personal computer and a ControlLogix, CompactLogix, PLC 5, or SLC 500, the switch settings part of the message does not pertain to the driver you are trying to configure.

19. _____ After fixing the problem, click OK (#3 in Figure 21-17) and the Auto Configuration Successful! message should display.
20. _____ This should return you to the Configure Drivers window like the one in Figure 21-19.

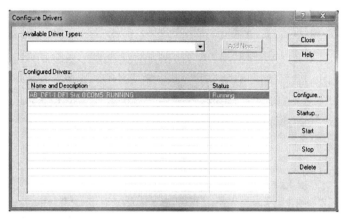

Figure 21-19 Configure Drivers window showing driver as "Running."
Used with permission Rockwell Automation, Inc.

21. _____ Under the heading Name and Description, notice the text "Running." In our exercise, there is only one driver configured and listed in the list. In reality, there may be a number of different drivers listed. Some drivers may be running, stopped, or even in error.
22. _____ If the text "Running" is displayed, as in the figure, close the Configure Drivers window.
23. _____ To verify communication has been established with the PAC, open RSWho. Refer to Figure 21-13 to review which icon opens RSWho.
24. _____ The RSWho window should look similar to Figure 21-20.

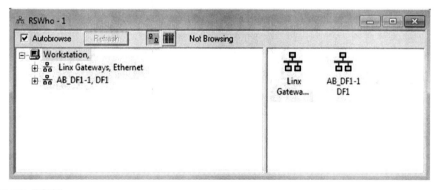

Figure 21-20 RSWho screen.
Used with permission Rockwell Automation, Inc.

25. _____ In the left pane, expand the AB_DF1-1 driver by clicking the +, which is A in Figure 21-21.
26. _____ In this example, we are communicating with a CompactLogix controller. Callouts B and D in the figure display the CompactLogix controller icon. If you are configuring a modular ControlLogix controller, the icon will resemble a modular controller instead of a CompactLogix.
27. _____ Callout C identifies the node address of this controller as 01. Note, serial communications is a direct, or point-to-point, connection; so actually there are no nodes as this is not a network connection. Controllers using serial communications typically default to node 01.

550 CONFIGURING AN RSLINX SERIAL DRIVER

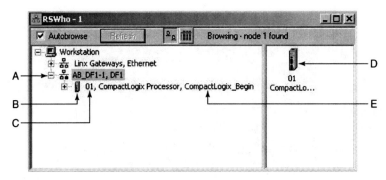

Figure 21-21 Verifying communications using RSWho.
Used with permission Rockwell Automation, Inc.

28. _____ The name of the project residing in the controller is CompactLogix_Begin as identified by E in Figure 21-21.
29. _____ The right pane of the RSWho window, callout E, displays the same information as in the left pane. If your RSWho window looks similar to the figure, you have successfully configured the serial driver.
30. _____ This completes the serial driver setup.

Identifying Problems

If the RSWho screen has an X through the controller, as in Figure 21-22, communications was established at one point but has currently been lost.
Check the following:

1. Is the PAC still powered?
2. Is the cable disconnected?
3. Is the cable damaged?
4. Has the PAC serial channel configuration changed?

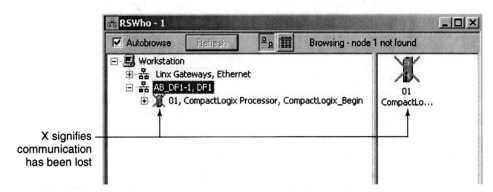

Figure 21-22 Communications, even though established at one point, is currently not available.
Used with permission Rockwell Automation, Inc.

CHAPTER

22

Installing and Configuring a USB Driver for 1756-L7 Series Controllers

Follow these procedures if you want to install the USB driver on your personal computer and use the RSLinx USB driver to establish communications with the 1756-L7 controllers using the controller's integrated USB port. Keep in mind RSLogix 5000 software version 18.11 is the minimum in order to incorporate the 1756-L73 or L75 controllers into a project and configure USB communications. RSLogix 5000 version 19.11 is the minimum to incorporate the 1756-L72 or L74 controllers into a project RSLogix 5000 software version 20 is the minimum required to use the L71 controller in a project. The 1756-L71, L72, L73, L74, and L75 controllers were the first ControlLogix modular controllers to have a USB port replacing the older standard 9-pin D-shell serial port. If you have 1756-L6 series controllers or earlier versions, this chapter does not pertain to these controllers because they do not have a USB port.

The first section of this chapter steps you through the initial install of the USB driver on a personal computer. This has to be completed before you can go into RSLinx and configure the driver. This only has to be completed once. The second section guides you through configuring RSLinx drivers, setting up the communication path in an RSLogix 5000 project, and then downloading the project.

INITIAL INSTALL OF USB DRIVERS ON PERSONAL COMPUTER

1. _____ Power up your ControlLogix.
2. _____ Connect a standard USB 2.0 A/B cable between the 1756-L7 controller and your personal computer. Refer to Figure 22-1 to identify the 1756-L7 controller's USB port.

Figure 22-1 1756-L7 Controller with USB port identified.
Used with permission Rockwell Automation, Inc.

If you have a Rockwell Automation 9300-USBS USB to serial adapter, shown in Figure 22-2, you can use the USB cable included in the kit. This kit is used when your personal computer has a standard serial port rather than a USB port. This adapter would be used to communicate between a personal computer with a standard serial port and a programmable automation controller (PAC) also with a standard serial port. Refer to Chapter 20 for the steps to configure a serial driver using the 9300-USBS.

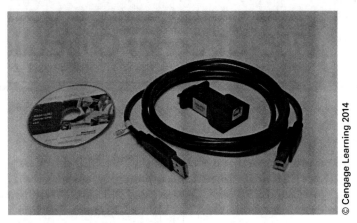

Figure 22-2 Rockwell Automation 9300-USBS USB to serial adapter.

3. _____ The Found New Hardware Wizard should open, as illustrated in Figure 22-3.

Figure 22-3 The Found New Hardware Wizard.

4. _____ Refer to Figure 22-4. Select No, not this time, regarding connecting to Windows Update.
5. _____ Click Next to continue.
6. _____ The next window is displayed as shown in Figure 22-5.
7. _____ Select Install the software automatically (Recommended).
8. _____ Click Next.
9. _____ The software will install as illustrated in Figure 22-6. The figure shows only one part of the process. Multiple messages will be displayed in the bottom of the window. Wait until the installation completes.

INSTALLING AND CONFIGURING A USB DRIVER FOR 1756-L7 SERIES CONTROLLERS 553

Figure 22-4 Select No regarding connecting to Windows Update.

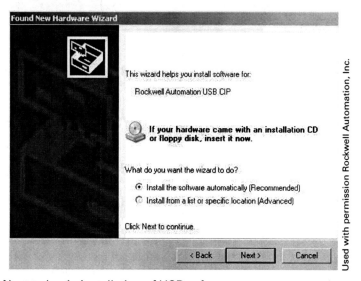

Figure 22-5 Click Next to begin installation of USB software on your personal computer.

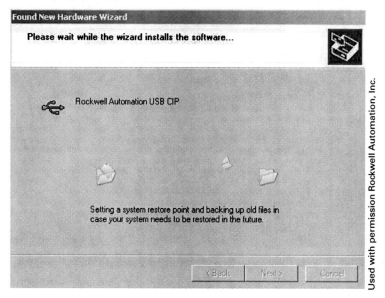

Figure 22-6 USB software being installed on your personal computer.

554 INSTALLING AND CONFIGURING A USB DRIVER FOR 1756-L7 SERIES CONTROLLERS

10. _____ Refer to Figure 22-7 and click Finish when the software is installed.

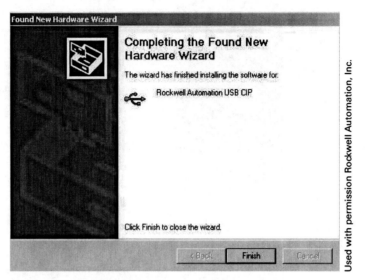

Figure 22-7 Software installation completed.

11. _____ After you click Finish to close the Wizard, your personal computer will have the USB driver installed. Next, you will open RSLinx to view the drivers installed and get ready to configure communications for download.

RSLINX DRIVER SETUP

1. _____ Open RSLinx software. Make sure RSLinx is current for the version of RSLogix 5000 software you are using.
2. _____ Open the RSLinx Configure Drivers window.
3. _____ When you open the RSLinx Configure Drivers window, you should see the AB_VBP-1 driver running, as illustrated in Figure 22-8. Note that we have deleted all other drivers from RSLinx for ease of viewing this configuration.

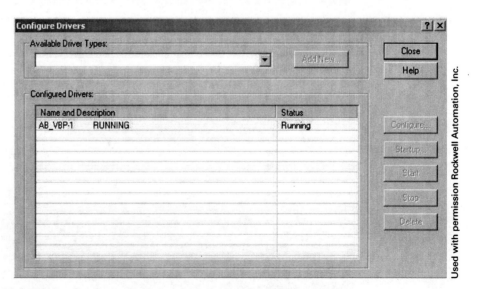

Figure 22-8 RSLinx Configure Drivers window showing driver running.

4. _____ Open RSWho. Refer to Figure 22-9; you should see two drivers, the Virtual Chassis driver and the USB driver.

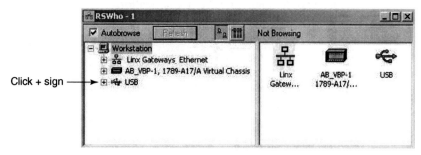

Figure 22-9 RSWho showing configured drivers. Click + to expand.
Used with permission Rockwell Automation, Inc.

5. _____ Click + to the left of the USB driver.
6. _____ The controller should be displayed in both the right and left panes, as in Figure 22-10.

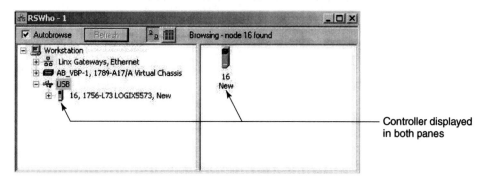

Figure 22-10 RSWho window showing communication has been established with the controller.
Used with permission Rockwell Automation, Inc.

The controller is displayed directly below the USB driver in the left pane and also in the right pane of RSWho, which shows that communications have been established with the controller. The same information is displayed in both panes. The name of the project currently in the controller is New. If the controller is not displayed, that tells you that communications was never established. If the controller icon has an X through it, that signifies that communications was established at one time, but not currently.

CONFIGURING COMMUNICATIONS PATH AND DOWNLOAD OF A RSLOGIX 5000 PROJECT

Before you can upload, download, or go online with a controller, you must verify whether the correct communications path is displayed in the RSLogix 5000 project Path toolbar. If the correct communications path is displayed, the project can perform the desired action. If the path is incorrect, you must create the correct path before you can continue. Keep in mind that the path being displayed is referred to as the current path. That does not necessarily mean that it is the correct path for what you wish to do at this moment. It is your responsibility to verify the correctness of the path to ensure that you do not download to the incorrect controller. To download an RSLogix 5000 project.

1. _____ Open RSLogix 5000 software.
2. _____ Open the project to be downloaded.

3. _____ Review the path displayed in the Path toolbar.
4. _____ The current path displayed in Figure 22-11 shows an example of an Ethernet path. Because we wish to use USB communications, this path must be replaced with the proper path to enable USB communications.

Figure 22-11 Path toolbar displaying an Ethernet communications path.
Used with permission Rockwell Automation, Inc.

5. _____ There are a number of ways to configure the desired communications path. One way is to click the network icon on the left side of the Path toolbar, as illustrated in Figure 22-12.

Figure 22-12 One way to continue configuring a new path is to click the network icon.
Used with permission Rockwell Automation, Inc.

6. _____ Another method is to click Communications from the RSLogix 5000 software Menu Bar and then click Who Active, as illustrated in Figure 22-13. Either method takes you to the same window.

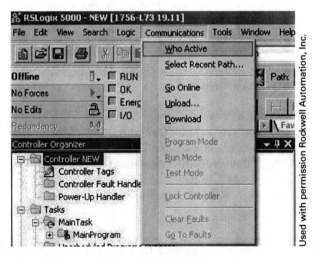

Figure 22-13 Click Communications and then Who Active.

7. _____ Using either method, the Who Active window should open as shown in Figure 22-14. The RSWho window in the figure has the USB driver in addition to a standard Ethernet IP driver.
8. _____ Note the Go Online, Upload …, and Download buttons are grayed out because a valid communications path has yet to be configured.
9. _____ To configure a communications path using the USB driver, click + to the left of the USB driver, as illustrated in Figure 22-14.
10. _____ The controller should display below the USB driver, as in Figure 22-15.

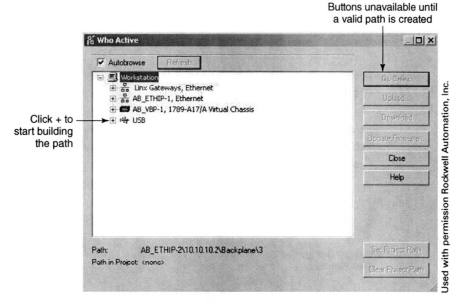

Figure 22-14 Who Active window showing available configured drivers.

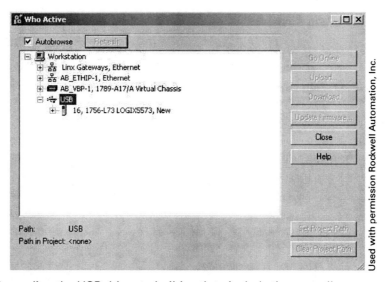

Figure 22-15 Expanding the USB driver to build path to include the controller.

11. _____ Click the controller. See Figure 22-16, #1.
12. _____ Refer to #2 in Figure 22-16 to see the Go Online, Upload ..., and Download buttons are now active because a valid communications path has been configured.
13. _____ Clicking the Set Project Path button, #3 in the figure, displays the new path in the RSLogix 5000 toolbar. The USB driver is now displayed, replacing the original Ethernet path from Figure 22-11. When this project is saved, the new path will be saved with the project. The USB path will then be the current path, as seen in Figure 22-17.
14. _____ On the right side of the window in Figure 22-18, click Download.

558 INSTALLING AND CONFIGURING A USB DRIVER FOR 1756-L7 SERIES CONTROLLERS

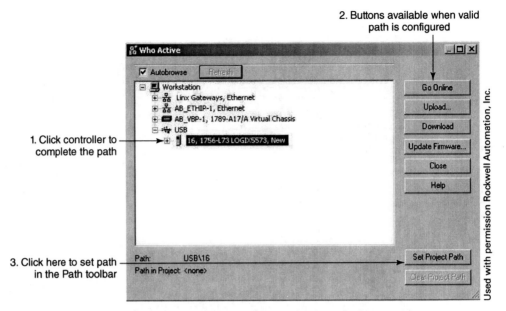

Figure 22-16 Select the controller to complete the communications path.

Figure 22-17 Path toolbar displaying the USB driver as the current path.
Used with permission Rockwell Automation, Inc.

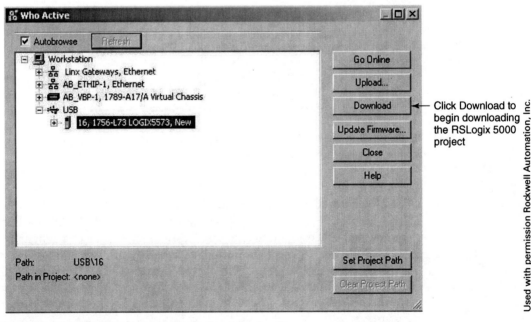

Figure 22-18 Click Download to begin the download.

15. _____ The Download window, as illustrated in Figure 22-19, should display.

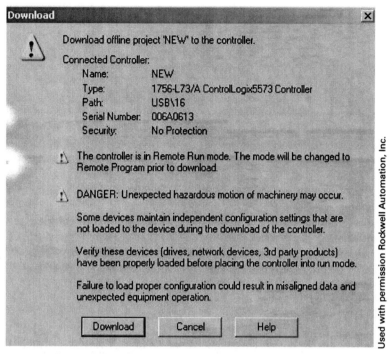

Figure 22-19 Click Download to continue.

16. _____ Click Download to begin the download.
17. _____ When the download is complete, put the controller in Run mode.
18. _____ The Online toolbar, as illustrated in Figure 22-20, should say Rem(ote) Run, and the path should display the USB driver. The gears to the left of the Path toolbar should be turning, signifying you are online.

Figure 22-20 RSLogix 5000 project showing controller in Remote Run mode, online, and using the USB driver.
Used with permission Rockwell Automation, Inc.

CHAPTER

23

Determine and Modify a Personal Computer's IP Address

Complete this lesson if you wish to determine a personal computer's IP address; determine, modify, or assign a personal computer's IP address using Windows 2000 or XP; determine, modify, or assign a personal computer's IP address using Windows Vista or Windows 7; or employ the MS-DOS ipconfig command to verify personal computer's IP address.

INTRODUCTION

Our personal computer is running RSLogix 5000 software with our PAC project, which needs to be downloaded, into the ControlLogix controller, using Ethernet communications. Before the personal computer and the ControlLogix controller can communicate using RSLinx software, the personal computer's IP address must be compatible with the IP address in the ControlLogix Ethernet communications module.

This lesson steps you through determining, modifying, or assigning a personal computer's IP address for Windows 2000, XP, Vista, and Windows 7. The steps for Windows 2000 and XP are very similar, although there are some differences; Vista and Windows 7 are somewhat similar. After you have configured your personal computer's IP address, you will use the ipconfig command to verify the address you set up is correct. This chapter contains three sections:

- Configuring an IP address for Windows 2000 or XP
- Configuring Vista IP address
- Configuring Windows 7 IP address

Turn to the specific section for the task you wish to complete.

CONFIGURING A PERSONAL COMPUTER'S ETHERNET ADDRESS USING WINDOWS 2000 OR XP

This section guides you through configuring the personal computer's IP address using Windows XP Professional. If you are using Windows 2000 software, the steps will be similar. Skip to the next sections if you are using Vista or Windows 7.

To view and/or modify the personal computer's IP address, follow the steps listed here.

1. _____ Start your personal computer.
2. _____ Refer to Figure 23-1 as you click Start.
3. _____ Click Network Connections.

DETERMINE AND MODIFY A PERSONAL COMPUTER'S IP ADDRESS   **561**

Figure 23-1 Click Start and then Network Connections in Windows XP.

4. _____ The Network Connections window, similar to the one shown in Figure 23-2, should display.

Figure 23-2 Select the proper network connection.

5. _____ If there are multiple selections in the right pane, ask your instructor which selection to make. If you are using a notebook personal computer and its integrated Ethernet port, the Local Area Connection selection, as illustrated in the figure, is typically used. Double-click the appropriate connection.
6. _____ The Local Area Connection Status window should open, similar to the one shown in Figure 23-3.
7. _____ Click Properties.
8. _____ The Local Area Network Connection Properties window, similar to that shown in Figure 23-4, should display.
9. _____ You need to scroll through the list to find Internet Protocol (TCP/IP), as it is typically at the end of the list. See #1 in Figure 23-4.
10. _____ Refer to #2 in the figure as you select Internet protocol (TCP/IP).
11. _____ Select Properties, #3 in the figure.
12. _____ The Internet Protocol (TCP/IP) Properties window, similar to the one in Figure 23-5, should display. At this point, the information displayed in the window may be different from that shown in the figure, as it depends on the current configuration of your personal computer.

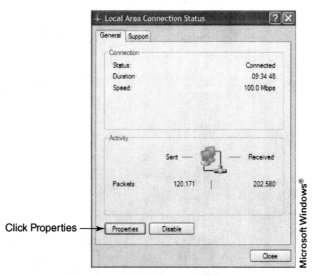

Click Properties

Figure 23-3 Local Area Connection Status window.

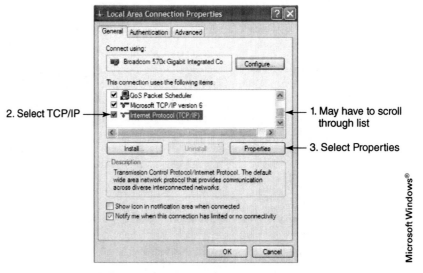

2. Select TCP/IP

1. May have to scroll through list

3. Select Properties

Figure 23-4 Local Area Network Connection Properties window.

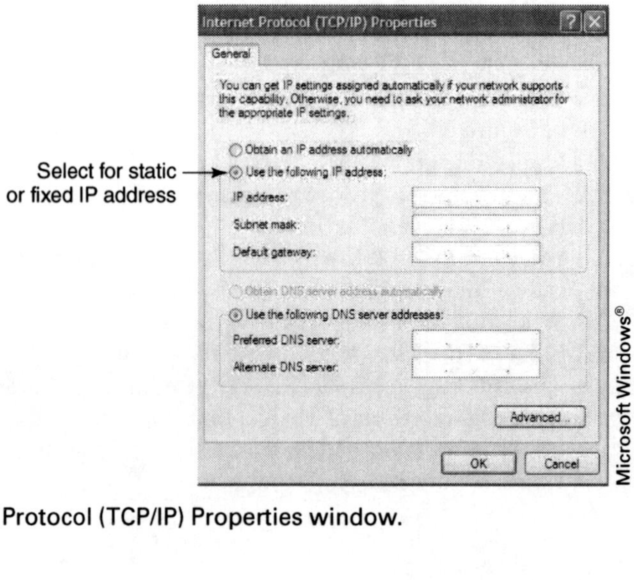

Select for static or fixed IP address

Figure 23-5 Internet Protocol (TCP/IP) Properties window.

13. _____ In an industrial environment, a fixed, also referred to as a static, Ethernet IP address is typically assigned to each device on the network. To enter a static IP address, select radio button labeled Use the following IP address, as noted in Figure 23-5.
14. _____ Check with your instructor to see what IP address to assign to your personal computer.
15. _____ For future reference, record your personal computer's IP address here: _____ . _____ . _____ . _____
16. _____ For this example, we use 192.168.1.3 as the IP address for our personal computer, the IP address you use may be different. We have selected this IP address because 192.168.x.x is a common private address used in many industrial environments. This class C Ethernet IP address has two parts: the network address and the Host Name, or node, address. In a class C IP address, the left three numbers, or octets, 192.168.1, identify the network address. Because every device we wish to communicate with is in many cases on the same network, in this example each device has the 192.168.1 as the network address. This network address is so common in industrial environments that the module properties General Tab in the RSLogix 5000 project I/O configuration for newer ControlLogix Ethernet modules has a selection with the network address of 192.168.1.x already entered for the user, as illustrated in Figure 23-6. The user can modify this network address by selecting the IP Address radio button and entering the desired network IP address.

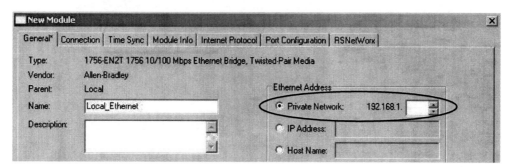

Figure 23-6 Ethernet Module properties showing common class C private network address of 192.168.1 entered into module properties by default.
Used with permission Rockwell Automation, Inc.

The rightmost octet in our IP address, identifies the node, or Host Name, address. Because every device on any network must have a unique node address, if we assign our personal computer the node address of 3, our PAC and any other devices we wish to communicate with on this network must have a different node value in its IP address. Figure 23-6 shows the node, or Host Name, address box empty, so the user can enter the desired value.

17. _____ Returning to assigning your personal computer's IP address, refer to #1 in Figure 23-7, and select Use the following IP address. Refer to #2 in the figure to enter each number, or octet, on your personal computer's Internet Protocol (TCP/IP) Properties window.
18. _____ In a similar manner, enter your personal computer's subnet mask as 255.255.255.0 or as specified by your instructor. Refer to #1 in Figure 23-8.
19. _____ For reference, record the personal computer's subnet mask here: _____ . _____ . _____ . _____
20. _____ If you are using the same IP address and subnet mask as the example, the Internet Protocol (TCP/IP) Properties window should look similar to the one shown in Figure 23-8.

564 DETERMINE AND MODIFY A PERSONAL COMPUTER'S IP ADDRESS

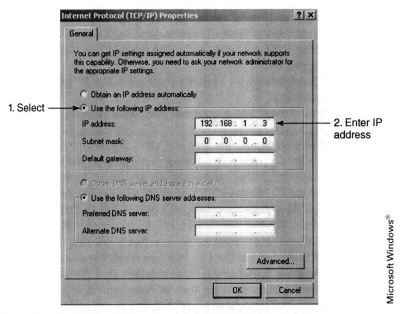

Figure 23-7 Enter the IP address values as assigned by your instructor.

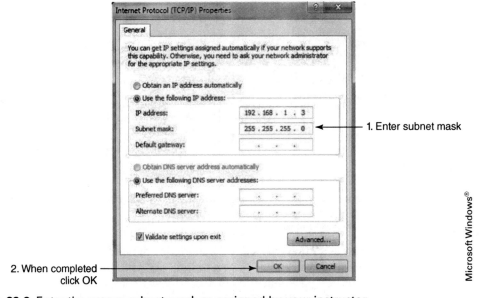

Figure 23-8 Enter the proper subnet mask as assigned by your instructor.

21. _____ Click OK, #2 in the figure, when completed.
22. _____ Close the Local Area Connection Properties window.
23. _____ Close the Local Area Communication Status window.
24. _____ Close the Network Connections window.
25. _____ Your personal computer's IP address should now be assigned. This address is fixed, or static. The address should not change when you shut down your computer.
26. _____ Figure 23-9 illustrates our personal computer with the IP address assigned. The figure also shows an Ethernet crossover cable connected directly between the personal computer and ControlLogix Ethernet communications module. We will assign a compatible IP address and subnet mask to your ControlLogix Ethernet communications module in a future chapter.

DETERMINE AND MODIFY A PERSONAL COMPUTER'S IP ADDRESS 565

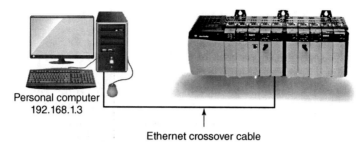

Figure 23-9 Personal computer's IP address assigned.
Used with permission Rockwell Automation, Inc.

VERIFY THE PERSONAL COMPUTER'S IP ADDRESS USING WINDOWS 2000 OR XP AND THE MS-DOS IPCONFIG COMMAND

Complete this section if you wish to use Windows 2000 or XP and execute the ipconfg command to verify the IP address of your personal computer. If you are using Windows 2000, you need to set up a network to successfully execute the ipconfig command. Use an Ethernet crossover cable, and connect the personal computer and ControlLogix together, as illustrated in Figure 23-9. Power up the ControlLogix. The ipconfig command should then successfully execute. If the computer is not connected to a network with Windows 2000, an error message, Media State — Cable Disconnected, will be returned when executing the ipconfig command. At this point, the IP address of the ControlLogix Ethernet communications module should not matter.

1. _____ To use ipconfig to verify the personal computer's IP address, click Start from your Windows desktop.
2. _____ Select Programs with Windows 2000 or All Programs with XP.
3. _____ Select Accessories.
4. _____ Select Command Prompt.
5. _____ Refer to Figure 23-10 as you enter the ipconfig command. The figure is from Windows 2000; Windows XP will be similar.

Figure 23-10 Entering the ipconfig command.

6. _____ Press Enter when the command has been entered.
7. _____ Figure 23-11 displays the personal computer's IP address and subnet mask. The figure is from Windows 2000; Windows XP will be similar.
8. _____ This information should verify that the configuration just completed.

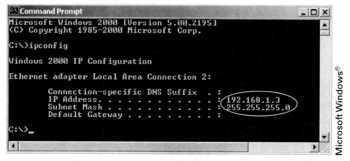

Figure 23-11 IPconfig command results in displaying the personal computer's IP address as well as the subnet mask.

CONFIGURING A PERSONAL COMPUTER'S ETHERNET ADDRESS USING WINDOWS VISTA

Complete this section if you wish to configure or monitor your personal computer's IP address using Windows Vista Premium. Our personal computer is running our RSLogix 5000 project that must be downloaded into the ControlLogix controller using Ethernet communications. Before the personal computer and the ControlLogix controller can communicate through the Ethernet communications module, both must have compatible IP addresses and subnet masks. This section steps you through configuring the personal computer's IP address using Windows Vista Premium. We will assign the ControlLogix Ethernet module IP address and subnet mask in a later chapter. To view or modify the personal computer's IP address, follow the steps listed here.

1. _____ Start your personal computer.
2. _____ From the Windows Vista desktop, click Start.
3. _____ Refer to Figure 23-12 as you click Connect To.

Figure 23-12 After clicking Start, select Connect To.
Microsoft Windows®

4. _____ The Connect to a network window should open, similar to the one shown in Figure 23-13.

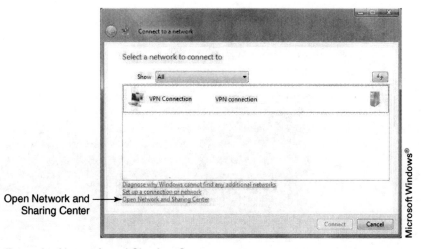

Figure 23-13 Open the Network and Sharing Center.

DETERMINE AND MODIFY A PERSONAL COMPUTER'S IP ADDRESS   567

5. _____ In the bottom-left-hand corner, click Open Network and Sharing Center.
6. _____ The Network and Sharing Center window should open as illustrated in Figure 23-14.

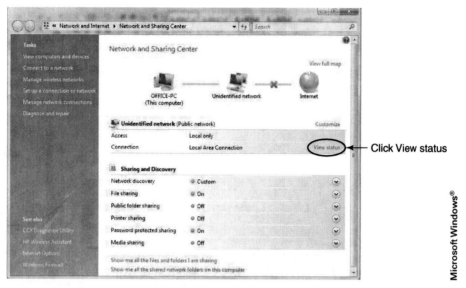

Figure 23-14 Network and Sharing Center.

7. _____ Near the center of the window is the Unidentified network section. To the far right of the word Connection, click View status, as identified in Figure 23-14.
8. _____ The Local Area Connection Status window should display, similar to the one in Figure 23-15.

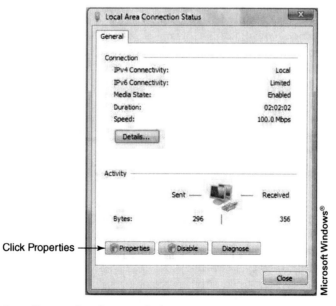

Figure 23-15 Local Area Connection Status window.

9. _____ Select Properties in the lower-left corner.
10. _____ The Local Area Connection Properties window, similar to that shown in Figure 23-16, should open.

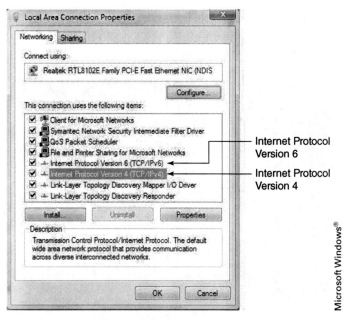

Figure 23-16 Local Area Connection Properties window.

11. _____ The Local Area Connection Properties window displays two selections for the Internet protocol versions: Internet Protocol Version 4 (TCP/IPv4) and Internet Protocol Version 6 (TCP/IPv6). Currently, the Internet is based on Internet Protocol Version 4. Internet Protocol Version 4 IP addresses are based on 32 bits. The IP address we will assign to our personal computer is 192.168.1.3. Each of the four numbers, referred to as octets, is actually represented internally in binary. Each octet is represented in eight bits. Eight bits times four numbers equals 32 bits.

Because of the popularity of the Internet, IP addresses are being exhausted rapidly. To resolve the future shortage of addresses, IP addresses will be migrating to the next generation, Internet Protocol Version 6 (TCP/IPv6). Although there are many advantages to version 6, one of the most notable is that the IP address will be 128 bits.

12. _____ For this example, we use 192.168.1.3 as the IP address for our personal computer. Ask your instructor if your IP address is different, and if so, record your personal computer's IP address here for future reference: _____ .
_____ . _____ . _____

We have selected this IP address because it is a common private address used in many industrial environments. This class C Ethernet IP address has two parts, the network address and the host, or node, address. In a class C address, the left three octets, 192.168.1 identify the network address. Because the devices we wish to communicate with are in many cases on the same network, for our example, each device has 192.168.1 as the network address. This network address is so common in industrial environments that the module properties General Tab in the RSLogix 5000 project I/O configuration for newer ControlLogix Ethernet modules has a selection with the network address of 192.168.1.x already entered for the user, as illustrated in Figure 23-17. This network address can be modified by the user by selecting the IP Address radio button and entering the desired IP network address.

The rightmost octet in our IP address, identifies the node, or host, address. If we assign our personal computer the node address of 3, our PAC and other devices we wish to communicate with on this network must all have a unique number representing the node, or host, portion of the IP address. Figure 23-17 shows the node, or host, address box empty, so the user can enter the desired value.

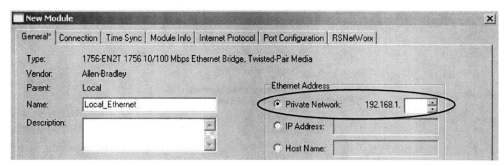

Figure 23-17 Ethernet Module properties showing common class C private network.
Used with permission Rockwell Automation, Inc.

13. _____ Returning to configuring our personal computer's IP address, refer to Figure 23-18, #1, and select TCP/IPv4.

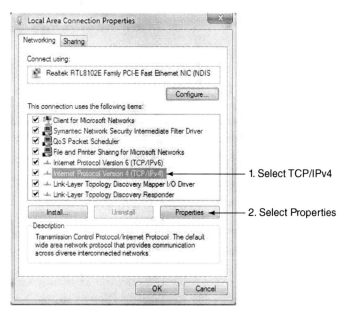

Figure 23-18 Local Area Connection Properties window.
Microsoft Windows®

14. _____ Click Properties, #2 in the figure.
15. _____ The Internet Protocol Version 4 (TCP/IPv4) Properties window should open, similar to the one shown in Figure 23-19. At this point, the information displayed in your window may be different from that in the figure, as it depends on the current configuration of your personal computer.
16. _____ Refer to Figure 23-19, #1, and click the radio button labeled Use the following IP address.
17. _____ To assign your personal computer's IP address of 192.168.1.3, enter each number, or octet, on your personal computer's Internet Protocol (TCP/IP) Properties window. Refer to #2 in Figure 23-19.
18. _____ Enter your personal computer's subnet mask as 255.255.255.0 or as specified by your instructor. See #3 in Figure 23-19.
19. _____ For reference, record the personal computer's subnet mask here: _____. _____ . _____ . _____
20. _____ If using the same IP address and subnet mask as seen in the example, the Internet Protocol (TCP/IP) Properties window should look similar to the one in Figure 23-19.

570 DETERMINE AND MODIFY A PERSONAL COMPUTER'S IP ADDRESS

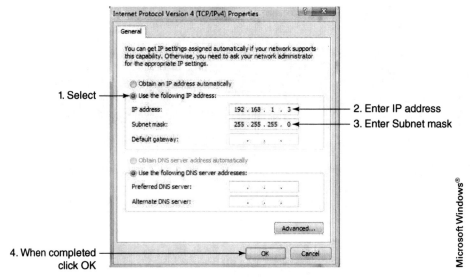

Figure 23-19 The Internet Protocol Version 4 Properties window.

21. _____ Click OK, #4 in the figure, to close the Internet Protocol Version (TCP/IPv4) Properties window, when completed.
22. _____ Close the Local Area Connection Properties window.
23. _____ Close the Local Area Connection Status window.
24. _____ Close the Network and Sharing center.
25. _____ Close the Connect to a network window.
26. _____ At this point, all network setup windows should be closed.
27. _____ Your personal computer IP address should now be assigned. This address is fixed, or static. The address should not change when you shut down your computer.
28. _____ Figure 23-20 illustrates our personal computer with the IP address assigned. The figure also shows an Ethernet crossover cable connected directly between the personal computer and ControlLogix Ethernet communications module. We will assign a compatible IP address and subnet mask to our ControlLogix Ethernet communications module in a future chapter.

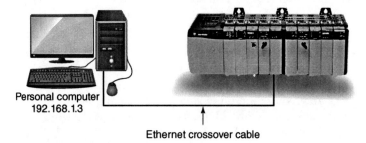

Figure 23-20 Personal computer's IP address assigned.
Used with permission Rockwell Automation, Inc.

VERIFY A PERSONAL COMPUTER'S IP ADDRESS USING THE IPCONFIG COMMAND

Complete this section if you wish to use Windows Vista to execute the ipconfig command and check or verify the personal computer's IP address and subnet mask configuration.

1. _____ From the Windows desktop, click Start.
2. _____ Enter the word "prompt" in the search box, #1 in Figure 23-21.
3. _____ Refer to Figure 23-21, #2, as you click Command Prompt near the top of the window.

DETERMINE AND MODIFY A PERSONAL COMPUTER'S IP ADDRESS 571

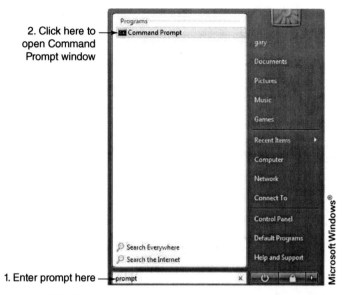

Figure 23-21 Enter "prompt" in the search box.

4. _____ The Administrator Command Prompt window, as shown in Figure 23-22 should display.
5. _____ Enter ipconfg on the command line, as shown in Figure 23-22.

Figure 23-22 Enter ipconfig.

6. _____ Press the Enter key on your computer. Your personal computer's IP address and subnet mask should be displayed, similar to that shown in Figure 23-23.

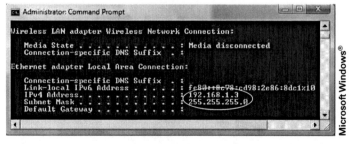

Figure 23-23 IP address displayed after execution of ipconfig command.

7. _____ The information in the figure verifies the setup of our IP address and subnet mask. This command can be used whenever this information needs to be checked.
8. _____ When you have finished, close the Administrator Command Prompt window.

9. _____ You have now assigned and verified your Windows Vista personal computer's static IP address.
10. _____ This completes this lab exercise.

CONFIGURING PERSONAL COMPUTER'S ETHERNET ADDRESS USING WINDOWS 7

Complete this section if you wish to configure or monitor your personal computer's IP address using Windows 7. RSLogix 5000 version 19 was the first release of the software that would run under Windows 7. Our personal computer is running our RSLogix project, which needs to be downloaded into the ControlLogix controller, using Ethernet communications. Before the personal computer and the ControlLogix controller can communicate through the Ethernet communications module, each must have compatible IP addresses and subnet masks. This section steps you through configuring the personal computer's IP address, using Windows 7. We will configure the ControlLogix Ethernet communications module IP address and subnet mask in a later chapter.

To view or modify the personal computer's IP address, follow the steps listed here.

1. _____ Connect the Ethernet crossover cable between your PAC Ethernet connection and personal computer. Figure 23-24 shows a crossover cable connected between our personal computer and a modular ControlLogix Communications module such as a 1756-ENBT.

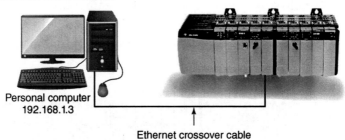

Figure 23-24 Connect crossover cable between the personal computer and the Ethernet communications module.
Used with permission Rockwell Automation, Inc.

2. _____ Power up your PAC.
3. _____ Power up your personal computer.
4. _____ From the Windows 7 desktop, click Start.
5. _____ Refer to Figure 23-25 as you click Control Panel.

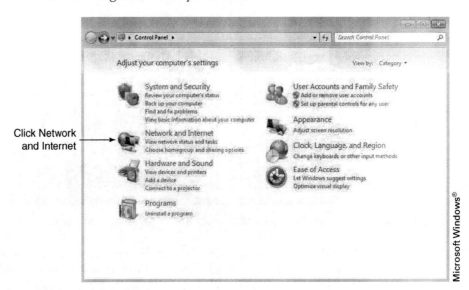

Figure 23-25 Select Control Panel.

DETERMINE AND MODIFY A PERSONAL COMPUTER'S IP ADDRESS   573

6. _____ The Adjust your computer's settings window should open, similar to the one in Figure 23-26.
7. _____ Refer to Figure 23-26 as you click Network and Sharing Center.

Figure 23-26 Network and Sharing Center.

8. _____ The View your basic network information and set up connections window shown in Figure 23-27 should display.

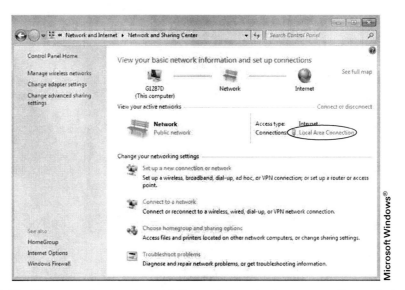

Figure 23-27 The View your basic network information and set up connections window.

9. _____ Click Local Area Connection as identified at the right in the figure.
10. _____ The Local Area Connection Status window, as shown in Figure 23-28, should display.
11. _____ Click Properties to display the Local Area Connection Properties window seen in Figure 23-29.
12. _____ The Local Area Connection Properties window displays two selections for the Internet protocol versions, Internet Protocol Version 4 (TCP/IPv4) and Internet Protocol Version 6 (TCP/IPv6). Currently, the Internet is based on Internet Protocol Version 4. Internet Protocol Version 4 IP addresses are based on 32 bits.

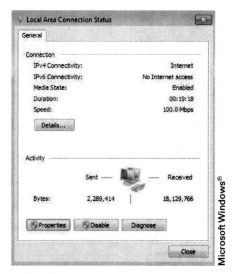

Figure 23-28 Local Area Connection Status.

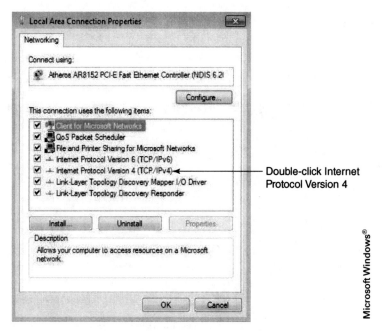

Figure 23-29 Local Area Connection Properties.

The IP address we will assign to our personal computer is 192.168.1.3. Each of the four numbers, referred to as octets, is actually represented internally in binary. Each octet is represented in eight bits. Eight bits times four numbers equals 32 bits.

Because of the popularity of the Internet, IP addresses are being exhausted rapidly. To resolve the future shortage of addresses, IP addresses will be migrating to the next generation, Internet Protocol Version 6 (TCP/IPv6). Even though there are many advantages to version 6, one of the most notable is that the IP address is 128 bits.

13. _____ For this example, we are going to use 192.168.1.3 as the IP address for our personal computer. Ask your instructor whether your IP address is different; if so, record your personal computer's IP address here for future reference:

_____ . _____ . _____ . _____

We have selected this IP address as it is a common private address used in many industrial environments. This class C Ethernet address has two parts: the network address and the host, or node, address. In a class C IP address, the left three octets, 192.168.1, identify the network address. Because the devices we wish to communicate with are in many cases on the same network, each device in this example has 192.168.1 as the network address. This network address is so common in industrial environments that the module properties General Tab in the RSLogix 5000 project I/O configuration for newer ControlLogix Ethernet modules has a selection with the network address of 192.168.1.x already entered for the user, as illustrated in Figure 23-30. This network address can be modified by the user by selecting the IP Address radio button and entering the desired network IP address.

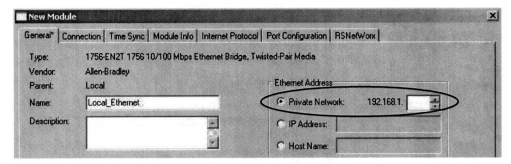

Figure 23-30 Ethernet Module properties showing common class C private network address of 192.168.1 entered into module properties by default.
Used with permission Rockwell Automation, Inc.

Figure 23-30 Ethernet Module properties shows the common class C private network address of 192.168.1 entered into module properties by default.

The rightmost octet in our IP address identifies the node, or host, address. If we assign our personal computer the node address of 3, our PLC and other devices we wish to communicate with on this network must all have a unique number representing the node, or host, portion of the IP address. Figure 23-30 shows the node, or host, address box empty so that the user can enter the desired value.

14. _____ Returning to configuring our personal computer's IP address, refer to Figure 23-31, #1, and select TCP/IPv4.

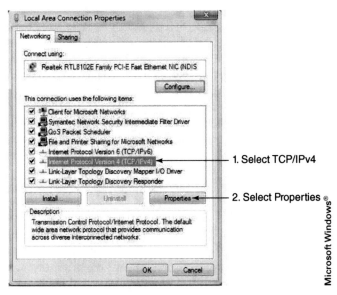

Figure 23-31 Local Area Connection Properties window.

576 DETERMINE AND MODIFY A PERSONAL COMPUTER'S IP ADDRESS

15. ____ Refer to #2 in the figure as you click Properties.
16. ____ The Internet Protocol Version 4 (TCP/IPv4) Properties window should open, similar to that shown in Figure 23-32. At this point, the information displayed in your window may be different form that shown in the figure, as it depends on the current configuration of your personal computer.

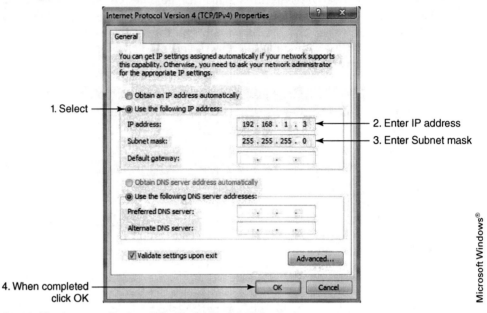

Figure 23-32 The Internet Protocol Version 4 Properties window.

17. ____ Refer to Figure 23-32, #1, and click the radio button labeled Use the following IP address.
18. ____ To assign your personal computer's IP address, enter each number, or octet, on your personal computer's Internet Protocol (TCP/IP) Properties window. Refer to #2 in Figure 23-32.
19. ____ Enter your personal computer's subnet mask as 255.255.255.0 or as specified by your instructor. See #3 in Figure 23-32.
20. ____ For reference, record your personal computer's subnet mask here: _____.
    _____ . _____ . _____
21. ____ If you are using the same IP address and subnet mask as in the example, the Internet Protocol (TCP/IP) Properties window should look similar to that shown in Figure 23-32.
22. ____ Click OK, #4 in the figure, to close the Internet Protocol Version (TCP/IPv4) Properties window, Figure 23-32, when completed.
23. ____ Close the Local Area Connection Properties window.
24. ____ Close the Local Area Connection Status window.
25. ____ Close the View your basic network information and set up communications window.
26. ____ At this point, all network setup windows should be closed.
27. ____ Your personal computer's IP address should now be assigned. This address is fixed, or static. The address should not change when you shut down your computer.
28. ____ Figure 23-33 illustrates our personal computer with the IP address assigned. The figure also shows an Ethernet crossover cable connected directly between the personal computer and ControlLogix Ethernet communications module. We will assign a compatible IP address and subnet mask to your ControlLogix Ethernet communications module in a future chapter.

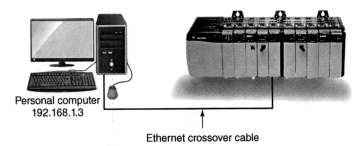

Figure 23-33 Personal computer's IP address assigned.
Used with permission Rockwell Automation, Inc.

VERIFY A PERSONAL COMPUTER'S IP ADDRESS USING THE IPCONFIG COMMAND

Complete this section if you wish to use Windows 7 to execute the ipconfig command and check or verify the personal computer's IP address and subnet mask configuration.

1. _____ From the Windows desktop, click Start.
2. _____ Enter the word "prompt" in the search programs and files box, as illustrated in Figure 23-34.

Figure 23-34 Enter "prompt" in the search box.

3. _____ Click Command Prompt near the top of the window, as in Figure 23-35, to open the Command Prompt window.

Figure 23-35 Click Command Prompt.

4. _____ Refer to Figure 23-36 as you enter ipconfg on the command line.

Figure 23-36 Enter ipconfig.

5. _____ Press the Enter key on your computer. Your personal computer's IP address and subnet mask should be displayed, similar to that shown in Figure 23-37.

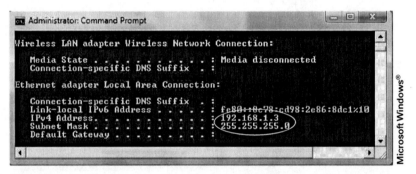

Figure 23-37 IP address displayed after execution of ipconfig command.

6. _____ The information in the figure verifies the setup of our IP address and subnet mask. This command can be used whenever this information needs to be checked.
7. _____ When you have finished, close the Command Prompt window.
8. _____ You have now assigned and verified your Windows 7 personal computer's static IP address.
9. _____ This lab exercise is completed.

CHAPTER

24

Configuring 1756 ControlLogix Modular Ethernet Hardware

Complete this lesson if you wish to identify ControlLogix modular Ethernet communication modules; assign an IP address to a new Ethernet module; determine whether an Ethernet module has an IP address configured; configure or modify a 1756-ENET, 1756-ENBT, 1756-EN2T, or 1756-EN2F modular Ethernet communications module; use a RSLinx serial driver to monitor, configure, or modify a 1756 modular Ethernet modules configuration; or execute the Ping command to test a network connection.

INTRODUCTION

As you remember from Chapter 1, there are a number of Ethernet modules for the modular ControlLogix. The older version of Ethernet uses the 1756-ENET module with a speed of 10 million bits per second (mbps), and half duplex. Currently, there are still facilities with older 1756-ENET modules in use. The newer modules include the 1756-ENBT, 1756-EN2T, and 1756-EN2F.

The newer Ethernet/IP modules can interface to the older 1756-ENET modules and the slower Ethernet network at 10 mbps. Keep in mind that the data rate for any network must be the same for all devices on that particular network. With that said, the newer IP modules can be configured on an existing network containing 1756-ENET modules, but the network must run at the slower speed. Likewise, a network containing the newer IP modules and running at 100 mbps cannot support the addition of an older 1756-ENET module, as this module can only run at the slower speed. The newer Ethernet is referred to as Ethernet/IP, where the "IP" stands for "industrial protocol." Ethernet/IP runs at 100 mbps full duplex. The table in Figure 24-1 provides an overview of selected ControlLogix Ethernet bridge modules.

\	\	\	CONTROLLOGIX ETHERNET BRIDGE MODULES	\
Part Number	Media	Minimum RSLogix 5000 Software	Communication Rate	Platform
1756-ENET	Cat 5 twisted pair	1.18	10 Half	1756 ControlLogix
1756-ENBT	Cat 5 twisted pair	11.11	10/100 Half or Full	1756 ControlLogix
1756-EN2T	Cat 5 twisted pair	15.0	100 M Full	1756 ControlLogix
1756-EN2F	Fiber	15.0	100 M Full	1756 ControlLogix
1768-ENBT	Cat 5 twisted pair	1.11	100 M Full	1768 CompactLogix module

Figure 24-1 ControlLogix Ethernet Bridge module overview.
© Cengage Learning 2014

Ethernet/IP is an open industrial network standard that supports real-time I/O control as well as messaging between different pieces of hardware such as programmable automation controllers (PACs), variable frequency drives, and operator interface. Today the most popular network for new applications for information transferring among computers, controllers, and human interface devices or variable frequency drives is Ethernet/IP. Figure 24-2 shows the ENET as well as an ENBT communications module. It is easy to identify which module is currently used in your system by looking at each module's features. The ENET module has no window near the top of the module, like the newer ENBT module. The window displays the module's IP address as well as other module status information. The ENET not only has the 10 Base T-RJ45 Ethernet connector on the front of the module behind the module's door, but also has the older AUI connector found on older Ethernet systems including many earlier PLC 5 Ethernet processors. There is no module door on the ENBT module and the RJ45 connector is on the bottom of the module.

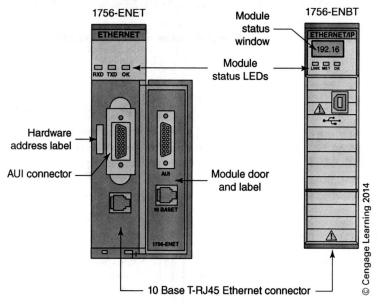

Figure 24-2 ControlLogix Ethernet communications modules.

Figure 24-3 illustrates the features of 1756-EN2T or 1756-EN2F module. The latest generation of Ethernet modules have a USB port for temporary connectivity to a personal computer to access, monitor, upload or download, and online edit the RSLogix 5000 project residing in the associated controller. Modules such as the 1756-EN2T and EN2F will also support twice as many connections as the ENBT. The EN2F connects to a fiber optic Ethernet network.

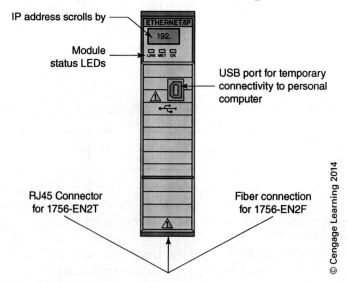

Figure 24-3 1756-EN2T or EN2F module.

Figure 24-4 shows the top view of 1756-EN2T or 1756-EN2F module rotary switches. There are three sets of rotary switches on the top of these modules. The switches can be used to manually configure the IP address for the module. Factory default for the switches is 999 so on power-up the module will check and see if DHCP or BOOTP is enabled. If it is, the module will request its IP address from the server. Switches set to 888 reset the module to initial factory settings. If using the switches to configure the IP address, the upper three octets are preconfigured as 192.168.1.xxx, whereas the lower octet is configured using the switches. If the switches are configured for a valid IP, 001 through 254, the switches will be read on power-up; the module's IP address will be the value 192.168.1. and the lower octet will be the switch values. For example, switches set to 123 result in an IP address of 192.168.1.123. The subnet mask is 255.255.255.0. For more information, consult the module's installation instruction publication.

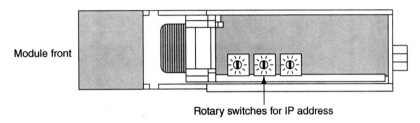

Figure 24-4 Top view of 1756-EN2F or EN2T ControlLogix Ethernet module rotary switches for IP address setting.
© Cengage Learning 2014

CONFIGURING CONTROLLOGIX ETHERNET MODULE IP ADDRESS USING RSLINX SOFTWARE

In this section, we determine, assign, or possibly modify the IP address currently stored in a ControlLogix modular Ethernet communication module. Keep in mind the following important points regarding these modules:

- A new module comes from the factory configured as BOOTP.
- The user will configure a static IP address if not using BOOTP.
- We will not be able to communicate with a device on the Ethernet network using Ethernet communications until we know or configure a compatible IP address in that device.
- When configuring an Ethernet network, the IP addresses in all devices on the network must be compatible. IP addresses that are not compatible will be unable to communicate on the network.

In Chapter 23, we assigned an IP address to our personal computer. Here we will view the module configuration and determine whether the configuration is compatible for the network we are going to configure. If the address is not compatible or not assigned, we will enter a compatible address. If you have a module with a window, the module's configuration, such as BOOTP or the current IP address, can be viewed by looking in the window on the front of the module. If using a 1756-ENET module or 1769 CompactLogix, which have no viewing window, we could determine the module's IP address by communicating with the module using communications other than Ethernet. If using a Series 7 controller or newer CompactLogix controller, the USB port is an easy way to access the current configuration of an Ethernet module.

For this exercise, we use serial communications to the controller, then to the backplane and on to the Ethernet module, so we can view and possibly modify the module's IP address. Configuring of serial drivers was introduced in earlier chapters. If you had a 1756-DH/RIO, 1756-CNB, or another Ethernet module in the ControlLogix chassis, those methods of communications could be used to access our module's configuration as well.

USING A SERIAL DRIVER TO VIEW AND MODIFY AN ETHERNET MODULE'S IP ADDRESS

For this lab, we use the serial driver previously configured to view and possibly modify our Ethernet module's IP address. If you have a controller with a USB port and the USB driver configured, the procedure to access the Ethernet module's configuration will be similar to the following steps. You may have to reconfigure the serial driver if it has been set up for a non ControlLogix product or deleted from the driver list in RSLinx. We are going to assume you are using a 1756-ENBT. If using a different module, the steps should be similar.

1. _____ Open RSLinx software.
2. _____ Reconfigure the serial driver if necessary.
3. _____ Open the RSWho screen. Refer to Figure 24-5 as you open RSWho.

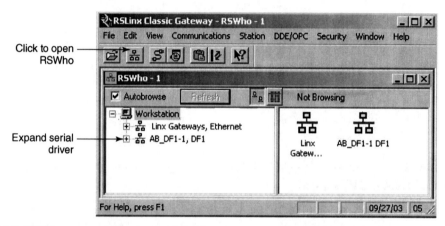

Figure 24-5 RSWho screen showing serial driver.
Used with permission Rockwell Automation, Inc.

4. _____ Expand the serial driver by clicking + to the left of the AB_DF1-1, DF1 identified in the figure.
5. _____ There should now be a + to the left of the controller in Figure 24-6.

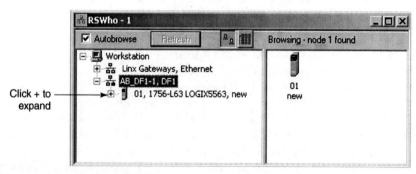

Figure 24-6 Serial driver expanded to display controller.
Used with permission Rockwell Automation, Inc.

6. _____ Click + to expand the tree.
7. _____ The tree should have expanded to show the backplane as illustrated in Figure 24-7.

CONFIGURING 1756 CONTROLLOGIX MODULAR ETHERNET HARDWARE   583

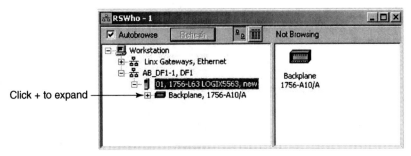

Figure 24-7 Tree expanded to show chassis backplane.
Used with permission Rockwell Automation, Inc.

8. _____ Continue to expand the tree by clicking + in front of the backplane. The tree should expand and display all modules in the chassis as illustrated in Figure 24-8.

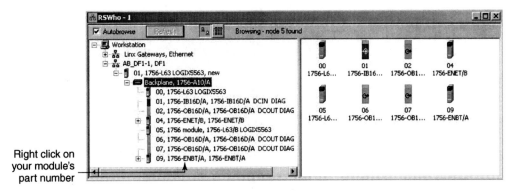

Figure 24-8 Backplane expanded to display chassis modules.
Used with permission Rockwell Automation, Inc.

9. _____ Our 1756-ENBT module resides in chassis slot 9. Right-click your Ethernet module's part number text, as identified in Figure 24-8. Remember your module be a different part number and may be in a different chassis slot.
10. _____ The drop-down box as illustrated in Figure 24-9 should appear.
11. _____ Double-click Module Configuration, as illustrated in Figure 24-9.

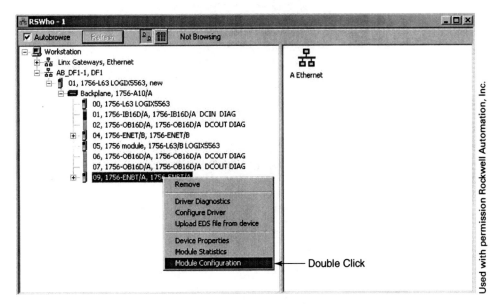

Figure 24-9 Double-click module information and select Module Configuration.

584 CONFIGURING 1756 CONTROLLOGIX MODULAR ETHERNET HARDWARE

12. ____ The Ethernet module Configuration General tab should display as in Figure 24-10. Click the Port Configuration tab as shown in the figure.

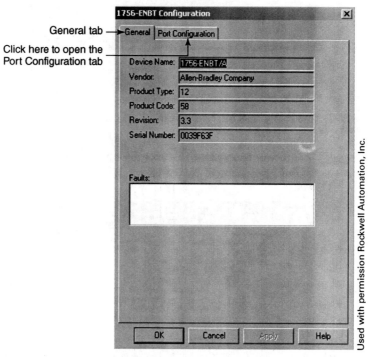

Figure 24-10 1756-ENBT Configuration General tab.

13. ____ By default, a new Ethernet module will be configured as Dynamic network configuration and have Use BOOTP selected, as illustrated in Figure 24-11. If using BOOTP, this module will request its IP address from a BOOTP server.

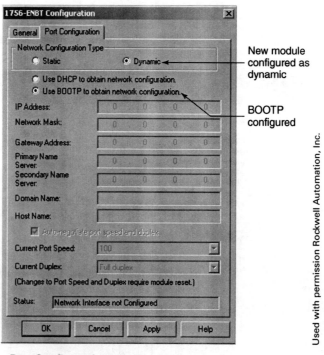

Figure 24-11 Ethernet Port Configuration tab.

14. _____ In many industrial networks, the module's IP address is configured as static. Static configuration signifies that the IP address of the device is fixed and will not change. To assign a static IP address, verify the Port Configuration tab is selected, as illustrated as A in Figure 24-12.

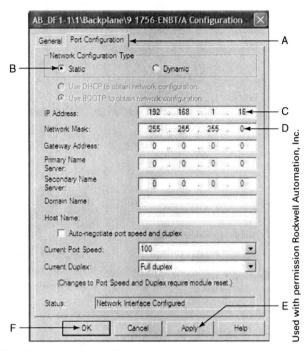

Figure 24-12 Static IP address configuration.

15. _____ Select Static configuration (B).
16. _____ As you remember from earlier studies the four number groups in an IP address are referred to as octets. To change the IP address octet value, click in the area and enter a new number for each octet. Click by 1, 2, 3, and 4 in the Figure 24-13 and enter each octet one at a time. Check with your instructor for the proper IP address to use for your lab exercise.

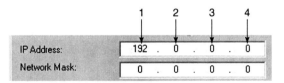

Figure 24-13 Enter IP address data for each octet.
Used with permission Rockwell Automation, Inc.

17. _____ Enter IP address 192.168.1.16 (or as directed by your instructor). Refer to C in Figure 24-12.
18. _____ In a similar manner, enter the subnet mask as 255.255.255.0 or as directed by your instructor. Refer to D in Figure 24-12.
19. _____ Refer to E in the figure as you click Apply.
20. _____ Click OK, F in the figure. Your Ethernet module's IP address should be assigned.
21. _____ For future reference, enter the following information in the table shown in Figure 24-14.

Ethernet Configuration				
Personal computer's IP address				
Personal computer's subnet mask				
ControlLogix Ethernet module IP				
ControlLogix Ethernet module subnet				

Figure 24-14 Your Ethernet setup for reference.
© Cengage Learning 2014

Figure 24-15 illustrates the directly connected Ethernet/IP network we configured.

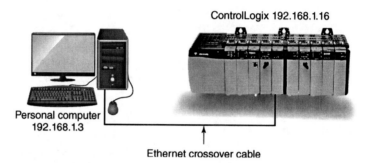

Figure 24-15 Lab exercise Ethernet network.
Used with permission Rockwell Automation, Inc.

USING THE PING COMMAND TO VERIFY COMMUNICATION TO YOUR PAC

Now that we have our Ethernet IP addresses assigned to our computer and ControlLogix communication module, we can use the ping command, an MS-DOS command, to test for a valid connection.

For the following lab exercise, we will modify our direct connection between our personal computer and ControlLogix and incorporate an Ethernet switch into our network.

Hardware Needed to Complete the Lab

- Personal computer with IP address configured
- Ethernet switch
- Two Ethernet patch cables
- ControlLogix with Ethernet communication module installed
- ControlLogix Ethernet communications model IP address configured

1. _____ Connect an Ethernet patch cable between the personal computer and Ethernet switch. Figure 24-16 illustrates the Ethernet network we will configure for this lab exercise. The switch you use in the lab may be different. The cable can be plugged into any of the switches ports.
2. _____ Connect an Ethernet patch cable between the Ethernet switch and the ControlLogix Ethernet communications module. Refer to Figure 24-16.

CONFIGURING 1756 CONTROLLOGIX MODULAR ETHERNET HARDWARE 587

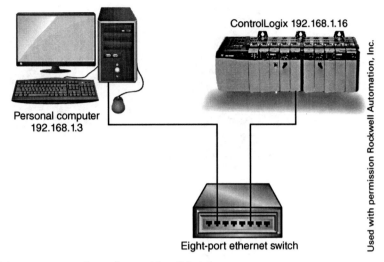

Figure 24-16 Ethernet network configured for this lab.

USING WINDOWS 2000 OR XP AND THE PING COMMAND TO VERIFY COMMUNICATION TO YOUR PLC

Complete this section if you wish to use Windows 2000 or XP and the MS-DOS Ping command to ping the ControlLogix Ethernet communications module to test the network connection using the configured IP address. If using Windows Vista to ping the network, skip to the next section in this chapter.

1. _____ Click Start.
2. _____ Select programs with Windows 2000 or All Programs if using XP.
3. _____ Select Accessories.
4. _____ Select Command Prompt.
5. _____ Refer to Figure 24-17 as you enter "ping" and the address of the ControlLogix communication module on the command line. Enter the IP address of your Ethernet communications module. For our example enter "ping 192.168.1.16."

Figure 24-17 Entering the Windows Ping command at the Command Prompt.

6. _____ Press the Enter key on your computer.
7. _____ The ping command should execute and display the results. In Figure 24-18, notice the text "Pinging 192.168.1.16 with 32 bits of data."
8. _____ The text on the next lines verifies the ControlLogix communications module responded to our communications request.
9. _____ Since the communications module responded we have established communications between our personal computer and ControlLogix Ethernet module.
10. _____ This completes this lab exercise. When completed close the Command Prompt window.

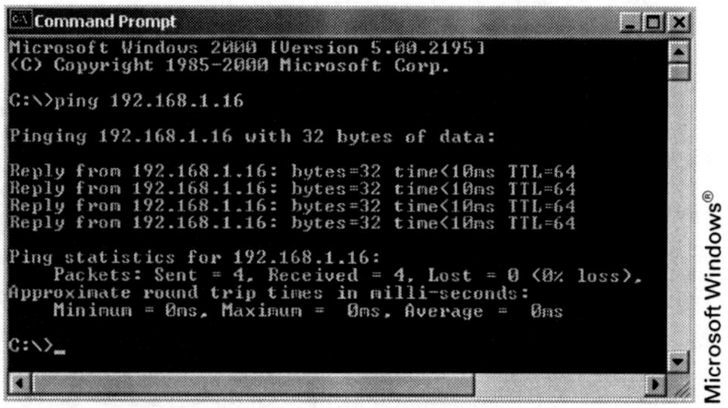

Figure 24-18 Ping command response from device at IP address 192.168.1.16.

USING WINDOWS VISTA AND THE PING COMMAND TO VERIFY COMMUNICATION TO YOUR PAC

Complete this section if you wish to use Windows Vista to ping the ControlLogix Ethernet communications module to test the network connection using the configured IP address. If using Windows 7 skip to the next section in this chapter.

1. _____ Using your personal computer, click Start.
2. _____ Click Command Prompt. See Figure 24-19.

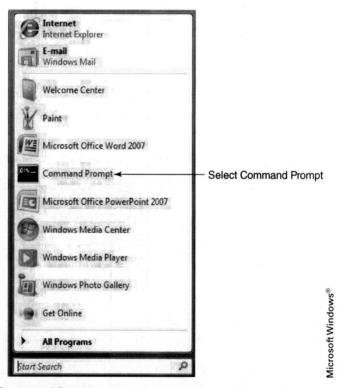

Figure 24-19 Click Command Prompt.

3. _____ Refer to Figure 24-20 as you enter ping and the address of the ControlLogix communications module on the command line.
4. _____ Press the Enter key on your computer.

CONFIGURING 1756 CONTROLLOGIX MODULAR ETHERNET HARDWARE 589

Figure 24-20 Enter the Ping command and ControlLogix Ethernet communications module address.

5. _____ The ping command should execute and display the results. In Figure 24-21, notice the text "Pinging 192.168.1.16 with 32 bits of data."

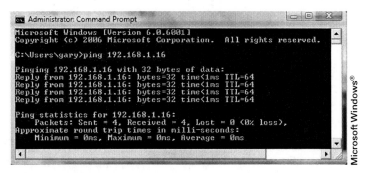

Figure 24-21 Ethernet communications module reply.

6. _____ The text on the next lines verifies that the ControlLogix communications module responded to our communications request.
7. _____ Because the communications module responded, we have established communications between our personal computer and the ControlLogix Ethernet module.
8. _____ This completes this portion of the lab. Close the Administrator: Command Prompt window.

USING THE PING COMMAND TO VERIFY COMMUNICATION TO YOUR PAC AND A PERSONAL COMPUTER USING WINDOWS 7

Complete this section if you wish to use Windows 7 and the MS-DOS ping command to ping the ControlLogix Ethernet communications module in order to test the network for a valid connection.

1. _____ From the Windows 7 desktop, click Start.
2. _____ Enter the word "prompt" in the search programs and files box, as illustrated in Figure 24-22.

Figure 24-22 Enter prompt in the search box.
Microsoft Windows®

590 CONFIGURING 1756 CONTROLLOGIX MODULAR ETHERNET HARDWARE

3. _____ Click Command Prompt near the top of the window in Figure 24-23 to open the Command Prompt window.

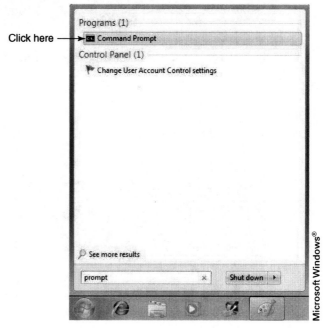

Figure 24-23 Click Command Prompt.

4. _____ Refer to Figure 24-24 as you enter "ping 192.168.1.16" on the command line in the Command Prompt window. Remember to enter the IP address of your communications module if it is different from our example.
5. _____ Press the Enter key on your computer.

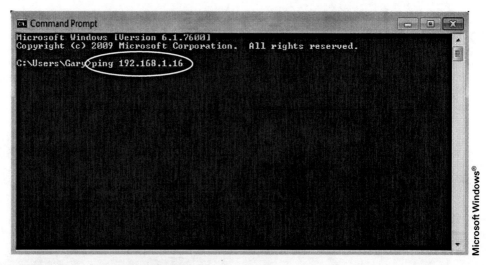

Figure 24-24 Enter "ping 192.168.1.16."

6. _____ The ping command should execute and display results similar to Figure 24-25. Notice the text "Pinging 192.168.1.16 with 32 bits of data."

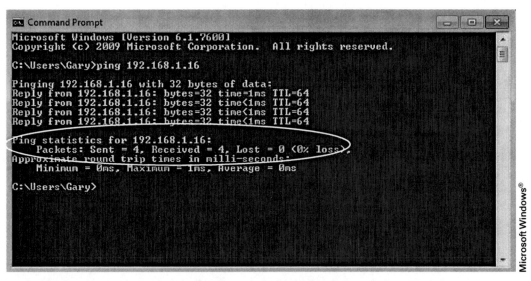

Figure 24-25 IP address displayed after execution of ipconfig command.

7. _____ The text displayed in the highlighted area verify that the ControlLogix communications module at address 192.168.1.16 responded.
8. _____ Because the communications module responded we have established communication between our personal computer and ControlLogix communications module.
9. _____ When you have finished, close the Command Prompt window.
10. _____ This completes this lab exercise.

CHAPTER 25

Configuring Ethernet IP Address for a CompactLogix 1769-L23E, 1769-L32E, or 1769-L35E

Complete this lesson if you need to identify CompactLogix Ethernet communication connections; assign an IP address to a 1769-L23E, 1769-L32E, or 1769-L35E controller; determine whether a CompactLogix controller has an IP address configured, or configure or modify a CompactLogix controller IP address.

CONFIGURING COMPACTLOGIX ETHERNET MODULE IP ADDRESS USING RSLINX SOFTWARE

In this chapter, we determine and assign or possibly modify the IP address for a CompactLogix Ethernet controller. Keep in mind the following important points regarding CompactLogix controller communications:

- The 1769 CompactLogix controller is supplied with two serial ports, or one serial and one Ethernet port, or one serial and one ControlNet communications port. The newest members of the CompactLogix family, released at the end of 2011 with version 20 of RSLogix 5000 Software, replace the standard serial port with a USB port. DeviceNet communications require the separate purchase of the communications modules for insertion into a CompactLogix slot.
- The 1768 CompactLogix also requires that separate communications modules be purchased for chassis slot insertion, as the 1768 CompactLogix controller is only available with a serial port.
- The 1769-L23s are not available with embedded ControlNet communications.
- A new Ethernet device comes from the factory configured as BOOTP. If not using BOOTP, the user configures a static IP address, as illustrated in this section.
- We will not be able to communicate with an Ethernet device using Ethernet communications until we configure and/or know the IP address for the device. For a new 1769 CompactLogix device, or if you are unsure of the Ethernet port configuration, use a different communications method, such as serial or USB, to view, modify, or initially configure the Ethernet port.
- When configuring a basic Ethernet network, the IP addresses and subnet masks in all devices on the same network must be compatible.

CONFIGURING ETHERNET IP ADDRESS FOR A COMPACTLOGIX 1769-L23E, 1769-L32E, OR 1769-L35E

INTRODUCTION

In a previous chapter, we assigned an IP address to our personal computer. In this chapter, we view the Ethernet port configuration for the CompactLogix and determine whether the address is compatible for the network we are going to connect. Once communications is established with the controller, we can view the port configuration and see how communications are configured in order to make the necessary modifications.

For this exercise, we use serial communications to communicate with the 1769-L23E, 1769-L32E, or 1769-L35E Ethernet controller so we can view and possibly modify its IP address.

USING A SERIAL DRIVER TO VIEW OR MODIFY A COMPACTLOGIX CONTROLLER'S ETHERNET IP ADDRESS

This lab uses the serial driver configured in Chapter 19 to view and possibly modify a CompactLogix Ethernet port configuration. Refer to Chapter 19, if necessary, to reconfigure an RSLinx serial driver.

1. _____ Connect the 1756-CP3, or null modem cable, between the personal computer and CompactLogix controller serial port. Refer to Figure 25-1 for connection ports on the controller. Even though the figure shows a 1769-L35E, the L23E and L32E are very similar.

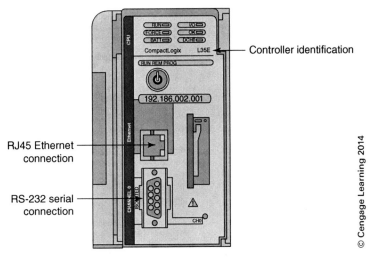

Figure 25-1 CompactLogix L35E module features.

2. _____ Power up the CompactLogix.
3. _____ Open RSLinx software. RSLinx should launch as shown in Figure 25-2.
4. _____ Refer to Figure 25-3 and verify that the serial driver is configured and running.
5. _____ Refer to Figure 25-4 and open RSWho.
6. _____ Expand the serial driver by clicking + to the left of the serial driver, as shown in Figure 25-5.
7. _____ As the driver expands, there should now be a + to the left of the controller, as in Figure 25-6.
8. _____ Click + to the left of the 01 CompactLogix controller to expand the tree.

594 CONFIGURING ETHERNET IP ADDRESS FOR A COMPACTLOGIX 1769-L23E, 1769-L32E, OR 1769-L35E

Figure 25-2 Open RSLinx software.

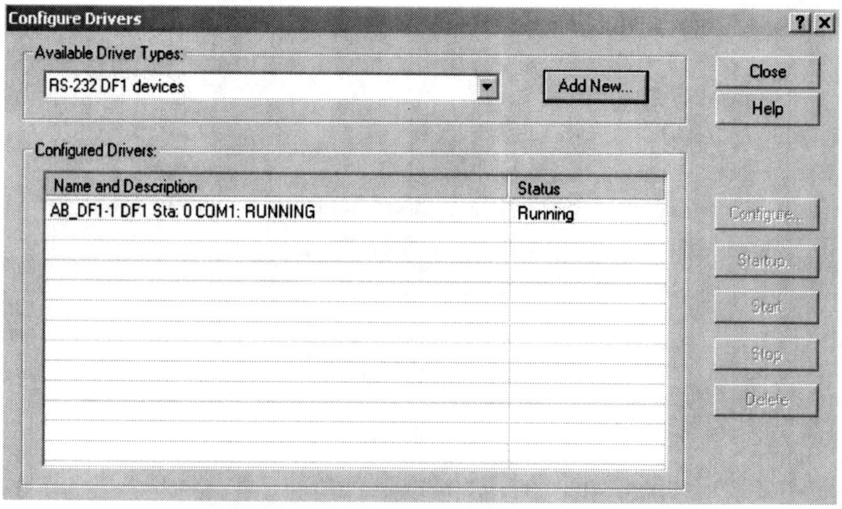

Figure 25-3 Verify that the serial driver is running.

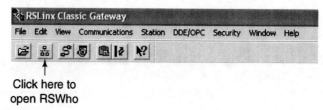

Click here to open RSWho

Figure 25-4 Open RSWho.
Used with permission Rockwell Automation, Inc.

CONFIGURING ETHERNET IP ADDRESS FOR A COMPACTLOGIX 1769-L23E, 1769-L32E, OR 1769-L35E

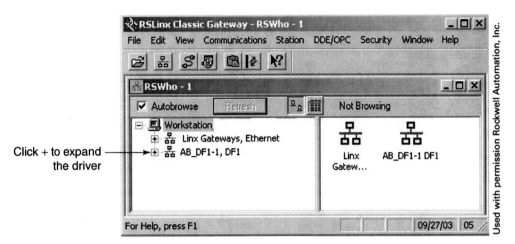

Figure 25-5 Expand the serial driver.

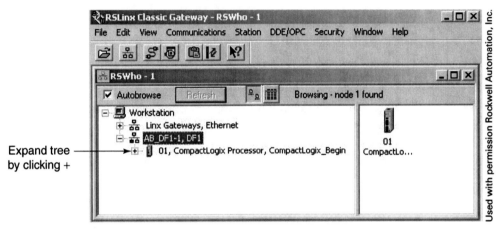

Figure 25-6 Serial driver expanded to display the controller.

9. _____ The tree should have expanded to show the backplane, as illustrated in Figure 25-7.

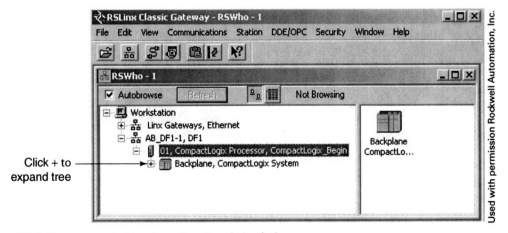

Figure 25-7 Tree expanded to show the chassis backplane.

10. _____ Continue to expand the tree by clicking + in front of the backplane. The tree should expand and display the CompactLogix controller, Ethernet port, and 1769 bus adapter, as illustrated in Figure 25-8.

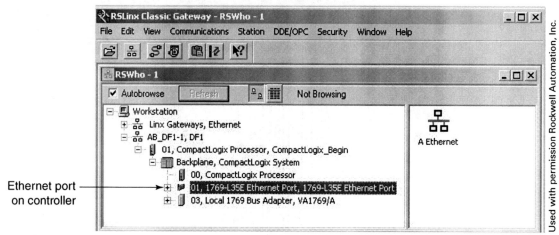

Figure 25-8 Expand to view the controller Ethernet port.

11. _____ Right-click the Ethernet port, the highlighted area in Figure 25-8.
12. _____ The drop-down right-click menu, as illustrated in Figure 25-9, should appear.

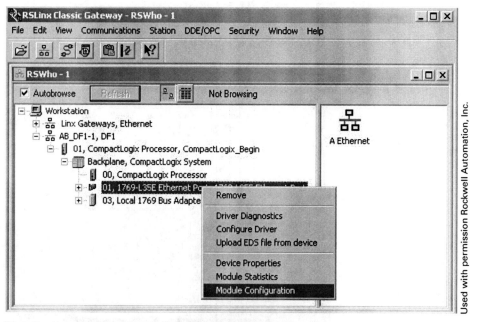

Figure 25-9 Backplane expanded to display chassis modules.

13. _____ Double-click Module Configuration.
14. _____ The 1769-L35E Ethernet Bridge Configuration General tab should display, as in Figure 25-10.

CONFIGURING ETHERNET IP ADDRESS FOR A COMPACTLOGIX 1769-L23E, 1769-L32E, OR 1769-L35E

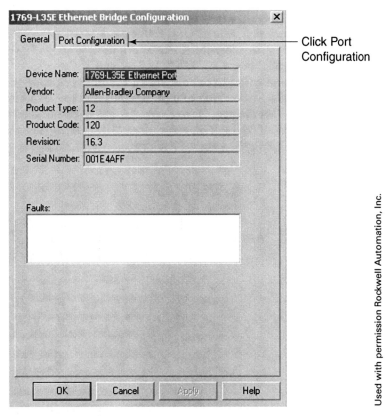

Figure 25-10 Click the Port Configuration tab.

15. _____ Click the Port Configuration tab to see Figure 25-11.

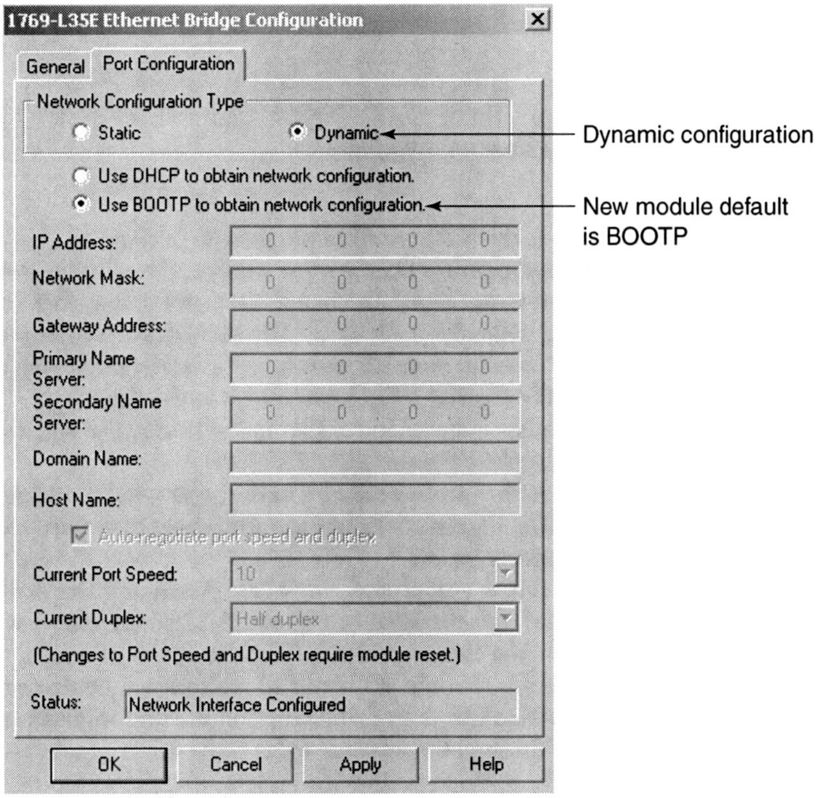

Figure 25-11 Ethernet Port Configuration tab.

598 CONFIGURING ETHERNET IP ADDRESS FOR A COMPACTLOGIX 1769-L23E, 1769-L32E, OR 1769-L35E

16. _____ By default, a new module is configured as a Dynamic network and has BOOTP selected. If using BOOTP, this module will request its IP address from a BOOTP server. In many cases, BOOTP is used to assign IP addresses during initial system installations.
17. _____ In many industrial networks, the network configuration type is configured as static. Static configuration signifies that the IP address of the device is fixed and will not change. Once a network has been installed and communication to other devices configured, rarely does the IP address change. Refer to Figure 25-12 as we configure a static IP address for our CompactlLogix.

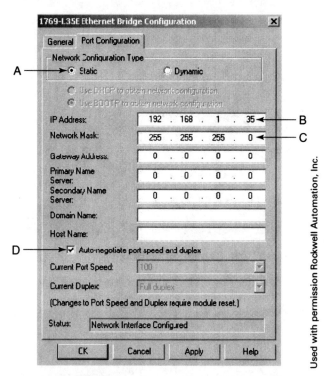

Figure 25-12 Static IP address configuration.

18. _____ Select A, the Static configuration radio button.
19. _____ As you remember from earlier studies, the number groups in an IP address are referred to as octets. To change the octet value, click in each area and enter the desired value for each octet. Check with your instructor for the proper IP address to use for your lab exercise. For our example, we use 192.168.1.35.
20. _____ Enter the desired IP address, into B in the figure.
21. _____ In a similar manner, enter the desired subnet mask, in area C in the figure. For our example, we enter 255.255.255.0.
22. _____ Check with your instructor whether you need to change the current port speed or duplex. Figure 25-12 has the Auto-negotiate port speed and duplex box, D in the figure, selected.
23. _____ When this is completed, click OK. Assuming the port configuration has been modified, the message in Figure 25-13 should display. The message states that by changing the controller's IP address, the previously configured communications to this module at the old IP address will be disrupted or stopped. Rockwell Automation refers to this disruption in communication as the connections to or routed through this module to be broken.

CONFIGURING ETHERNET IP ADDRESS FOR A COMPACTLOGIX 1769-L23E, 1769-L32E, OR 1769-L35E

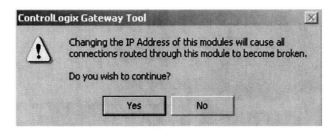

Figure 25-13 Changing the IP address will disrupt current communications (connections) through this module at the old IP address.
Used with permission Rockwell Automation, Inc.

24. _____ In this situation, we do wish to assign and use the new IP address, so click Yes.
25. _____ Click OK to exit the Port Configuration window.
26. _____ For future reference, enter the following information in the table shown in Figure 25-14.

Ethernet Configuration				
Personal computer's IP address				
Personal computer's subnet mask				
CompactLogix Ethernet IP				
CompactLogix Ethernet subnet				

Figure 25-14 Your Ethernet setup for reference.
© Cengage Learning 2014

We have our newly configured CompactLogix Ethernet/IP address compatible with our personal computer's. If we had used an Ethernet crossover cable to connect between our personal computer and CompactLogix controller, Figure 25-15 illustrates the Ethernet network we would have just configured.

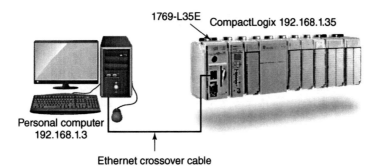

Figure 25-15 Lab exercise Ethernet network.
Used with permission Rockwell Automation, Inc.

With our personal computer and CompactLogix IP addresses configured, the next step before we can communicate between our personal computer and CompactLogix is to configure an Ethernet/IP RSLinx driver. RSLinx driver configuration is covered in Chapter 27.

CHAPTER

26

Configuring a 1756-ENET Ethernet Driver Using RSLinx

Complete this lab exercise if you wish to configure an RSLinx Ethernet driver for the older 1756-ENET module. If you wish to configure an RSLinx Ethernet/IP driver for a 1756-ENBT, 1756-EN2T, or 1756-EN2F, continue on to Chapter 27.

The pictures for this chapter are from RSLinx version 2.57 and 2.59. If you have a different version of RSLinx, your screens might be a bit different. Refer back to Figure 24-2 if you need help identifying which Ethernet module you have. In Chapter 23 the IP address of a personal computer was configured, and in Chapters 24 and 25 ControlLogix and CompactLogix hardware Ethernet ports were configured. This chapter steps you through using RSLinx software to configure an RSLinx driver for the 1756-ENET module.

THE LAB

1. _____ Verify that the IP address of your personal computer is properly configured.
2. _____ Make sure you are using a 1756-ENET Ethernet module.
3. _____ Select an Ethernet crossover cable and connect it between your personal computer and the ControlLogix 1756-ENET Ethernet module.
4. _____ Power up the ControlLogix.
5. _____ Open RSLinx by clicking the icon on your desktop. The icon should look similar to that shown in Figure 26-1. If there is no icon on your desktop, go to Start, Programs, Rockwell Software, RSLinx, RSLinx Classic.

Figure 26-1 RSLinx Classic icon on your desktop.
Used with permission Rockwell Automation, Inc.

The RSLinx splash screen, similar to the one shown in Figure 26-2, should display.

6. _____ After RSLinx is launched, click the icon to configure a driver, as illustrated in Figure 26-3.
7. _____ The Configure Drivers window should open, as seen in Figure 26-4. Note that there are two Ethernet drivers listed. The Ethernet devices selection, which is currently highlighted, can be used either to configure a driver for the 1756-ENET or to manually configure the newer modules. The Ethernet/IP Driver can only be used to configure modules newer than the 1756-ENET.
8. _____ Select the Ethernet devices driver, as highlighted in the figure.

CONFIGURING A 1756-ENET ETHERNET DRIVER USING RSLINX 601

Figure 26-2 RSLinx Classic splash screen.

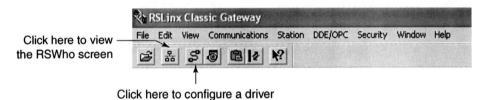

Figure 26-3 Select the configure driver icon.
Used with permission Rockwell Automation, Inc.

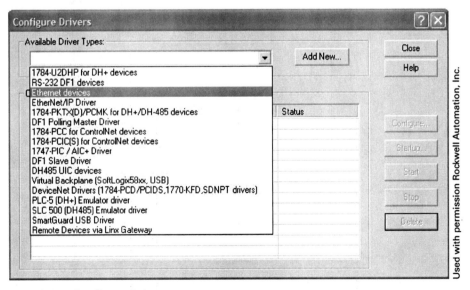

Figure 26-4 RSLinx Configure Drivers screen.

9. _____ Click Add New.
10. _____ The Add New RSLinx Classic Driver window should display, as illustrated in Figure 26-5. The default driver name, AB_ETH-1, is automatically inserted. You could change the name by entering a new name. The best option is to accept the default driver name and click OK.

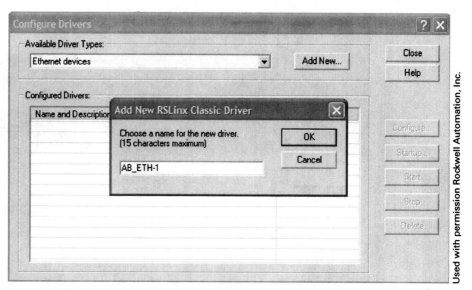

Figure 26-5 Select the name for your RSLinx driver.

11. _____ The Configure driver: AB_ETH-1 window should open, as shown in Figure 26-6.

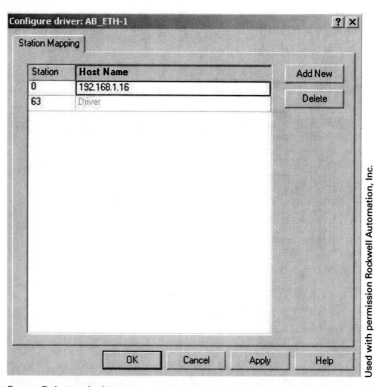

Figure 26-6 Configure Driver window.

12. _____ Enter the IP address for the device you wish to communicate with, which is the 1756-ENET module. Do not enter the IP address of your personal computer here. Enter the IP address 192.168.1.16 (or the address specified by your instructor), as shown in Figure 26-6. Remember, the personal computer was assigned the IP address of 192.168.1.3 in Chapter 23, and the ENET module was assigned the IP address of 192.168.1.16 in Chapter 24.

CONFIGURING A 1756-ENET ETHERNET DRIVER USING RSLINX 603

13. _____ When this is completed, click OK.
14. _____ Refer to Figure 26-7 as the Configure Drivers window should reopen, showing the newly configured driver in the Configured Drivers list. Note the word "Running" is listed twice. This text only states that the software driver just configured is running. Do not confuse these words as stating that you definitely have communication with the target device.

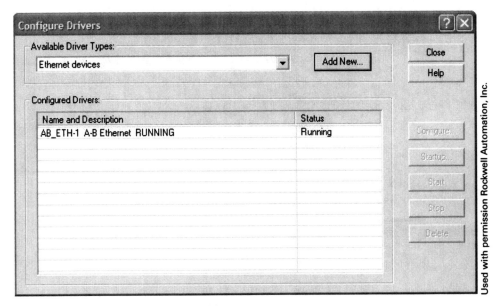

Figure 26-7 Ethernet driver shown as running.

15. _____ If Running is displayed in both positions as illustrated, click close. If something other than Running is being displayed, the driver was not configured correctly. Verify compatible IP addresses in both the personal computer and 1756-ENET module. Delete the RSLinx driver and try again. An RSLinx driver cannot be deleted if it is in use anywhere. Make sure all windows applications using the driver are closed before attempting to delete the driver. If you continue to have problems, delete the RSLinx driver, and shut down RSLinx by going to File and Exit and Shutdown. Reopen RSLinx and try again.
16. _____ Refer back to Figure 26-3 and click the RSWho icon.
17. _____ Referring to Figure 26-8, there should be a + in the box to the left of the driver. Click + to expand the display.

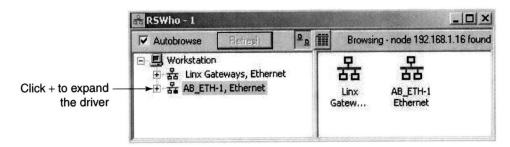

Figure 26-8 Click + to expand the driver.
Used with permission Rockwell Automation, Inc.

18. _____ Hopefully, your driver has been successfully configured and the RSWho screen looks as it does in Figure 26-9. Note the following information:
   A. Driver name
   B. IP address of Ethernet module
   C. Module part number
   D. That RSLinx is browsing the network

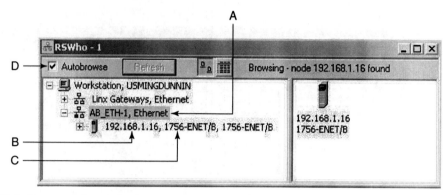

Figure 26-9 Successfully configured Ethernet driver.
Used with permission Rockwell Automation, Inc.

COMMUNICATIONS PROBLEMS

19. _____ If an X is displayed on the Ethernet Module icon (see A in Figure 26-10), and an X in the right pane, as illustrated by B, the network device is currently unrecognized. To solve this problem, verify that the personal computer and the Ethernet module have compatible IP addresses. Also, make sure the RSLinx driver is correctly configured. The driver may need to be deleted and reconfigured.

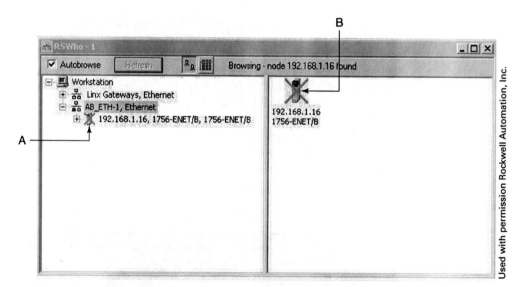

Figure 26-10 RSWho showing an unrecognized device on the network.

20. _____ This completes this portion of the lab. Your driver has been successfully configured.

CHAPTER

27

Configuring Ethernet/IP Drivers Using RSLinx Software

Complete this lab exercise if you wish to configure an RSLinx Ethernet/IP driver for the 1756-ENBT, 1756-EN2T, or 1756-EN2F module. The pictures of RSLinx for this chapter are from RSLinx version 2.57 and 2.59. If you have a different version of RSLinx, your screens might be a bit different. Refer back to Figure 24-2 if you need help identifying which Ethernet module you have. In Chapter 23 a personal computer's IP address was configured, and in Chapters 24 and 25 the ControlLogix hardware Ethernet ports were configured. This chapter steps you through using RSLinx software to configure an RSLinx driver for the 1756-ENBT module. Whether you are using the 1756-EN2T or the 1756-EN2F module, the steps are the same.

THE LAB

1. _____ Verify that the IP address of your personal computer is properly configured.
2. _____ Make sure you are using the proper Ethernet communications module.
3. _____ Select an Ethernet crossover cable and connect it between your personal computer and the ControlLogix Ethernet module.
4. _____ Power up the ControlLogix.
5. _____ Open RSLinx by clicking the RSLinx Classic icon on your desktop. The icon should look similar to that shown in Figure 27-1. If there is no icon on your desktop, go to Start, Programs, Rockwell Software, RSLinx, RSLinx Classic.

Figure 27-1 RSLinx Classic icon on your desktop.
Used with permission Rockwell Automation, Inc.

The RSLinx splash screen, similar to the one shown in Figure 27-2, should display.

6. _____ After RSLinx is launched, click the Configure Drivers icon, as illustrated in Figure 27-3.
7. _____ The Configure Drivers window should open. See Figure 27-4. Note there are two Ethernet drivers listed. The Ethernet/IP Driver selection is currently highlighted, and can only be used to configure a driver for the Ethernet/IP modules such as 1756-ENBT, EN2T, EN2F, or newer modules. Even though the older Ethernet devices driver could be used to configure most Ethernet modules, it is typically used for the 1756-ENET module. The older driver, as set up in the last chapter, can be

605

606 CONFIGURING ETHERNET/IP DRIVERS USING RSLINX SOFTWARE

Figure 27-2 RSLinx Classic splash screen.

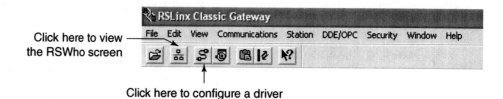

Figure 27-3 Select the Configure Drivers icon.
Used with permission Rockwell Automation, Inc.

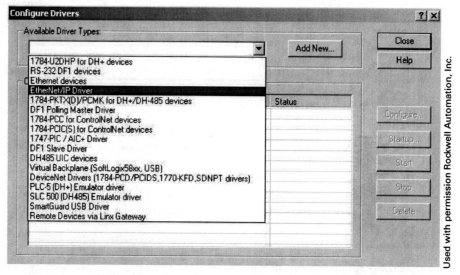

Figure 27-4 Configure Drivers window.

configured as illustrated in that chapter to manually configure an RSLinx driver to communicate with the newer Ethernet/IP modules. When you have finished configuring the Ethernet/IP driver, all recognized devices on the network will be displayed in RSWho. Manually configuring the driver using the Ethernet devices driver can be used to display only the desired Ethernet/IP devices in RSWho.

8. _____ Select the Ethernet/IP driver.

9. _____ After the driver to be configured has been selected, refer to Figure 27-5 and click Add New.

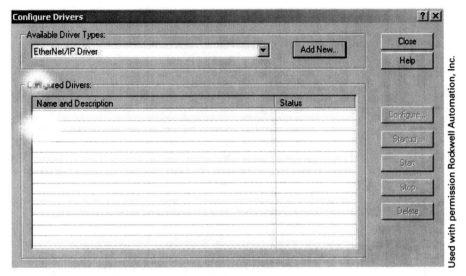

Figure 27-5 With the driver selected, click Add New.

10. _____ The Add New RSLinx Classic Driver window should display, as illustrated in Figure 27-6. The default driver name of AB_ETHIP-1 is automatically inserted. The name can be changed by entering a new name. For this lab accept the default name and click OK.

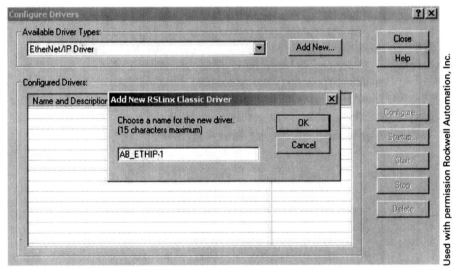

Figure 27-6 Accept the default driver name and click OK.

11. _____ The Configure Drivers window, as illustrated in Figure 27-7, should display. The selections displayed in the description list are computer specific. The Ethernet port and the currently configured IP address are highlighted in the figure. Note that the computer IP address assigned as 192.168.1.3 in Chapter 23 is displayed. You need to select your specific interface and personal computer IP address from the Description list. If your personal computer's IP address is not displayed, verify that the personal computer's IP address has been properly configured and that it is connected to the ControlLogix Ethernet communications module.

608 CONFIGURING ETHERNET/IP DRIVERS USING RSLINX SOFTWARE

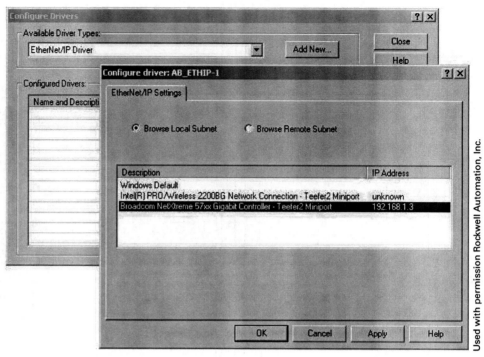

Figure 27-7 Select your specific Ethernet interface.

12. _____ Notice that the Browse Local Subnet radio button is selected. One feature of the newer Ethernet/IP driver is that RSLinx browses the local subnet, automatically finds devices with compatible IP addresses, and then finishes configuring the driver for you.
13. _____ Click OK to configure the driver.
14. _____ The Configure Drivers window should reopen, similar to the one in Figure 27-8 showing the newly configured driver in the Configured Drivers list. Note the word "Running" is listed twice. This text states only that the software driver just configured is running. Do not confuse these words as stating that you definitely have communication with the target device.

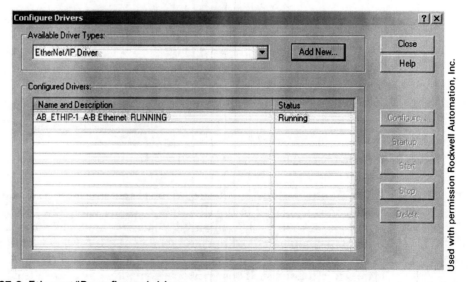

Figure 27-8 Ethernet/IP configured driver.

CONFIGURING ETHERNET/IP DRIVERS USING RSLINX SOFTWARE 609

15. _____ If "Running" is displayed in both positions, as illustrated, click close. If something other than "Running" is being displayed, the driver was not configured correctly. Verify compatible IP addresses in both the personal computer and Ethernet module. Delete the RSLinx driver and try again. An RSLinx driver cannot be deleted if it is in use anywhere else. Make sure all windows using the driver are closed before attempting to delete the driver. If you continue to have problems, delete the RSLinx driver, and shut down RSLinx by going to File and Exit and Shutdown. Reopen RSLinx and try again.
16. _____ Refer back to Figure 27-3 and click the RSWho icon.
17. _____ Figure 27-9 displays the RSWho window showing the configured Ethernet IP driver. Double-click the driver name in either the left or right pane to expand the driver.

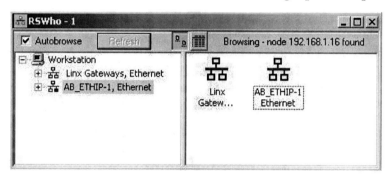

Figure 27-9 RSWho screen. Double-click the driver to expand.
Used with permission Rockwell Automation, Inc.

18. _____ Your driver should have successfully configured if the RSWho screen looks as it does in Figure 27-10. Note the following information:
    A. Driver name
    B. IP address of Ethernet module
    C. Module part number
    D. That RSLinx is browsing the network

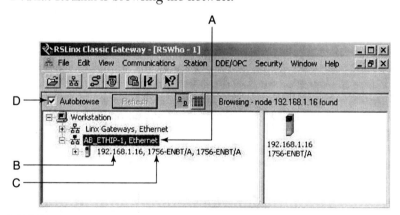

Figure 27-10 RSWho screen showing the configured driver.
Used with permission Rockwell Automation, Inc.

UNDERSTANDING THE RSLINX DISPLAY

Figure 27-10 shows our Ethernet driver and the Ethernet communications port with its IP address in the right and left of the RSWho window. This view illustrates a successfully configured Ethernet driver. Figure 27-11 displays a very similar view; however, note that there is no icon of the Ethernet communications module in the right pane. This illustrates a failed attempt to configure the driver. Start troubleshooting by checking for incompatible IP addresses or subnet masks. In many cases, the error is a simple typing error when entering the IP address or subnet mask.

610 CONFIGURING ETHERNET/IP DRIVERS USING RSLINX SOFTWARE

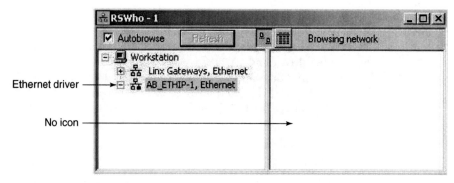

Figure 27-11 RSWho showing Ethernet driver but no icon in the right pane.
Used with permission Rockwell Automation, Inc.

If communication was lost between your computer and the Ethernet communications module after the driver has been successfully configured, an X will display on top of the module's icon, as illustrated in Figure 27-12. This signifies that communication was established at one time; however, there is currently no communication with this specific device.

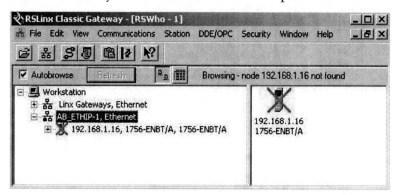

Figure 27-12 Loss of communication between personal computer and Ethernet communications module.
Used with permission Rockwell Automation, Inc.

In some cases, a yellow question mark could be displayed, as illustrated in Figure 27-13. Note the left pane with the question mark and text stating unrecognized device. In most situations where the question mark is displayed, RSLinx has identified a device on the network at the address displayed; however, RSLinx does not know what the device is. Until the problem is fixed, communications with the device will not be possible. Most likely the problem here is that the firmware of the Ethernet communications module is newer than the current revision of RSLinx software can recognize. Updating RSLinx to the latest software version typically takes care of the problem. Also check for proper configuration.

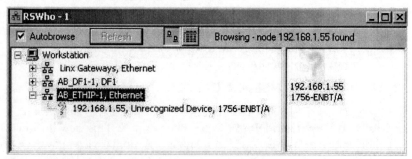

Figure 27-13 Unrecognized device on network.
Used with permission Rockwell Automation, Inc.

21. _____ This completes this portion of the lab. If your RSLinx RSWho window looks similar to the one in Figure 27-10, your driver has been successfully configured.

CHAPTER

28

Configuring a CompactLogix 1769-L23E, 1769-L32E, or 1769-L35E Ethernet/IP Driver Using RSLinx

Complete this lab exercise if you wish to configure an Ethernet/IP driver for a Bulletin 1769 CompactLogix controller. The illustrations for this chapter are from RSLinx version 2.57 and 2.59. If you have a different version of RSLinx, your screens might be a bit different. In Chapter 23 a personal computer's IP address was configured and in Chapter 25 the CompactLogix hardware Ethernet port was configured. This chapter steps you through using RSLinx software to configure an RSLinx Ethernet/IP driver for the bulletin 1769 CompactLogix controllers.

THE LAB

1. _____ Verify whether your personal computer's IP address is properly configured.
2. _____ Verify whether you are using a CompactLogix Ethernet controller. The serial port and Ethernet/IP port will be on the controller.
3. _____ Select an Ethernet crossover cable, and connect it between your personal computer and the CompactLogix Ethernet port.
4. _____ Power up the CompactLogix.
5. _____ Open RSLinx by clicking the RSLinx Classic icon on your desktop. The icon should look similar to that shown in Figure 28-1. If there is no icon on your desktop, go to Start, Programs, Rockwell Software, RSLinx, RSLinx Classic.

Figure 28-1 RSLinx Classic icon on your desktop.
Used with permission Rockwell Automation, Inc.

The RSLinx splash screen, as illustrated in Figure 28-2, should display.

611

Figure 28-2 RSLinx Classic splash screen.

6. _____ After RSLinx is launched, click the Configure Drivers icon as illustrated in Figure 28-3.

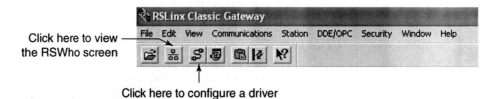

Figure 28-3 Select the Configure Drivers icon.
Used with permission Rockwell Automation, Inc.

7. _____ The Configure Drivers window should open. See Figure 28-4. Note that there are two Ethernet drivers listed. The Ethernet/IP Driver selection is currently highlighted and can only be used to configure a driver for the newer modular Ethernet/IP communications modules for the modular ControlLogix chassis and CompactLogix with Ethernet/IP. Even though the older Ethernet devices driver could be used to configure most Ethernet modules, it is typically used for the 1756-ENET module. The older driver, as set up in Chapter 26, could be configured as illustrated in that chapter to manually configure an RSLinx driver to communicate with the newer Ethernet/IP CompactLogix controller Ethernet port. When completed configuring the Ethernet/IP driver, all recognized devices on the network will be displayed in RSWho. Manually configuring the driver using the Ethernet devices driver can be used to display only the desired Ethernet/IP devices in RSWho.

8. _____ Select the Ethernet/IP driver from Figure 28-4.

9. _____ After the driver to be configured has been selected, refer to Figure 28-5 and click Add New.

10. _____ The Add New RSLinx Classic Driver window should display as illustrated in Figure 28-6. The default driver name of AB_ETHIP-1 is automatically inserted. The name can be changed by entering a new name. For this lab, refer to Figure 28-6 and accept the default name and click OK.

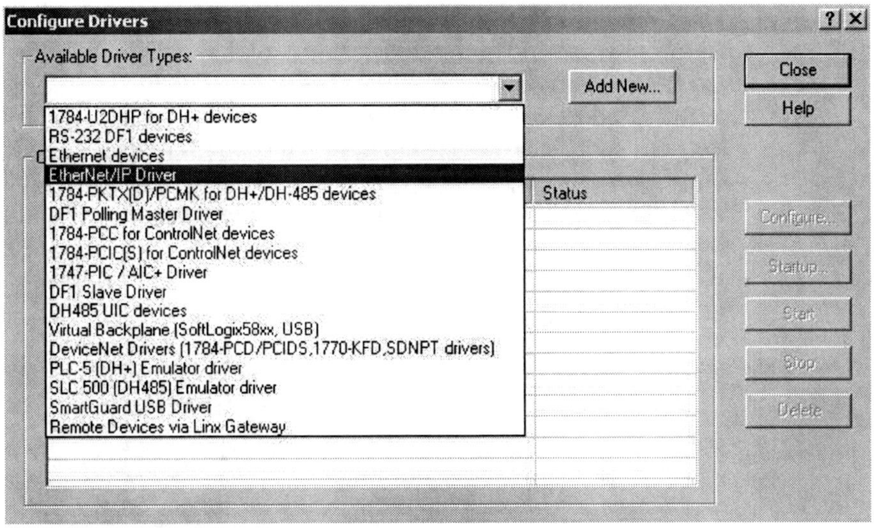

CONFIGURING A COMPACTLOGIX 1769-L23E, 1769-L32E, OR 1769-L35E ETHERNET/IP DRIVER USING RSLINX 613

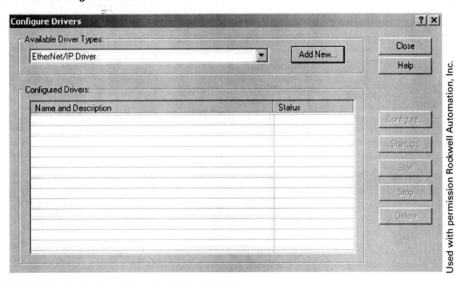

Figure 28-4 The Configure Drivers window.

Figure 28-5 With the driver selected, click Add New.

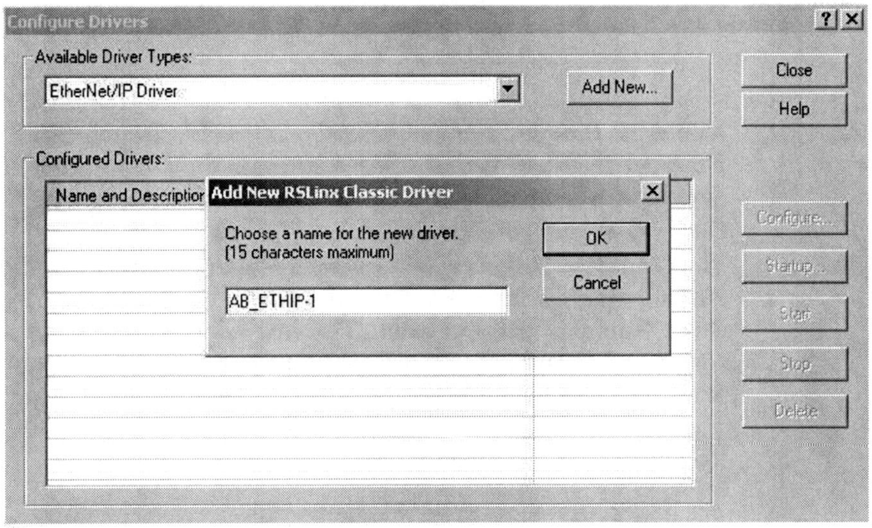

Figure 28-6 Accept the default driver name and click OK.

11. _____ The Configure Drivers window, as illustrated in Figure 28-7, should display. The selections displayed in the description list are computer specific. The Ethernet port and the currently configured IP address are highlighted in the figure. Note the computer's IP address assigned as 192.168.1.3 in Chapter 23 is displayed. You need to select your specific interface and personal computer IP address from the Description list. If your personal computer's IP address is not displayed, verify whether the personal computer's IP address has been properly configured and is connected to the ControlLogix Ethernet communications module.

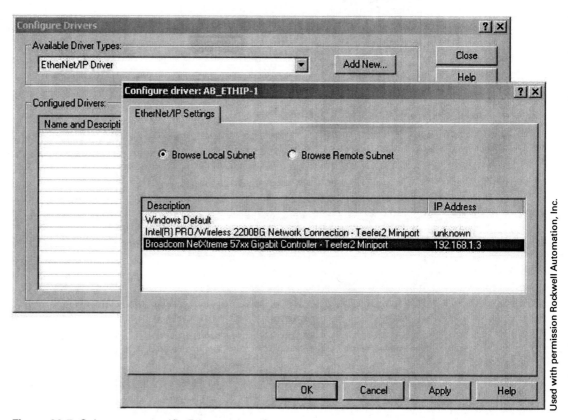

Figure 28-7 Select your specific Ethernet interface.

12. _____ Notice the Browse Local Subnet radio button is selected. One feature of the newer Ethernet/IP driver is that RSLinx browses the local subnet, automatically finds devices with compatible IP addresses, and finishes configuring the driver for you.
13. _____ Click OK to configure the driver.
14. _____ The Configure Drivers window should reopen, similar to the one shown in Figure 28-8 showing the newly configured driver in the Configured Drivers list. Note the word "Running" is listed twice. This text only states that the software driver just configured is running. Do not confuse these words as stating you definitely have communication with the target device.

CONFIGURING A COMPACTLOGIX 1769-L23E, 1769-L32E, OR 1769-L35E ETHERNET/IP DRIVER USING RSLINX **615**

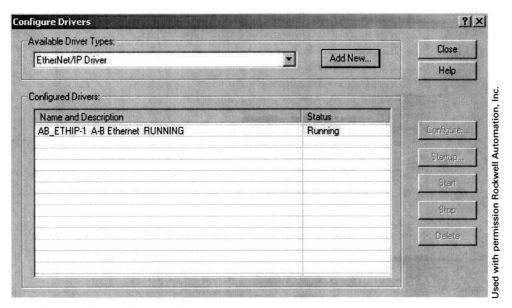

Figure 28-8 Ethernet/IP configured driver.

15. _____ If Running is displayed in both positions as illustrated, click close. If something other than Running is being displayed, the driver was not configured correctly. Verify compatible IP addresses in both the personal computer and Ethernet module. Delete the RSLinx driver and try again. An RSLinx driver cannot be deleted if it is in use anywhere else. Make sure all windows using the driver are closed before attempting to delete the driver. If you continue to have problems, delete the RSLinx driver, and shut down RSLinx by going to File and Exit and Shutdown. Reopen RSLinx and try again.
16. _____ Refer back to Figure 28-3 and click the RSWho icon.
17. _____ Figure 28-9 displays the RSWho window showing the configured Ethernet IP driver. Double-click the driver name in either the left or right pane to expand the driver.

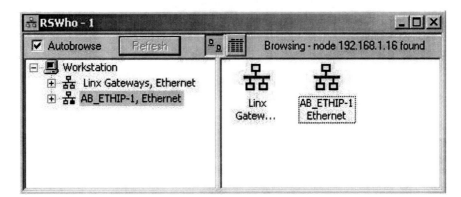

Figure 28-9 RSWho screen. Double-click the driver to expand.
Used with permission Rockwell Automation, Inc.

18. _____ Your driver should have successfully configured if the RSWho screen looks as in Figure 28-10. Note the following information.
   A. Driver name
   B. IP address of Ethernet module
   C. Module part number
   D. RSLinx is browsing the network

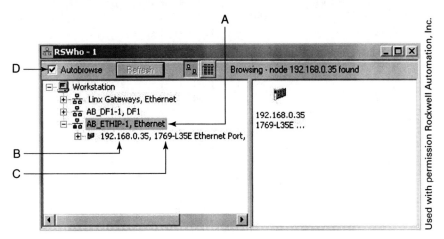

Figure 28-10 RSWho screen showing configured driver.

UNDERSTANDING THE RSLINX DISPLAY

Figure 28-10 shows our Ethernet driver and the Ethernet communications port with its IP address in the right and left pane of the RSWho window. This view illustrates a successfully configured Ethernet driver. Figure 28-11 is a very similar view; however, note that there is no icon of the Ethernet communications module in the right pane. This illustrates a failed attempt to configure the driver. Start troubleshooting by checking for incompatible IP addresses or subnet masks. In many cases, the error is a simple typing error when entering the IP address or subnet mask.

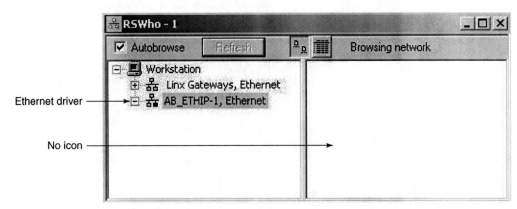

Figure 28-11 RSWho showing Ethernet driver, but no icon in the right pane.
Used with permission Rockwell Automation, Inc.

If communication was lost between your computer and the Ethernet communications module after the driver has been successfully configured, an X will display on top of the module's icon, as illustrated in Figure 28-12. This signifies that communication was established at one time; however, there is currently no communication with this specific device.

CONFIGURING A COMPACTLOGIX 1769-L23E, 1769-L32E, OR 1769-L35E ETHERNET/IP DRIVER USING RSLINX 617

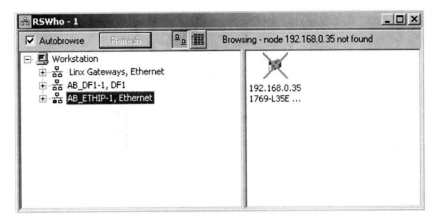

Figure 28-12 Loss of communication between personal computer and L35E.
Used with permission Rockwell Automation, Inc.

In some cases, a yellow question mark could be displayed as illustrated in Figure 28-13. Even though the figure illustrates an RSWho window for a modular 1756-ENBT Ethernet communications module, the view would be very similar for other devices including CompactLogix. Note that the left pane with the question mark and text stating unrecognized device. In most situations where the question mark is displayed, RSLinx has identified a device on the network at the address displayed; however, RSLinx does not know what the device is. Until the problem is fixed, communications with the device will not be possible. Most likely, the problem here is that the firmware of the Ethernet communications module is newer than the current revision of RSLinx software can recognize. Updating RSLinx to the latest software version typically takes care of the problem. Also check for proper configuration.

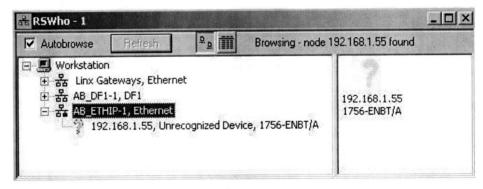

Figure 28-13 Unrecognized device on network.
Used with permission Rockwell Automation, Inc.

21. _____ This completes this portion of the lab. If your RSLinx RSWho window looks similar to the one shown in Figure 28-10, your driver has been successfully configured.

CHAPTER

29

Configuring a USB Driver for a 1756 Ethernet Communications Module

Complete this exercise if you wish to configure USB communications between a personal computer and a modular ControlLogix Ethernet/IP communications module containing an integrated USB port. For this chapter, we have selected to demonstrate configuring the 1756-EN2T. Examples of Ethernet modules containing integrated USB ports include the 1756-ENBT, 1756-EN2T, and 1756-EN2F. Personal computer and Ethernet port configuration is basically the same among the three modules. Module features are illustrated in Figure 29-1.

Examples of additional features supported by newer modules include the following:

- Personal computer with USB ports instead of the traditional serial port that has traditionally been used for module configuration
- Direct USB connection to Ethernet module versus connecting to the controller and accessing the Ethernet module over the backplane
- 1756-EN2T or EN2F module, which provides better performance of up to 10,000 packets per second versus 5,000 for the ENBT
- 1756-EN2T and EN2F, each supporting twice as many connections

MODULE CONFIGURATION

The modules have three rotary switches located on the top of the module on the circuit board that can be used for IP address configuration. Because the switches are inaccessible after the module has been inserted in the chassis, modifications must be made before installing the module. With the module lying on its side and the front of the module to the left, the switches are identified from left to right as switches X, Y, and Z. Refer to Figure 24-4 for the location of the switches. For easy setup, a fixed or static IP address can be configured using the rotary switches. The IP address will be 192.168.1.XYZ. The value set on the three rotary switches contains the value of the far right octet. Each rotary switch can be set between 0 and 9. If the switches were set to 123, for example, the IP address of the module would be 192.168.1.123. See Figure 29-2 for switch setting definitions. The module's IP address can also be monitored and configured using RSLogix 5000 or RSLinx software. We cover RSLinx software IP address configuration in this lesson. The module's IP address and other properties can also be viewed using Windows Explorer and by entering the module's IP address.

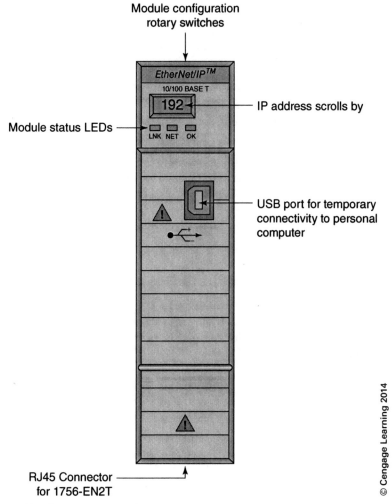

Figure 29-1 1756-EN2T module identification.

XYZ Switch Setting	Resultant Module Behavior
001 to 254	• IP address = 192.168.1.XYZ • Subnet mask = 255.255.255.0 • Gateway address = 0.0.0.0
888	• Resets module to initial factory settings, which is BOOTP • Used only to reset module, not for normal operation
Any other value	IP address configured using RSLogix 5000 or RSLinx software Available choices: • BOOTP (factory default) • User configured • DHCP

Figure 29-2 Ethernet port configuration rotary switches functionality.
© Cengage Learning 2014

Points to Keep in Mind for Using USB Communications between a Personal Computer and a ControlLogix Communications Module

- RSLinx 2.51 or later must be installed on your personal computer.
- The personal computer needs USB communications port connectivity.
- USB port communication is intended for temporary local programming or monitoring.
- USB cable must not exceed 9.8 feet (3 m).
- Connection must not contain any USB hubs.
- Connecting or disconnecting the module's communication cable with power applied could cause an explosion in hazardous locations.
- Refer to Rockwell Automation literature for cable requirements to maintain product certification.

COMPONENTS NEEDED TO COMPLETE THE LAB

Even though any of the three modules could be configured with the instructions given in this lab, we assume you are using a 1756-EN2T with an integrated USB port. The following equipment is required to complete this lab for initial USB personal computer port configuration and Ethernet module port configuration:

- Modular ControlLogix
- 1756-ENBT, 1756-EN2T, or 1756-EN2F with integrated USB port
- Personal computer running RSLinx software version 2.51or later version
- USB cable 9.8 feet (3 m) long or shorter with a type A and a type B connector. Refer to Rockwell Automation module installation instructions for cable selection considerations.

INITIAL PERSONAL COMPUTER USB DRIVER CONFIGURATION LAB

1. _____ Power up the personal computer.
2. _____ Insert the 1756-EN2T module into your ControlLogix chassis. Verify the module's address switches are set to a user-configured setting.
3. _____ Connect the USB cable between the personal computer and 1756-EN2T.
4. _____ Power up the ControlLogix. If this is the first time you have used a ControlLogix module with an integrated USB port with this personal computer, the required USB drivers must be configured. These drivers will enable us to communicate with the ControlLogix module's USB port. The RSLinx USB driver configuration that enables us to communicate with our controller is also included in this setup. No additional RSLinx driver configuration is required. This personal computer USB driver configuration typically only has to be completed once. If your personal computer already has the USB drivers configured, skip to the next section, "Configuring Ethernet Communications Using the Module's USB Port and RSLinx Software."
5. _____ After a short time, the Found New Hardware Wizard dialog appears, as illustrated in Figure 29-3.
6. _____ Select Install the software automatically (Recommended), as shown in the figure.
7. _____ Click Next.

CONFIGURING A USB DRIVER FOR A 1756 ETHERNET COMMUNICATIONS MODULE 621

Figure 29-3 The Found New Hardware Wizard dialog.

8. _____ The Found New Hardware Wizard searches for the software, as illustrated in Figure 29-4.

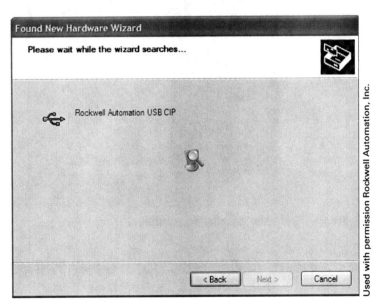

Figure 29-4 Hardware Wizard searching.

9. _____ A software installation window opens similar to the one in Figure 29-5. Multiple messages will be displayed as the driver configures.
10. _____ When the Wizard has completed installing the software, a dialog window opens, as illustrated in Figure 29-6.
11. _____ Click Finish to complete. The necessary USB drivers to enable communication with ControlLogix modules with an integrated USB port are complete.

622 CONFIGURING A USB DRIVER FOR A 1756 ETHERNET COMMUNICATIONS MODULE

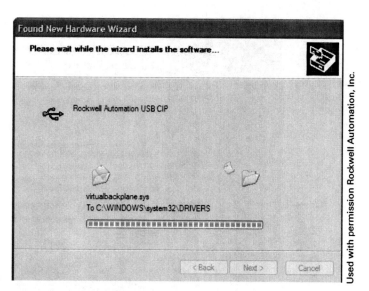

Figure 29-5 New software installation.

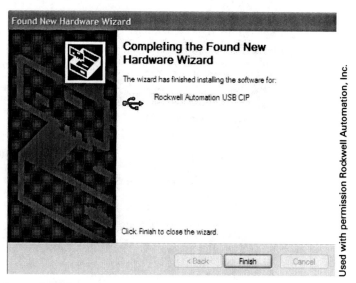

Figure 29-6 Software installation complete.

CONFIGURING ETHERNET COMMUNICATIONS USING THE MODULE'S USB PORT AND RSLINX SOFTWARE

This section views the automatically configured RSLinx driver and the use of the USB port to monitor or modify the Ethernet communication module's Ethernet port configuration. The pictures for this chapter are from RSLinx versions 2.57. If you have a different version of RSLinx, your screens might be a bit different.

1. _____ Open RSLinx software by clicking the icon on your desktop. The icon should look similar to that shown in Figure 29-7. If there is no icon on your desktop, go to Start, Programs, Rockwell Software, RSLinx, RSLinx Classic.
2. _____ The RSLinx splash screen, similar to the one shown in Figure 29-8, should display.
3. _____ After RSLinx is launched, click the RSWho icon, as illustrated in Figure 29-9.

CONFIGURING A USB DRIVER FOR A 1756 ETHERNET COMMUNICATIONS MODULE **623**

Figure 29-7 RSLinx Classic icon.
Used with permission Rockwell Automation, Inc.

Figure 29-8 RSLinx Classic splash screen.

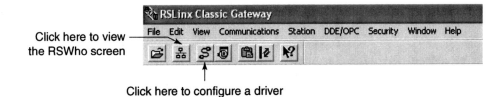

Figure 29-9 Open RSWho.
Used with permission Rockwell Automation, Inc.

4. _____ The RSLinx USB driver was automatically configured as part of the personal computer USB port driver installation. The RSWho window should look similar to the one in Figure 29-10 displaying the USB driver. Note that all other RSLinx drivers have been removed for ease of viewing the RSWho window. In the plant environment, there are typically multiple drivers listed.
5. _____ Refer to Figure 29-10 and expand the USB driver by clicking + to the left of the USB text in the left pane.
6. _____ Click + to the left of the module part number to expand the view and show the backplane. The RSWho screen should look similar to the one displayed in Figure 29-11.
7. _____ Refer to Figure 29-12 as you click + to the left of the backplane (# 1 in the figure) to expand the list to display all chassis modules.
8. _____ Right-click the 1756-EN2T module text, #2 in the figure.
9. _____ From the right-click menu, select Module Configuration, # 3 as illustrated in Figure 29-12.

624 CONFIGURING A USB DRIVER FOR A 1756 ETHERNET COMMUNICATIONS MODULE

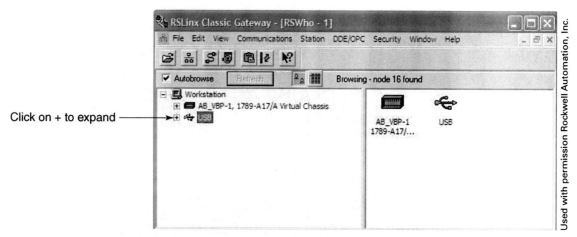

Figure 29-10 RSWho showing the newly configured USB driver.

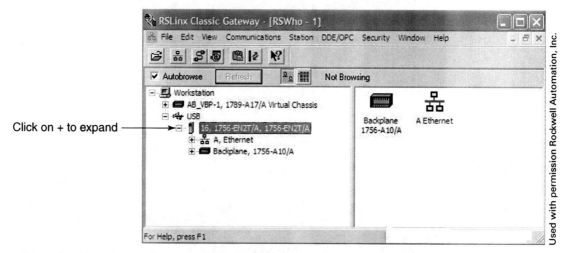

Figure 29-11 RSWho window showing the chassis backplane.

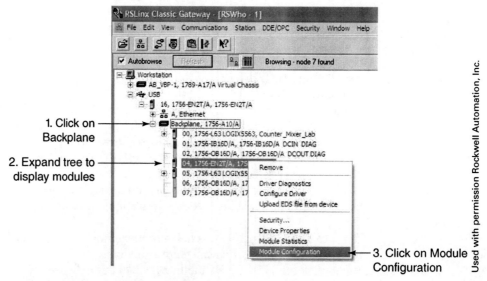

Figure 29-12 Click Module Configuration.

10. _____ See Figure 29-13 as the 1756-EN2T Configuration screen General tab should display.

CONFIGURING A USB DRIVER FOR A 1756 ETHERNET COMMUNICATIONS MODULE 625

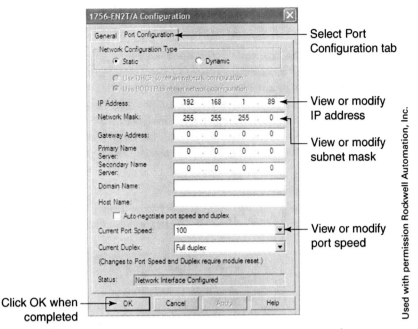

Figure 29-13 Port configuration.

11. _____ Click the Port Configuration tab.
12. _____ Figure 29-13 displays the Port Configuration tab.
13. _____ To modify the IP address, enter a new address. The module switches must be set for user configuration. If you attempt to change the IP address manually when the switches are set for a specific address, this will cause an error to be displayed. Modify Network Mask if necessary.
14. _____ To modify port speed or duplex, make new selections from the drop down.
15. _____ Click OK when you have completed the steps above.
16. _____ The message displayed in Figure 29-14 should display. The message states that by changing the module's IP address, other devices on the network expecting to communicate with that address will be unable to communicate without their configuration being modified.

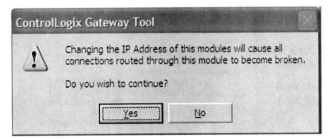

Figure 29-14 Message stating communications will be lost if IP address is modified.
Used with permission Rockwell Automation, Inc.

17. _____ Click Yes to continue.
18. _____ This completes the Ethernet port configuration or modification.

RSLOGIX 5000 SOFTWARE COMMUNICATIONS USB CONNECTION TO UPLOAD, DOWNLOAD, OR GO ONLINE WITH THE CONTROLLER

If you have a personal computer with no standard serial port but with USB ports, and a ControlLogix controller with a traditional 9-pin D-shell serial port, direct serial communication between the personal computer and the ControlLogix controller will be difficult. With the RSLinx driver configured, you could download your RSLogix 5000 project from the personal computer to the ControlLogix controller by way of the 1756-EN2T's USB port. The USB port of the Ethernet module could also be used to upload, download, or go online with the controller in a similar manner to that of any other RSLinx driver. To use the USB port to communicate with the controller, follow these steps.

1. _____ Open RSLogix 5000 software.
2. _____ If downloading, open project to be downloaded.
3. _____ Select Communications from the Windows Menu bar, as shown in Figure 29-15.

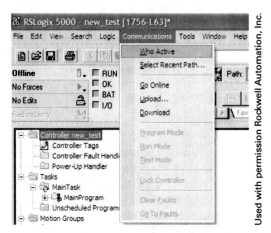

Figure 29-15 Select Who Active to configure the communications path using USB.

4. _____ From the drop down, select Who Active.
5. _____ The Who Active window should display, similar to the one seen in Figure 29-16.
6. _____ To configure communications, refer to Figure 29-16 and click on + to the left of the USB driver.

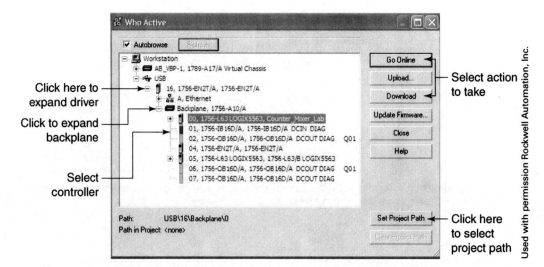

Figure 29-16 Who Active window.

CONFIGURING A USB DRIVER FOR A 1756 ETHERNET COMMUNICATIONS MODULE 627

7. _____ Click + to the left of the backplane.
8. _____ Click once to select the controller.
9. _____ Click Set Project Path to set the path in the RSLogix 5000 software toolbar. The Path toolbar in RSLogix 5000 project should update and look like the Path toolbar seen in Figure 29-17.

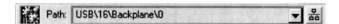

Figure 29-17 RSLogix 5000 Path toolbar.
Used with permission Rockwell Automation, Inc.

10. _____ Do you want to upload, download, or go online? Select the desired action to take and proceed as normal.

RSLINX DRIVER CONFIGURATION OR MODIFICATION

One last note: Because the setup of this USB driver was part of the USB configuration of the personal computer, we did not configure the driver in RSLinx as we would normally configure our RSLinx drivers. Because the USB driver was configured in conjunction with the USB port configuration of our personal computer, the driver cannot be reconfigured in RSLinx. Although the driver is listed in the RSLinx Configure Drivers list, if there is an attempt to modify the driver from that screen, an error message is displayed like that shown in Figure 29-18.

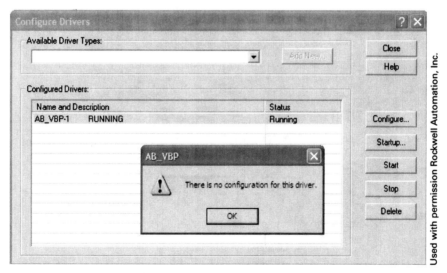

Figure 29-18 Configuration error when attempting to reconfigure USB driver in the RSLinx Configure Drivers window.

GLOSSARY

I/O Digital signals are represented by 1 for on and 0 for off. The binary number system is used to identify the two states.

l5k text file RSLogix 5000 software project windows file extension is .acd. This Windows file can be compressed into a text file, which has a ".l5k" extension. An example of a Windows file might be "Conveyor.acd," whereas a text file would be "Conveyor.l5k."

actual packet interval (API) This is the actual data transfer time in milliseconds for information across a ControlNet or Ethernet/IP network. *See also* requested packet interval.

add-on instruction Add-on instructions were available starting with RSLogix 5000 Software version 16. These instructions provide a programmer with the option of creating executable code and encapsulating it in a single instruction. Sometimes a machine function such as a startup sequence could be programmed as a series of many ladder rungs. For ease of use, these rungs could be incorporated into and programmed as a single instruction. This is extremely useful when the code may be reused multiple times within the project or stored for reuse in future projects in a library file.

address The address of an input or saved data is the unique identification of the location where the data are stored. Computer memory is arranged in blocks of locations, much like a grouping of post office boxes, each of which has its own unique address.

alias tag An alias can be assigned as another name of a tag. A ControlLogix base tag could be something like Local:2:I.Data.6. An alias tag for this base tag could be simply Start_Conveyor. Start_Conveyor is easier to work with and remember than the base tag. An alias tag simply points to another tag, another alias tag, bit, array element, or member of a user-defined data type. A ControlLogix alias tag is similar to a symbol in an SLC 500 or a PLC 5.

analog input module An analog input module converts DC analog incoming signals to digital values that can be manipulated by the controller.

analog output module Analog output modules convert digital signals from the controller into analog output voltage or current signals to operate analog output hardware devices.

analog signal A voltage or current signal that continuously changes in a smooth gradual progression over a specific range is an analog signal.

API *See* actual packet interval.

array A stored consecutive block of data of the same type. All information within a particular array must be entirely of the same data type. A programmer could create an array of BOOL, SINT, INT, DINT, or real numbers. Arrays can be one, two, or three dimensions. Structures such as timers and counters and user-defined data types can be stored only as a single-dimensional array. One member of an array is called an *element*.

ASCII *American standard code for information interchange* is a seven-bit code used for digital communications. ASCII is a binary code used to represent letters and characters. Seven-bit ASCII can represent 128 different binary combinations that represent letters, symbols, and decimal numbers found on a standard computer keyboard. There are 26 alphanumeric capital characters and 26 lowercase characters, 10 numerals (0 through 9), punctuation, mathematical and special symbols, and control characters.

asynchronous ControlLogix I/O updates are not synchronized to the scan of the logic. ControlLogix I/O updates are determined by the *requested packet interval* (RPI) as programmed by the programmer and are not synchronized to the scan of the logic as in the SLC 500 or PLC 5. The SLC and PLC 5 input and updates are separate independent steps in the scan sequence. ControlLogix I/O updates and logic scan occur independently, or asynchronous to each other.

attach To attach is to establish communication between a PLC and an outside device.

backplane The printed circuit board that runs along the back of a modular rack, base, or chassis. This board accepts each I/O module's signal through a connector into which each module plugs. Signals are transferred between the modules and the controller via the backplane.

base tag The base tag is the basic or starting tag that refers directly to the area of memory where the data are stored. An example of a base tag could be the I/O address—for example, Local:2:I. Data.0. This is the tag assigned to this particular I/O point in the I/O configuration. Programmers could alias off of this base tag and call it Start_Conveyor.

battery backup A battery is used on a controller and some specialized modules as a way of providing power to keep volatile memory chips energized to ensure that memory will be retained even if the main power is lost. Newer memory chips are more energy-efficient, and many newer PLCs thus use internal capacitors in place of batteries for memory retention.

baud rate The speed of bits-per-second transmission when using serial communication. PLCs typically transmit serial data from 1,200 to 19,200 bits per second.

binary-coded decimal (BCD) BCD is a simple, four-bit code number system developed as an easy way to convert decimal numbers 0 through 9 into a binary format and help humans interface with the computer.

binary number system A number system in which only two digits (1 and 0) are used to represent numerical values. Also known as the base-2 number system.

bit The abbreviation for a *binary digit*. A single 1 or 0 is a bit.

Boolean (BOOL) The Boolean data type is composed of one bit and thus is also known as a bit. BOOL is assigned as a data type for a ControlLogix tag.

branch A logical, parallel path within a rung of ladder logic.

byte Eight bits, or binary digits. An example of a byte would be the series of bits 11001010.

central processing unit This microprocessor inside any computer controls system activities. In a modular programmable controller, the controller is the module that contains the microprocessor. The controller controls the execution of the user program, I/O updates, and associated housekeeping chores.

chassis The hardware assembly (sometimes also called a *rack* or *base*, depending on the manufacturer) that holds the power supply, controller, and I/O together as a unit. The chassis contains the backplane, which transfers control signals and data between the controller and I/O modules.

checksum An error-checking routine used for verifying the validity of transmitted data.

clear To reset a bit, memory location, or entire memory to a 0 logical status is to clear the bit or memory location. Each data table location must contain either an on or an off signal level. The on signal level stored in memory is called a 1, whereas an off signal level is associated with a 0. Thus, a data table that has been cleared contains only 0s.

compatible module When setting up ControlLogix I/O configuration for each module, the programmer must determine and program how close a replacement I/O module must be to the module that is being replaced. A compatible module would be one with the same part number, same manufacturer, and same major revision; the minor revision must be equal to or greater than the original module. If the replacement module is not compatible, then the connection to the replacement module will be rejected by the controller.

consumed tag Data can be shared between two or more devices over either a ControlNet or Ethernet/IP network. As an example, a controller can send or produce data and send it over the network to the controller that uses or consumes the data. The producing controller will have a tag that is of the produced type, whereas the consumer will have a tag with the exact same name that is of the consumed type.

consumer controller When the consumed tag is configured, the consuming controller will be programmed to receive or consume the data being sent from the producing controller across the network. The producing controller will have a tag that is of the produced type, whereas the consumer will have a tag with the exact same name that is of the consumed type.

continuous task A task that continuously runs but can be interrupted by other higher-priority tasks such as a periodic or event task. When interrupted, the continuous task stops executing and waits until it receives permission from the interrupting task to continue. The continuous task has no priority. Even though an RSLogix 5000 project can have one continuous task, it is not required.

ControlBus The backplane of a ControlLogix chassis in a ControlNet network. *See also* backplane.

controller *See* central processing unit.

controller organizer The RSLogix 5000 controller organizer displays the project organization in a tree format, showing tasks, programs, routines, data types, trends, I/O configuration, and both controller scoped and program scoped tags.

controller scoped tag ControlLogix tags are grouped into two general classifications: either local or global. Controller scoped tags are global tags and can be used anywhere within the project, whereas program scoped tags are local and can only be used within the current program. Most tags can be assigned by the programmer as either controller scoped or program scoped. Some tags such as I/O tags from the I/O configuration are controller scoped by default. *See also* program scoped tags.

coordinated system timer (CST) The CST master is the controller that contains the master clock or coordinated system timer to which all I/O modules and controllers in the chassis are synchronized.

CST *See* coordinated system timer.

current path When uploading, downloading, or going online with a controller from your personal computer directions, a path or route to send the data between the computer and the controller must be configured. The current path is displayed in the RSLogix 5000 Path toolbar.

data type The kind of information or data represented by a tag. When creating a tag, its name is assigned first, and then its type is identified, such as BOOL, SINT, INT, DINT, timer, or counter.

DDE *See* dynamic data exchange.

debounced (input signal) An input signal that has had either intermediate mechanical noise or multiple input signals removed from it as a result of mechanical contact or bounce that is inherent when two contacts are brought together.

default value The starting value, or beginning settings, provided to the user by the software or hardware.

diagnostic I/O modules Newer I/O modules have the ability to send diagnostic information back to the PLC. Diagnostic output modules that have electronic fusing can alert the controller that the fuse has tripped. Diagnostic input and output modules can also detect an open output circuit and send a diagnostic bit back to the controller. Diagnostic bits are interpreted by the ladder logic and the appropriate action taken.

digital Digital signals can be only on or off. In a PLC, information is stored, transferred, or processed in a two-state numerical representation—that is, on or off, open or closed, true or false, or 1 or 0.

DINT *See* double integer.

disable keying When setting up ControlLogix I/O configuration for each module, the programmer must determine and program how close a replacement I/O module must be to the module that is being replaced. When disable keying is selected, the keying feature is basically unused. Any module of the same type can be used in the chassis slot.

discrete signals Two-state signals, usually on or off, true or false, yes or no, or 1 or 0. *Discrete signal* is another term used for a digital signal.

disk operating system (DOS) The operating system that makes a personal computer work.

double integer (DINT) Two 16-bit words, or 32 bits total, used to represent integer values. The minimum memory allocation and the basic format of data or a tag is stored in the ControlLogix.

double word A double word is two 16-bit words used together.

download To transfer a program from a computer into a programmable controller's memory.

drop. *See* node.

drop lines Flexible cables that drop from a tap box in a communication network. The drop line is used to connect a hardware device to the main (trunk) line of a communications network.

dynamic data exchange (DDE) Process that provides the capability to link data from one application such as a running PLC program to another DDE-compliant program, such as Microsoft Excel. As an example, the previous day's production data contained in the PLC's data tables can be used to populate an Excel spreadsheet on a scheduled basis before the morning's production meeting.

edit To change a ladder diagram or program.

electrically erasable programmable read-only memory (EEPROM) A nonvolatile memory chip used in PLCs to store a controller's firmware or the user ladder program in a memory chip that can be read but not written to. The EEPROM can be erased electrically and reprogrammed.

enable Either energize or make capable of being energized under proper signal conditions.

end instruction Instruction preprogrammed on the last rung by the programming software. The end rung cannot be deleted, and logic cannot be programmed on it. When executing the end rung, the controller is directed to return to the previous ladder file, routine, or subroutine. If currently executing logic in the main ladder or routine, the controller is directed to move to the next step in the program scan.

error message A visual indication (such as "E 008") either displayed in software or on a module that alerts the user to an improper software instruction entry attempt, incomplete instruction sequence, or hardware malfunction.

Ethernet network The TCP/IP Ethernet network is a local area network designed for high-speed exchange of information between computers, programmable controllers, and other devices. Ethernet has a high bandwidth between 10 million bits per second (10 Mbps) and 100 Mbps for high-speed communication between computers and PLCs over vast distances. An Ethernet network is desirable because it allows plant floor PLC data to be accessed by office or corporate mainframe databases.

event task An interrupt based on a predetermined trigger. When the trigger is seen, the event task interrupts any lower-priority tasks currently executing. When done, control is given back to the interrupted task. Priority assigned is between 1 and 15, with 1 being the highest. The event task is similar to a DII or PII in the SLC 500 or PLC 5s.

exact match When setting up ControlLogix I/O configuration, for each module configured the programmer must determine how close a replacement I/O module must be to the module being replaced. Choices are exact match, compatible module, and disable keying. An exact match would require that a replacement module be an exact match: same part number, same manufacturer, same major revision, and same minor revision as the original module. If the replacement module is not an exact match, then the connection to the replacement module will be rejected by the controller. *See also* compatible module, disable keying.

examine if closed (XIC) An input instruction that is logically true when the input status bit associated to its address is a 1 and logically false when it is a 0.

examine if open (XIO) An input instruction that is logically true when the input status bit associated with its address is a 0 and logically false when it is a 1.

executable code RSLogix 5000 routines contain the executable code or programming language that is used to control the application. RSLogix 5000 or ControlLogix can be programmed in any or all of four programming languages: ladder logic, sequential function chart, function block diagram, and structured text.

false An instruction is false when not on, true, or failing to provide a continuous logical path on a ladder rung.

fault routine A special routine used to determine what the controller is to do when a fault occurs. A programmer can create a local fault routine to handle instruction execution problems. A global fault handler called the *controller fault handler* is available to handle other types of faults.

field bus architecture A control architecture that uses serial, digital, multidrop, and two-way communication between intelligent field devices.

file A collection of like information organized into one group. As an example, timer data must be stored somewhere, so the area of memory that stores such data is called a *timer file*. Think of a file as a file folder storing like data in the file cabinet of the controller memory.

firmware The set of software commands that defines the personality of a system such as a PLC or variable-frequency drive main computer board. The firmware in a PLC defines its personality as a PLC and not a variable-frequency drive or other piece of hardware. Firmware is typically stored in flash memory.

fixed I/O A fixed-style PLC's I/O screw terminals are built into the unit and not changeable. A fixed I/O PLC has no removable modules. All I/O points are built-in or in the form of fixed screw terminals that are nonchangeable.

flash erasable programmable read-only memory (EPROM) Combines the versatility of the EEPROM with the security provided by a UVPROM.

floating point data file File type used to store integers and other numerical values that cannot be stored as an integer or DINT. Any number outside the range of an INT or DINT, or any fractional number, must be stored in a floating point file. Numbers such as 0.333 or 0.25 are not whole numbers, so they will be stored in a floating point file. Floating point data is also referred to as a REAL number.

function block programming language A graphical programming language in which control elements are represented by blocks connected together with lines called *wires*. In some cases, a single function block can replace 20 to 30 or more equivalent ladder rungs, making function block more suitable for process control or to drive control applications. The function block programming language is only available in RSLogix 5000 software and the ControlLogix family of Allen-Bradley PLCs.

function keys Keys on an electronic operator device that are labeled F1, F2, F3, and so on. The operation of each key is defined by the software. Function keys can be user-defined on many electronic operator interface devices.

hardware Hardware includes the physical PLC with its chassis, modules, power supply, and controller.

hexadecimal number system Hexadecimal (hex) numbers are an extension of the BCD number system. Hex is base-16, meaning it has 16 digits: 0, 1, 2, 3, 4, 5, 6, 7, 8, 9, A, B, C, D, E, F. Digits 10 through 15 inclusive are represented by the first six letters of the English alphabet. Hex is typically used for error codes, masking, and EDS files.

human–machine interface (HMI) Graphical display hardware in which machine status, alarms, messages, diagnostics, and data entry are available to the operator in graphical display format. The graphical hardware can be a personal computer or industrial computer running software such as Rockwell Automation's RSView 32, RSView ME, or RSView SE, or a panel view display terminal.

IEC 1131 International standard for machine control programming tools such as PLCs and their associated programming software. The standard comprises five programming languages with standard commands and data structures.

industrial computer An industrial computer is a personal computer that has been built specifically to withstand a harsh industrial environment.

inhibit Turns a feature or task off. An I/O module that is inhibited will not communicate or connect to the controller. Inhibiting is typically used for startups or testing.

input Incoming signals to the PLC from outside hardware devices such as limit switches, sensors, push buttons, and so forth. These incoming signals are stored in memory locations in the input tag.

input module The removable input section of a modular PLC. Each input module typically has 8, 16, or 32 terminal-block screw connections where incoming signal wires are attached. Each input module transforms and isolates incoming signal levels from outside hardware devices into signal levels that the controller can understand and process.

instruction A rung of logic contains input contacts or other symbols representing action the controller is to take depending on whether these contacts or symbols are found to be true or false. Each symbol represents an instruction. Instructions in the user program direct the controller how to react to an on or off input signal seen during the program scan.

instruction set A list of available instructions that a particular controller will understand and execute when running the user program.

integer A positive or negative whole number such as $-2, 1, 0, 1, 2, 3, 4, 5$, and so on.

integer (INT) data type Data type made up of 16 consecutive bits. An integer is typically used to store signed whole numbers such as $-2, -1, 0, 1, 2, 3, 4, 5$. An INT is assigned as a data type for a ControlLogix tag. The data range of an integer is $-32,768$ to $+32,767$.

intelligent field devices Microprocessor-based devices used to provide process variables, performance, and diagnostic information to the PLC. These devices are able to execute their assigned control functions with little interaction, except communication, with their host controller.

intelligent I/O modules I/O modules with their own microprocessor intelligence to process input values and the ability to decide how to control their associated output devices.

Internet A global collection of industrial, personal, commercial, academic, and government computers connected on one large network, or on a collection of smaller networks, for the purpose of exchanging information.

interoperability When a product from one vendor can be substituted for a similar product by another vendor. The DeviceNet, ControlNet, or Ethernet/IP network allows the user to mix and match products from many vendors on the same network. The user is not locked into the product offerings from a single vendor.

I/O Input/output.

I/O interface A hardware device that enables the PLC and external hardware devices to work together. The PLC needs to communicate with or control outside devices with incompatible signal levels. The interface device, typically an I/O module, contains circuitry that converts and isolates signals. The interface allows hardware devices to send each other understandable signals.

IP address The software IP or Internet protocol address assigned to a device on an Ethernet/IP network. The IP address comprises the network address as well as the device's node or host address. An IP address is made of four groups of three numbers called *octets*. An example of an IP address would be 192.168.4.2.

jump-to-subroutine (JSR) instruction When true, the JSR instruction redirects program flow from the current ladder file to another ladder file or routine called a *subroutine*. The SLC 500, MicroLogix, and PLC 5 subroutine files start with ladder file 3. The ControlLogix does not have ladder files with numbers. Because ControlLogix is a name-based PLC, all routines are assigned a name rather than a file number. A ControlLogix subroutine is any routine other than the main routine or a fault routine.

keeper When using a ControlNet network, the 1756-CNB(R) module with the lowest node number holds or keeps the network configuration information. This module is known as the keeper. A ControlNet network is configured using RSNetWorx for ControlNet software. Once the software is configured, the configuration is downloaded to the keeper. This is called *scheduling the network*.

ladder diagram A shorthand representation of a circuit in which symbols to represent the actual on or off status of input hardware devices is a ladder diagram. In addition to input and output symbols, the ladder diagram contains internal instructions including timer, counter, sequencer, math, and other data-manipulation instructions.

light-emitting diodes (LEDs) Semiconductor diodes that emit light when energized. LEDs are used in input and output displays in electronic and electrical equipment.

listen only connection When a module is set up in the I/O configuration, the communications format or ownership is determined. An input module can have multiple owners, whereas an output module can be assigned only one owner. A second communications format option is a listen only connection. A listen only connection allows another controller to listen to the status and I/O data. Such a controller does not have the module's I/O configuration, so it is not the owner and cannot control the module. Each I/O module must have one owner.

local I/O The chassis that the controller resides that a particular project is going to be downloaded into is called the local chassis. I/O modules in the local chassis are considered local I/O. Modules residing in a remote chassis are remote I/O modules.

logic When the controller is running the user program, the microprocessor is solving the program logic. Logic is the set of rules for interconnecting discrete ladder program instructions to arrive at conclusion. The logic represented by each rung's output instruction is the solution of the rung. In digital electronics, complex problems are solved using logic gates such as AND, OR, and NOT. Programmable controllers employ logic in the same manner, using ladder logic.

long integer (LINT) Data type is currently only used in the ControlLogix family of PACs and comprising 64 bits. Currently the LINT is used only as the coordinated system timer in ControlLogix.

lower byte The lower 8 bits in a 16-bit word.

lower nibble The lower 4 bits of a byte. Bits 0 through 3 make up the lower nibble of a byte. Bits 4 through 7 make up the upper nibble of a byte.

main operand The tag being operated on in a programming instruction. In the case of a Move instruction, there is a source tag and a destination tag. The destination tag is the main operand because it is the tag that is being operated on or changed. When adding documentation to a programming instruction, the description text that can be entered following the main operand is the main operand description.

main routine Each program must have a main routine that executes automatically when the program is executed. In the controller organizer, the main routine can be identified by a 1 on a piece of paper on top of the routine's language icon. Additional routines can be subroutines or a single fault routine. The number of routines in a program is only limited by controller memory. A routine can be in any of the four programming languages.

man–machine interface (MMI) Refers to the graphic terminals used to display status or alarm information about the process being controlled. Operator display terminals can be touchscreen or keypad units. Keypad units use function keys that are separate from the display screen for operator data input. MMI devices allow the operator to enter process parameter information or view status data on numerous screens. The operator interface device screens are developed using a screen development software package on a PC and then downloading into the operator interface device.

mask A 16-bit word or 32-bit double word used to filter or mask out selected data bits. The mask stops the selected bits from being transferred from a source word, or location, to a destination word. Often the mask is represented as a hexadecimal value.

member One piece of a timer, counter, or user-defined data-type structure. A counter, for example, is made up of a preset value and the accumulated value, both DINTs. A counter also has a DINT representing the counter's status bits. Members within a structure can be the same or of differing data types.

memory Where data are stored in a PLC in an orderly manner. Data are typically stored in a file by address. A file cabinet full of file folders storing information in an orderly manner serves as a metaphor for how the PLC stores and organizes data.

micro PLCs A term used to identify a new generation of physically smaller and more powerful PLCs. Their small size and increased capabilities are a result of advances in smaller and more powerful microprocessor and in solid-state components.

microprocessor A single integrated circuit chip that is the *central processing unit controller,* or "brain," in computerized hardware such as a PLC. A modular PLC has the controller as a separate modular piece of hardware that is either inserted into a rack or chassis or clipped onto a rack. Fixed PLCs have the controller built into the hardware housing that also contains the power supply and I/O screw terminals.

MMI *See* man–machine interface.

modular I/O Modular PLCs have removable assemblies called *I/O modules,* with each typically having 8, 16, or 32 terminal-block screw connections to which signal wires are attached. The advantage of a modular PLC I/O system is the flexibility to mix and match module signal levels and input and output designations to suit a particular application.

Move Data are moved from one location to another with a Move instruction. Although a Move instruction typically places the data in a new location, the original data still reside in their original location. The Move instruction is deceiving in that data are really copied to the destination rather than physically moved. When using a Move instruction, some PLCs allow the use of a mask to filter out specific data bits from being copied to the destination.

multicast ControlNet and Ethernet/IP are referred to as producer–consumer networks. These networks can handle programming uploads, downloads, program monitoring, I/O control, and messaging across the same network. Time-critical data have an associated RPI and are multicast. Multicast information is produced once and received simultaneously by multiple destinations or consumers. A tag is produced to the backplane once by the producing controller. Any controllers that have the same tag name but are of a consumed tag type are able to consume the tags from the backplane and network simultaneously.

multiple owners An input module can have multiple owners, meaning that one or more controllers hold the I/O configuration for the module. The owner or owners can be in the same or different chassis. When configuring multiple owners for input modules, the I/O configuration must match or the connection will be rejected. Copy and paste the module's I/O configuration between the different projects. *See also* owner.

network Hardware devices connected through a communication link to enable communication among multiple devices. A network can be used to (1) streamline system operation by sharing available hardware resources or (2) share operator interface data among multiple PLCs. A central personal computer can exchange programs, program data, and monitor any station on the network.

network update time (NUT) The time required for a ControlNet network to make one cycle, or update, of the network.

nibble Four bits make up a nibble.

node A physical device on a network. When incorporating a network into PLC system architecture, the main network cable makes up the network trunk line. All devices are connected to the trunk line cable using some type of junction box, link coupler, or station connector. The cable segment from the trunk line connector box to the hardware device being connected is a *drop line*. Each piece of hardware has a unique address called a node address. Typically, only one PLC, programming terminal, or operator interface device is allowed on a single node.

node address Unique address much like a house on a street. Also called a *station address*.

nonvolatile memory Memory that does not lose its contents after the power is lost. Usually this memory is called *nonvolatile ROM*.

octal number system Base-8 system that uses the numbers 0–7, 10–17, 20–27, 30–37, 40–47, 50–57, 60–67, 70–77, 100–107, and so on. (There are no 8s or 9s in the octal number system.) Octal numbers are used for addressing on older PLCs. Many newer PLCs use decimal numbering for their addressing assignments.

OEMs Abbreviation for *original equipment manufacturers*.

offline When a programming device and its associated PLC are not communicating, the devices are considered to be offline. A personal computer is typically used stand alone offline while the user develops or edits programs.

online A computer establishes communication by going online with the controller. Being online, the programming device and PLC are able to communicate with each other. Devices that are online can exchange data, files, or programs between each other's memory.

open system An open system is one in which the user has interchangeability and connectivity choices.

output The resulting on or off signal from solving the ladder rung instructions is sent out, or *output*, from the controller and stored in the output tag. During the portion of the scan when the controller updates its outputs, the on or off signals residing in the output tags are output to each output module's screw terminals.

output energize (OTE) An Allen-Bradley PLC output instruction. The resulting true or false logical status of all input instructions on a particular rung is reflected in the on or off status of the OTE output instruction. If the logical resultant of all input instructions is true, the rung will become true and the OTE instruction will go true. With the OTE instruction true, the associated output hardware device will energize. If the rung is evaluated as false, then the OTE will be false and output devices will either stay off or turn off.

output module The output section of a PLC. The output module isolates and converts the low-voltage output signals from the controller to the proper voltage or current levels needed to control output circuits.

output tag Memory location in which the on–off status of each output is stored. The controller sends the logical on–off status for each output, which is the result of solving the user program, to the output tag. The on–off signals from the output tags ultimately control each corresponding output screw terminal.

owner I/O module ownership is determined by the I/O configuration. When setting up each module's I/O configuration, the programmer configures the communication format for each module. Ownership is determined at this point. When the project is downloaded to the controller, the

I/O configuration is also downloaded. When going into Run mode, the controller sends the I/O configuration to the I/O modules it owns. Until a module receives its configuration, the module is considered in a waiting state. On receiving its correct I/O configuration, a module attempts to enter Run mode. An input module can have one or more owner controllers, whereas an output module can have only one owner. Because an output module can have only one owner, only one controller can control the outputs on a particular module. Module ownership can be either in the local chassis or in a remote chassis. *See also* listen only connection.

owner controller The controller that holds the I/O configuration for the I/O module. An input module can have multiple owners, whereas an output module can have only one owner. If a particular module is not in the I/O configuration, then this controller is not the owner. If this controller has the module in its I/O configuration, then this controller is the sole owner or one of the owners, has a listen only connection, or has the module inhibited. *See also* owner.

PAC *See* programmable automation controller.

parity bit The bit is added to a binary array to make the sum of the bits always odd or even. Parity is used for error checking during data transmission.

path When uploading, downloading, or going online with a controller from your personal computer directions, the route that must be configured to send the data between the computer and the controller. The current path is displayed in the RSLogix 5000 Path toolbar.

PC control Sometimes referred to as *soft control*, a control system in which the traditional PLC has been replaced with a personal or industrial computer running under Windows and software control. In many cases, soft PLC control is incorporated into systems requiring high degrees of data collection and processing or connectivity to multiple networks.

periodic task A periodic task is an interrupt based on a period of time. A periodic task is triggered on a time interval determined by a programmer. When the programmed time expires, the periodic task interrupts any lower priority tasks currently executing. When completed control is given back to the interrupted task. Priority assigned is between 1 and 15. Highest priority is 1. The periodic task is similar to a selectable timed interrupt or STI in the SLC 500 or PLC 5s.

physical chassis Houses the controller, the input and output modules, and, in some cases, the power supply. Some PLC manufacturers use the term *chassis*, some use *rack*, and others refer to a *base*. Although each manufacturer's hardware device may have a different name and look different, all are used to hold together the pieces of a modular PLC.

PID *See* proportional, integral, derivative control.

priority Periodic and event tasks are interrupts. These two task types can be assigned a priority between 1 and 15. A higher-priority task can interrupt a currently executing lower-priority task. Priority 1 is the highest priority, and 15 is the lowest. The continuous task has no priority.

processor The *central processing unit* (controller).

processor configuration When setting up software to begin developing ladder logic, the software needs to be told what controller will be used to run the program. Typically, there is a menu in the software from which the controller to be used in this application is selected.

produced tag A controller scoped tag that can be shared with or consumed by other controllers across either ControlNet or Ethernet/IP is said to be produced by the sending controller. A produced tag type is assigned by the programmer as the tag is created. Produced information is time-critical information and has an RPI associated with it. Non–time critical information is transferred using a message instruction. *See also* multicast.

producer controller The controller that originates or produces information from one controller to one or more consumer controllers across ControlNet or Ethernet/IP. A produced tag must be controller scoped.

program A set of related routines and their associated program scoped tags. The program is the second level of scheduling within an RSLogix 5000 project. Below the task are the programs. RSLogix 5000 projects starting with software version 15 can contain as many as 100 programs per task. There is no executable code within a program. The main reason for programs is to provide the programmer the opportunity to determine which program executes first, second, third, and so on. Programs execute in the order listed below their associated task in the controller organizer. The list of programs, or order of execution, can be modified by a programmer. Modifying the order of how programs execute is referred to as *adjusting* the program schedule. *See also* unscheduled programs.

Program mode A PLC is either in Run mode, solving the user program, or in Program mode. In Program mode, the controller scan is stopped. The programmer develops or edits the user program while the controller is in Program mode.

program scan The program scan is one part of the PLC scan operating cycle. During the controller scan, the controller scans each rung of the user program. This is the scan of the program, or the program scan. During the scan and execution of the ladder program, each instruction is executed. The resulting true or false logical state derived from solving all of a rung's instructions results in the output instruction status. The output instructions' on or off statuses are stored in the output tag.

program scoped tag ControlLogix tags are grouped into two general classifications: local or global tags. Controller scoped tags are global tags and can be used anywhere within the project, whereas program scoped tags are local and only used within the current program. Most tags can be assigned by the programmer as either controller scoped or program scoped. A programmer might use program scoped tags to help organize tags associated with a specific program rather than having to work with a larger alphabetical controller scoped tags collection. Program scoped tags are sometimes created by a programmer as a collection of tags that can be easily copied and pasted into another similar program. *See also* controller scoped tags.

programmable automation controller (PAC) The designation given to ControlLogix because of its additional functionality.

programmable logic controller (PLC) Usually called a *programmable controller*, a PLC is a digitally operated, electronic industrial computer. It has a programmable memory for internally storing instructions and data. Programmed instructions execute control logic functions such as timing, counting, sequencing, arithmetic, communication, and data-manipulation instructions. The PLC is typically used when referring to a SLC 500 or PLC 5.

project file RSLogix Software stores a controller's programming and configuration information in a file called a *project*. The project file contains all information relating to the project. Information contained within a project includes tasks, programs, routines, configuration information, trends, documentation, and so on. RSLogix 5000 project RSLogix 5000 window files have an .acd file extension.

prompt A symbol used to inform the user that a response is required.

proportional, integral, derivative (PID) control Control executed through either an intelligent I/O module or a program instruction, providing automatic closed-loop operation of process control loops.

protocol The set of rules that defines the format and timing of data between data communications devices—for example, Allen-Bradley PLCs communicating via RS-232 serial communications using Allen-Bradley's DF-1 protocol.

radix Another way to describe the base of a number system. The decimal number system is base-10 or radix-10, whereas binary is base-2 or radix-2.

random access memory (RAM) This term is a misnomer. RAM is more accurately described as read–write memory. This means that the controller can *write,* or place data, into memory locations. The controller can also *read,* or take data out of, a memory location. The *random access* part of this term simply means that the controller can access data by going directly to the desired address rather than going through each and every address in a serial fashion.

read To acquire a copy of the desired data from a storage area such as a hard disk, floppy disk, RAM, or data stored in another PLC. When read, a copy of the original data is made and transferred to the target memory's storage location.

read-only memory (ROM) Memory information that can only be read. Under most circumstances, the average user cannot write any data to memory that is read-only. ROM may be programmed by an OEM and may contain the operating system, or personality, of the computer system. Specifically relating to a PLC, an OEM may develop and load the user program into some type of ROM as a way to restrict end users from modifying it.

real number A number assigned as a data type for a ControlLogix tag. Real numbers allow for a decimal point, whereas an integer is a whole number. *See also* floating-point.

real-time sampling (RTS) The input sampling rate used in analog type input modules. The RTS is assigned by the programmer to set up how often an analog input module samples the incoming field data. These data are stored in a buffer on the module until the RPI expires and the data are transferred to the module's input tags. *See also* requested packet interval.

remote I/O (RIO) Chassis of I/O can be remotely distributed around the plant floor, close to the machine or process, as a way to provide short wiring runs between I/O devices and their respective I/O modules. Communication using either ControlNet or Ethernet/IP will connect the remote chassis back to their local controller.

Remote Run mode Changing a controller from Program to Run mode by using the personal computer and the mouse while concurrently running programming software. This action puts the PLC into Remote Run mode.

requested packet interval (RPI) The requested data-transfer time in milliseconds for information across a ControlNet or Ethernet/IP Network. *See also* actual packet interval.

retentive data Data that are not lost when power is interrupted to the PLC and its memory. Typically, a battery is used to keep power on for volatile RAM chips after a power interruption or shutdown of the PLC. If there is power to RAM chips, then memory will be retained. If the battery is dead or missing, then data in the user program will be lost during a power interruption.

return (RET) instruction The Return instruction is an output instruction which can be conditioned in a subroutine as a way to exit the subroutine, if certain conditions exist, without executing the remaining rungs. Also, when using a PLC 5 or ControlLogix, input parameters can be used to transfer data into a subroutine where the data is to be manipulated. When data is transferred into a subroutine for manipulation, an answer is expected by the calling routine. The return instruction is the vehicle used to return the answer from the data manipulated within the subroutine to the calling routine.

ROM *See* read-only memory.

routine Where RSLogix 5000 projects store executable code. Each program has a main routine, an optional fault routine, and the desired number of subroutines. Subroutines are accessed using the jump to subroutine (JSR) instruction. Using a JSR to jump into a routine of another programming language is OK. The number of routines in a project is only limited by the amount of memory in the controller. Any routine must be entirely in the same programming language: ladder logic, sequential function chart, function block diagram, or structured text.

RPI *See* requested packet interval.

RSLogix 5 software Software used to program, edit, or monitor the Allen-Bradley PLC 5 family of PLCs.

RSLogix 500 software Software used to program, edit, or monitor the Allen-Bradley SLC 500 family of PLCs.

RSLogix 5000 software Software used to program, edit, or monitor the Allen-Bradley ControlLogix family of PLCs. The family includes the ControlLogix, CompactLogix, FlexLogix, DriveLogix, GuardLogix, and SoftLogix.

RTS. *See* real-time sampling.

Run mode The operating cycle in which the controller is running the project and actively controlling the machine or process.

rung comment Text that is included with each PLC ladder rung and used to help individuals understand how the program operates or how the rung interacts with the rest of the program. Up to 1,000 lines of text can be entered as an RSLogix 5000 rung comment.

SBR instruction *See* Subroutine instruction.

self-test Hardware that uses its firmware to monitor, test, and verify a device such as a PLC and detect any faults or errors in operation.

sequential function chart (SFC) Another programming language available only on certain PLCs. SFC is similar to flowchart-type programming and uses step-type boxes containing structured text and transition steps, which also contain structured text. Each step is executed until it is completed. When the transition step is true, the preceding step is completed and the flow can proceed to the next step. SFC programming is an alternative to ladder logic if the process can be broken into logical steps. The sequential function chart language is only available in the PLC 5 and ControlLogix families of Allen-Bradley PLCs.

short integer (SINT) Data type made up of 8 bits. This is sometimes called a *byte*. A SINT is assigned as a data type for a ControlLogix tag.

sign bit The leftmost bit in a 16- or 32-bit binary number representing the positive or negative status of the value represented. If the bit is a 1, then the value is negative. A 0 in the sign bit position represents a positive number.

signed integer A signed whole number such as $-2, -1, 0, +1, +2, +3, +4, +5$, and so on. Numerical data are represented in binary format within the PLC. A 16-bit signed integer uses the leftmost bit as the sign bit and the lower 15 bits as the numerical value. Fifteen bits can represent the values 0 to 32,767. The range of integers represented by a 16-bit signed integer is $-32,768$ to $+32,767$. Likewise, a 32-bit PLC using 32-bit signed integers can represent data ranging from $-2,147,483,648$ to $+2,147,483,647$.

single integer One 16-bit word used to represent integer values.

sinking input A sinking input point switches the negative DC current side of its physical input field device. A sourcing inductive proximity switch is interfaced to a sinking 24-volt DC input module. The sourcing proximity sensor switches the positive side, whereas the input point sinks the current to ground.

sinking output A sinking output module switches the negative DC current side in relation to the physical output field device. A sinking output point is interfaced to a sourcing 24-volt DC output load. The sinking output sinks the current to ground after current has been seen by the load.

SINT *See* short integer.

software The program on a disk or CD-ROM purchased for use on a personal computer to create your user program. Also refers to the program a user develops and stores in the programmable controller's memory.

sourcing input A sourcing input point switches the positive DC current side of its physical input field device. A sinking inductive proximity switch is interfaced to a sourcing 24-volt DC input module. The proximity sensor sinks current to ground, or the negative side of the power supply.

sourcing output A sourcing output module switches the positive DC current side in relation to the physical output field device. A sourcing output point is interfaced to a sinking 24-volt DC output load. The sourcing output point switches the positive side, whereas the output field device sinks the current to ground.

station A PLC, computer, operator interface, or other hardware device connected to a network. This hardware device is then called a *station*, or *node*. Each node has a unique address on the network.

structure An RSLogix 5000 data type that can contain members of different data types. A timer or counter structure, for example, comprises three double integers: the preset value, the accumulated value, and the instructions' BOOL status bits. There are several predefined structures in RSLogix 5000 software. Users can create their own user-defined data type that could incorporate multiple members of differing data types.

structured text A text- or statement-based programming language similar to the Basic programming language. Uses for this programming language could range from motion-control applications to number crunching. The structured text language is only available in the PLC 5 and ControlLogix families of Allen-Bradley PLCs.

subroutine A subprogram contained within the main program. Subroutine program logic typically will not be solved during every scan of the PLC. A jump to subroutine instruction is included at the proper point in the PLC ladder logic to direct the controller to jump out of the current routine and jump to and scan a different routine dictated by specific inputs or conditions. Subroutines are used to save scan time. The number of subroutines that could be configured in a ControlLogix project is limited only by controller memory. An example of subroutine usage is alarm logic. Under normal conditions, alarm conditions are not present. Alarm logic can be put into a subroutine because it is unnecessary to scan all alarm logic during every program scan. Incorporating a rung into your program directs the controller to execute this subroutine, or separate program, only under alarm conditions. ControlLogix routines contain the project's executable code.

subroutine (SBR) instruction The first instruction programmed within the subroutine. When using a PLC 5 or ControlLogix family PLC, the SBR instruction is optional unless input and return parameters are incorporated into the JSR, SBR, and RET instructions.

system overhead time Time slice in a ControlLogix system; the percentage of time taken from the continuous task for controller communications to I/O and sending messages across a network connection.

tag ControlLogix is a name-based, or tag-based, machine. A tag is simply a name. When something is created using the RSLogix 5000 Software, such as a timer or counter, the programmer assigns its name or tag. For example, a timer may be called Mixing_Timer, and a counter might be called Parts_Counter. Each object created must have a unique tag. Once the tag has been created, its data type must be identified. For example, Mixing_Timer is a timer, and Parts_Counter is a counter data type. Start_Machine could be a single bit or a BOOL tag. There are four types of tags: base, alias, produced, and consumed.

task The first level of scheduling within an RSLogix 5000 project. The project organization from top to bottom is task, program, and routine. A task is a collection of scheduled programs. When a task is executed, the associated programs are executed in the order listed. This list of programs is called the *program schedule*. There are four types of tasks: continuous, periodic, event, and safety. There is no executable code in any task. Executable code is reserved for the routines. Newer versions of RSLogix 5000 can have as many as 32 tasks.

Test mode A PLC operating mode used to test a PLC program. Test mode is typically used in troubleshooting a system or system installation. When the PLC is in Test mode, the ladder program operates as normal except that output module outputs are by default not energized. Output behavior for ControlLogix is user configurable in the I/O configuration. Test mode is for testing input interaction with the ladder program and observing the resulting output status without physical field outputs being energized.

throughput The time it takes for an input signal to be processed and seen at an output point.

truth table Represents how outputs are expected to behave as the result of specific input signals, which are, for computer purposes, usually in digital format.

two's complement The system used to represent negative numbers while executing math operations in a digital computer. The leftmost bit is the *sign bit*, whereas the remaining bits represent the number itself. If the leftmost bit is a 1, then the number is negative. If the leftmost bit is a 0, then the number is positive.

unscheduled program Programs that are currently not needed can be unscheduled. For the controller, unscheduled programs do not exist and are not executed. Unscheduled programs typically fall under debug or installation programs that are only needed when troubleshooting or when removing undesired options when the customer purchases the machine.

unsigned integer When all 16 bits in a 16-bit word are used to represent a numerical value with no sign bit, there is no sign; therefore, there are only positive numbers. Sixteen bits can represent a value from 0 to 65,535.

upper byte The upper 8 bits in a 16-bit word form the upper byte.

upper nibble The upper 4 bits of any specific byte form the upper nibble.

user-defined data type (UDT) A structure created by a programmer to store unlike data types. This structure could contain several related tags for a variable-frequency drive—for example, related but mixed data relating to a tank or vessel. Maybe you need to know the drive heat sink temperature (a real number), commanded motor speed (an integer), and drive status information (bits). Because an array can only store data of the same data type, a user-defined data type is used to store this drive data as a group.

volatile memory Memory, usually identified as RAM, that will not retain its original contents if the power is removed.

watchdog timer A hardware timer used in PLCs to ensure that the program scan is completed in a timely manner. The watchdog timer ensures that the program has not been caught in an endless loop or for some reason become hung up and unable to complete its program scan. ControlLogix watchdog timers are task based. The controller resets the watchdog timer at the end of each task's execution to ensure continuous operation.

Who An RSLinx utility used with the Allen-Bradley PLCs that enables the programmer to see which devices are on a network.

word Sixteen bits is defined as a word. The ControlLogix integer (INT) data type is a word.

write To transfer a copy of data from the originator's memory to the specified storage device. As an example, current output update data are written to the output tags. A message instruction could be used to write data from a tag or group of tags in one controller to a tag or group of tags in a remote controller. Data written to a specified storage area write over current data stored there.

XIC *See* examine if closed.

XIO *See* examine if open.

CPSIA information can be obtained
at www.ICGtesting.com
Printed in the USA
FFOW04n1142201215
19740FF